THE METAL-RICH UNIVERSE

Metal-rich stars accumulate their metals from previous generations of stars, and so they contain the history of their galaxy. By studying these stars we can gain valuable insights into how metals change the formation and evolution of stars, and why giant exoplanets seem to be found around metal-rich stars, and explain the extraordinary massive-star populations observed in the metal-rich region of our own galaxy.

Until now this topic has received much less attention than very metal-poor stars, which give clues to the early chemical evolution of galaxies. Recent observations of metal-rich regions have shown that stars hosting giant planets are generally metal-rich, which has triggered further observations of metal-rich stars. This has led to the discovery of new exoplanets, and advances in the study of planet formation and the late chemical evolution of galaxies.

This book is the first on this topic, and it covers many aspects, from spectral-line formation to stellar formation and evolution in high-metallicity regimes. It is invaluable to researchers and graduate students in stellar evolution, extragalactic astronomy, and planet formation.

GARIK ISRAELIAN is Staff Astronomer at the Instituto de Astrofísica de Canarias. His areas of research include stellar physics, stellar spectroscopy, extra-solar planets, massive stars, and metal-poor stars.

GEORGES MEYNET is a Professor in the Astronomy Department at the University of Geneva. His research focuses on stellar evolution and stellar nucleosynthesis.

THE METAL-RICH UNIVERSE

Edited by

GARIK ISRAELIAN
Instituto de Astrofísica de Canarias
and
GEORGES MEYNET
University of Geneva

CAMBRIDGE UNIVERSITY PRESS
Cambridge, New York, Melbourne, Madrid, Cape Town, Singapore, São Paulo, Delhi

Cambridge University Press
The Edinburgh Building, Cambridge CB2 8RU, UK

Published in the United States of America by Cambridge University Press, New York

www.cambridge.org
Information on the title.www.cambridge.org/9780521879989

© Cambridge University Press 2008

This publicaion is in copyright. Subject to statutory exception
and to the provisions of relevant collective licensing agreements,
no reproduction of any part may take place without
the written permission of Cambridge University Press.

First published 2008

Printed in the United Kingdom at the University Press, Cambridge

A catalogue record for this publication is available from the British Library

ISBN 978-0-521-87998-9 hardback

Cambridge University Press has no responsibility for the persistence or accuracy of URLs for external or third-party internet websites referred to in this publication, and does not guarantee that any content on such websites is, or will remain, accurate or appropriate.

Contents

List of contributors	page ix
Preface	xvii

Part I   Abundances in the Galaxy: field stars — **1**

1 Metal-rich stars and stellar populations: a brief history and new results — 3
   R. M. Rich

2 The metal-rich nature of stars with planets — 17
   N. C. Santos

3 Solar chemical peculiarities? — 30
   C. Allende Prieto

4 Kinematics of metal-rich stars with and without planets — 36
   A. Ecuvillon, G. Israelian, F. Pont, N. C. Santos & M. Mayor

5 Elemental abundance trends in the metal-rich thin and thick disks — 41
   S. Feltzing

6 Metal-rich massive stars: how metal-rich are they? — 53
   D. J. Lennon and C. Trundle

7 Hercules-stream stars and the metal-rich thick disk — 62
   T. Bensby, M. S. Oey, S. Feltzing & B. Gustafsson

8 An abundance survey of the Galactic thick disk — 69
   B. E. Reddy, D. L. Lambert & C. Allende Prieto

Part II   Abundances in the Galaxy: Galactic stars in clusters, bulges and the centre — **75**

9 Galactic open clusters with supersolar metallicities — 77
   S. Randich

10 Old and very-metal-rich open clusters in the BOCCE project — 88
   A. Bragaglia, E. Carretta, R. Gratton & M. Tosi

11	Massive-star versus nebular abundances in the Orion nebula *S. Simón-Díaz*	94
12	Abundance surveys of metal-rich bulge stars *J. P. Fulbright, R. M. Rich & A. McWilliam*	100
13	Metal abundances in the Galactic Center *F. Najarro*	112
14	Light elements in the Galactic bulge *A. Lecureur, V. Hill, M. Zoccali & B. Barbuy*	126
15	Metallicity and ages of selected G–K giants *L. Pasquini, M. Döllinger, J. Setiawan, A. Hatzes, L. Girardi, L. da Silva, J. R. de Medeiros, A. Weiss & O. Von Der Lühe*	132

Part III Observations – abundances in extragalactic contexts   **139**

16	Stellar abundances of early-type galaxies *S. C. Trager*	141
17	Measuring chemical abundances in extragalactic metal-rich H II regions *F. Bresolin*	155
18	On the maximum oxygen abundance in metal-rich spiral galaxies *J. M. Vílchez, L. Pilyugin & T. X. Thuan*	168
19	Starbursts and their contribution to metal enrichment *D. Kunth*	173
20	High metallicities at high redshifts *M. Pettini*	186
21	Evolution of dust and elemental abundances in quasar DLAs and GRB afterglows as a function of cosmic time *B. E. Penprase, W. Sargent & E. Berger*	199
22	Dust, metals and diffuse interstellar bands in damped Lyman-alpha systems *S. Ellison*	205
23	Tracing metallicities in the Universe with the James Webb Space Telescope *R. Maiolino, S. Arribas, T. Böker, A. Bunker, S. Charlot, G. de Marchi, P. Ferruit, M. Franx, P. Jakobsen, H. Moseley, T. Nagao, L. Origlia, B. Rauscher, M. Regan, H. W. Rix & C. J. Willott*	212

Part IV Stellar populations and mass functions   **225**

24	The stellar initial mass function of metal-rich populations *P. Kroupa*	227
25	Initial-mass-function effects on the metallicity and colour evolution of disc galaxies *P. Westera, M. Samland, R. Buser & K. Ammon*	249

26	The metallicity of circumnuclear star-forming regions *A. I. Díaz, E. Terlevich, M. Castellanos & G. Hägele*	255
27	The stellar population of bulges *P. Jablonka*	261
28	The metallicity distribution of the stars in elliptical galaxies *A. Pipino & F. Matteucci*	270
29	Wolf–Rayet populations at high metallicity *P. A. Crowther*	276
30	The stellar populations of metal-rich starburst galaxies: the frequency of Wolf–Rayet stars *J. R. Souza Leão, C. Leitherer, F. Bresolin & R. Cid Fernandes*	288
Part V	**Physical processes at high metallicity**	**293**
31	Stellar winds from Solar-metallicity and metal-rich massive stars *J. Puls*	295
32	On the determination of stellar parameters and abundances of metal-rich stars *Y. Takeda*	308
33	Are WNL stars tracers of high metallicity? *G. Gräfener & W.-R. Hamann*	321
34	The observable metal-enrichment of radiation-driven-plus-wind-blown H II regions in the Wolf–Rayet stage *G. Hensler, D. Kroeger & T. Freyer*	327
35	Metal-rich A-type supergiants in M31 *N. Przybilla, K. Butler & R.-P. Kudritzki*	332
Part VI	**Formation and evolution of metal-rich stars and stellar yields**	**339**
36	Massive-star evolution at high metallicity *G. Meynet, N. Mowlavi & A. Maeder*	341
37	Supernovae in Galactic evolution: direct and indirect metallicity effects *C. Fröhlich, R. Hirschi, M. Liebendörfer, F.-K. Thielemann, G. Martínez Pinedo & E. Bravo*	354
38	Progenitor evolution of Type-I supernovae: evolution and implications for yields *Sung-Chul Yoon*	369
39	Star formation in the metal-rich Universe *I. A. Bonnell*	380
40	Metallicity of Solar-type main-sequence stars: seismic tests *S. Vauclair*	393
41	Chemical-abundance gradients in early-type galaxies *P. Sánchez-Blazquez*	403

42	Oxygen-rich droplets and the enrichment of the interstellar medium *G. Stasińska, G. Tenorio-Tagle, M. Rodríguez & W. J. Henney*	409
Part VII	**Chemical and photometric evolution beyond Solar metallicity**	**413**
43	Models of the Solar vicinity: the metal-rich stage *L. Carigi*	415
44	Chemical-evolution models of ellipticals and bulges *F. Matteucci*	428
45	Chemical evolution of the Galactic bulge *B. Gibson, A. J. MacDonald, P. Sánchez-Blázquez & L. Carigi*	441
46	How do galaxies become metal-rich? An examination of the yield problem *M. G. Edmunds*	447
47	Abundance patterns: thick and thin disks *C. Chiappini*	454
48	Formation and evolution of the Galactic bulge: constraints from stellar abundances *S. K. Ballero, F. Matteucci & L. Origlia*	460
49	Summary *B. E. J. Pagel*	465

Contributors

Karin Ammon, Astronomical Institute, Department of Physics and Astronomy, Universität Basel, Venusstrasse 7, CH-4102 Binningen, Switzerland

S. Arribas, CSIC – Departamento de Astrofísica Molecular e Infrarroja, Madrid, Spain

Silvia K. Ballero, Dipartimento di Astronomia, Università di Trieste, Via G. B. Tiepolo 11, I-34124 Trieste, Italy

B. Barbuy, Universidade de São Paulo, IAG, Rua do Matão 1226, São Paulo 05508-900, Brazil

T. Bensby, European Southern Observatory, Alonso de Cordova 3107, Vitacura, Casilla 19001, Santiago 19, Chile

Edo Berger, Observatories of the Carnegie Institution, Pasadena, California, USA

T. Böker, European Space Agency – ESTEC, Noordwijk, The Netherlands

Ian A. Bonnell, School of Physics and Astronomy, University of St Andrews, St Andrews KY16 9SS, UK

Angela Bragaglia, INAF – Osservatorio Astronomico di Bologna, Via Ranzani 1, I-40127 Bologna, Italy

Eduardo Bravo, Departament de Física i Enginyeria Nuclear, Universitat Politècnica de Catalunya, Barcelona, Spain

Fabio Bresolin, Institute for Astronomy, University of Hawaii, 2680 Woodlawn Drive, Honolulu, HI 96822, USA

A. Bunker, School of Physics, University of Exeter, Exeter, Devon EX4 4QJ, UK

Roland Buser, Astronomical Institute, Department of Physics and Astronomy, Universität Basel, Venusstrasse 7, CH-4102 Binningen, Switzerland

Keith Butler, Universitätssternwarte München, Scheinerstrasse 1, D-81679 München, Germany

Leticia Carigi, Instituto de Astronomía, Universidad Nacional Autónoma de México, Apartado Postal 70-264, México 04510, D.F., Mexico

Eugenio Carretta, INAF – Osservatorio Astronomico di Bologna, Via Ranzani 1, I-40127 Bologna, Italy

Marcelo Castellanos, Universidad Autónoma de Madrid, Madrid, Spain

S. Charlot, Institute d'Astrophysique de Paris, Paris, France

Cristina Chiappini, INAF – Osservatorio Astronomico di Trieste, Via Tiepolo 11, I-34131 Trieste, Italy

Roberto Cid Fernandes, Universidade Federal de Santa Caterina, Florianópolis, SC, Brazil

Paul A. Crowther, Department of Physics and Astronomy, University of Sheffield, Hounsfield Road, Sheffield S3 7RH, UK

L. da Silva, Observatorio Nacional, R. Gal. José Cristino, 77, 20921-400, São Cristóvão, Rio de Janeiro, Brazil

G. de Marchi, European Space Agency – ESTEC, Noordwijk, The Netherlands

J. R. de Medeiros, Universidade Federal do Rio G. Norte, CEP 59072-970 Natal, Brazil

Ángeles I. Díaz, Universidad Autónoma de Madrid, Madrid, Spain

M. Döllinger, European Southern Observatory, Karl-Schwarzschild-Strasse 2, D-85748 Garching bei München, Germany

Alexandra Ecuvillon, Instituto de Astrofísica de Canarias, E-38200 La Laguna, Tenerife, Spain

M. G. Edmunds, School of Physics and Astronomy, Cardiff University, Queens Buildings, 5 The Parade, Cardiff CF24 3AA, UK

Sara L. Ellison, Department of Physics and Astronomy, University of Victoria, 3800 Finnerty Road, Victoria, BC, Canada

Sofia Feltzing, Lund Observatory, Box 43, SE-221 00 Lund, Sweden

P. Ferruit, CRAL – Observatoire de Lyon, 9 Avenue Charles André, Saint-Genis Laval, France

M. Franx, Leiden Observatory, Leiden, The Netherlands

Tim Freyer, Institut für Theoretische Physik und Astrophysik, Universität Kiel, D-24098 Kiel, Germany

Carla Fröhlich, Department of Physics and Astronomy, Universität Basel, Switzerland

Jon P. Fulbright, Department of Physics and Astronomy, The Johns Hopkins University, Baltimore, MD 21218, USA

Brad K. Gibson, University of Central Lancashire, Centre for Astrophysics, Preston PR1 2HE, UK

L. Girardi, INAF – Osservatorio Astronomico di Trieste, Via Tiepolo 11, I-34131 Trieste, Italy

G. Gräfener, Institut für Physik, Universität Potsdam, Am Neuen Palais 10, D-14469 Potsdam, Germany

Raffaele Gratton, INAF – Osservatorio Astronomico di Padova, Vicolo dell'Osservatorio 5, I-35122 Padova, Italy

B. Gustafsson, Department of Astronomy and Space Physics, University of Uppsala, SE-751 20 Uppsala, Sweden

Guillermo Hägele, Universidad Autónoma de Madrid, Madrid, Spain

W.-R. Hamann, Institut für Physik, Universität Potsdam, Am Neuen Palais 10, D-14469 Potsdam, Germany

A. Hatzes, Tautenburg, Germany

William J. Henney, Centro de Radioastronomía y Astrofísica, Universidad Nacional Autónoma de México, Campus Morelia, Apartado Postal 3-72, 58090 Morelia, Mexico

Gerhard Hensler, Institut für Astronomie, Universität Wien, A-1180 Wien, Austria

V. Hill, Observatoire de Paris-Meudon, GEPI et CNRS UMR 8111, 92125 Meudon Cedex, France

Raphael Hirschi, Department of Physics and Astronomy, Universität Basel, Basel, Switzerland

Garik Israelian, Instituto de Astrofísica de Canarias, E-38200 La Laguna, Tenerife, Spain

P. Jablonka, Observatoire de Genève, Chemin des Maillettes 51, CH-1290 Sauverny, Switzerland

P. Jakobsen, European Space Agency – ESTEC, Noordwijk, The Netherlands

Danica Kroeger, Institut für Theoretische Physik und Astrophysik, Universität Kiel, D-24098 Kiel, Germany

Pavel Kroupa, Argelander Institut für Astronomie, Universität Bonn, Auf dem Hügel 71, D-53121 Bonn, Germany

Rolf-Peter Kudritzki, Institute for Astronomy, 2680 Woodlawn Drive, Honolulu, HI 96822, USA

Daniel Kunth, Institut d'Astrophysique de Paris, F-75014 Paris, France

David L. Lambert, McDonald Observatory and Department of Astronomy, University of Texas, Austin, TX 78712, USA

A. Lecureur, Observatoire de Paris-Meudon, GEPI et CNRS UMR 8111, F-92125 Meudon Cedex, France

Claus Leitherer, Space Telescope Science Institute, Baltimore, MD, USA

Daniel J. Lennon, Isaac Newton Group of Telescopes, E-8700 Santa Cruz de La Palma, Tenerife, Spain

Matthias Liebendörfer, Department of Physics and Astronomy, Universität Basel, Switzerland

Angela J. MacDonald, University of Central Lancashire, Centre for Astrophysics, Preston PR1 2HE, UK

André Maeder, ISDC, Observatoire de Genève, Université de Genève, Chemin d'Ecogia 16, CH-1290 Versoix, Switzerland

R. Maiolino, INAF – Osservatorio Astronomico di Roma, Rome, Italy

Francesca Matteucci, Dipartimento di Astronomia, Università di Trieste, Via G. B. Tiepolo 11, I-34100 Trieste, Italy

Michael Mayor, Observatoire de Genève, 51 Chemin des Maillettes, CH-1290, Sauverny, Switzerland

A. McWilliam, Observatories of the Carnegie Institute of Washington, 813 Santa Barbara Street, Pasadena, CA 91101, USA

Georges Meynet, Observatoire de Genève, Université de Genève, CH-1290 Sauverny, Switzerland

H. Moseley, NASA Goddard Space Flight Center, Greenbelt, MD 20771, USA

Nami Mowlavi, ISDC, Observatoire de Genève, Université de Genève, Chemin d'Ecogia 16, CH-1290 Versoix, Switzerland

T. Nagao, National Astronomical Observatory of Japan, Osawa, Japan

Francisco Najarro, Instituto de Estructura de la Materia, CSIC, Serrano 121, E-29006 Madrid, Spain

M. S. Oey, Department of Astronomy, University of Michigan, Ann Arbor, MI 48109-1042, USA

Livia Origlia, INAF – Osservatorio Astronomico di Bologna, Via G. Ranzani 1, I-40127 Bologna, Italy

B. E. J. Pagel, Astronomy Centre, University of Sussex, Brighton BN1 9QH, UK

Luca Pasquini, European Southern Observatory, Karl-Schwarzschild-Strasse 2, D-85748 Garching bei München, Germany

Bryan E. Penprase, Pomona College, Department of Physics and Astronomy, Claremont, CA, USA

Max Pettini, Institute of Astronomy, Madingley Road, Cambridge CB3 0HA, UK

Leonid Pilyugin, Main Astronomical Observatory, National Academy of Sciences of Ukraine, 03680 Kiev, Ukraine

Gabriel Martínez Pinedo, Gesellschaft für Schwerionenforschung, Darmstadt, Germany

Antonio Pipino, Dipartimento di Astronomia, Università di Trieste, I-34127 Trieste, Italy

Frédéric Pont, Observatoire de Genève, 51 Chemin des Maillettes, CH-1290, Sauverny, Switzerland

Carlos Allende Prieto, McDonald Observatory and Department of Astronomy, University of Texas, Austin, TX 78712, USA

Norbert Przybilla, Dr. Karl Remeis-Sternwarte Bamberg, Sternwartstrasse 7, D-96049 Bamberg, Germany

Joachim Puls, Universitätssternwarte München, Scheinerstrasse 1, D-81679 München, Germany

Sofia Randich, INAF – Osservatorio di Arcetri, Largo E. Fermi 5, I-50125 Firenze, Italy

B. Rauscher, NASA Goddard Space Flight Center, Greenbelt, MD 20771, USA

Bacham E. Reddy, Indian Institute of Astrophysics, Bangalore, India

M. Regan, Space Telescope Science Institute, Baltimore, MD, USA

R. Michael Rich, Department of Physics & Astronomy, UCLA, Los Angeles, CA 90095-1547, USA

H. W. Rix, Max-Planck-Institut für Astronomie, Heidelberg, Germany

Mónica Rodríguez, Instituto Nacional de Astrofísica Óptica y Electrónica, AP 51, 72000 Puebla, Mexico

Nuno C. Santos, Centro de Astronomia e Astrofísica da Universidade de Lisboa, Observatorio Astronomico de Lisboa, Tapada da Ajuda, 1349-018 Lisboa, Portugal

Markus Samland, Astronomical Institute, Department of Physics and Astronomy, Universität Basel, Venusstrasse 7, CH-4102 Binningen, Switzerland

Patricia Sánchez-Blázquez, University of Central Lancashire, Centre for Astrophysics, Preston PR1 2HE, UK

Wallace Sargent, California Institute of Technology, Pasadena, CA 91125, USA

J. Setiawan, Max-Planck-Institut für Astronomie, Heidelberg, Germany

S. Simón-Díaz, Instituto de Astrofísica de Canarias, E-38200 La Laguna, Tenerife, Spain

João Rodrigo Souza Leão, Space Telescope Science Institute, Baltimore, MD, USA

Grażyna Stasińska, LUTh, Observatoire de Paris-Meudon, 5 Place Jules Jansen, F-92195 Meudon, France

Yoichi Takeda, National Astronomical Observatory of Japan, 2-21-1 Osawa, Mitaka, Tokyo 181-8588, Japan

Guillermo Tenorio-Tagle, Instituto Nacional de Astrofísica, Óptica y Electrónica, AP 51, 72000 Puebla, Mexico

Elena Terlevich, Instituto Nacional de Astrofísica, Óptica y Electrónica, AP 51, 72000 Puebla, México

Friedrich-Karl Thielemann, Department of Physics and Astronomy, Universität Basel, Switzerland

Monica Tosi, INAF – Osservatorio Astronomico di Bologna, Via Ranzani 1, I-40127 Bologna, Italy

Trinh X. Thuan, Astronomy Department, University of Virginia, Charlottesville, VA 22904, USA

S. C. Trager, Kapteyn Astronomical Institute, Rijksuniversiteit Groningen, Postbus 800, NL-9700 AV Groningen, The Netherlands

Carrie Trundle, Astrophysics Research Centre, The Queen's University of Belfast, Belfast BT7 1NN, Northern Ireland, UK

Sylvie Vauclair, Laboratoire d'Astrophysique de Toulouse Tarbes, Observatoire Midi–Pyrénées, Université Paul Sabatier, Toulouse, France

Jose M. Vílchez, Instituto de Astrofísica de Andalucía (CSIC), Apartado Postal 3004, E-18080 Granada, Spain

O. Von der Lühe, Kipenheuer Institut für Sonnenphysik, Schoeneckerstrasse 6, D-79104 Freiburg, Germany

A. Weiss, Max-Planck-Institut für Astrophysik, Karl-Schwarzschild-Strasse 1, Postfach 1317, D-85741 Garching bei München, Germany

Pieter Westera, Astronomical Institute, Department of Physics and Astronomy, Universität Basel, Venusstrasse 7, CH-4102 Binningen, Switzerland

C. J. Willott, Herzberg Institute of Astrophysics, Victoria, BC, Canada V9E 2E7

Sung-Chul Yoon, Astronomical Institute "Anton Pannekoek", Universiteit Amsterdam, Kruislaan 403, 1098 SJ Amsterdam, The Netherlands

M. Zoccali, Popular Universidad Católica de Chile, Departamento de Astronomía y Astrofísica, Casilla 306, Santiago 22, Chile

Preface

Even though metals constitute only a few per cent of the total mass fraction of stars, they have a huge impact on the way stars and galaxies evolve. In that respect, metallicity in the Universe is, like the salt in a dish, a small amount that can completely change its flavour!

The metal-rich stars have never attracted as much attention as the metal-poor halo stars, which tell us about the first supernovae and the early chemical evolution of our Galaxy. However, metal-rich stars are of interest in their own right and can shed new lights on very topical subjects. For instance, it is now well established that stars rich in metals are more likely to harbour giant planets. This awareness has elicited careful and detailed abundance studies of ever more metal-rich stars. As a byproduct, trends of the abundances of many elements at high metallicity are now available and await an interpretation in terms of stellar nucleosynthesis and chemical-evolution models. The extent to which these observed trends are in line with what is expected from the current stellar and chemical-evolution models largely remains to be checked and this is one of the main topics of these proceedings.

Putting the subject into a larger context, let us recall that the attainment of adequate models of the high-metallicity regime is of great interest for the study of the central regions of galaxies, which are thought to have higher-than-solar metallicity. Also, it appears that many quasar environments are metal-rich out to redshifts of at least 5. A better knowledge of star formation and evolution in central regions of galaxies would thus appear crucial to enhancing our understanding of these fascinating objects.

In these proceedings, the reader will find the latest observations of metal-rich stars (in the field, clusters, bulge, planet hosts, etc.), as well as presentations of models of atmospheres and spectral-line formation, models of stellar evolution

The Metal-rich Universe, eds. G. Israelian and G. Meynet. Published by Cambridge University Press.
© Cambridge University Press 2008.

and nucleosynthesis at high [Fe/H], discussions on the contribution of metal-rich stars to nucleosynthesis, and models for the chemical evolution of galaxies in the high-metallicity regime.

Many very interesting questions are addressed, for instance the following.

- How do stars form in metal-rich regions?
- What can be said on the possible variation of the initial mass function at high metallicity?
- Is the upper mass limit lower at higher metallicity?
- Is high metallicity a necessary condition for planet formation around stars?
- Is there a minimum metallicity for planets to form around stars?
- How different is the evolution of stars with higher-than-solar metallicity from that of their solar-metallicity counterparts?
- What are the consequences for the stellar populations expected and the chemical enrichment of the interstellar medium?

About 100 participants from 19 countries took part and, it is hoped, enjoyed the week they spent in La Palma. Their written contributions contained in the present book will help to make the metal-rich Universe a topical subject for the next few years. Some of the figures communicate little or no information in black and white, so they have been made available in colour on the book's website.

This conference could be organised thanks to the financial help of Cabildo Insular de La Palma, Patronato de turismo de La Palma, DISA Corporacion Petrolifera S.A. and Banco Bilbao Vizcaya Argentaria.

We would like to thank also the Scientific Organising Committee, which consisted of S. Feltzing (Sweden), D. Garnett (USA), G. Gilmore (UK), A. Herrero (Spain), D. Lambert (USA), F. Matteucci (Italy), A. McWilliam (USA), S. Randich (Italy), N. Santos (Portugal), Y. Takeda (Japan) and the two undersigned.

Our warmest thanks go to Judith de Araoz for her efficient work in managing the secretary's office before, during and after the conference.

Garik Israelian and Georges Meynet

Part I

Abundances in the Galaxy: field stars

1

Metal-rich stars and stellar populations: a brief history and new results

R. Michael Rich
Department of Physics and Astronomy, UCLA, Los Angeles CA 90095-1547, USA

The subject of metal-rich stars has been controversial for over 40 years, and I review some of the major developments in the subject area during that period, emphasizing those papers that set the subject on its present-day course. Metals emerge in the Universe at very high redshift, and galaxies with roughly Solar metallicity are documented even at redshift 3. In the local Universe, disks and bulges are often metal-rich, but metal-rich stars can also be found in distant halo populations, likely ejected into those environments by merger events. The Galactic bulge has a mean abundance of slightly subsolar but contains stars as metal-rich as [Fe/H] $\sim+0.5$; these stars have a complicated enhancement of light elements.

1 Introduction

As a graduate student in the 1980s, I was warned by senior colleagues to stay clear of the issue of high metallicity. That subject, it was said, had been controversial, and careers had foundered on claims of metallicity greater than Solar. My thesis work was on the Galactic bulge, and, in the long run, it would help to return the subject of super-metal-rich stars to respectability. Still, it is surprising that this is the first meeting on the metal-rich Universe. Spheroidal populations, which have elevated metallicity, account for some 50%–70% of the stellar mass in the local Universe (Fukugita *et al.* 1998), and are also known as the hosts of black holes (Tremaine *et al.* 2002). The era of metal production keeps getting pushed back to earlier and earlier epochs; metal lines are found in the most distant quasars, with Fe II clearly appearing at $z = 6.4$ (Barth *et al.* 2003), less than 1 Gyr after the Big Bang. In the same quasar, $\sim 10^{10} M_\odot$ of molecular gas (CO-line emission) is observed (Walter

The Metal-rich Universe, eds. G. Israelian and G. Meynet. Published by Cambridge University Press.
© Cambridge University Press 2008.

et al. 2004). The association of quasars with galaxy bulges in formation marks them as evidently prodigious sources of metals in the early Universe.

By 2.4 Gyr after the Big Bang, at $z = 2.77$, in an exquisite study of a lensed starforming (so-called Lyman-break) galaxy, MS 1512-cB58, Pettini *et al.* (2000) find a total metallicity of 0.25 Solar – quite respectable by the standards of the present-day Universe. Galaxies clearly built up their metals early and rapidly. Early spectroscopy of this distant starburst galaxy was of such high signal-to-noise ratio (SNR) that it was necessary to obtain better UV spectra of nearby starbursts in order to have a local comparison sample. The agents of metal buildup are massive stars, and this galaxy is a snapshot of chemical evolution in action. Since the pioneering work of Steidel *et al.* (1996) it has been found that the Lyman-break galaxy population is surprisingly metal-rich, with evidence of metal-enriched outflows. The work of Erb *et al.* (2006) uses the classical strong optical lines, shifted to the infrared, to derive abundances in a population of $z \sim 2$ galaxies. They find supersolar effective yields in their galaxy population and a gas outflow rate of approximately four times the star-formation rate. Although the data are less secure, the abundances derived from the broad lines of quasars have also been claimed to be high (e.g. Hamann *et al.* 2002).

The detection of high metallicity in this high-redshift galaxy should not come as a surprise to those who have followed the study of the Galactic bulge in recent years. Zoccali *et al.* (2003) show that the turnoff age for the bulge is comparable to that of a metal-rich halo globular cluster and estimate an age for the bulge in excess of 10 Gyr, a secure demonstration of the great age hinted at in prior studies (Ortolani *et al.* 1995, Kuijken & Rich 2002). The observation of nearly Solar metallicity at high redshift should come as an expectation, not a surprise.

Yet another means of quantitative measurement of metals in the high-redshift Universe is offered by damped Lyman-alpha systems in quasars. These are gas clouds of sufficient H I column that the associated Lyman-alpha lines have damping wings; associated with these clouds are also metal lines, and it is possible to derive a surprisingly accurate metal abundance for these systems. Dessauges-Zavadsky *et al.* (2006) analyze systems over a redshift range of $1.8 < z < 2.5$, ranging from 1/55 to 1/5 the Solar iron abundance.

Metals at redshift 6 are not confined to quasars or distinct bodies like Lyman-limit or damped Lyman-alpha systems. Metals are distributed widely; Sargent, Simcoe, and collaborators (e.g. Becker *et al.* 2006) have used statistical methods to find C, O, and Si in the intergalactic medium at $z \sim 6$, presumably placed there by wind outflow from star-forming galaxies. There are metals in the Universe as far as the eye can see.

Returning closer to home, metal-rich populations are found in surprising venues. The well-studied open cluster NGC 6791 is found to have [Fe/H] $= +0.4$ even with modern abundance determinations (Gratton *et al.* 2006; Origlia *et al.* 2006). The

Sagittarius dwarf spheroidal galaxy has abundances up to Solar (McWilliam *et al.* 2003; Sbordone *et al.* 2006). Even the stellar halos of luminous galaxies have metallicities approaching Solar (Mouhcine *et al.* 2005b), as does the outer disk of M31 some 30 kpc from the nucleus (Brown *et al.* 2006). Regions and stars of high metallicity provide insight into the star-formation process and nucleosynthesis, and it is now clear that regions of high metallicity must be considered to be of great importance in the formation of galaxies and that they are widespread, not only confined to the nuclear regions.

2 A brief history of supermetallicity

In the 1950s, spectroscopists were aware that the nucleus of M31 has strong CN lines, with strengths similar to the line strengths found for stars in the Solar vicinity (Morgan 1958). Baade (1963) stated that "After the first generation of stars has been formed, we can hardly speak of a generation, because the enrichment takes place so soon, and there is probably very little time difference. So the CN giants that contribute most of the light in the nuclear region of the Nebula must also be called old stars; they are not young." This is a remarkably prescient insight that largely describes the present-day picture of the chemical evolution of spheroids.

The work of Spinrad & Taylor (1969) is considered the seminal work on supermetallicity, but it was predated by an interest in the spectroscopy of galaxies (Spinrad 1961) that resulted in the identification of strong Na lines in the nucleus of M31, a finding that was to be a subject of debate for over two decades. Hindsight finds the most extreme metallicities ([Fe/H] = +0.75 for NGC 6791) to be skirting the bounds of credibility, and this no doubt has contributed to the general atmosphere of skepticism; but one must consider the state of abundance analysis at that time. No doubt NGC 6791 remains today the most metal-rich open cluster; Spinrad and Taylor at least succeeded in getting the correct ranking. Was the supermetallicity in fact a real phenomenon? On the one hand, Gustafsson *et al.* (1974) argued for its reality, using narrow-band *photometry* (with a pulse-counting photometer) of spectral regions selected to have clumps of weak iron lines adjacent to clean continuum; this was an innovative method for its time and turns out to have given the correct answer. On the other hand, Peterson (1976) argued that supermetallicity was spurious and arises from the temperature profile of the stellar atmosphere (boundary cooling). The argument stated that excess CN causes a steepening of the boundary temperature gradient; the anomalous cooling strengthens the lines of neutral metals. This in turn masquerades as supermetallicity. The dispute was largely settled by two papers. The first was that of Branch *et al.* (1978), which was the first modern analysis of the prototype super-metal-rich (SMR) star μ-Leonis, using a reticon detector that yielded spectra with very high SNR (Figure 1.1). The application of a modern detector with a linear response and the capability to produce high-SNR

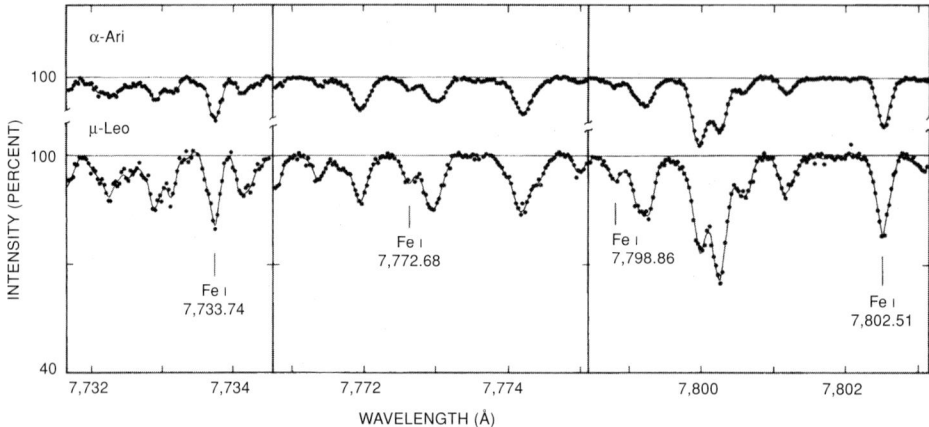

Figure 1.1. One of the first high-resolution digital reticon spectra of μ-Leo, the comparison with Solar-metallicity α-Ari shows the clear enhancement of the weak iron lines relative to the continuum (Branch *et al.* 1978). The enhancement of metals in the super-metal-rich star μ-Leo ([Fe/H] = +0.3) is obvious.

data represented a breakthrough in the subject. While the abundance of μ-Leonis is debated at the 0.1-dex level, that study finds what is essentially the modern value of [Fe/H] for μ-Leonis. The second was that of Deming & Butler (1979), who considered the abundances of binary-star companions of SMR stars; these authors also find a genuinely elevated iron abundance in SMR stars. Deming (1980) argued that Peterson's work had been affected by a subtle misplacement of the continuum in her photographic spectra (a continuum misplacement of 1% leading to a 0.1-dex effect on metallicity).

A sidebar to this debate was Spinrad's (1961) claim that enhanced Na D lines in the M31 nucleus argued for dwarf enhancement in that population. The resulting controversy continued through studies by Whitford (1977), on the FeH Wing–Ford bands, and Faber & French (1980), on the Na 8190 lines in the nucleus of M31, while ultimately settling on giant-dominated light on the basis of infrared studies (e.g. Frogel & Whitford 1987). Studies of the Galactic-bulge luminosity function (Zoccali *et al.* 2000) do not find an abnormally bottom-heavy mass function; the issue is resolved in favor of giant-dominated light.

The effect of the debate over supermetallicity has been, nonetheless, to cast aspersions on the subject. That is in part why we had to wait until 2006 for a meeting on the metal-rich Universe.

2.1 Supermetallicity and stellar evolution

The rich globular cluster systems of the Milky Way and the Magellanic Clouds have provided a set of stellar-population templates spanning from very low to Solar

metallicity. However, only the open cluster NGC 6791 is metal rich at +0.4 dex.
The Galactic bulge is a complex population with a range in abundance with even
some age spread possible; see e.g. Zoccali *et al.* (2003). At high metallicity, we
observe the RGB to become dominated by M giants; metal-line and TiO opacity
in those stars can cause the V magnitude of the RGB tip to be as faint as the
horizonal branch. The I band is also affected; the first giant-branch tip appears to
curve downward or descend. The AGB becomes populated with Miras and OH/IR
stars; these are numerous in the bulge, and the early work of Blanco found large
numbers of M giants toward the Galactic Center. The progression to cool, luminous
giant branches with Mira and OH/IR stars is the dominant effect observed at high
metallicity.

The helium-burning stars are generally confined to the red clump, with a blue
horizontal-branch extension seen only in old, metal-poor populations – at least
in the classical view of stellar populations. However, blue EHB populations are
observed in NGC 6388 and 6441. These are not super-metal-rich clusters, but are
metal-rich enough that such blue HBs should not exist (Rich *et al.* 1997). Blue
HB stars are also found in NGC 6791 (Peterson & Green 1998). In red elliptical
galaxies with no sign of star formation, a hot component (now known as the UVX)
has been detected since the early days of satellite astronomy (Code 1969) and
remains unexplained (O'Connell 1999). Burstein *et al.* (1988) in an influential
paper found that ellipticals and bulges exhibit a correlation between the UVX and
the metallicity-sensitive Mg_2 index; this has not been confirmed in a larger sample
of GALEX-selected quiescent early-type galaxies from the SDSS (Rich *et al.* 2005)
and there remains only a weak correlation in a larger sample of nearby elliptical
galaxies (Rich *et al.*, work in preparation). While UV light is present in many early-
type galaxies, its cause remains mysterious. A full review is beyond the scope of
this paper, but additional factors (subpopulations with enhanced helium abundance)
might be at play.

3 Supermetallicity in stellar populations

The major paradigm in the chemical evolution of stellar populations is that the
chemical enrichment due to massive-star SNe relative to Type Ia SNe reflects the
rate of star formation; the emergence of Solar abundance ratios is a reflection of
the point at which the Type Ia SNe begin to produce substantial iron (Wheeler *et al.*
1989; McWilliam 1997).

Metal-rich populations have turned up in unexpected locations, not only in
the Galactic bulge or elliptical galaxies. In Baade's original population model,
Population II was thought to be older and more globular-cluster-like, and therefore more metal-poor. These ideas persisted well into the 1950s, aided by the

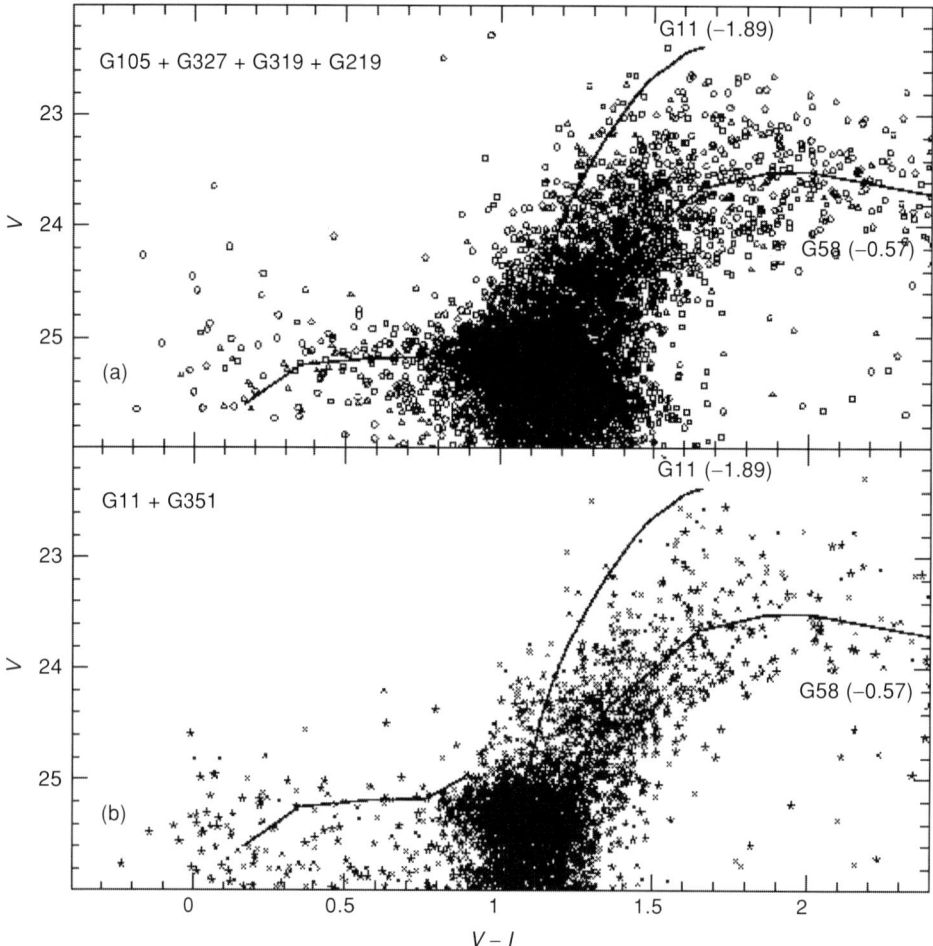

Figure 1.2. Color–magnitude diagrams of the field population in the halo of M31, derived from the field populations of the globular clusters indicated, obtained using WFPC2 on the HST (Bellazini *et al.* 2003). The blue HB ridgeline is that for M68 while the RGB ridges are for G11 ([Fe/H] = -2) and G58 ([Fe/H] = -0.6). Notice the large numbers of fainter, redder stars that must clearly be more metal-rich. These populations may reflect either an extension of the spheroid or the debris of merger events.

difficulties of actually making quantitative measurements in the Galactic bulge. At present, metal-rich stars are found in the old (10–11 Gyr) disk (Castro *et al.* 1997; Pompeia *et al.* 2002, 2003) and in NGC 6791, as mentioned earlier. In fact, the metallicities of these populations reach extremes as high as those found in the Galactic bulge. One important difference, however, is that these disk populations generally have scaled Solar abundances for the light elements, in contrast with the bulge, for which the levels of light elements (especially Mg) remain enhanced to

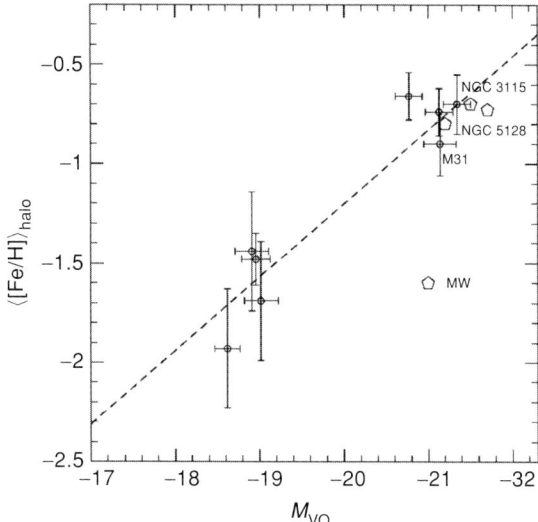

Figure 1.3. Correlation between the photometric metallicity (from the red giant-branch color) of halo populations roughly 10 kpc distant from the plane, and luminosity, for spiral and S0 galaxies (Mouhcine *et al.* 2005a). At this time, it is not clear whether we are observing a trend, or a bimodal distribution, where the presence of metal-rich stars arises from the extension of massive bulges. In some of the most luminous galaxies, it appears that the halo is indeed an outer extension of the spheroid. However, the distant-field population in M31 is polluted by material from merger remnants; this process likely also populates the halo with metal-rich stars.

metallicities even above the Solar value. The galaxy NGC 6791 has, in fact, subsolar abundances.

There is now evidence for resolved metal-rich populations in the halo and disk of M31. Bellazzini *et al.* (2003) find descending metal-rich giant branches in the M31 halo (Figure 1.2), and numerous studies since the pioneering work of Mould & Kristian (1986) find a metal-rich halo extended to 30 kpc (e.g. Durrell *et al.* 2004). Brown *et al.* (2003, 2006) have undertaken deep HST/ACS imaging of M31 halo fields and find evidence for suprasolar metallicities both in the disk (32 kpc distant from the nucleus) and in the spheroid (12 kpc from the nucleus). In the case of the M31 fields, the case for the super-metal-rich populations is based on the modeling of the main-sequence turnoff. Keck spectroscopy of the stars in these populations (Koch *et al.*, work in preparation) now in progress will further test whether super-metal-rich stars are present in these low-density environments.

The galaxy M31 is not alone in possessing an extended population of metal-rich stars. Mouhcine *et al.* (2005a) find a trend of halo metallicity with parent-galaxy luminosity (Figure 1.3), with the field populations spanning 1.5 dex in metallicity. However, M31 does remain as having among the most metal-rich halos in our sample of nine galaxies. Is the complex interaction history of M31 an anomaly, or does it suggest a mechanism for populating the stellar content of the halos of massive galaxies? Since it is difficult to form metal-rich stars in the very-low-density environments of halos, one might instead suspect that these stars form in dense star clusters or in bulges and inner disks. It is more likely that the presence of these metal-rich populations at great distances is the result of their ejection via one or more significant mergers; it will be interesting to model this mechanism in detail.

4 The Galactic bulge

The metal-rich populations in NGC 6791 and the old disk have Solar or subsolar scaled alpha abundances, suggesting that those populations have enriched over timescales longer than 1 Gyr, so that Type Ia SNe stars had time to contribute substantial iron. Bulges form more rapidly. The Galactic bulge's formation timescale is likely ~1 Gyr or less (Ortolani *et al.* 1995; Zoccali *et al.* 2003). When disk stars in the foreground of the bulge are excluded by proper motion (Kuijken & Rich 2002) the remaining bulge population bears a strong resemblance to the main-sequence turnoff of an old globular cluster; the constraint on the numbers of stars brighter than the turnoff (even blue stragglers) is remarkable. In short, there appears to be very little room for an extended star-formation history in the bulge, arguing both from the standpoint of the main-sequence turnoff and from the chemical-enrichment perspective (see below).

In the case of the bulge, the Blancos produced an R, I color–magnitude diagram that for the first time revealed a clear red giant branch. Armed with the new pulse-counting detectors developed by Shectman at Las Campanas, Whitford and I took the first digital spectra of bulge giants in the 1980s (Whitford & Rich 1983). At the telescope, the bulge giants looked remarkable, especially their strong Na D and Mg lines, which exceeded dramatically anything from the standard stars. My original abundance scale was high (another pang in the supermetallicity controversy). Two factors likely contributed to this. First, I derived an abundance scale based on iron, but using the Mg b 5170 (Mg2) index (and not accounting for selective Mg enhancement). A second factor is more subtle: the standard stars observed in Rich (1988) and Rich (1990) were very bright and were observed behind heavy neutral-density filters. Even so, they frequently came close to or exceeded the coincidence count limits, diminishing the measured depth of the Mg index in the standard stars. Nonetheless, the Rich (1990) abundance scale was only 0.3 dex higher than the McWilliam & Rich (1994) scale based on high-resolution echelle spectra. The present-day iron-abundance scale of Fulbright *et al.* (2006a), derived from $R = 67{,}000$, SNR > 50 Keck echelle spectra, is very close to the original McWilliam &

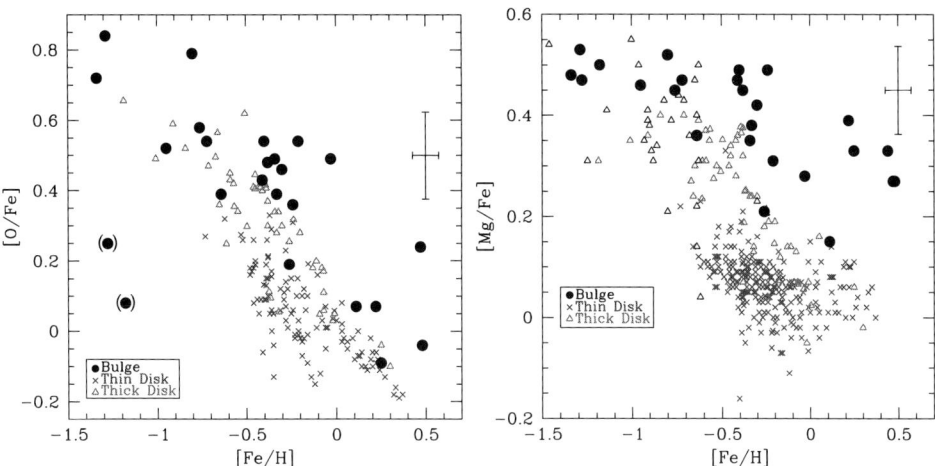

Figure 1.4. Left panel: [O/Fe] versus [Fe/H] for Galactic-bulge giants (filled symbols) and the thin/thick-disk population (Fulbright *et al.* 2006b). Notice that the bulge [O/Fe] is only mildly elevated relative to the disk. Right panel: the trend of [Mg/Fe] versus [Fe/H] for the bulge stars (filled symbols), also from Fulbright *et al.* (2006b). Notice the very strong enhancement of Mg that is carried through to the highest metallicities. This is not characteristic of the SMR disk population. The enhancement in Mg is also seen in the integrated light of elliptical galaxies (Worthey *et al.* 1992).

Rich (1994) scale. The mean [Fe/H] is slightly subsolar, extending to [Fe/H] = +0.5 (which is also, coincidentally, the upper limit of the old disk stars). With the problem of the bulge iron-abundance scale settled (Fulbright *et al.* 2006a) we may turn to the determination of the alpha abundances (Fulbright *et al.* 2006b). However, we should emphasize that the bulge is not extremely metal-rich; rather, ⟨[Fe/H]⟩ is somewhat subsolar. It is in the alpha elements than one can observe striking differences in composition relative to the disk. Figure 1.4 shows a fundamental result that has been established since McWilliam & Rich (1994): that Mg remains elevated in the bulge to [Fe/H] = +0.5. It has been known for more than a decade that Mg is also elevated in massive elliptical galaxies (Worthey *et al.* 1992). However, oxygen, which is also believed to be formed in hydrostatic burning, follows a trend very similar to that of the disk, with only marginal elevation relative to the Solar vicinity. Because O and Mg are believed to be produced in the hydrostatic burning envelopes of massive stars, the disconnect between these two elements is not understood. Fortunately, both our results and that of Zoccali *et al.* (2006), derived on a different sample of bulge giants, find this trend for oxygen. It is possible that the early generations of massive stars underwent substantial mass loss via a Wolf–Rayet phase; much of the outer envelope was lost to the interstellar medium before the nucleosynthesis of substantial oxygen (McWilliam & Rich 2004).

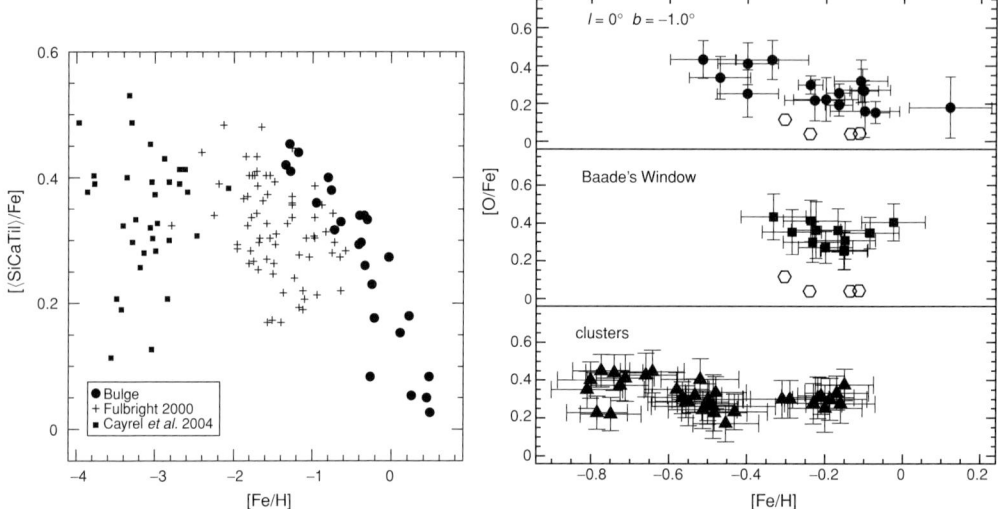

Figure 1.5. Left panel: [⟨SiCaTiI⟩/Fe] for the Fulbright *et al.* (2006b) bulge giants (filled circles) compared with the halo stars of Fulbright (2000) (plus signs) and Cayrel *et al.* (2004) (filled squares). Notice that the bulge stars occupy a locus that defines an upper envelope relative to the halo; the bulge locus is also tighter than that of the halo (from Fulbright *et al.* 2006b). Right panel: the trend of [O/Fe] versus [Fe/H] for bulge fields close to the Galactic Center, the well-studied Baade's Window field at $b = -4°$, and for Galactic-bulge globular clusters (Origlia & Rich 2007). The open symbols correspond to four disk giants in the Solar vicinity. See Rich & Origlia (2005) for details. The oxygen is measured from the 1.6-μm OH lines in the infrared. Notice the absence of metal-rich stars in the bulge fields; the wider abundance range toward the Galactic center continues to be a topic of investigation.

Finally, considering the run of the explosive alphas (those light elements synthesized during the explosion rather than in hydrostatic burning, [⟨SiCaTiI⟩/Fe] versus [Fe/H]), the bulge locus defines an envelope that is always more alpha-enhanced than the halo (Figure 1.5). Earlier notions that the bulge is an extension of the halo abundance distribution must be revised; even those bulge stars with [Fe/H] overlapping that of the halo define an upper envelope in the explosive alphas that is not seen in the halo. In the scheme of metal-rich populations, the bulge has clearly experienced the chemical fingerprinting of its unique enrichment process.

4.1 New directions

The bulge provides both a background screen for microlensing events and a rich population of stars for investigating phenomena like planetary transits. The recent Sagittarius Window Exoplanet Probe Survey (SWEEPS) (Sahu *et al.* 2006) has

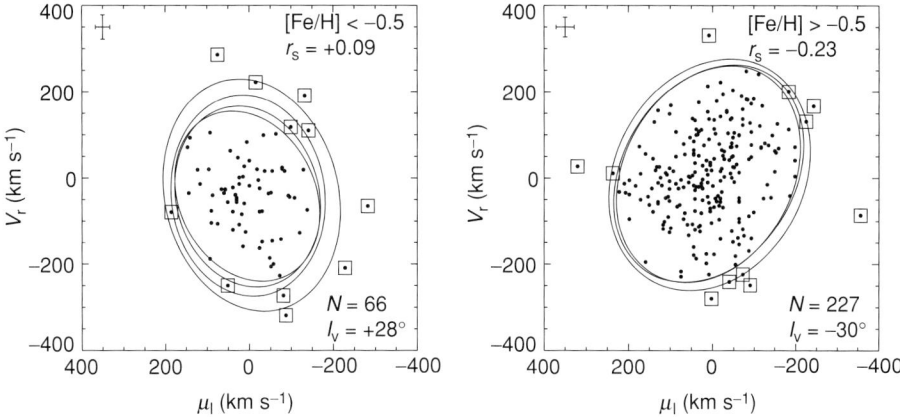

Figure 1.6. Detection of vertex deviation in the bulge/bar population as a function of abundance (Soto et al. 2006). The radial velocity (y-axis) is plotted against the transverse proper motion (x-axis); [Fe/H] is from Sadler et al. (1996) but using the scale of Fulbright et al. (2006a). The rotation of the axes for the metal-rich subset reflects the streaming of stars along the bar. This larger sample confirms the effect first noted by Zhao et al. (1994)

identified 16 transit candidates, some of which have periods <1 day. These ultra-short-period planets are the analogs of hot Jupiters, but they orbit M dwarfs. There is a hint that this class of object is preferentially associated with metal-rich stars, but considerably more work is required before we know how the incidence of bulge-transiting planets depends on metallicity. Although it is beyond the scope of this review, note that there have been several microlensing planet detections toward the bulge, yielding terrestrial-mass planets. A proposed spacecraft, the Microlensing Planet Finder (Bennett et al. 2004) would image continuously a several-square-degree area of the bulge from space, potentially discovering tens of terrestrial planet lenses and 50,000 transit events. Such a mission would settle the question of planetary incidence as a function of galactocentric distance and metallicity.

The advent of the nirspec high-dispersion infrared echelle spectrograph (McLean et al. 1998) at Keck II made possible the determination of detailed abundances for cool and obscured stars. This enables a push toward the Galactic Center. Rich & Origlia (2005) find that bulge M giants have [Fe/H] ∼ Solar with [α/Fe] = +0.3, while Rich & Origlia (2007) find no gradient in both [Fe/H] and [α/Fe] within 4° of the Galactic center (Figure 1.5). Using these powerful infrared techniques but at lower spectral resolution, we plan to extend our work to the nucleus of our nearest spiral galaxy, M31. The OSIRIS spectrograph at Keck is capable of integral field spectroscopy and, in several-hour exposures, should be able to measure abundances for red giants in the bulge of M31, at $R = 2,000$–$3,000$. These observations might unveil for the first time the stellar populations that resemble those in the centers of giant elliptical galaxies.

Combining proper-motion studies with abundances and kinematics yields the surprising result (Figure 1.6) (Soto *et al.* 2006) that the more metal-rich stars in the bulge exhibit a strong vertex deviation (principal-axis rotation) of the velocity ellipsoid. If the metal-rich subset of stars in the bulge have a different formation history from the bulge as a whole, this would seem problematic for any "simple model" of chemical evolution, in which the chemical enrichment has taken place rapidly, without infall or outflow of material. The issue now becomes that of understanding the chemical/dynamical evolution of populations as they enrich past Solar metallicity.

5 Conclusions

The subject of metal-rich stellar populations arose surprisingly early, in the 1950s. When the linear detectors became available, Spinrad and colleagues discovered the first metal-rich populations, but their early work, while ranking abundances correctly, found [Fe/H] values that pushed the bounds of credibility (+0.75 dex for NGC 6791). Peterson (1976) then argued that supermetallicity was an artifact due to boundary cooling in the stellar atmosphere. A series of subsequent studies used a variety of innovative methods to confirm that some stars do indeed have more metals than does the Sun. These included the use of groups of weak iron lines adjacent to continuum points (Gustafsson *et al.* 1974), analysis of the hotter components of binary stars that included CN strong giants (Deming & Butler 1979), and high-resolution, high-SNR, spectroscopy of μ-Leonis itself (Branch *et al.* 1978). The confirmation of supermetallicity extends into the present day (e.g. McWilliam & Rich 1994; Fulbright *et al.* 2006a). The bulge has super-metal-rich stars with selective alpha enhancements, whereas NGC 6791 has subsolar alphas. The emphasis now must be on using the study of metal-rich populations to gain insight about the formation and evolution of galaxies.

Acknowledgments

I wish to thank Bernard Pagel for pointing out the Deming & Butler (1979) paper. I acknowledge support from NSF grant AST-0307931.

References

Baade, W. (1963), in *The Evolution of Galaxies and Stellar Populations*, ed. C. Payne-Gaposchkin (Cambridge MA, Harvard University Press), p. 256
Barth, A. J., Martini, P., Nelson, C. H., & Ho, L. C. (2003) *ApJ* **594**, L95
Becker, G. D., Sargent, W. L. W., Rauch, M., & Simcoe, R. A. (2006) *ApJ* **640**, 69
Burstein, D. *et al.* (1988), *ApJ* **328**, 440

Bellazzini, M. et al. (2003), *A&A* **405**, 867
Bennett, D. P., Bond, I., Cheng, E. et al. (2004), *Proc. SPIE* **5487**, 1453
Branch, D., Bonnell, J., & Tomkin, J. (1978), *ApJ* **225**, 902
Brown, T. M., Ferguson, H. C., Smith, E. et al. (2003), *ApJ* **592**, L17
Brown, T. M., Smith, E., Ferguson, H. C. et al. (2006), *ApJ* **652**, 323
Castro, S., Rich, R. M., Grenon, M., Barbuy, B., & McCarthy, J. K. (1997), *AJ* **111**, 2439
Cayrel, R., Depagne, E., Spite, M. et al. (2004), *A&A* **416**, 1117
Code, A. D. (1969), *PASP* **81**, 475
Deming, D. (1980), *ApJ* **236**, 230
Deming, D., & Butler, D. (1979), *AJ* **84**, 839
Dessauges-Zavadsky, M., Prochaska, J. X., D'Odorico, S., Calura, F., & Matteucci, F. (2006), *A&A* **445**, 93
Durrell, P. R., Harris, W. E., & Pritchet, C. J. (2004), *AJ* **128**, 260
Erb, D. K., Shapley, A. E., Pettini, M., Steidel, C. C., Reddy, N. A., & Adelberger, K. L. (2006), *ApJ* **644**, 813
Faber, S. M., & French, H. B. (1980), *ApJ* **235**, 405
Frogel, J. A., & Whitford, A. E. (1987), *ApJ* **320**, 199
Fulbright, J. P. (2000), *AJ* **120**, 1841
Fulbright, J. P., McWilliam, A., & Rich, R. M. (2006a), *ApJ* **636**, 821
Fulbright, J. P., McWilliam, A., & Rich, R. M. (2006b), ArXiv Astrophysics e-prints, arXiv:astro-ph/0609087
Fukugita, M., Hogan, C. J., & Peebles, P. J. E. (1998), *ApJ* **503**, 518
Gratton, R., Bragaglia, A., Carretta, E., & Tosi, M. (2006), *ApJ* **642**, 462
Gustafsson, B., Kjærgaard, P., & Andersen, S. (1974), *A&A* **34**, 99
Hamann, F., Korista, K. T., Ferland, G. J., Warner, C., & Baldwin, J. (2002), *ApJ* **564**, 592
Kuijken, K., & Rich, R. M. (2002), *ApJ* **124**, 2054
McLean, I. S., et al. (1998), *Proc. SPIE* **3354**, 566
McWilliam, A. (1997), *ARAA* **35**, 503
McWilliam, A., & Rich, R. M. (1994), *ApJ* **91**, 749
  (2004), in *Origin and Evolution of the Chemical Elements*, eds. A. McWilliam & M. Rauch (Pasadena, CA, Carnegie Observatory), p. 38
McWilliam, A., Rich, R. M., & Smecker-Hane, T. A. (2003), *ApJ* **592**, L21
Morgan, W. W. (1958), *Ricerche Astronomiche* **5**, 325
Mouhcine, M., Ferguson, H. C., Rich, R. M., Brown, T. M., & Smith, T. E. (2005a), *ApJ* **633**, 821
Mouhcine, M., Rich, R. M., Ferguson, H. C., Brown, T. M., & Smith, T. E. (2005b), *ApJ* **633**, 828
Mould, J., & Kristian, J. (1986), *ApJ* **305**, 591
O'Connell, R. W. (1999), *ARAA* **37**, 603
Origlia, L., Valenti, E., Rich, R. M., & Ferraro, F. R. (2006), *ApJ* **646**, 499
Ortolani, S., Renzini, A., Gilmozzi, R. et al. (1995), *Nature* **377**, 701
Peterson, R. (1976), *ApJS* **30**, 61
Peterson, R. C., & Green, E. M. (1998), *ApJ* **502**, L39
Pettini, M., Steidel, C. C., Adelberger, K. L., Dickinson, M., & Giavalisco, M. (2000), *ApJ* **528**, 96
Pompeia, L., Barbuy, B., & Grenon, M. (2002), *ApJ* **566**, 845
  (2003), *ApJ* **592**, 1173
Rich, R. M. (1988), *AJ* **95**, 828
  (1990), *ApJ* **362**, 604
Rich, R. M., Sosin, C., Djorgovski, S. G. et al. (1997), *ApJ* **484**, L25

Rich, R. M., & Origlia, L. (2005), *ApJ* **634**, 1293
 (2007), *ApJ* **665**, L119
Rich, R. M., Salim, S., Brinchmann, F. *et al.* (2005), *ApJ* **619**, L107
Sadler, E. M., Rich, R. M., & Terndrup, D. M. (1996), *AJ* **112**, 171
Sahu, K. C., Casertano, S., Bond, H. E. *et al.* (2006), *Nature* **443**, 534
Sbordone, L., Bonifacio, P., Buonanno, R., Marconi, G., Monaco, L., & Zaggia, S. (2006), ArXiv Astrophysics e-prints, arXiv:astro-ph/0612125
Soto, M., Rich, R. M., & Kuijken, K. (2006), ArXiv Astrophysics e-prints, arXiv:astro-ph/0611433
Spinrad, H. (1961), *PASP* **73**, 336
Spinrad, H., & Taylor, B. J. (1969), *ApJ* **157**, 1279
Steidel, C. C., Giavalisco, M., Pettini, M., Dickinson, M., & Adelberger, K. L. (1996), *ApJ* **462**, L17
Tremaine, S., Gebhardt, K., Bender, R. *et al.* (2002), *ApJ* **574**, 740
Walter, F., Carilli, C., Bertoldi, F. *et al.* (2004), *ApJ* **615**, L17
Wheeler, J. C., Sneden, C., & Truran, J. W. Jr. (1989), *ARAA* **27**, 279
Whitford, A. E. (1977), *ApJ* **211**, 527
Whitford, A. E., & Rich, R. M. (1983), *ApJ* **274**, 723
Worthey, G., Faber, S. M., & González, J. J. (1992), *ApJ* **398**, 69
Zhao, H., Spergel, D. N., & Rich, R. M. (1994), *AJ* **108**, 2154
Zoccali, M., Cassis, S., Frogel, J. A. *et al.* (2000), *ApJ* **530**, 418
Zoccali, M., Renzini, A., Ortolani, S. *et al.* (2003), *A&A* **399**, 931
Zoccali, M., Lecureur, A., Barbuy, B. *et al.* (2006), *A&A* **457**, L1

2

The metal-rich nature of stars with planets

Nuno C. Santos[1,2,3]

[1]*Centro de Astronomia e Astrofísica da Universidade de Lisboa, Observatorio Astronomico de Lisboa, Tapada da Ajuda, 1349-018 Lisboa, Portugal*
[2]*Centro de Geofísica de Évora, Rua Romão Ramalho 59, 7000 Évora, Portugal*
[3]*Observatoire de Genève, 51 Chemin des Maillettes, CH-1290 Sauverny, Switzerland*

Several spectroscopic studies have shown that stars with giant planets are particularly metal-rich compared with average field stars. In this paper we review the most recent results concerning the study of the chemical abundances of planet-host stars. Abundance distributions for several elements are presented or discussed, including those of iron-peak and alpha-elements, and the light elements lithium (both ^{7}Li and ^{6}Li) and beryllium. The impact of these results on the theories of planet formation and evolution is discussed.

1 Introduction

Following the discovery of a giant planet orbiting the Solar-type star 51 Peg (Mayor & Queloz 1995), planet hunters have unveiled the presence of about 200 exoworlds.[1] Most of these discoveries were made using the radial-velocity technique, and include ~20 multi-planetary systems, several confirmed transiting planets (e.g. Charbonneau *et al.* 2000; Konacki *et al.* 2003; Bouchy *et al.* 2004), as well as the first Neptune-mass planets (e.g. Santos *et al.* 2004b; McArthur *et al.* 2004).

Globally, these findings brought to light the existence of planets with a huge variety of characteristics, eliciting unexpected questions about the processes of giant-planet formation. For instance, some of the planets are on eccentric orbits (Naef *et al.* 2001), which would be more typical of some comets in the Solar System, while some behemoths have more than 15 times the mass of Jupiter (Udry *et al.* 2002). As a result, the definition of a planet has itself been put into question; see e.g. the discussion in Santos *et al.* (2005a).

---

[1] See the table at http://obswww.unige.ch/exoplanets for continuous updates. Before these discoveries, only planets around a pulsar had been detected (Wolszczan & Frail 1992). Given the violent supernova explosion that gave rise to the pulsar, however, it is believed that these are probably second-generation planets.

The Metal-rich Universe, eds. G. Israelian and G. Meynet. Published by Cambridge University Press.
© Cambridge University Press 2008.

With the number of detected exoplanets increasing very fast, current results are already giving us the chance to undertake the first statistical studies of their properties, as well as of their host stars (e.g. Cumming *et al.* 1999; Zucker *et al.* 2002; Udry *et al.* 2003; Santos *et al.* 2003; Eggenberger *et al.* 2004; Halbwachs *et al.* 2005). These studies, together with theoretical models, are now helping us to better our understanding of the way planets are formed and evolve. In this paper we will review one of the major observational constraints regarding this issue: the study of the metallicity of stars with giant planets.

2 The metallicity of stars with planets

The discussion about the origin of the metal-rich stars in our Galaxy is nowadays intimately related to the study of stars with giant planets. Soon after the discovery of the first extra-Solar planets, it was noticed that planet-host stars are particularly metal-rich compared with "single" field dwarfs (González 1998; González *et al.* 2001; Santos *et al.* 2001, 2004a, 2005b; Fischer & Valenti 2005), i.e. on average they present a metal content greater than that found in stars not known to be orbited by any planetary-mass companion. This result, which has clearly been confirmed by a uniform spectroscopic analysis of large samples of stars with and without detected giant planets (Santos *et al.* 2001) is obtained by using various kinds of techniques to derive the stellar metallicity (e.g. Reid 2002). Furthermore, it is found both for Solar-neighborhood planet-hosts and for their most distant counterparts (Santos *et al.* 2006b), such as the ones found by the OGLE photometric transit campaign.

It was shown that the metallicity excess observed cannot be explained by any sampling or observational biases (Santos *et al.* 2003). Planet-host stars are indeed significantly more metal-rich than stars without known giant planets. The average metallicity difference between the two samples is $\sim$0.25 dex.

Furthermore, and most importantly, the results show that the probability of finding a planet is proportional to the metallicity of the star: more metal-rich stars have a higher probability of harboring a planet than do objects of lower metallicity (e.g. Santos *et al.* 2001, 2003, 2004a; Reid 2002; Fischer & Valenti 2005) – see the lower right panel of Figure 2.1. About 3% of Solar-metallicity stars seem to harbor a planetary-mass companion, while more than 20% of stars with twice the Solar metallicity have been detected to have orbiting planets.

This result is probably telling us that the probability of forming a giant planet depends strongly on the metallicity of the cloud of gas and dust that gave rise to the star and planetary system. Although it is unwise to draw any strong conclusions on the basis of just one point, it is also worth noticing that our own Sun is in the "metal-poor" tail of the planet-host [Fe/H] distribution.

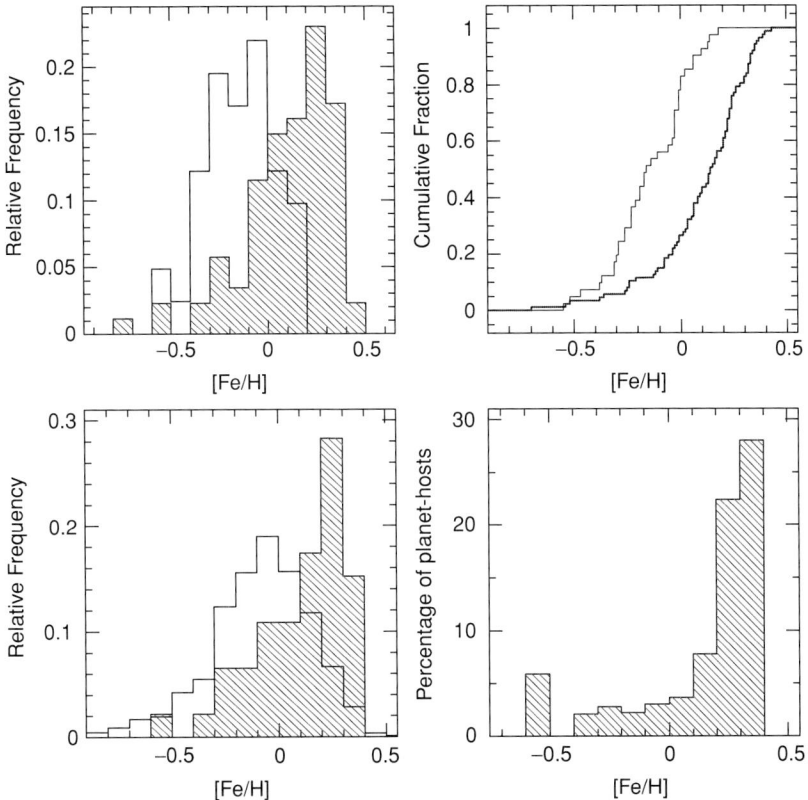

Figure 2.1. Upper panels: [Fe/H] distributions for planet-host stars (hashed histogram) and for a volume-limited comparison sample of stars not known to harbor any planetary-mass companion (open bars). The average difference between the [Fe/H] of the two samples is ∼0.25 dex. A Kolgomorov–Smirnov test showed that the probability that the two samples are parts of the same population is of the order of 10^{-8}. Lower panel, left: [Fe/H] distributions for planet-host stars (hashed histogram) included in the CORALIE planet-search sample, compared with the same distribution for ∼900 stars in the whole CORALIE program (solid-line open histogram). Lower panel, right: the percentage of planet-hosts found among the stars in the CORALIE sample as a function of stellar metallicity. From Santos et al. (2004a).

It is important to remember at this point that the metallicities for the two samples of stars plotted in Figure 2.1 were derived using exactly the same techniques, and are thus to the same scale.

2.1 The origin of the high metallicity

All the conclusions discussed above are true under the assumption that the metallicity excess observed is original to the cloud of gas and dust that gave rise to the star and its planetary system. In other words, we are supposing that the higher prevalence of planets around metal-rich stars is reflecting a higher probability of forming a planet around such a star before the disk dissipates.

However, one other interpretation has been discussed in the literature to explain the [Fe/H] excess observed for stars with planets. In fact, it has been suggested that the high metal content is the result of the accretion of planets and/or planetary material into the star (e.g. González 1998). In such a case, the observed metallicity excess would itself be a by-product of planet formation.

There are multiple ways of deciding between the two scenarios above, and in particular to try to see whether "pollution" might indeed have played an important role in increasing the metal content of the planet-host stars relative to that of their non-planet-host counterparts. Probably the clearest argument is based on stellar internal structure. Material falling onto a star's surface would induce a different increase in [Fe/H] depending on the depth of its convective envelope, which is where mixing can occur. However, no correlation between the metallicity of the planet-host stars and their convective-envelope mass has been found (e.g. Pinsonneault *et al.* 2001; Santos *et al.* 2003).

Some doubts have recently been advanced against the contention that this lack of correlation is a good reason to exclude the possibility that stellar pollution could have caused the observed [Fe/H] "excess" (Vauclair 2004). Furthermore, it has been shown that in a few cases stellar pollution may have played some role (Israelian *et al.* 2001, 2003; Laws & González 2001), although not a strong enough role for it to be responsible for a large variation in [Fe/H]. However, the evidence for an "original" source is further supported by the huge quantities of "pollution" by hydrogen-poor (planetary) material needed to explain the metallicity excess observed for a few late-type very-metal-rich dwarfs known to harbor giant-planets, as well as for a few sub-giant-planet-host stars (Santos *et al.* 2003). Recent asteroseismological measurements were also not conclusive regarding this issue (Bazot *et al.* 2005). In other words, the bulk metallicity "excess" observed most probably has a "primordial" origin.

The study of kinematic properties of planet-host stars has also given some interesting information. In particular, recent results suggest that the planet-host stars have kinematics typical of metal-rich stars in the Solar neighborhood (Ecuvillon *et al.* 2007). This comes as no surprise if we accept that pollution did not play an important role in defining the metallicities for these stars. The results also indicate that planet-hosts may have originated in the inner regions of the Galaxy.

2.2 Implications for the models

These conclusions have many important implications for theories of planet formation. In this respect, two main proposals are now being debated in the literature.

On the one side, the traditional core-accretion scenario (e.g. Pollack *et al.* 1996; Alibert *et al.* 2004) tells us that giant planets are formed as the result of the runaway accretion of gas around a previously formed icy core with about 10–20 times the mass of the Earth. In contrast to this idea, some authors have proposed that giant planets may form by a disk-instability process (Boss 1997).

According to the instability model, the efficiency of planet formation should not be dependent on the metallicity of the star/disk (Boss 2002). This is opposite to what is expected from the traditional core-accretion scenario (Ida & Lin 2004), since the higher the grain content of the disk, the easier it should be for the "metal" cores that will later on accrete gas to form before the gas disk dissipates.

The results presented above, showing that the probability of finding a planet is a strong function of the stellar metallicity, thus favor the core-accretion model as the main mechanism responsible for the formation of giant planets, although they do not completely exclude the disk-instability model). Indeed, it has even been shown that according to the core-accretion model it is possible to predict the observed [Fe/H] distribution of planet-host stars (e.g. Ida & Lin 2004).

3 The situation for other elements

The authors of the majority of the spectroscopic studies on the chemical properties of stars with planets have concentrated their efforts on measuring the abundances of iron as a metallicity proxy. However, the number of studies regarding other elemental abundances is increasing. These concerned the abundances of a variety of elements, from Li and Be (see Section 4), up to alpha and Fe-group elements (e.g. Sadakane *et al.* 1999, 2002; Santos *et al.* 2000, 2006b; González *et al.* 2001; Smith *et al.* 2001; Bodaghee *et al.* 2003; Beirão *et al.* 2005; Ecuvillon *et al.* 2004a, 2004b, 2006a; Gilli *et al.* 2006; Luck & Heiter 2006) and have unveiled a few interesting trends. The main conclusion, though, is that the stars with planets seem to be the metal-rich component of the Solar-neighborhood population (Figure 2.2).

The study of other elements might be of great importance. If stellar pollution is a common occurrence among planet-host stars, or any field star, e.g. Murray *et al.* (2001), Quillem (2002), Wilden (2002), Laws & González (2003), and Shen *et al.* (2005), we can expect to find these to be more enriched in refractory elements, since volatiles could evaporate from infalling bodies before being accreted (Smith *et al.* (2001; Ecuvillon *et al.* 2006b).

Furthermore, elements like C, O, and N (e.g. Ecuvillon *et al.* 2004a, 2006a; Robinson *et al.* 2006) may be essential to form the cores of giant planets. An overabundance of these species could thus increase the efficiency of planet formation. Up to now, however, no clear global evidence of differences between elements of different condensation temperatures has been found regarding any of these matters (Smith *et al.* 2001; Takeda *et al.* 2001; Ecuvillon *et al.* 2006b).

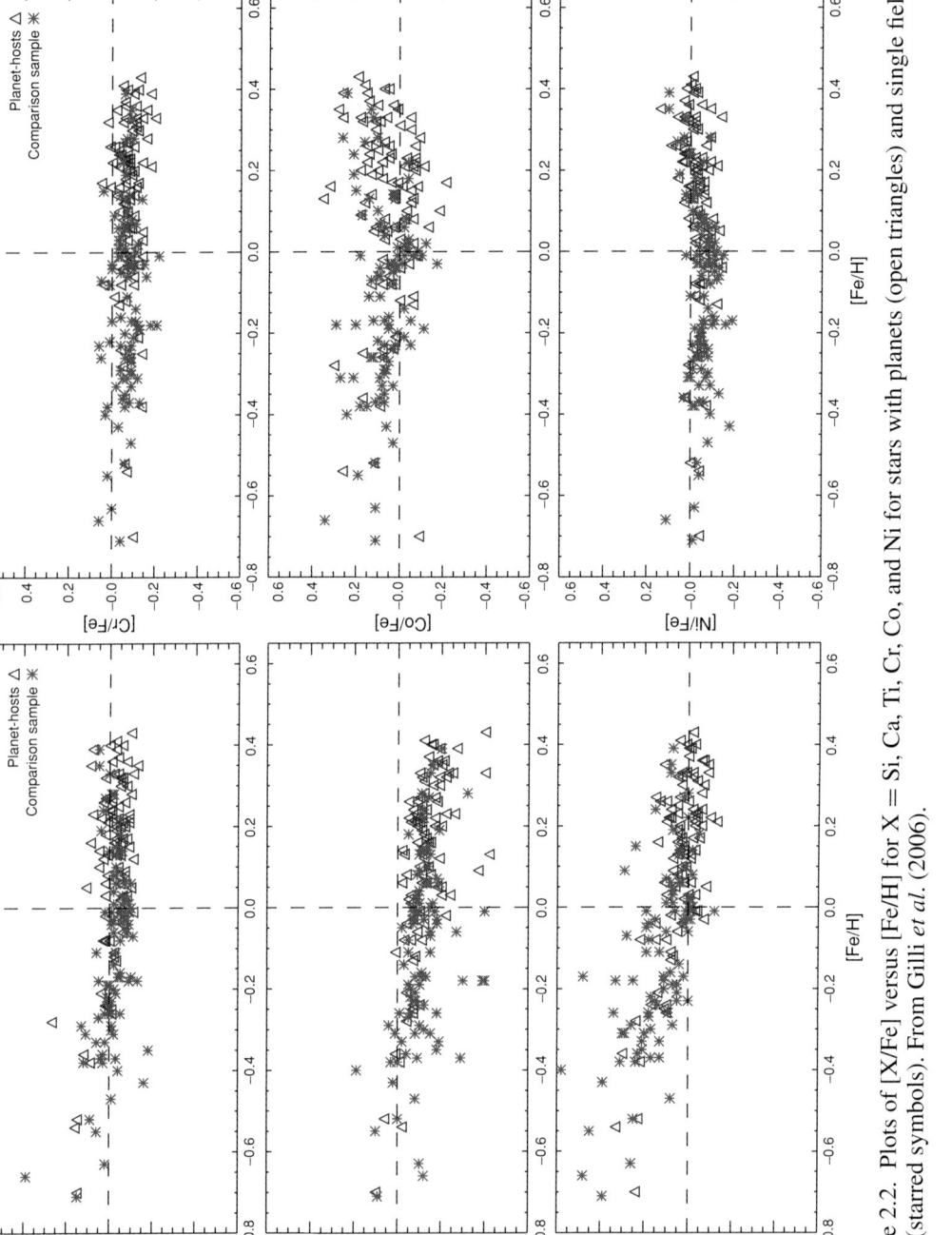

Figure 2.2. Plots of [X/Fe] versus [Fe/H] for X = Si, Ca, Ti, Cr, Co, and Ni for stars with planets (open triangles) and single field stars (starred symbols). From Gilli *et al.* (2006).

4 Light elements

The study of the light elements has an enormous potential for the understanding of planet formation. First, it is well known that light elements and their abundance ratios are good tracers of stellar internal mixing and rotation (e.g. Stephens *et al.* 1997). From the several mixing mechanisms that have been referred to in the literature as responsible for the depletion of light elements in Solar-type stars, rotation and angular-momentum loss are among the leading processes. The study of light-element abundances may thus probably tell us much about processes related to the angular-momentum evolution of planet-host stars. If the formation of giant planets needs the presence of massive proto-planetary disks, we can consequently expect that planet-hosts and "single" stars might have had a different angular-momentum history, thus presenting different light-element abundances. Finally, an angular-momentum variation can also be induced by the accretion of planetary-mass bodies by the star (Siess & Livio 1999; Israelian *et al.* 2003).

Furthermore, if at least part of the metal "enrichment" found for planet-host stars is due to stellar "pollution" effects, we should also be able to observe an enhancement in the abundances of the light elements in planet-hosts. This enhancement should be of at least the same order of magnitude as the excess metallicity observed, although subsequent Li and Be depletion could mask the "pollution" effect.

A few studies of the chemical abundances of planet-host stars have now investigated the abundances of the light elements ^{6}Li (Israelian *et al.* 2001, 2003; Reddy *et al.* 2003), ^{7}Li (González & Laws 2000; Ryan 2000; Israelian *et al.* 2004), and ^{9}Be (García López & Pérez de Taoro 1998; Deliyannis *et al.* 2000; Santos *et al.* 2002, 2004c). Overall, and putting aside a few exceptions (see Section 4.1), the results of these studies suggest that stars with planets have in general normal light-element abundances, i.e. those typical of field stars, even though a few interesting correlations may have been found. Among these, one of the most interesting is probably the fact that, in the temperature interval between ∼5,600 and 5,850 K, planet-hosts seem to have systematically lower Li abundances than do "average" field stars (e.g. Israelian *et al.* 2004; Chen & Zhao 2006) – see Figure 2.3.

4.1 Hints of pollution?

Although, as we have seen in the previous section, the bulk stellar metallicity "excess" observed seems to have a "primordial" origin, some hints of stellar "pollution" related to planet-host stars have been discussed in the literature (e.g. Laws & González 2001; Israelian *et al.* 2001, 2003). One of the clearest of these has to do with the detection of significant amounts of ^{6}Li in the atmosphere of the metal-rich Solar-type dwarf HD 82943.

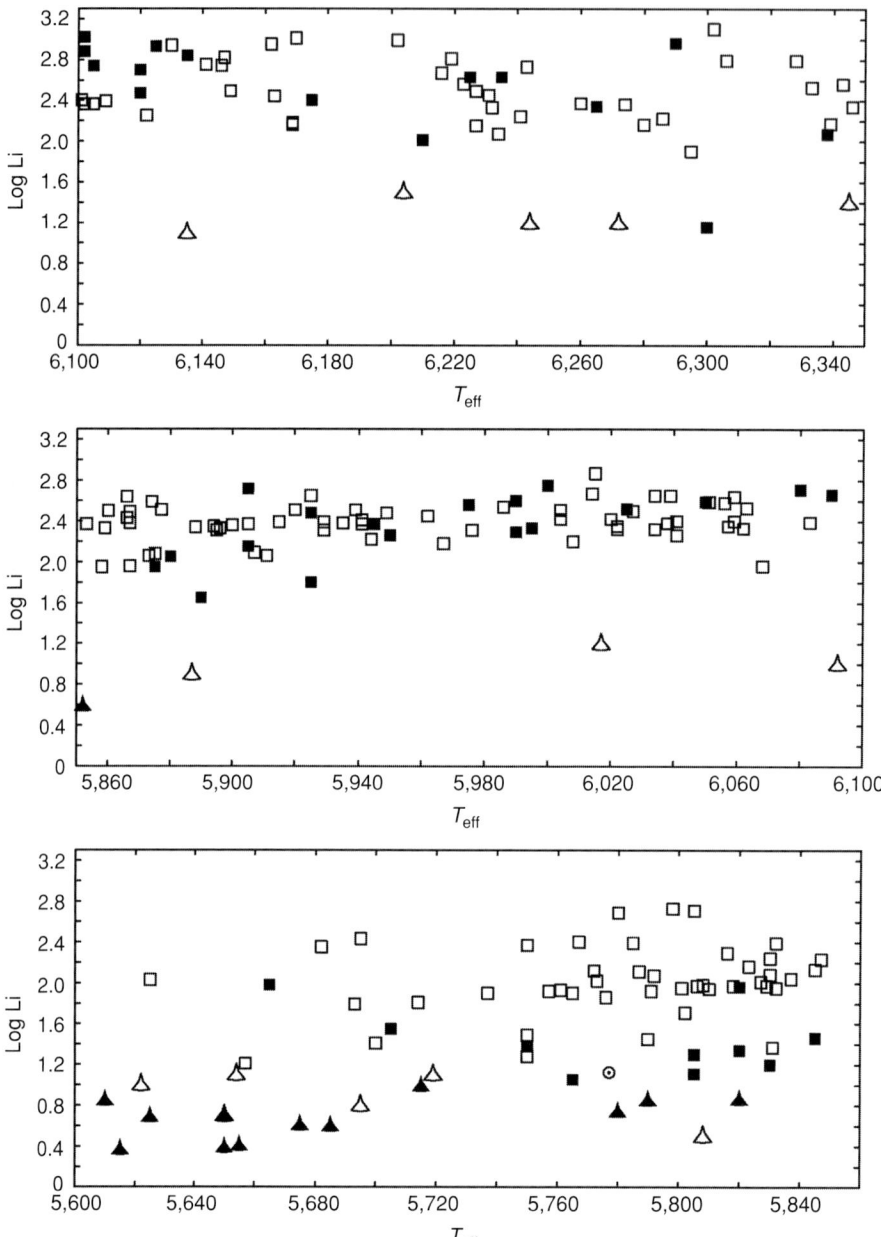

Figure 2.3. Lithium (Li) abundances as a function of effective temperature for planet-host stars (filled symbols) and field stars in three different temperature regimes. In the temperature interval between ∼5,600 and 5,850 K, planet-hosts seem to have systematically lower Li abundances. From Israelian *et al.* (2004).

The rare isotope ^{6}Li provides a unique way of looking for traces of "pollution." Until recently, this isotope had been detected in only a few metal-poor halo and disk stars, but never with a high level of confidence in any metal-rich or even Solar-metallicity star. Standard models of stellar evolution predict that ^{6}Li nuclei are efficiently destroyed during the early evolution of Solar-type (and -metallicity) stars and disappear from their atmospheres within a few million years. Planets, however, do not reach high enough temperatures to burn ^{6}Li nuclei, and fully preserve their primordial content of this isotope. A planet engulfed by its parent star would boost the star's atmospheric abundance of ^{6}Li.

In fact, as discussed in Israelian *et al.* (2003), planet (or planetesimal) engulfment seems to be the only convincing and the least speculative way of explaining the presence of this isotope in the atmosphere of HD 82943, a planet-host star known to harbor a system of two resonant giant planets.

It is important to note, however, that the quantity of material that we need to add to the atmosphere of HD 82943 in order to explain the lithium isotopic ratio would not be able to change the [Fe/H] of the star by more than a few hundreaths of a dex. Furthermore, the results presented above, supporting a "primordial" source for the high [Fe/H] of planet-host stars, are not dependent on these probably isolated cases of "pollution".

Note also that we are referring to the infall of planets or planetary material after the star has reached the main-sequence phase and fully developed a convective envelope; if engulfment happens before that, all planetary material will be deeply mixed, and no traces of "pollution" might be found. It is interesting to note that the whole giant-planet-formation phase must take place when a disk of gas (and debris) is present. Massive gas disks may not exist at all when a star like the Sun reaches the main-sequence phase; although it is still not known definitively, inner disks seem to disappear after $\sim$10 Myr – e.g. Haisch *et al.* (2001). Thus, all the "massive" infall that would be capable of changing the measured elemental abundances if the star were already in the main sequence might simply occur too early, explaining why we do not see strong traces of "pollution" (in particular, concerning iron).

5 Metallicity and orbital parameters

Some hints of trends relating the metallicity of the host stars and the orbital parameters of the planets have been discussed already in the literature.

Several authors (e.g. González 1998; Queloz *et al.* 2000; Santos *et al.* 2003; Sozzetti 2004) have considered the possibility that stars with short-period planets may be particularly metal-rich, even among planet-hosts. This result could be supported by theoretical models, since migration mechanisms could induce stellar "pollution" and/or depend on the quantity of planetesimals in the disk (e.g. Lin *et al.* 1996; Murray *et al.* 1998). The excess metallicity observed could, for example, be inherent to planets that have migrated sufficiently to become detectable in the baseline of measurements of the current radial-velocity surveys. However, current results are not statistically significant. Furthermore, the most recent models suggest that "Type II" migration should not be very dependent on the metallicity of the disk (Livio & Pringle 2003).

However, we must note that there exist other possibilities that could explain a putative metallicity–period correlation; see e.g. the discussion in Santos *et al.* (2006a). For instance, increasing the metallicity may decrease the formation timescales of the giant planets (e.g. Ida & Lin 2004), giving them more time to migrate. Furthermore, if giant planets formed around metal-richer stars have bigger cores (Guillot *et al.* 2006), they may be less fragile against evaporation.

A relation between the stellar metallicity and other planetary orbital parameters has also been considered, namely for the eccentricity and planetary minimum mass (Santos *et al.* 2003; Laws *et al.* 2003). Again, no significant trends have been found, although hints for some possible correlations have been pointed out. The same is true concerning the frequency of planets as a function of the stellar mass (Laws *et al.* 2003), although caution is due here, since, as shown by Santos *et al.* (2003), planet-search samples may be strongly biased in stellar mass. Furthermore, it now seems clear that lower-mass stars have a lower frequency of giant planets (X. Bonfils, private communication), something that may be expected from current models (Laughlin *et al.* 2004).

Finally, in the mass regime below $\sim 20 M_{\rm Jup}$, no clear metallicity differences seem to exist between stars having companions with masses above and below ten times the mass of Jupiter (e.g. Santos *et al.* 2003). Such a difference could be expected if these two populations had undergone different formation processes (Rice *et al.* 2003).

6 Particular cases

6.1 Stars with low-mass planets

Interestingly, the well-known correlation between the presence of planets and the stellar metallicity that exists for stars hosting giant planets (e.g. González *et al.* 2001; Santos *et al.* 2001) does not seem to exist for stars hosting their lower-mass counterparts (e.g. Udry *et al.* 2005). The stars with Neptune-mass planets found so far have a rather flat metallicity distribution.

The reason for this lack of correlation may have a theoretical background (Benz *et al.* 2006). Indeed, according to the core-accretion model, stars with high metallicity are supposed to form planetary cores faster. These thus have time to accrete gas (and become giant planets). For planets forming around stars with lower metallicity, the cores never grow to the critical mass above which considerable gas accretion takes place. This way, it is even likely that very-low-mass planets (rocky or icy) are more prevalent around lower-metallicity stars, although some threshold must exist below which no material is available for planets at all.

6.2 Intermediate-mass stars

A similar lack of metallicity–planet correlation may now have been found for intermediate-mass stars hosting giant planets. As recently noted by Da Silva *et al.* (2006), for the known giant stars hosting a planet the metallicity correlation may no longer be valid. The reason for this possible lack of correlation is still under discussion, but the higher mass of the target stars may have something to do with it (e.g. Laughlin *et al.* 2004).

7 Concluding remarks

As we have seen in this review, the study of the stars that have planetary companions is greatly helping us to understand the processes of giant-planet formation and evolution. In particular, these stars were found to be particularly metal-rich (for all the metals studied so far) compared with the average local field dwarfs. While this metallicity excess is in general common to all the elements studied, a few interesting cases may have been found, including the detection of traces of infall of planetary material into the stellar convective zones. Globally speaking though, the excess metallicity seems to reflect the higher metallicity of the gas clouds that gave rise to these stars and their planetary systems. About 25% of the stars in the planet-search samples having [Fe/H] between 0.3 and 0.4 dex have a planetary companion. This proportion falls to a value of less than $\sim$3% for stars with Solar metallicity. These numbers probably reflect the higher probability of forming a planet around a metal-rich star.

Current data lend some support to the core-accretion scenario against the disk-instability model as the "main" mechanism of giant-planet formation. We note, however, that the current data do not allow us to discard the possibility that both situations can occur; the disk-instability model could be, for example, responsible for the formation of the companions (brown dwarfs or planets) around the more metal-deficient stars.

As new planets are added to the lists, new hints of correlation are being found. It is thus extremely important to continue to monitor the trends of chemical abundances of planet-host stars. The analysis of stars for which real Solar-System analogs have been discovered will be of particular interest, since the Sun seems to occupy a region in the low-metallicity tail of the [Fe/H] distribution of stars with giant planets. The study of stars hosting very-low-mass planets, of intermediate-mass stars with planets, and of planet-hosts in different regions of the Galaxy should also merit particular attention.

Acknowledgment

Support from the Fundação para a Ciência e a Tecnologia (Portugal) in the form of a fellowship (reference SFRH/BPD/8116/2002) and a grant (reference POCI/CTE-AST/56453/2004) is gratefully acknowledged.

References

Alibert, Y., Mordasini, C., & Benz, W. (2004), *A&A* **417**, L25
Bazot, M., Vauclair, S., Bouchy, F., & Santos, N. C. (2005), *A&A* **440**, 615
Beirão, P., Santos, N. C., Israelian, G., & Mayor, M. (2005), *A&A* **438**, 251
Benz, W., Mordasini, C., Alibert, Y., & Naef, D. (2006), in L. Arnold, F. Bouchy, & C. Moutou (eds.), *Tenth Anniversary of 51 Peg-b: Status of and Prospects for Hot Jupiter Studies*, Paris: Frontier Group
Bodaghee, A., Santos, N. C., Israelian, G., & Mayor, M. (2003), *A&A* **404**, 715
A. P. Boss (1997), *Science* **276**, 1836
 (2002), *ApJ* **567**, L149
Bouchy, F., Pont, F., Santos, N. C. *et al.* (2004), *A&A* **421**, L13
Charbonneau D., Brown, T. M., Latham, D. W., & Mayor, M. (2000), *ApJ* **529**, L45
Chen, Y. Q., & Zhao, G. (2006), *AJ* **131**, 1816
Cumming, A., Marcy, G. W., & Butler, R. P. (1999), *ApJ* **526**, 890
Da Silva, L., Girardi, L. Pasquini, L. *et al.* (2006), *A&A* **458**, 609
Deliyannis, C. P., Cunha, K., King, J. R., & Boesgaard, A. M. (2000), *AJ* **119**, 2437
Ecuvillon, A., Israelian, G., Santos, N. C. *et al.* (2004a), *A&A* **426**, 619
 (2004b), *A&A* **418**, 703
Ecuvillon, A., Israelian, G., Shchukina, N. G. *et al.* (2006a), *A&A* **445**, 633
Ecuvillon, A., Israelian, G., Santos, N. C. *et al.* (2006b), *A&A* **449**, 809
Ecuvillon, A., Israelian, G., Pont, F., Santos, N. C., & Mayor, M. (2007), *A&A*, **461**, 171
Eggenberger, A., Udry, S., & Mayor, M. (2004), *A&A* **417**, 353
Fischer, D., & Valenti, J. (2005), *ApJ* **622**, 1102
Gilli, G., Israelian, G., Ecuvillon, A., Santos, N. C. & Mayor, M. (2006), *A&A* **449**, 723
García López, R. J., & Pérez de Taoro, M. R. (1998), *A&A* **334**, 599
González, G. (1998), *A&A* **334**, 221
González, G., & Laws, C. (2000), *ApJ* **119**, 390
González, G., Laws, C., Tyagi, S, & Reddy, B. (2001), *AJ* **121**, 432
Guillot, T., Santos, N. C., Pont, F. *et al.* (2006), *A&A* **453**, L21
Haisch K. E. Jr., Lada, E. A., & Lada, C. J. (2001), *ApJ* **553**, L153
Halbwachs, J. L., Mayor, M., & Udry, S. (2005), *A&A* **431**, 1129
Ida, S., & Lin, D. (2004), *ApJ* **616**, 567
Israelian, G., Santos, N. C., Mayor, M., & Rebolo, R. (2001), *Nature* **411**, 163
 (2003), *A&A* **405**, 753
 (2004), *A&A* **414**, 601
Konacki, M., Torres, G., Jha, S., & Sasselov, D. (2003), *Nature* **121**, 507
Laughlin, G., Bodenheimer, P., & Adams, F. (2004), *ApJ* **612**, L73
Laws, C., & González, G. (2001), *ApJ* **553**, 405
 (2003), *ApJ* **595**, 1148

Lin, D. N. C., Bodenheimer, P., & Richardson, D. C. (1996), *Nature* **380**, 606
Livio, M., & Pringle, J. E. (2003), *MNRAS* **346**, L42
Luck, R., & Heiter, U., (2006), *AJ* **131**, 3069
Mayor, M, & Queloz, D. (1995), *Nature* **378**, 355
McArthur, B. E., Endl, M., Cochran, W. D. *et al.* (2004), *ApJ* **614**, L81
Murray, N., Hansen, B., Holman, M., & Tremaine, S. (1998), *Science* **279**, 69
Murray, N., Chaboyer, B., Arras, P., Hansen, B., & Noyes, R. W. (2001), *ApJ* **555**, 801
Naef, D., Latham, D. W., Mayor, M. *et al.* (2001), *A&A* **375**, L27
Pinsonneault, M., DePoy, D., & Coffee, M. (2001), *ApJ* **556**, L59
Pollack, J., Hubickyj, O., Bodenheimer, P. *et al.* (1996), *Icarus* **124**, 62
Queloz, D., Mayor, M., Weber, L. *et al.* (2000), *A&A* **354**, 99
Quillen A. C. (2002), *AJ* **124**, 400
Reddy, B., Lambert, D., Laws, C., González, G., & Covey, K. (2003), *MNRAS* **335**, 1005
Reid, I. N. (2002), *PASP* **114**, 306
Rice, W. K. M., Armitage, P. J., Bonnel, I. A. *et al.* (2003), *MNRAS* **346**, L36
Robinson, S. E., Laughlin, G., Bodenheimer, P., & Fischer, D. (2006), *ApJ* **643**, 484
Ryan, S. (2000), *MNRAS* **316**, L35
Sadakane K., Honda, S., Kawanomoto, S. *et al.* (1999), *PASJ* **51**, 505
Sadakane K., Ohkubo M., Takada Y. *et al.* (2002), *PASJ* **54**, 911
Santos, N. C., Israelian, G., & Mayor, M. (2000), *A&A* **363**, 228
   (2001), *A&A* **373**, 1019
Santos, N. C., García López, R. J., Israelian, G. *et al.* (2002), *A&A* **386**, 1028
Santos, N. C., Israelian, G., Mayor, M. *et al.* (2003), *A&A* **398**, 363
   (2004a), *A&A* **415**, 1153
Santos, N. C., Bouchy, F., Mayor, M. *et al.* (2004b), *A&A* **426**, L19
Santos, N. C., Israelian, G., García López, R. J. *et al.* (2004c), *A&A* **427**, 1085
Santos, N. C., Benz, W., & Mayor, M. (2005a), *Science* **310**, 251
Santos, N. C., Israelian, G., Mayor, M. *et al.* (2005b), *A&A* **437**, 1127
Santos, N. C., Pont, F., Melo, C. *et al.* (2006a), *A&A* **450**, 825
Santos, N. C., Ecuvillon, A., Melo, C. *et al.* (2006b), *A&A* in press
Siess, L., & Livio, M. (1999), *MNRAS* **308**, 1133
Shen, Z.-X., Jones, B., Lin, D. N. C., Liu, X.-W., & Li, S.-L. (2005), *ApJ* **635**, 608
Smith V. V., Cunha K., & Lazzaro D. (2001), *AJ* **121**, 3207
Sozzetti, A. (2004), *MNRAS* **354**, 1194
Stephens, A., Boesgaard, A. M., King, J. R., & Deliyannis, C. P. (1997), *ApJ* **491**, 339
Takeda, Y., Sato, B., Kambe, E. *et al.* (2001), *PASJ* **53**, 1211
Udry, S., Mayor, M., Naef, D. *et al.* (2002), *A&A* **390**, 267
Udry, S., Mayor, M., & Santos, N. (2003), *A&A* **407**, 369
Udry, S. *et al.* (2005), *A&A* **447**, 361
Vauclair, S. (2004), *ApJ* **605**, 874
Wilden, B. S., Jones, B. F., Lin, D. N. C., & Soderblom, D. R. (2002), *AJ* **124**, 2799
Wolszczan, A., & Frail, D. A. (1992), *Nature* **355**, 145
Zucker, S., & Mazeh, T. (2002), *ApJ* **568**, L113

3
Solar chemical peculiarities?

Carlos Allende Prieto

McDonald Observatory and Department of Astronomy, University of Texas, Austin, TX, USA

Results of several investigations of FGK stars in the Solar neighborhood have suggested that thin-disk stars with an iron abundance similar to that of the Sun appear to have higher abundances of other elements, such as silicon, titanium, and nickel. Offsets could arise if the samples contain stars with ages, mean Galactocentric distances, or kinematics that differ on average from the Solar values. They could also arise due to systematic errors in the abundance determinations, if the samples contain stars that are different from the Sun regarding their atmospheric parameters. We re-examine this issue by studying a sample of 80 nearby stars with Solar-like colors and luminosities. Among these Solar *analogs*, the objects with Solar iron abundances exhibit Solar abundances of carbon, silicon, calcium, titanium, and nickel.

1 Introduction

Under the assumption that low-mass dwarf stars with convective envelopes have a surface chemical composition that simply reflects that of their natal clouds, one can use such stars to trace the chemical evolution of the Galaxy. The Sun is then a convenient reference, but are there nearby stars with Solar composition? Or, in other words, is the Solar abundance pattern the norm in the local thin disk?

Inspection of some of the most recent synoptic studies of nearby stars suggests that the Sun's metallicity is slightly off from average (metallicity is here equated to the iron abundance [Fe/H][1]). Nordström *et al.* (2004) obtained metallicities from Strömgren photometry for nearly 14,000 F- and G-type stars within 70 pc,

---

[1] $[Fe/H] = \log_{10}(N(Fe)/N(H)) + 12$, where N is the number density.

The Metal-rich Universe, eds. G. Israelian and G. Meynet. Published by Cambridge University Press.
© Cambridge University Press 2008.

finding that their distribution could be approximated by a Gaussian with a mean of [Fe/H] = −0.14 and a σ of 0.19 dex. Allende Prieto et al. (2004) studied spectroscopically the stars more luminous than $M_V = 6.5$ ($M > 0.76 M_\odot$) within 14.5 pc of the Sun and concluded that their metallicity distribution is centered at [Fe/H] = −0.11 and has a σ of 0.18 dex. Luck & Heiter (2005) derived spectroscopic metallicities for a sample of 114 FGK stars within 15 pc similar to that analyzed by Allende Prieto et al. (2004), finding a metallicity distribution with a consistent width ($\sigma = 0.16$ dex), but centered at a value slightly closer to Solar (−0.07 for the complete sample, and −0.04 when thick-disk stars are excluded). Haywood (2002) has argued that sample selection based on spectral type discriminates against high-metallicity stars, proposing a metallicity distribution (based on photometric indices) for the Solar neighborhood that is centered at the Solar value.

Inevitably, one must ask whether there is any reason to expect the local metallicity distribution to be centered at the Solar value. Chemical differences among the Sun and its neighbors may be reasonable if the age or the Galactic orbit of the Sun is somewhat off from the average for nearby stars. The age distribution or, equivalently, the star-formation history of the Solar neighborhood is an unsolved problem, judging from the discrepant results obtained from analyses of the Hipparcos *HR* diagram (Bertelli & Nasi 2001; Vergely et al. 2002) and studies of stellar activity (e.g. Rocha-Pinto et al. 2000).

What about abundance ratios? Should we expect the ratios such as C/Fe to be fairly uniform at any given iron abundance? Chemical uniformity requires the interstellar medium to be extremely well mixed, but that is precisely what local spectroscopic studies find. Reddy et al. (2003) examined this issue by analyzing high-dispersion spectra of a few hundred stars and were unable to detect any cosmic scatter. The dispersion was as small as 0.03–0.04 dex for many elements, and could be entirely accounted for by considering the uncertainties in the atmospheric parameters. The immediate implication is that the local interstellar medium is well mixed and has been well mixed for many Ga. Such a conclusion is not contradicted by the results of studies of interstellar gas toward bright stars within and beyond the local bubble (e.g. Oliveira et al. 2005).

In this situation it seems only natural to expect the Sun to have abundance ratios similar to those of other low-mass dwarfs in the Solar vicinity with similar metallicity. That is indeed the case for most elements, but there are some striking offsets. The landmark study by Edvardsson et al. (1993) found nearby FGK-type stars with Solar iron abundance to be, on average, richer than the Sun in Na, Al, and Si. Part of this trend, but not all, could be linked to biases in other stellar parameters, such as mean Galactocentric distance and age. More recent studies of nearby low-mass stars kept finding offsets between the abundance ratios of stars with Solar iron abundances and the Sun. For example, Reddy et al. (2003) found small offsets, in the same sense as Edvardsson *et al.* for the ratios C/Fe, N/Fe, K/Fe, S/Fe, Al/Fe, and Si/Fe (and perhaps Na/Fe), but opposite trends for Mn/Fe and V/Fe. Allende Prieto *et al.* (2004) also found similar patterns in their sample for O/Fe, Si/Fe, Ca/Fe, Sc/Fe, Ti/Fe, Ni/Fe, and some neutron-capture elements (Na was not studied).

The lack of consistency among results regarding the existence and size of these chemical offsets is worrisome. Local samples of stars span variable ranges in spectral type, which may be associated with different systematic errors. In order to investigate further the nature of the observed offsets, we have observed a sample of Solar analogs selected from the Hipparcos color–magnitude diagram. We describe the results below.

2 Data and analysis

To select Solar analogs we used the Johnson M_V absolute magnitudes and the $(B-V)$ color indices compiled by Allende Prieto & Lambert (1999) for 17,219 nearby stars ($d < 100$ pc) included in the Hipparcos catalog. Were selected stars within 0.07 mag of the adopted values for the Sun $(B-V, M_V) = (0.65, 4.85)$ accessible to the 9.2-m Hobby–Eberly Telescope (HET) at McDonald Observatory during the first observing period of 2005 (December 2004–March 2005), when the observations were obtained. A list of 130 stars was placed on the HET queue, and 94 were spectroscopically observed.

The observations employed the High Resolution Spectrograph (HRS) (Tull 1998), a fiber-coupled spectrograph, using the first-order diffraction grating g316 as cross disperser to give almost continuous coverage between 407.6 and 783.8 nm. A fiber with a diameter of 2 arcsec fed the 0.625-arcsec-wide slit of the spectrograph, providing a FWHM resolving power of $R \sim 120,000$. The data reduction was carried out with an automated pipeline within IRAF, performing bias correction, flat-fielding, scattered-light correction, extraction, and wavelength calibration based on Th–Ar hollow-cathode spectra.

The stellar effective temperatures, surface gravities, and overall metallicity were derived by a χ^2 fitting of the spectral order containing Hβ (Allende Prieto 2003). First, the procedure was applied to the spectra of FG dwarfs included as part of the Elodie library at a resolving power of $R \sim 10,000$, then the residuals were fit by linear trends. After applying the linear corrections, the rms scatter between our results and those in the Elodie catalog was found to be 1.5%, 0.16 dex, and 0.07 dex for $T_{\rm eff}$, $\log g$, and [Fe/H], respectively.

The HRS spectra were processed in exactly the same manner, after smoothing them to a resolution $R = 10,000$, and the resulting parameters were subjected to the linear corrections inferred from the comparison with the Elodie library.

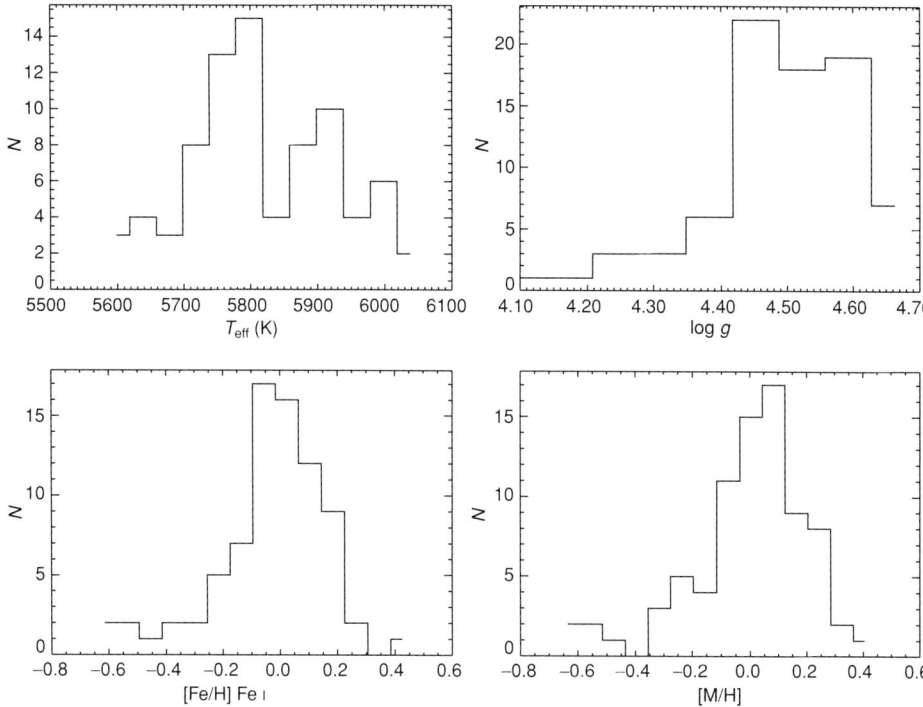

Figure 3.1. Stellar parameters for the sample. Two *metallicity* distributions are shown: [M/H] indicates the values derived from the analysis of the spectral order that includes Hβ (used to select the model atmosphere), and [Fe/H] indicates the values subsequently derived from the analysis of equivalent widths of Fe I lines. The surface gravities shown here correspond to the spectroscopic values derived from the Hβ order, but the true gravities are likely tightly concentrated around the Solar value ($\log g \simeq 4.437$), given the narrow distribution of the sample stars in M_V.

Figure 3.1 shows the distribution of the final atmospheric parameters. Once the basic atmospheric parameters had been constrained, we measured and made use of the equivalent widths of 46 Fe I lines to derive the appropriate value of the microturbulence and the iron abundance using MOOG (Sneden 2002). The iron linelist is a subset of that described in Ramírez *et al.* (2007), and for other elements we used the same lines as Allende Prieto *et al.* (2004). Abundances of C, Si, Ca, Ti, and Ni were also determined assuming LTE.

3 Results and discussion

Figure 3.2 shows our results for silicon and titanium. The thin- and thick-disk membership can easily be decided from these plots. The offset from [Si/Fe] = 0 at [Fe/H] = 0 found in several previous surveys is not apparent in the left-hand panels

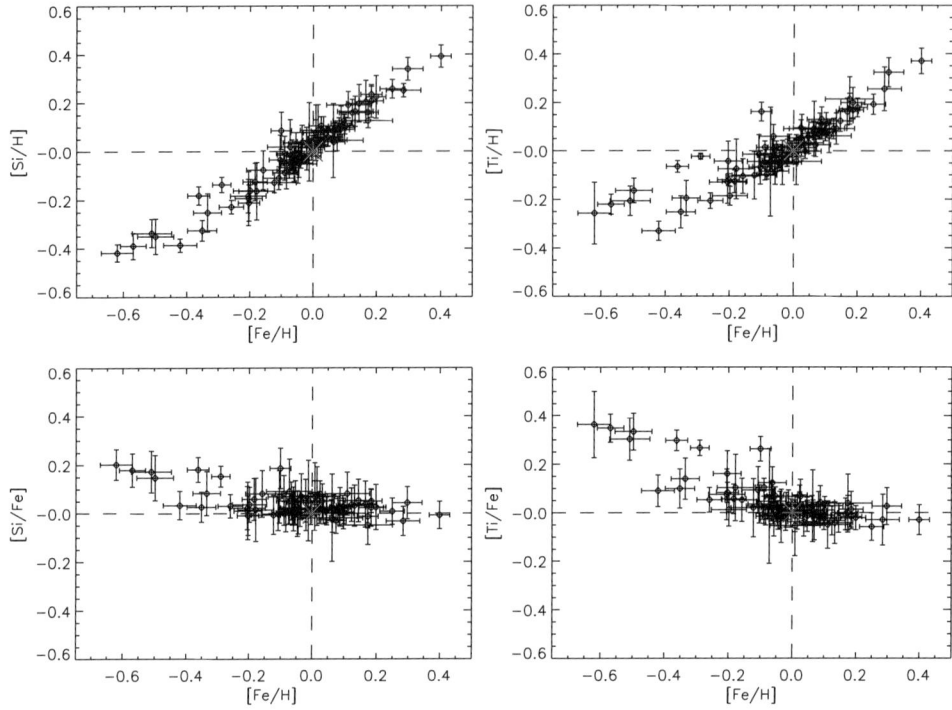

Figure 3.2. Abundances of Si and Ti, and their ratios to the iron abundances for the sample stars. The broken lines mark the location of the Solar reference values.

of Figure 3.2. Similarly, the right-hand panels do not confirm the offset found by Allende Prieto *et al.* (2004) for titanium, and no significant offsets were found for carbon, calcium, titanium, and nickel either. This result suggests that the offsets in previous analyses were likely the result of systematic errors.

Samples of stars spanning a narrow range in atmospheric parameters are well suited for carrying out differential studies of chemical evolution. With such samples, we can potentially minimize the impact of shortcomings in the theory of stellar atmospheres and line formation on the derived abundances, as the small scatter in the abundance ratios shown in Figure 3.2 suggests.

References

Allende Prieto, C. (2003), *MNRAS* **339**, 1111
Allende Prieto, C., Barklem, P. S., Lambert, D. L., & Cunha, K. (2004), *A&A* **420**, 183
Allende Prieto, C., & Lambert, D. L. (1999), *A&A* **352**, 555
Bertelli, G., & Nasi, E. (2001), *AJ* **121**, 1013
Edvardsson, B., Andersen, J., Gustafsson, B., Lambert, D. L., Nissen, P. E., & Tomkin, J. (1993), *A&A* **275**, 101
Haywood, M. (2002), *MNRAS* **337**, 151
Nordström, B., Mayor, M., Andersen, J., Holmberg, J., & Pont, F. (2004), *A&A* **418**, 989

Luck, R. E., & Heiter, U. (2005), *AJ* **129**, 1063
Oliveira, C. M., Dupuis, J., Chayer, P., & Moos, H. W. (2005), *ApJ* **625**, 232
Ramírez, I., Allende Prieto, C., & Lambert, D. L. (2007), *A&A* **465**, 271
Reddy, B. E., Tomkin, J., Lambert, D. L., & Allende Prieto, C. (2003), *MNRAS* **340**, 304
Rocha-Pinto, H. J., Scalo, J., Maciel, W. J., & Flynn, C. (2000), *A&A* **358**, 869
Sneden, C. (2002), http://verdi.as.utexas.edu/moog.html
Tull, R. G. (1998), *Proc. SPIE* **3355**, 387
Vergely, J.-L., Köppen, J., Egret, D., & Bienaymé, O. (2002), *A&A* **390**, 917

4

Kinematics of metal-rich stars with and without planets

Alexandra Ecuvillon,[1] Garik Israelian,[1] Frédéric Pont,[2] Nuno C. Santos[2,3,4] & Michel Mayor[2]

[1]*Instituto de Astrofísica de Canarias, E-38200 La Laguna, Tenerife, Spain*
[2]*Observatoire de Genève, 51 Chemin des Maillettes, CH-1290 Sauverny, Switzerland*
[3]*Observatorio Astronomico de Lisboa, 1349-018 Lisboa, Portugal*
[4]*Centro de Geofísica de Évora, Rua Romão Ramalho 59, 7000 Évora, Portugal*

We present a detailed study on the kinematics of metal-rich stars with and without planets, and their relation with the Hyades, Sirius and Hercules dynamical streams in the Solar neighbourhood. We compare the kinematic behaviour of known planet-host stars with that of the remaining targets belonging to the CORALIE volume-limited sample, in particular its metal-rich population. The high average metallicity of the Hyades stream is confirmed. The planet-host targets exhibit a kinematic behaviour similar to that of the metal-rich comparison subsample, rather than to that of the comparison sample as a whole, thus supporting the hypothesis of a primordial origin for the metal excess observed in stars with known planetary companions. According to the scenarios proposed as an explanation for the dynamical streams, systems with giant planets could have formed more easily in metal-rich inner Galactic regions

1 Introduction

The kinematic distribution of stars in the Solar neighbourhood is far from smooth. According to the standard Eggen scenario, stars in a moving group were formed simultaneously in a small phase-space volume, so that we can still observe a stream of young stars with similar velocities. However, several results have raised problems with this interpretation (e.g. Chereul *et al.* 1999; Dehnen 1998). Among the alternative explanations proposed, the most likely is that the substructures observed in the velocity distribution are caused by purely dynamical mechanisms. Authors of several works (e.g. Dehnen 2000; Sellwood & Binney 2002; De Simone *et al.* 2004) found that non-axysimmetric components in the Galactic disc, such as the rotating Galactic bar and spiral waves, may cause radial migration. In this framework, the

The Metal-rich Universe, eds. G. Israelian and G. Meynet. Published by Cambridge University Press.
© Cambridge University Press 2008.

Hyades and Sirius streams would be the outward- and inward-moving streams of stars formed in inner/outer Galactic regions (Famaey *et al.* 2005).

It is not unlikely that known exoplanetary systems could also have suffered such radial displacements. The typical contributors to the kinematic structures in the Solar neighbourhood are metal-rich old stars (Raboud *et al.* 1998). The well-known metal-rich nature of planet-host stars (e.g. Santos *et al.* 2005 and references therein) might thus be related to dynamical streams. It is probable that planet-host stars formed in protoplanetary clouds with a high metal content, such as those located in inner regions of the Milky Way, and were then brought closer to the Sun by dynamical streams (Famaey *et al.* 2005). Famaey *et al.* (2005) have found an indicative clump of stars hosting planets from the data of Santos *et al.* (2003) in the Hyades streams. Several works have investigated the kinematics of stars with extrasolar planets (e.g. Barbieri & Gratton 2002; Santos *et al.* 2003), without finding any significative peculiarity.

In this work we study the kinematics of planet-host stars and their relation to the dynamical streams in the Solar neighbourhood and investigate whether this relation corresponds to the behaviour of the metal-rich population. The idea is that, if the metal-rich nature of planet-host stars is intrinsic, then they should be kinematically identical to the metal-rich comparison sample, whereas if the metallicity excess is due to planet-related enrichment, then they should be kinematically identical to the total comparison sample.

2 Data and analysis

We used precise radial-velocity measurements and cross-correlation-function parameters from the CORALIE database, and parallaxes, photometry and proper motions from the HIPPARCOS and Tycho-2 catalogues, to derive accurate kinematics for all the targets belonging to the volume-limited CORALIE planet-search survey. The segregation between thin- and thick-disc members was done according to the procedure described by Bensby *et al.* (2003). Chemical abundances for CORALIE stars already spectroscopically studied were extracted from Ecuvillon *et al.* (2004a, 2004b, 2006) and Gilli *et al.* (2006) and combined with the kinematic information in order to analyse the relation of the trends [X/Fe] versus [Fe/H] to the thin and thick discs.

We obtained the density distribution of space velocities by applying the fifth-nearest-neighbour method. A Gaussian background was removed in order to obtain the ellipsoids corresponding to the remnant structures, the Hyades, Sirius and Hercules *dynamical streams*. The Hyades stream was then analysed by the following two methods: (i) counting the number of stars of each group – stars with planets, comparison sample and metal-rich comparison subsample – in the Hyades and field

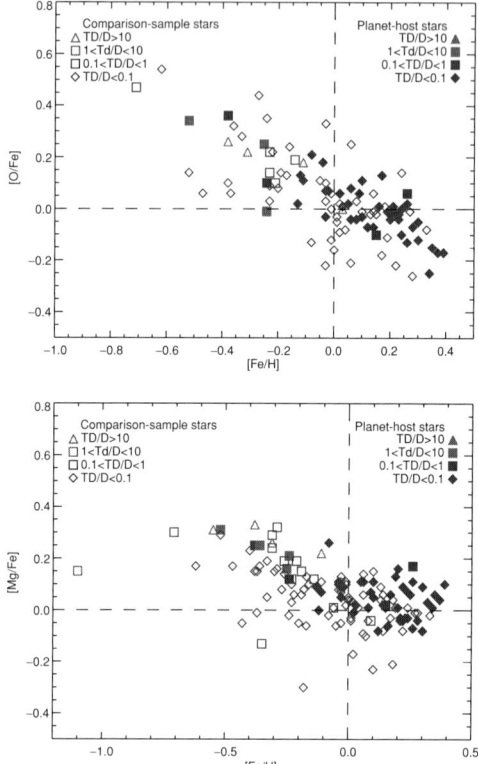

Figure 4.1. Left panels: [O/Fe] and [Mg/Fe] versus [Fe/H] for CORALIE targets with detailed chemical abundances. Full and intermediate members of the thick/thin-disc population are indicated with red triangles and squares/blue diamonds and squares, respectively. Right panel: the density distribution of Galactic space velocities in the (V, U) plane. The uncorrected, subtracted Gaussian and corrected distributions are presented in the upper, central and lower panels, respectively. The field, Hyades, Sirius and Hercules ellipsoids are indicated by the blue, red, blue green and yellow lines. (The right panel can be seen, in colour, only on the book's website.)

ellipsoids, and then subtracting the expected numbers corresponding to Gaussian distributions; and (ii) computing the average overdensities (from the fifth-nearest-neighbour method) in the Hyades and field ellipsoids, and subtracting the average in the field from that in the Hyades. We refer the reader to Ecuvillon *et al.* (2007) for a detailed description of the data and analysis.

3 Results and discussion

Concerning the thick/thin-disc analysis (see Figure 4.1, left panels), we found that the planet-host star HD 4308, with [Fe/H] = -0.27, exhibits a kinematic behaviour

Table 4.1. *Results for the two methods. The numbers of stars in the field and Hyades ellipsoids and the excess ratio in the Hyades stream are listed in the first columns, while the last column indicates the average overdensity in the Hyades ellipsoid. The assumed errors are Poissonian.*

Group	Counting method			Overdensity method
	Field	Hyades	Excess	Overdensity
Planet-hosts	47	10	0.16 ± 0.06	0.029 ± 0.018
Whole comparison	440	40	0.05 ± 0.01	0.004 ± 0.004
Metal-rich comparison	119	19	0.11 ± 0.04	0.020 ± 0.011
Solar comparison	321	21	0.03 ± 0.02	0.003 ± 0.005

typical of the thick-disc population. Moreover, we confirm the possibility of thick-disc members with supersolar metallicity, since one comparison-sample star out of 21 belonging to the thick disc was found to have [Fe/H] > 0. However, we did not observe the significant α enrichment reported by other authors for the thick disc.

From the density-distribution analysis (see Figure 4.1, right panels), the high average metallicity of the Hyades-stream members is confirmed, and this is in agreement with the scenario of non-axisymmetric perturbations as the source of radial displacement (Famaey *et al.* 2005). The two methods used in the comparison between planet-host stars and the comparsion sample gave consistent results (see Table 4.1): the kinematics of planet-host stars is different by 1.5σ from that of the whole comparison sample and similar to that of the comparison subsample of metal-rich stars.

This is a strong argument in favour of a primordial origin of the metal excess in planet-host stars. This result, linked with the mechanism proposed to explain the presence of the Hyades stream, suggests that stars with giant planets might have formed more easily at inner Galactic radii, in a metal-rich interstellar medium, and then suffered radial displacements taking them to the Solar neighbourhood.

4 Concluding remarks

The main outcomes of this work are the following.

- The CORALIE planet-host target HD 4308 has been found to be a very probable member of the Galactic thick disc, together with another 21 comparison stars.
- No significant differences or peculiar enrichment in α-elements are observed in the abundance patterns of the subset of stars with homogeneous abundance determinations from previous studies. However, the small number of thick-disc members prevents us from obtaining conclusive results.

- The tentative existence of thick-disc stars at supersolar metallicities is confirmed by the appearance of one thick-disc member out of 21 with high metallicity (HD 152391 with [Fe/H] = 0.03).
- The high average metallicity of the Hyades stream reported in previous works is confirmed. Our results are in agreement with the scenario of non-axysimmetric perturbations, such as transient spiral waves, which has been proposed on the basis of several results of simulations and observations. According to this model, the Hyades stream would be the outward-moving stream of high-metallicity stars born at inner Galactic radii and then pushed into the Solar neighborhood. We have also found a new clue confirming the Solar and subsolar average metallicity suggested for the Sirius stream.
- The group of planet-host stars exhibits a kinematic behaviour much more similar to that of the metal-rich comparison subsample than to that of the whole comparison sample, in the sense that the overdensity in the Hyades stream observed for stars with planets is much higher than that observed for the whole comparison group. This result has been reached by use of the two methods independently and strongly supports the hypothesis of a primordial origin for the metal excess observed in stars with giant planets.
- If we interpret our results within the scenario proposed to explain the presence and origin of dynamical streams in the Solar neighbourhood, stars with giant planets could have formed more easily at inner Galactic radii in a more metal-rich interstellar medium and then suffered radial displacements due to a non-axysimmetric component of the Galactic potential, pushing them into the Solar neighbourhood.

References

Barbieri, M., & Gratton, R. G. (2002), *A&A* **384**, 879–883
Bensby, T., Feltzing, S., & Lundström, I. (2003), *A&A* **410**, 527–551
Chereul, E., Creze, M., & Bienayme, O. (1999), *A&A Suppl.* **135**, 5–28
Dehnen, W. (1998), *AJ* **115**, 2384–2396
  (2000), *AJ* **119**, 800–812
De Simone, R., Wu, X., & Tremaine, S. (2004), *MNRAS* **350**, 627–643
Ecuvillon, A., Israelian, G., Santos, N. C. *et al.* (2004a), *A&A* **418**, 703–715
  (2004b), *A&A* **426**, 619–630
Ecuvillon, A., Israelian, G., Shchukina, N. G. *et al.* (2006), *A&A* **445**, 633–645
Ecuvillon, A., Israelian, G., Pont, F., Santos, N. C., & Mayor, M. (2007), *A&A* **461**, 171
Famaey, B., Jorissen, A., Luri, X. *et al.* (2005), *A&A* **430**, 165–186
Gilli, G., Israelian, G., Ecuvillon, A., Santos, N. C., & Mayor, M. (2006), *A&A* **449**, 723–736
Raboud, D., Grenon, M., Martinet, L. *et al.* (1998), *A&A* **335**, L61–L64
Santos, N. C., Israelian, G., Mayor, M. *et al.* (2003), *A&A* **398**, 363–376
  (2005), *A&A* **437**, 1127–1133
Sellwood, J. A., & Binney, J. J. (2002), *MNRAS* **336**, 785–796

5

Elemental abundance trends in the metal-rich thin and thick disks

Sofia Feltzing

Lund Observatory, Box 43, SE-221 00 Lund, Sweden

Thick disks are common in spiral and S0 galaxies and seem to be an inherent part of galaxy formation and evolution. Our own Milky Way is host to an old thick disk. The stars associated with this disk are enhanced in the α-elements relative to similar stars present in the thin disk. The Milky Way thin disk also appears to be younger than the thick disk. Elemental-abundance trends in stellar samples associated with the thin and thick disks in the Milky Way are reviewed. Special attention is paid to how such samples are selected. Our current understanding of the elemental abundances and ages in the Milky Way thick and thin disks is summarized and discussed. The need for differential studies is stressed. Finally, formation scenarios for the thick disk are briefly discussed in the light of the current observational picture.

1 Introduction

Thick disks appear to be ubiquitous in spiral and S0 galaxies (e.g. Schwarzkopf & Dettmar 2000; Dalcanton & Bernstein 2002; Davidge 2005; Mould 2005; Elmegreen & Elmegreen 2006) and the Milky Way is no different since it hosts a thick disk in addition to the thin disk (Gilmore & Reid 1983). The Milky Way thick disk has a scale height of about 1 kpc, which is three times that of the thin disk.

Available instrumentation and telescopes limit us to studying the most nearby stars if we wish to derive their elemental abundances and study the chemical history of the Milky Way. Several stellar populations overlap in the Solar neighborhood. The major components are the thin and thick disks and the halo. There is also a multitude of streams and so-called moving groups. The thick disk lags behind the local standard of rest by $\sim 46\,\mathrm{km\,s^{-1}}$ and the two disks have different velocity

The Metal-rich Universe, eds. G. Israelian and G. Meynet. Published by Cambridge University Press.
© Cambridge University Press 2008.

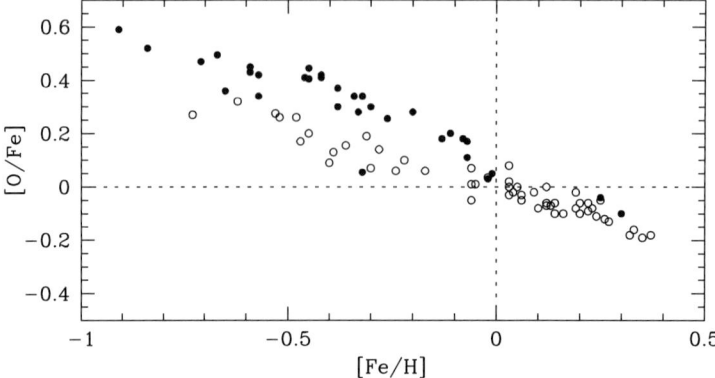

Figure 5.1. A plot of [O/Fe] versus [Fe/H] for stars with kinematics typical of the thick disk (marked with •) and stars with kinematics typical of the thin disk (marked with ○). The data are taken from Bensby *et al.* (2004a & 2005).

dispersions. We are thus able to distinguish the thin- and thick-disk stars from each other (at least on a statistical basis) by using kinematic information.

It has proved very fruitful to combine kinematic and chemical information for stars to derive the history of the disks in the Milky Way. Edvardsson *et al.* (1993) published one of the very first studies fully exploiting this technique. In subsequent studies utilizing the combination of kinematic and elemental-abundance information all thick-disk stars that have been studied to date have been found to be older than the thin-disk stars (e.g. Bensby *et al.* 2003; Fuhrmann 2004). It is also now well established that stars in the Solar neighborhood with kinematics typical of the thick disk are, at a given [Fe/H], enhanced in α-elements relative to the thin-disk stars (e.g. Bensby *et al.* 2004; Fuhrmann 2004) (Figure 5.1).

2 How to define a thick-disk star – selecting stars for spectroscopy

Stars in the thick disk rotate more slowly, in the plane, around the Galactic Center than do the thin-disk stars. The thick-disk stars also move higher above the Galactic plane than do the thin-disk stars. The velocity dispersions in all three Galactic velocities for the thick disk are also larger than the equivalent dispersions for the thin disk.

In the Solar neighborhood we see a mixture of stars from both disks and from the halo. In very rough numbers, we find that ten per hundred stars are thick-disk stars and the rest are thin-disk stars, and that one per thousand Solar-neighborhood stars is a halo star (Buser 2000). If we want to study the elemental abundances in stars that we believe belong to the thick disk we need to decide how to select appropriate targets. There are essentially two ways to do that: *position* – sufficiently high above the plane that the star is more likely a thick-disk than a thin-disk star; or *kinematics* – various kinematic criteria may be formulated to distinguish the disks from one another.

Various kinematic criteria have been used. The following three examples highlight some of the differences and similarities. It is interesting to note, however, that the resulting abundance trends essentially show the same results.

2.1 *Bensby* et al. *(2003, 2005)*

All their stars are from the Hipparcos catalog and data on their radial velocities as well as Strömgren photometry are available in the literature. For these stars a kinematic selection was done in the following way: assume that the velocity components have Gaussian distributions unique to each population (i.e. halo, thick disk, thin disk); allow for the different asymmetric drifts; calculate the probability that each star belongs to the halo, thick disk, and thin disk, respectively; then, for the thick-disk component they selected stars that were more likely to be thick-disk than thin-disk stars, and vice versa for the thin disk. It turns out that the selection is not very sensitive to the local normalizations of the number densities for the disks (Bensby *et al.* 2005). This also shows, as can be expected from the procedure itself, that two fairly extreme samples are selected.

2.2 *Gratton* et al. *(2003)*

Accurate parallaxes from the Hipparcos catalog and radial velocities from the literature were used to calculate the orbital parameters and space velocities for the stars. The stars were then subdivided into three categories: an inner rotating population; a second population containing non-rotating and rotating stars; and a third category containing the thin disk. The last category is confined to the Galactic plane as defined by the orbital parameters of the stars (i.e. maximum height reached above the plane and eccentricity). The second category is identified as the halo; and the first includes part of what is in the two other studies called the halo and all of their thick disk. This population is referred to as the dissipative component since the authors were not able, in their kinematic as well as abundance data, to find any discontinuity between what is generally called the halo and the thick disk.

2.3 *Reddy* et al. *(2003, 2006)*

All stars are from the Hipparcos catalog and have radial velocities available in the literature. A cutoff distance of 150 pc was imposed in order to avoid problems with reddening. An initial selection of stars belonging to the thin and thick disks was done by imposing cuts in V_{LSR} and W_{LSR}. These selections are also "verified" by computing probabilities akin to those in e.g. Bensby *et al.* (2003).

2.4 Discussion

These methods also impose criteria such that only stars within a fairly narrow range of effective temperature and log g are selected. Hence, it becomes possible to (1) carry out a differential study and (2) obtain ages for the stars on the basis of their positions in the H–R diagram.

Additionally, Klaus Fuhrmann has in a series of papers (Fuhrmann 1998; 2004) and further, unpublished work investigated a sample containing all mid-F to early-K dwarf stars within 25 pc and with $M_V = 6$ and location north of declination $\delta = 15°$. Since such a sample is mainly made up of thin-disk stars, he added stars at larger distances that are assumed representative of the thick disk and halo. However, the basic criteria used to assign a star to either thin or thick disk have evolved between the papers. In the first paper the chemical signatures (i.e. Mg abundances) were the major criteria, whereas in the second paper age is envisioned as the criterion that will distinguish a star as belonging to the thin or the thick disk. It also turns out that these assignments do agree with kinematic classifications based on e.g. V_{LSR} and total velocity.

However, a more robust and straightforward method appears to be first identifying stars according to reproducible kinematic criteria and then studying their abundances and ages, since we do not a priori have knowledge about what the thick disk is but want to find out.

3 The abundance trends in the thick and thin disks

Recent studies of the elemental abundance trends in the thin and thick disks include the following differential studies: Fuhrmann (1998, 2004), Chen *et al.* (2000), Mashonkina *et al.* (2003), Gratton *et al.* (2003), Bensby *et al.* (2003, 2004a, 2005), Mishenina *et al.* (2004), Bensby & Feltzing (2006), and Feltzing *et al.* (2006). These are complemented by studies that have focused on just one of the disks. The two most important studies of thick-disk stars along are Prochaska *et al.* (2000) and Reddy *et al.* (2006). For the thin disk Reddy *et al.* (2003) and Allende Prieto *et al.* (2004) are of particular interest.

The main findings from these studies may be summarized as follows.

(a) At a given [Fe/H] the stars with kinematics typical of the thick disk are more enhanced in α-elements than are the stars with kinematics typical of the thin disk (e.g. Bensby *et al.* 2003, 2005; Fuhrmann 1998, 2004; Gratton *et al.* 2003).
(b) There are also differences between the two disks in terms of other elements, e.g. Ba, Al, Eu, and Mn (Mashonkina *et al.* 2003; Bensby *et al.* 2005; Feltzing 2006; Feltzing *et al.* 2006).
(c) The elemental-abundance trends for the kinematically selected samples are tight (Bensby *et al.* 2004a; Reddy *et al.* 2006).

(d) In studies that follow stars with kinematics typical of the thick disk up to Solar metallicities a downward trend in e.g. [O/Fe] as a function of [Fe/H] has been noted. This is most easily interpreted as a contribution from SN Ia to the chemical enrichment (Figure 5.1) (Bensby *et al.* 2003).

An essential part of the studies cited in the list above is that they all employ a differential method. That is, in the study both stars with kinematics typical of the thin disk and stars with kinematics typical of the thick disk are included. Furthermore, the stars span only narrow ranges in effective temperature and $\log g$. This means that, to first order, any modeling errors in the abundance determination cancel out. It is important to note that when data from different studies are combined the distinct trends are often blurred because it is very difficult to put data from different studies on the same baseline insofor as the derived abundances are concerned.

It is interesting and important to note that, although authors apply various kinematic selection criteria, the results are robust and remain the same. This implies that the currently used criteria are selecting essentially the same stellar populations. The important fact we have learnt in the last decade is that stars that occupy the velocity space associated with the thick disk have elemental abundance trends that are distinct from the trends traced by stars with kinematics typical of the thin disk.

3.1 Vertical structures

Changes in the properties of the stellar populations as a function of height above the Galactic plane are important clues regarding the formation of the Galactic disk system. A slow, monolithic collapse would, for example, result in clear trends such as that the mean metallicity would increase with decreasing distance from the Galactic plane. If instead the thick disk formed from an originally thin disk that was later puffed up in a merger event, we should see no such trends. In that case we would also expect the abundance trends in the thick disk to be the same at all heights.

Gilmore *et al.* (1995) studied the metallicity distribution function at 1.5 and 2 kpc above the galactic plane. They found no differences between the two distributions. Davidge (2005) and Mould (2005) also find that there is no appreciable gradient in the colors of the stellar populations as a function of the height above the Galactic plane in nearby spiral galaxies. This is consistent with the result found for the Milky Way by Gilmore *et al.* (1995).

In Figure 5.2 we have divided the stars with kinematics typical of the thick disk into two samples on the basis of how far their $W_{\rm LSR}$ velocities will take them above the galactic plane (Bensby *et al.* 2005). The two samples exhibit exactly the same abundance trends. These findings appear to exclude the possibility of a monolithic collapse for the formation of the thick disk and favor a puffing-up scenario. Tentative

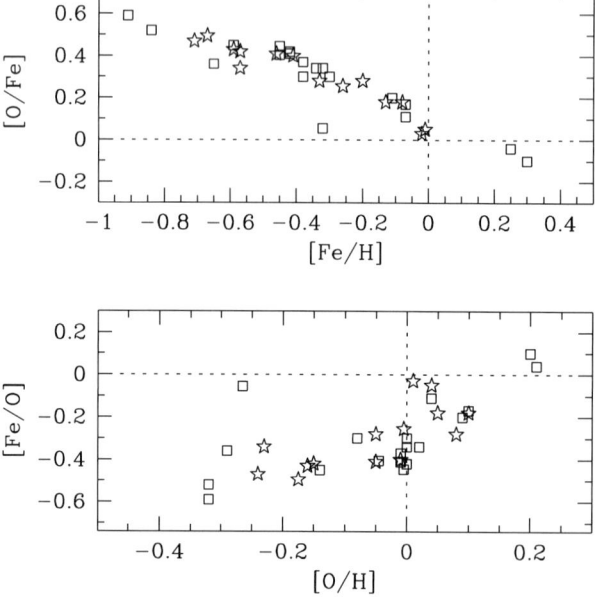

Figure 5.2. Plots of [O/Fe] versus [Fe/H] and [Fe/O] versus [O/H] for all stars with thick-disk kinematics in Bensby et al. (2003, 2004a, 2005). The stars have been divided according to how far they reach above the Galactic plane. Stars with $Z_{max} > 500$ pc are marked by open squares and those with $Z_{max} \leq 500$ pc are marked by open stars. See also Figure 5.3, which shows the W velocities and estimated distances from the Galactic plane for these stars as well as for the stars with kinematics typical of the thin disk from the same studies.

results from an abundance study of five dwarf stars situated above the Galactic plane at ∼1 kpc show that the elemental-abundance trends for these stars are the same as for the local, kinematically selected thick-disk stars (Feltzing et al. 2006). Figure 5.4 shows the results for Ba and Al. The [Ba/Fe] versus [Fe/H] trend for the local, kinematically selected thick-disk stars is well separated from that of the thin-disk stars at [Fe/H] ∼ 0. The five "*in situ*" dwarf stars clearly follow the local-thick-disk trend rather than the local-thin-disk trend. Also for Al we see a clear separation of the two trends and the stars at ∼1 kpc exhibit the same trend as that for the stars with kinematics typical of the thick disk.

3.2 How metal-rich can the thick disk be?

An interesting and unanswered questions is the following: how metal-rich are the most metal-rich stars in the thick disk? When selecting stars with kinematics typical of the thick disk we find stars with typical thick-disk kinematics at up to [Fe/H] = 0 and even a few stars of above-Solar metallicity (e.g. Bensby et al. 2005). That such

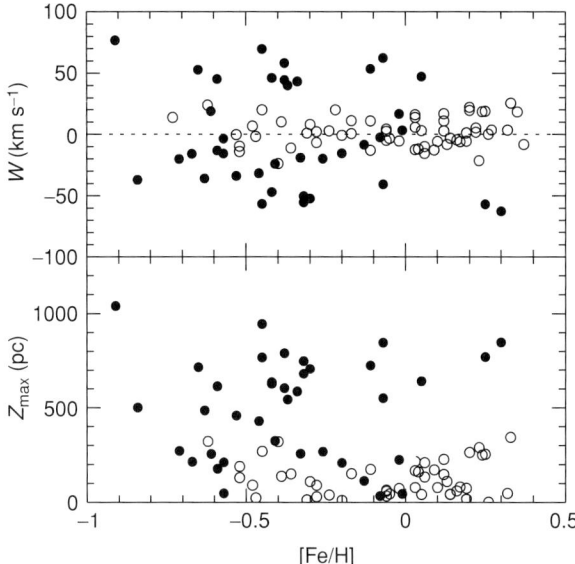

Figure 5.3. For all stars in Bensby *et al.* (2003, 2005) we show in the top panel the W velocity as a function of [Fe/H] and in the bottom panel an estimate based on their measured W velocities of the Z_{max} these stars will reach. For details see Bensby *et al.* (2005). Stars that have kinematics like that of the thick disk are shown as • and stars with kinematics like that of the thin disk are marked with ○. For the thick disk $\sigma_W \sim 35$ km s^{-1}.

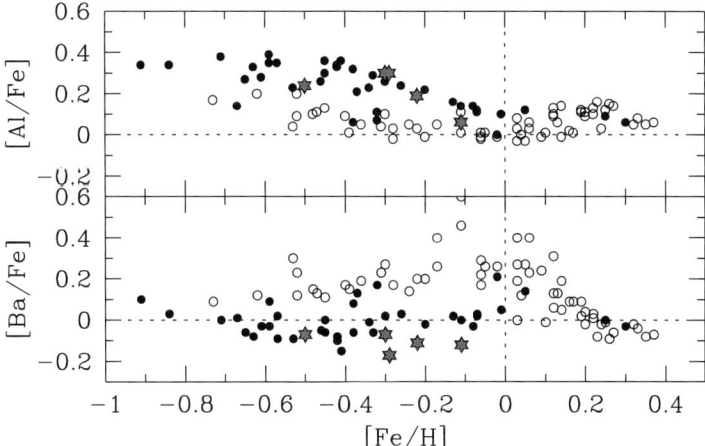

Figure 5.4. Plots of [Al/Fe] and [Ba/Fe] versus [Fe/H] for three stellar samples: • marks local Solar-neighborhood stars with kinematics typical of the thick disk and ○ marks stars with kinematics typical of the thin disk. Both samples are from Bensby *et al.* (2003, 2005). Stars mark the five dwarf stars at the South Galactic Pole for which we have obtained spectra (Feltzing *et al.* in preparation). The stars are on average ~ 1 kpc away from the Sun.

stars really belong to the thick disk has been questioned. Rather it could be argued that they belong to the tail of the velocity distribution of the thin disk (Mishenina *et al.* 2004). Figure 5.3 shows the relevant kinematic data and estimated distance reached above the Galactic plane for all stars in Bensby *et al.* (2003, 2005)

However, tentative results for stars $\sim$2 kpc above the Galactic disk show that a large portion of such stars also have Solar metallicities (Arnadottir, Feltzing *et al.* in preparation). Whether the number of metal-rich stars is compatible with the expected number of thin-disk stars at these heights remains to be confirmed. Also, when inspecting large kinematic samples, e.g. the sample collected in Bensby *et al.* (2004b), it is clear that stars with $|W_{\mathrm{LSR}}| > 35 \,\mathrm{km\,s^{-1}}$ (the velocity dispersion for the thick disk) are common also at Solar metallicities. However, also here detailed modeling is needed in order to draw any firm conclusions. For now, we feel that it might be most productive to keep an open mind on this particular issue and investigate the metal-rich thick disk further.

4 Ages

For dwarf stars that have evolved off the main sequence it is possible, albeit difficult, to derive their ages (Jørgensen & Lindegren 2005). Also for studies of age structure(s) in the disks it is important to carry out differential studies, i.e ones in which all stellar parameters are derived in the same way and all ages are derived using the same isochrones. With this approach modeling errors will cancel out (at least to first order) and we can say with confidence which stars are the older ones and find out whether there are any trends such as that e.g. more metal-rich stars are younger. It is also important to take α-enhancement into account, since that tends to make a star younger (e.g. Kim *et al.* 2002; Yi *et al.* 2001).

So far, in all abundance studies of the thin and the thick disk has been found that the stars with kinematics typical of the thick disk are older than the stars with kinematics typical of the thin disk (e.g. Fuhrmann 2004; Bensby *et al.* 2005). Results from recent studies of resolved stellar populations in nearby spiral galaxies show that their thick disks also appear to be populated exclusively by old stars (Davidge 2005; Mould 2005).

Whether there is a hiatus in star formation such that there is an age gap between the thin and the thick disk is debated (e.g. Gratton *et al.* 2003; Bensby *et al.* 2004b, 2005). Here exact selection criteria might play a role.

More comprehensive studies of the age properties of the thick-disk component have been carried out by Bensby *et al.* (2004b), Schuster *et al.* (2006), and Haywood (2006). In all of these studies it has been found that there appears to be an age–metallicity relation for the stellar population with kinematics typical of the thick disk. The change in ages might be as large as 3–4 Gyr (Bensby *et al.* 2004b). Thus, it does appear that the star formation in this stellar population has been extended over time.

5 Discussion

The two most important features of the disk that any model must be able to reproduce is the fact that tight and different trends are observed for elemental abundances (e.g. oxygen) for stars with kinematics typical of the thin and the thick disks, and that there is no evidence that either the elemental-abundance trends or the metallicity distribution functions in the thick disk vary with height above the Galactic plane (Bensby *et al.* 2005; Gilmore *et al.* 1995).

Additional constraints are provided by the stellar ages. It appears that the stellar population in the thick disk is older than that in the thin disk (e.g. Fuhrmann 2004; Bensby *et al.* 2004b). Also, there is evidence for an age–metallicity relation in the thick disk (Schuster *et al.* 2006; Haywood 2006; Bensby *et al.* 2004b).

Several scenarios for the formation of thick disks in spiral galaxies have been suggested. A comprehensive, and still valid, summary can be found in Gilmore *et al.* (1989). Earlier work focused on various versions of monolithic collapse. Although the observational evidence is still somewhat meagre, it does appear that the fact that to date no vertical gradients have been observed invalidates this scenario. Recent efforts have focused on ΛCDM realizations of the formation of Milky Way-like galaxies (e.g. Abadi *et al.* 2003; Brooks *et al.* 2004, 2005). These models suggest that the thick disk in the Solar neighborhood could be made up of a single accreted dwarf galaxy in which the stars had formed prior to the merger but that other parts of the thick disk, e.g. those closer to the Bulge, would have originated in other dwarf galaxies. If this is correct we should expect to see rather different abundance patterns in various parts of the Milky Way thick disk, since there is no reason to believe that the merged galaxies would have the same potential wells and hence produce identical abundance patterns. In general, if thick-disk stars are accreted to the Milky Way, they (at least the ones close to the Sun) must come from the same object, since otherwise it would be hard to imagine how to create the tight abundance trends that we observe in the local thick disk. Other models, e.g. Kroupa (2002), suggest that the Milky Way has gone through phases of enhanced star formation due to close encounters between the Milky Way and a passing satellite galaxy. Not only would the star-formation rate be enhanced in these models but also gas would be stirred and excited to higher latitudes in the Milky Way, thus creating the thick-disk stars "*in situ.*"

A widely advocated scenario is that what we today see as the thick disk was originally a thin-disk that was heated by a minor merger between the Milky Way and a satellite galaxy. However, it currently appears that the thick disk is all old.

This would mean that the last merger happened long ago (3–4 Gyr if we should believe the current age estimates). This in turn would not work well with the ΛCDM models. Another constraint is that the stars that make up our local thick disk must have formed in a fairly deep potential well, since the mean metallicity of the thick disk is high (Wyse 1995). This, in combination with the clear enhancement of the elemental abundances for the α-elements which indicates that SN II dominated the chemical enrichment, indicates that, if the thick-disk stars did form in a satellite, it must have been large (Wyse 2006).

6 Summary

Our current knowledge about the thin and thick disks can be summarized as follows. The summary has been divided into three categories.

Controversial findings/claims

- The thick-disk extends to [Fe/H] = 0.
- There is an age–metallicity relation for the thick disk.

The following are items that are fairly well agreed upon

- The thick disk exhibits evidence for extended star formation.
- No variations in abundance trends and/or metallicity distribution functions as a function of height above the Galactic plane have been found (yet).

Commonly acknowledged as well established are the following

- Abundance trends for kinematically selected samples differ.
- The elemental-abundance trends in the kinematically selected samples are very tight.
- Stars with kinematics typical of the thick disk are enhanced in α-elements relative to the stars with kinematics typical of the thin disk (at a given [Fe/H]).
- The Solar-neighborhood thick-disk stars that have been studied are all old.
- To date all stars with kinematics typical of the thick disk that have been studied with high-resolution spectroscopy appear to be older than those with kinematics typical of the thin disk.

The real challenge for models of galaxy formation is to explain the tight elemental-abundance trends found for kinematically selected populations of disk stars, i.e. the thin and the thick disk. There is also a need to be able to accommodate the observed age constraints, i.e. the fact that the stars in the thick disk are older than those in the thin disk and perhaps that there is an age–metallicity relation for the thick disk.

A wealth of information about the stars in the Milky Way disks has been collected so far. However, as the discussion above suggests, we are still quite far away from our goal of understanding how the two disks formed and evolved. The current observational evidence does not even appear to be quite strong enough to distinguish between some of the major formation scenarios that have been proposed so far.

Future progress will depend on access to large surveys (see e.g. various contributions to Joint Discussion 13 held at the IAU General Assembly in Prague in 2006, http://clavius.as.arizona.edu/vo/jd13/) but it will also, in equal measure, depend on the quality of the data (i.e. resolution and signal-to-noise ratio) and the treatment of the data (i.e. the modeling of stellar atmospheres and line formation). This includes obtaining improved atomic data, three-dimensional modeling of stellar atmospheres, and NLTE treatment of line formation.

Acknowledgment

S.F. is a Royal Swedish Academy of Sciences Research Fellow supported by a grant from the Knut and Alice Wallenberg Foundation.

References

Abadi, M., Navarro, J., Steinmetz, M., & Eke, V. (2003), *ApJ* **597**, 21–34
Allende Prieto, C., Barklem, P. S., Lambert, D. L., & Cunha, K. (2004), *A&A* **420**, 183–205
Bensby, T., & Feltzing, S. (2006), *MNRAS* **367**, 1181–1193
Bensby, T., Feltzing, S., & Lundström, I. (2003), *A&A* **384**, 879–883
    (2004a), *A&A* **415**, 155–170
    (2004b), *A&A* **421**, 969–976
Bensby, T., Feltzing, S. Lundström, I., & Ilyin, I. (2005), *A&A* **433**, 185–203
Brook, C. B., Kawata, D., Gibson, B. K., & Freeman, K. C. (2004), *ApJ* **612**, 894–899
Brook, C. B., Gibson, B. K., Martel, H., & Kawata, D. (2005), *ApJ* **630**, 298–308
Buser, R. (2000), *Science* **287**, 69–74
Chen, Y. Q., Nissen, P. E., Zhao, G., Zhang, H. W., & Benoni, T. (2000), *A&AS* **141**, 491–506
Dalcanton, J. J., & Bernstein, R. A. (2002), *AJ* **124**, 1328–1359
Davidge, T. J. (2005), *ApJ* **622**, 279–293
Elmegreen, B. G., & Elmegreen, D. M. (2006), *ApJ* **650**, 644–
Feltzing, S. (2006) astro-ph/0609264, to appear in the proceedings of Joint Discussion 13 held at the 2006 IAU General Assembly in Prague
Feltzing, S., Fohlman, M., & Bensby, T. (2006), *A&A*, accepted for publication
Fuhrmann, K. (1998), *A&A* **338**, 161–183
    (2004), *AN* **325**, 3–80
Gilmore, G., & Reid, N. (1983), *MNRAS* **202**, 1025–1047
Gilmore, G., Wyse, R. F. G., & Jones, J. J. (1995), *AJ* **109**, 1095–1111
Gilmore, G., Wyse, R. F. G., & Kuijken, K. (1989), *ARA&A* **27**, 555–627
Gratton, R. G., Carretta, E., Desidera, S., Lucatello, S., & Barbieri, M. (2003), *A&A* **406**, 131–140
Haywood, M. (2006), *MNRAS* **371**, 1760–1776
Jørgensen, B. R., & Lindegren, L. (2005), *A&A* **436**, 127–143
Kim, Y.-C., Demarque, P., Yi, S. K., & Alexander, D. R. (2002), *ApJS* **143**, 499–511
Kroupa, P. (2002), *MNRAS* **330**, 707–718

Mashonkina, L., Gehren, T., Travaglio, C., & Borkova, T. (2003), *A&A* **397**, 275–284
Mishenina, T. V., Soubiran, C., Kovtyukh, V. V., & Korotin, S. A. (2004), *A&A* **418**, 551–562
Mould, J. (2005), *AJ* **129**, 698–711
Prochaska, J. X., Naumov, S. O., Carney, B. W., McWilliam, A., & Wolfe, A. M. (2000), *ApJ* **120**, 2513–2549
Reddy, B. E., Tomkin, J., Lambbert, D. L., & Allende Prieto, C. (2003), *MNRAS* **340**, 304–340
Reddy, B. E., Lambbert, D. L., & Allende Prieto, C. (2006), *MNRAS* **367**, 1329–1366
Schwarzkopf, U., & Dettmar, R.-J. (2000), *A&A* **361**, 451–464
Schuster, W. J., A. Moitinho, A., Márquez, A., Parrao, L., & Covarrubias, E. (2006), *A&A* **445**, 939–958
Wyse, R. F. G. (2006), In *The Local Group as an Astrophysical Laboratory. Proceedings of the Space Telescope Science Institute Symposium, Baltimore, USA, 2003*, eds. M. Livio & T. M. Brown, pp. 33–46.
Wyse, R. F. G., & Gilmore, G. (1995), *AJ* **110**, 2771–2787
Yi, S., Demarque, P., Kim, Y.-C. *et al.* (2001), *ApJS* **136**, 417–437

6

Metal-rich massive stars: how metal-rich are they?

Daniel J. Lennon[1,2] & Carrie Trundle[3]

[1]*Isaac Newton Group of Telescopes, E-38700 Santa Cruz de La Palma, Tenerife, Spain*
[2]*Instituto de Astrofísica de Canarias, E-38200 La Laguna, Tenerife, Spain*
[3]*The Queen's University of Belfast, Belfast BT7 1NN, Northern Ireland, UK*

We discuss the metallicity of massive stars in the Solar neighbourhood, comparing new results with those for the Sun. We find that, despite there being small systematic differences between various NLTE determinations of [O/H] in hot stars, there is reasonable agreement among results from various studies of nearby stars, with a value of 8.60 ± 0.1 dex being implied. This is in good agreement with the latest Solar estimate based on three-dimensional models, and is in good agreement with recent estimates of the nebular oxygen abundance in Orion. We review the evidence for metal-rich massive stars in our own galaxy and in M31, concluding that there is little convincing evidence for supersolar [O/H] in massive stars in the Milky Way, while there is only limited evidence for mildly metal-rich regions in M31 with [O/H] relative to Solar of only $+0.2$. Discrepancies between stellar and nebular abundances at high metallicity can be traced to problems in calibrating the R_{23} index for H II regions in the metal-rich regime.

1 Introduction

Any discussion of metal-rich massive stars would be incomplete without first addressing the issue of the abundance scale in order to answer the questions 'What constitutes a metal-rich massive star?' We can then address the question 'Do we observe any metal-rich massive stars directly?'

Determining the abundance scale is important. For example, nearby massive stars are generally assumed to have 'Solar' composition, which is typically used in stellar-evolution calculations and formulations of mass-loss rate, and which in turn is scaled to other metallicities (Z). Historically, however, there have always

The Metal-rich Universe, eds. G. Israelian and G. Meynet. Published by Cambridge University Press.
© Cambridge University Press 2008.

been problems in resolving the chemical composition of nearby OB stars with that of the Sun, and to a large extent it was discrepancies such as this that drove much of the early development of NLTE model atmospheres for hot stars; see Mihalas (1972) for the case of magnesium. The expectation was that much of these apparent discrepancies was due to the neglect of NLTE and, in more recent times, to the neglect of line-blanketing and winds in the hotter and more luminous stars. So, while a differential analysis can result in precise relative abundances, even using LTE methods, NLTE techniques are required in order to set the zero-point of the abundance scale and enable us to compare with the Sun. In Section 2 we will review abundances for nearby B-type stars and compare these with the latest results from Solar three-dimensional models. It is interesting to note in this context that the accepted Solar composition has been changing as a result of the implementation of these new models and, as we will see, at least for oxygen, nearby massive stars and the Sun are converging to the same abundance.

Besides the Sun, H II regions provide another very important comparison for massive stars. In a sense this is a more relevant comparison since the chemical compositions of both massive stars and H II regions should reflect that of the current interstellar medium. Massive stars and their associated ionized nebulae should therefore have the same composition (though perhaps modified slightly by condensation onto dust grains). Comparing these two classes of objects is important because H II regions are widely used to infer the dependences of various observables on Z, such as the blue-to-red supergiant ratio, Wolf–Rayet populations and in general calibrate indirect methods of inferring Z. In Section 3 we will look at the comparison of OB stars in Orion with the nebular abundance. Then in Section 4 we will turn to the comparison of stellar with nebular abundances in those inner regions of the Milky Way and M31 thought to be metal-rich from studies of these galaxies' abundance gradients as derived from nebular analysis.

Finally, throughout his article when we refer to metallicity, or Z, we will in general mean the oxygen abundance (or [O/H]). It is difficult to derive accurate and precise iron abundances in massive stars, since most iron lines lie in the UV and, at Solar metallicity and above, the lines are saturated and insensitive to abundance, although it may well be possible to estimate metallicity from mass-loss rates. see Haser *et al.* (1998) for a detailed discussion on these points. Oxygen, on the other hand, is an excellent tracer of metallicity; there are many O II lines in the optical spectra of stars of spectral types B2–O9 and therefore we will use [O/H] as a proxy for Z unless specified otherwise. This has the further advantage that oxygen abundances are readily obtainable from H II regions, indeed often this is the only abundance readily derived from faint and distant H II regions.

Table 6.1. *Mean oxygen abundances ([O/H]) for B2–O9 stars within 500 pc of the Sun, the sample size (N) is also given. For comparison two Solar values are shown; the old value (Anders & Grevesse 1989), which is weighted towards analysis of the forbidden [O I] line using one-dimensional models, and the new results using a three-dimensional model (Asplund* et al. *2004).*

Source	[O/H]	N
Kilian (1992)	8.54 ± 0.14	21
Gies & Lambert (1992)	8.69 ± 0.14	22
Daflon *et al.* (1999, 2001)	8.61 ± 0.15	23
Solar, one-dimensional	8.93 ± 0.04	
Solar, three-dimensional	8.66 ± 0.05	

2 Nearby OB stars

Here we consider results for nearby massive stars, by which we mean OB stars within approximately 500 pc of the Sun. Within this distance we expect that the abundance dispersion due to the expected Galactic abundance gradient (~ 0.05 dex) can be neglected. We further restrict ourselves to results from NLTE calculations only, since we wish to compare the absolute abundances with Solar. Unfortunately most such results are based on NLTE line-formation calculations in which LTE line-blanketed model atmospheres have been adopted. There are currently very few results from full NLTE line-blanketed model atmospheres, but see Hunter *et al.* (2007). There have been essentially three studies matching these requirements; those of Kilian (1992), hereafter K92, Gies & Lambert (1992), GL92, and Daflon *et al.* (1999, 2001), D99. These samples have been filtered further to remove supergiants (from GL92) and the mean values of [O/H] are listed in Table 6.1.

Referring to Table 6.1, one can see that the various NLTE results for nearby B-type stars are rather similar, while the left-hand panels of Figure 6.1 confirm that there are small systematic differences between the various sets of results. Indeed, for five stars in common between K92 and GL92 one finds a mean difference in [O/H] of 0.10 dex, the K92 results being the lower ones. This is a substantial part of the 0.15-dex difference between the complete samples. Furthermore, when one considers the trend of [O/H] with effective temperature (right-hand panels of Figure 6.1) it is clear that both GL92 and D99 display strong trends, though in opposite senses; in GL92 [O/H] increases with effective temperature whereas in D99 [O/H] decreases with effective temperature. It is only in the latter two cases that one finds mildly 'metal-rich' stars, but, since these lie at the extremes of the temperature ranges, it is clear that their metallicities must be suspect. The reason for these trends with

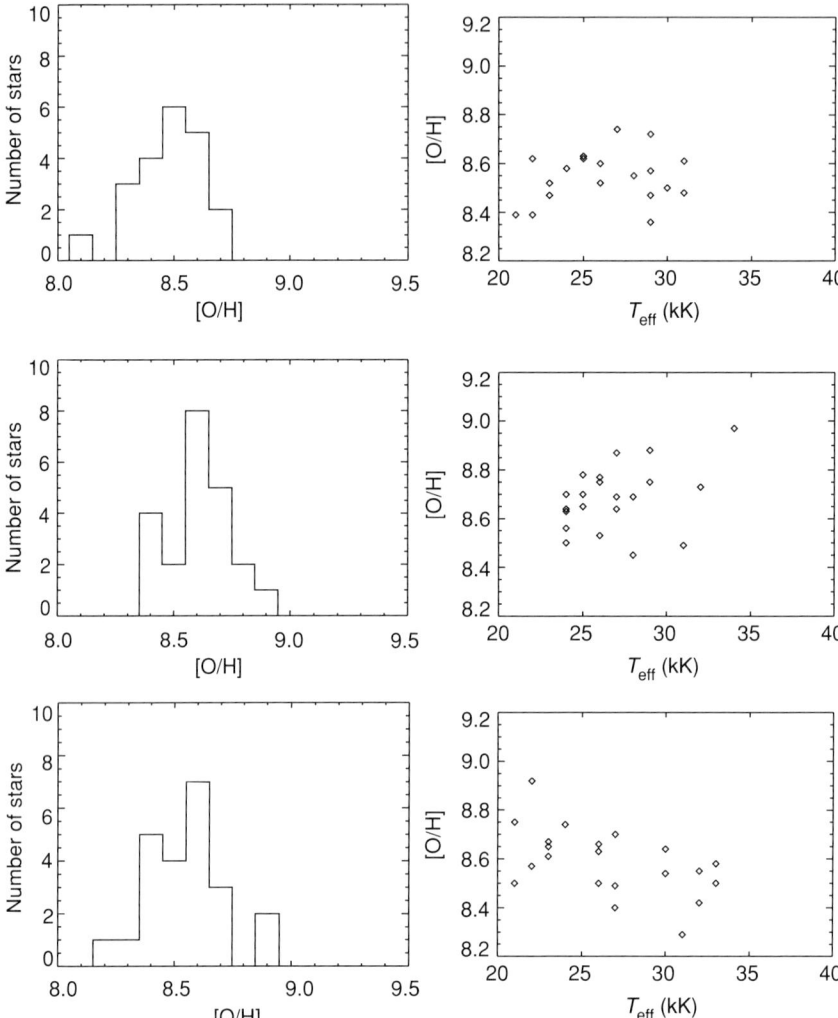

Figure 6.1. Oxygen abundance ([O/H]) of B-type stars in the Solar neighbourhood. Top panel: results from Kilian (1992) (K92) using NLTE line formation and LTE models with partial line-blanketing, mean [O/H] = 8.54 dex. Central panels: results from Gies & Lambert (1992; GL92) using NLTE line formation and LTE models with line-blanketing, mean [O/H] = 8.69 dex. Bottom panels: results from Daflon et al. (1999, 2001) (D99) using NLTE line formation and LTE line-blanketed models, mean [O/H] = 8.61 dex. Note the temperature trends in GL92 and D99; the trend in GL92 is further enhanced when one removes the two points above 25 000 K with the lowest [O/H], these being for a variable star and an SB2.

temperature probably lies in the methods used to derive effective temperatures; K92 used the silicon-ionization-balance method, GL92 used the Strömgren $[c_1]$ index with empirical corrections to bring the results into line with the empirical estimates of Code *et al.* (1972) and D99 used another photometric method based on the Q-value derived from UBV magnitudes. These temperature trends are clearly contributing to the dispersion of [O/H] results and it is to be hoped that fully line-blanketed models and line-formation calculations will resolve this issue and lead to a better zero-point for the composition of local B-type stars. It is certainly worth noting, however, that the long-standing discrepancy between oxygen abundances for the B-type stars and the Sun is largely removed by improvements in the Solar estimates, the latest three-dimensional results agreeing very nicely with those for the massive stars!

3 Orion

The Orion association, with its constituent nebulae and clusters, provides an important reference point for massive-star and nebular (H II region) abundances. The NLTE results for B-type stars in the Orion OB1 association tend to agree well with those for nearby B-type stars and with each other. For example, K92 derive [O/H] = 8.60 ± 0.11 from 7 stars, while Cunha & Lambert (1992, 1994) derive 8.65 ± 0.12 and 8.61 ± 0.17 from 18 stars in Orion OB1. This agrees well with the oxygen abundance derived for the M42 nebula of 8.65 ± 0.03 by Esteban *et al.* (2004), so all seems well. However, the three B-type stars of the M42 ionizing (Trapezium) cluster have much higher [O/H] abundances of 8.92, 8.76 and 8.97 dex, very different from that of the nebula. The solution, as hinted at above, is that these stars lie at the hot end of the effective temperature range and their temperatures are overestimated (also resulting in surface gravities higher than expected for main-sequence stars). Simón Díaz *et al.* (2006) have resolved this discrepancy in a very detailed analysis of the Trapezium cluster stars using NLTE models with line-blocking. They derive lower temperatures (and surface gravities) with resulting [O/H] abundances of 8.65, 8.59 and 8.64 dex, in good agreement with the nebular results and results for nearby B-type stars. This work is described further in Chapter 11 of these proceedings.

4 Galactic abundance gradients – the metal-rich end

Given the various systematic effects on massive-star abundances, some of which have been discussed above, it is clear that to investigate systematic differences between populations it is preferable to control the sample such that these

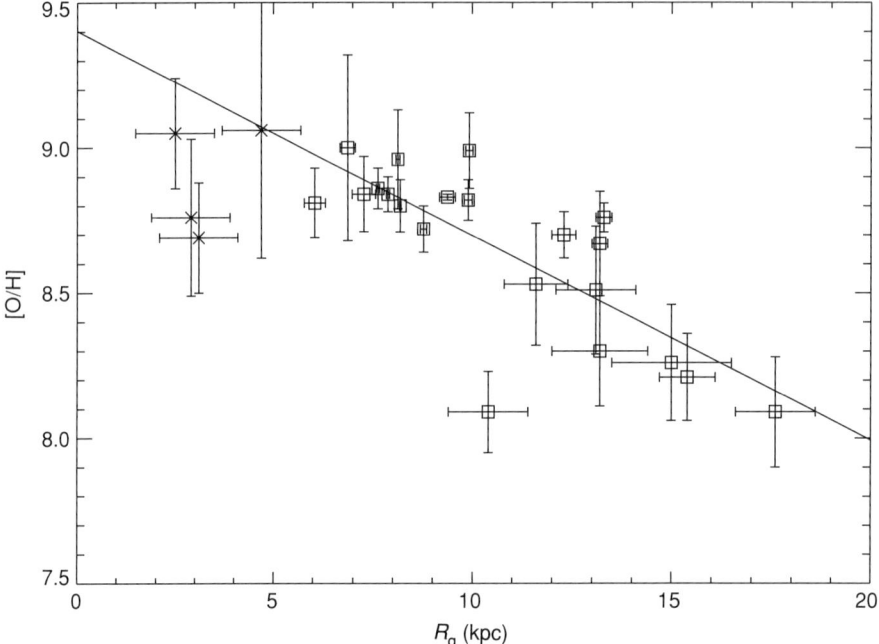

Figure 6.2. The oxygen-abundance gradient in the Milky Way from B-type stars; squares are points of Smartt & Rolleston (1997), asterisks the high-Galactic-latitude stars of Smartt et al. (2001b). Oxygen abundance is plotted as a function of the Galactocentric radial distance. Note the apparent flattening of the gradient within the solar circle ($R_g \sim 8.5$). The solid line is a fit to the stellar points of Smartt and Rolleston and is in good agreement with the gradient obtained from H II regions (~ -0.7 dex kpc^{-1}).

systematics are minimized. Smartt & Rolleston (1997) and Rolleston et al. (2000) attempted this in their study of the abundance gradient of the Milky Way using B-type stars. Their work, though based on LTE methods, was well controlled such that the differential abundances between objects were not greatly influenced by NLTE effects, differences in effective temperature or choice of absorption lines. They found good agreement with the abundance gradients derived from H II regions but crucially, from the point of view of this conference, did not sample the metal-rich end of the abundance gradient as it is extrapolated towards the Galactic Centre, similarly to the results of Daflon & Cunha (2004). The reason of course is the increasing extinction in this region. However, they attempted to rectify this (Smartt et al. 2001b) by observing four B-type stars at Galactic latitudes between 5 and 8 degrees above the Galactic plane, and Galactic longitudes between -40 and $+20$ degrees. Surprisingly all four objects, lying between 3 and 5 kpc from the Galactic Centre, turn out to have [O/H] similar to those of nearby B-type stars, rather than the expected supersolar [O/H] as predicted from the abundance gradient (Figure 6.2).

The interpretation of the results is somewhat unclear, however, since the abundances of magnesium and silicon are consistent with the expected abundance gradient, while the origin of these high-latitude stars is also uncertain.

The local-group galaxy M31 presents us with an excellent alternative possibility in which to study metal-rich massive stars; extinction is low and the generally accepted abundance gradient as derived from H II regions implies supersolar [O/H] in the inner regions of this galaxy. Its remoteness, however, constrains us to the study of supergiant stars, and A-type and B-type supergiants have been used for this purpose. Venn *et al.* (2000) used A-type supergiants to probe the iron abundance directly, also deriving [O/H]. However, early B-type supergiants are ideal tracers of [O/H] since they have many O II lines in the optical that one can use for determining the oxygen abundance, which is of special interest here because it can be used to compare directly with nebular results. The first attempts at this were those of Smartt *et al.* (2001a), one star, and Trundle *et al.* (2002), an additional six stars.

In general we find the surprising result that supergiant stars display a rather flat gradient for galactocentric distances between 5 and 25 kpc, whereas the H II results exhibit a strong gradient in this region. Smartt *et al.* (2001a) also demonstrated that the nebular results, which are based on calibrations of the R_{23} index, depend very sensitively on which calibration is used in the metal-rich domain (their object lying in the inner region of M31). They give the example of a nebula coincident with the position of the association (OB10) containing their star whose R_{23} ratio implies [O/H] values of 9.07 dex using Zaritsky *et al.* (1994) or 8.90 dex using McGaugh (1991) compared with their LTE abundance of 8.69 dex. Trundle *et al.* (2002) developed this theme further, showing that the calibration of Pilyugin (2001a, 2001b) implied even lower values of [O/H] at galactocentric distances of 5 kpc. Subsequently the stellar results have been improved by the application of NLTE methods including the effects of line-blanketing and winds (see elsewhere in these proceedings for a more complete description). The results are summarized in Figure 6.3, where one can see that, of the five stars analysed, three have subsolar [O/H] while two stars are mildly oxygen-rich. Furthermore, the two oxygen-rich stars belong to the same association, OB78, suggesting that this may be a characteristic of the association. In any case, it is clear that we have again failed to discover any massive star that could be said to be 'metal-rich'.

5 Concluding remarks

We have reviewed the literature on the direct determination of metallicities of massive stars, concentrating on establishing the metallicity scale of OB stars and its relationship with the solar metallcity and H II-region results, and investigated

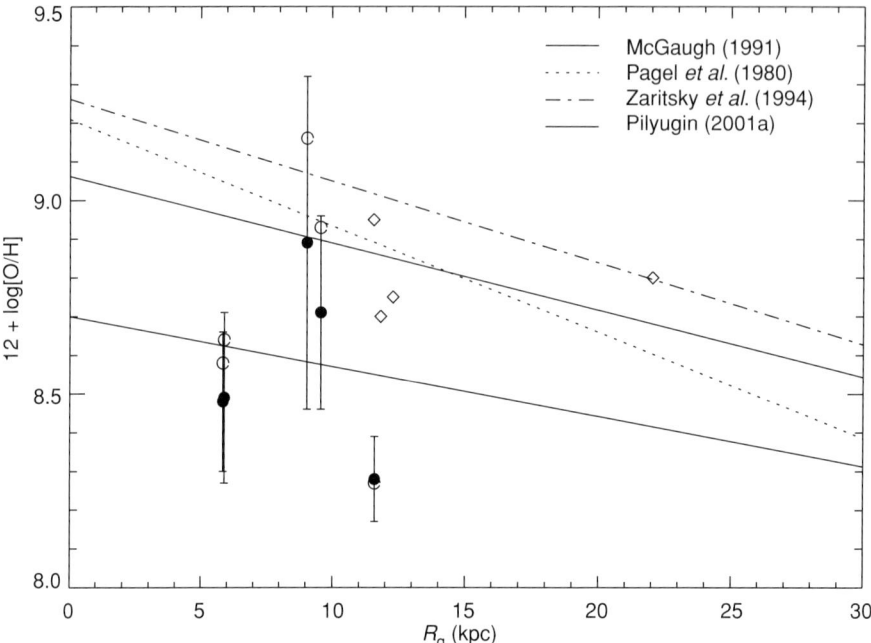

Figure 6.3. Oxygen abundances for supergiants in M31 as a function of galactocentric distance. The solid circles refer to NLTE results including line-blanketing and winds, while the open circles show the LTE results for the same stars. Also shown are LTE results for other B- and A-type supergiants in M31, while the nebular abundance gradients derived with various calibrations of R_{23} are shown as lines (see the key). Note the lack of supersolar stellar abundances in contrast with the nebular results at low galactocentric distances.

the oxygen abundance of massive stars in supposedly metal-rich environments of the inner Milky Way and M31. We draw the following conclusions.

- OB stars within approximately 500 pc of the Sun have similar metallicities with [O/H] ∼ 8.60 ± 0.1, in good agreement with the latest estimates of the Solar oxygen abundance derived using three-dimensional models of 8.66 dex.
- This value agrees well with the oxygen abundance of B-type stars in the Orion OB1 association and with the H II-region oxygen abundance of this region.
- Little evidence is found for superSolar oxygen abundance in the Galactic disc, even at small Galactocentric distances, the abundance gradient for this element being rather flat within the Solar circle.
- Massive stars in the inner regions of M31 are at best only mildly metal-rich, again displaying a rather flat gradient in [O/H], in disagreement with H II-region results.
- Estimates of [O/H] using the R_{23} index depend very sensitively on the calibration used, especially in the metal-rich regime. In addition, most calibrations predict very high [O/H] compared with that of the stars in the inner regions of M31.

References

Anders, E., & Grevesse, N. (1989), *Geochim. Cosmochim. Acta* **53**, 197–214
Asplund, M., Grevesse, N., Sauval, A. J. *et al.* (2004), *A&A* **417**, 751–768
Code, A. D., Davis, J., Bless, R. C., & Hanbury Brown, R. (1972), *ApJ* **203**, 417
Cunha, K. & Lambert, D. L. (1992), *ApJ* **399**, 586–598
  (1994), *ApJ* **426**, 170–191
Daflon, S., Cunha, K., & Becker, S. R. (1999), *ApJ* **522**, 950–959
Daflon, S., Cunha, K., Becker, S. R., & Smith, V. V. (2001), *ApJ* **552**, 309–320
Daflon, S., & Cunha, K. (2004), *ApJ* **617**, 1115–1126
Gies, D. R., & Lambert, D. L. (1992), *ApJ* **387**, 673–700
Haser, S. M., Pauldrach, A. W. A., Lennon, D. J. *et al.* (1998), *A&A* **330**, 285
Hunter, I., Dufton, P. L., Smartt, S. J. *et al.* (2007), *A&A* **466**, 277
Kilian, J. (1992), *A&A* **262**, 171–187
McGaugh, S. S. (1991), *ApJ* **505**, 793
Mihalas, D. (1972), *ApJ* **177**, 115
Pagel, B. E. J., Edmunds, M. G., & Smith, G. (1980), *MNRAS* **193**, 219
Pilyugin, L. S. (2001a), *A&A* **369**, 594
  (2001b), *A&A* **373**, 56
Rolleston, W. R. J., Smartt, S. J., Dufton, P. L., & Ryans, R. S. I. (2000), *A&A* **362**, 537–544
Simón-Díaz, S., Herrero, A., Esteban, C., & Najarro, F. (2006), *A&A* **448**, 351–366
Smartt, S. J., Crowther, P. A., Dufton, P. L. *et al.* (2001a), *MNRAS* **325**, 257–272
Smartt, S. J., & Rolleston, W. R. J. (1997), *ApJL* **481**, 47
Smartt, S. J., Venn, K. A., Dufton, P. L., & Lennon, D. J. (2001b), *A&A* **367**, 86–105
Trundle, C., Dufton, P. L., Lennon, D. J, Smartt, S. J., & Urbaneja, M. A. (2002), *A&A* **395**, 519–533.
Venn, K. A., McCarthy, J. K, Lennon, D. L. *et al.* (2000), *ApJ* **541**, 610
Zaritsky, D., Kennicutt, R. C., & Huchra, J. P. (1994), *ApJ* **420**, 87

7

Hercules-stream stars and the metal-rich thick disk

T. Bensby[1], M. S. Oey[2], S. Feltzing[3] & B. Gustafsson[4]

[1] *European Southern Observatory, Alonso de Cordova 3107, Vitacura, Casilla 19001, Santiago 19, Chile*
[2] *Department of Astronomy, University of Michigan, 830 Dennison Building, 500 Church Street, Ann Arbor, MI 48109-1042, USA*
[3] *Lund Observatory, Box 43, SE-221 00 Lund, Sweden*
[4] *Department of Astronomy and Space Physics, University of Uppsala, Box 515, SE-751 20 Uppsala, Sweden*

Using the MIKE spectrograph, mounted on the 6.5-m Magellan/Clay telescope at the Las Campanas observatory in Chile, we have obtained high-resolution spectra for 60 F and G dwarf stars, all likely members of a density enhancement in the local velocity distribution, referred to as the Hercules stream. By comparing with an existing sample of 102 thin- and thick-disk stars we have used space velocities, detailed elemental abundances, and stellar ages to trace the origin of the Hercules stream. We find that the Hercules-stream stars exhibit a wide spread in stellar ages, metallicities, and element abundances. However, the spreads are not random but separate the Hercules stream into the abundance and age trends outlined by either the thin disk or the thick disk. We hence claim that the major constituents of the Hercules stream actually are thin- and thick-disk stars. These diverse properties of the Hercules stream indicate a dynamical origin, probably caused by the Galactic bar. However, we can at the moment not entirely rule out the possibility that the Hercules stream could be the remnants of a relatively recent merger event.

1 Introduction

The stellar velocity distribution in the Solar neighborhood is not smooth, but exhibits lots of substructure (e.g. Dehnen 1998; Skuljan *et al.* 1999; Famaey *et al.* 2005; Arifyanto & Fuchs 2006; Helmi *et al.* 2006). The most prominent features are the Pleiades–Hyades super-cluster, the Sirius cluster, and the Hercules stream (also known as the *u*-anomaly). By studying nearby G and K giants Famaey *et al.* (2005) found that the Hercules stream makes up ∼6% of the stars in the Solar neighborhood, with its stars moving on highly eccentric orbits. On average these stars have a net

The Metal-rich Universe, eds. G. Israelian and G. Meynet. Published by Cambridge University Press.
© Cambridge University Press 2008.

drift of ~40 km s^{-1} directed radially away from the Galactic Center, and, just as for the thick disk, their orbital velocities around the Galaxy lag behind the local standard of rest (LSR) by ~50 km s^{-1}. See also Ecuvillon *et al.* (2007), who found similar properties for nearby F and G dwarf stars.

As several numerical simulations have shown, this excess of stars at ($U_{\rm LSR}$, $V_{\rm LSR}$) ≈ (−40, −50) km s^{-1} can be explained as a signature of the Galactic bar (e.g. Raboud *et al.* 1998; Dehnen 1999, 2000; Fux 2001). Whether it is a chaotic process, whereby stars are gravitationally scattered off the inner regions by the bar, or whether they have ordered orbits coupled to the outer Lindblad resonance of the bar is, however, unknown (Fux 2001). In any case, this explanation means that the stars in the Hercules stream should originate from the inner disk regions. So, is the Hercules stream a distinct Galactic stellar population with a unique origin and history or is it a mixture of the other populations? Or could it even be a remnant of an ancient merger event?

To trace the origin of the Hercules stream further we have observed a sample of 60 F and G dwarf stars. By performing a strictly differential detailed abundance analysis of the Hercules-stream stars relative to stars of the two disk populations previously studied by us (Bensby *et al.* 2003, 2005) we minimized uncertainties due to systematic errors in the analysis.

2 Observations, abundance analysis, and age determinations

High-resolution ($R \approx 65{,}000$), high-quality ($S/N \gtrsim 250$) echelle spectra were obtained for 60 F and G dwarfs by T.B. in January, April, and August 2006 with the MIKE spectrograph (Bernstein *et al.* 2003) on the Magellan/Clay 6.5-m telescope at the Las Campanas Observatory in Chile. A Toomre diagram for the sample can be seen in Figure 7.1(a).

For the abundance analysis we used the Uppsala MARCS stellar model atmospheres (Gustafsson *et al.* 1975; Edvardsson *et al.* 1993; Asplund *et al.* 1997). These models are one-dimensional, plane-parallel, and calculated under the assumption of local thermodynamic equilibrium (LTE). Their chemical compositions have been scaled with metallicity relative to the standard Solar abundances as given in Asplund *et al.* (2005), using enhanced abundances for the α-elements at subsolar metallicities. To determine effective temperatures we use excitation equilibrium of Fe I and to estimate the microturbulence parameter we require all Fe I lines to yield the same abundance independently of line strength. To estimate the surface gravities we utilize the fact that for all our stars we have accurate Hipparcos parallaxes (ESA 1997). Final abundances were first normalized on a line-by-line basis with our Solar values as reference and then averaged for each element.

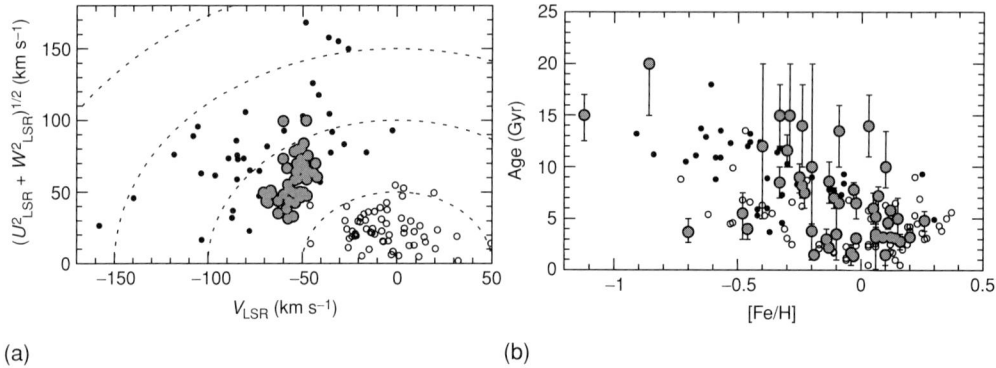

Figure 7.1. (a) A Toomre diagram. (b) Stellar ages versus [Fe/H]. Data for Hercules-stream stars are marked by gray stars, data for thin- and thick-disk stars from Bensby *et al.* (2003, 2005) by open and filled circles, respectively. Error bars are explained in the text.

Stellar ages were determined from the Yonsei–Yale (Y^2) α-enhanced isochrones (Kim *et al.* 2002; Demarque *et al.* 2004) in the T_{eff}–M_V plane. Lower and upper limits on the ages were estimated by taking the error bars arising from an assumed uncertainty of ± 70 K (Bensby *et al.* 2003) in T_{eff} and the uncertainty in M_V due to the error in the parallax.

3 Results and discussion

In Figure 7.1(b) we plot ages versus [Fe/H] together with the thin- and thick-disk samples from Bensby *et al.* (2003, 2005). Below [Fe/H] = 0 it appears that the Hercules stream divides into two (or maybe three) branches: one that follows the thick-disk age trend (the downward age–metallicity relation for the thick disk was seen in Bensby *et al.* (2004) and then also verified by Haywood (2006) and Schuster *et al.* (2006)); one that follows the thin-disk age trend; and a few stars (four or five) that tend to have high ages of ~ 15 Gyr in the interval $-0.4 \lesssim$ [Fe/H] $\lesssim 0$. Although tempting, it seems premature to conclude that we see a thin-disk and a thick-disk branch in the Hercules stream.

Abundance trends for nine elements are shown in Figure 7.2. On considering the α-elements (Mg, Si, Ca, Ti), the first impression is that most stars follow the trends outlined by the thick-disk stars. At [Fe/H] $\gtrsim -0.1$ the thin and thick disk are, however, too close to distinguish in the α-element trends, but the separation can be extended a few more tenths of a dex by looking at the Ba trend. Even here most Hercules stars follow the thick-disk trend. We note that the Al abundance also is a good criterion for distinguishing between the two disks; here we find that the trend for the Hercules stars again accompanies that of the thick-disk stars. The

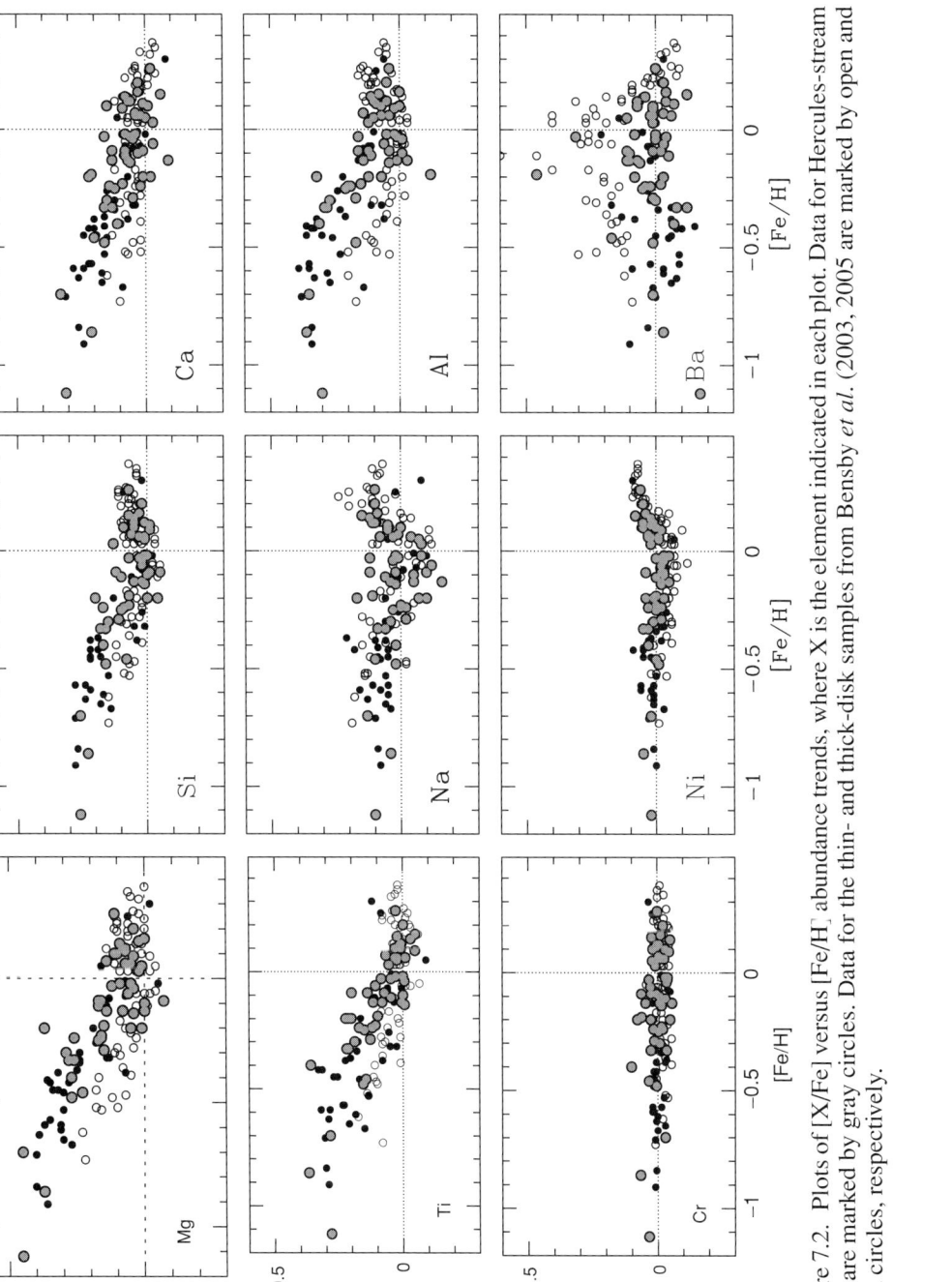

Figure 7.2. Plots of [X/Fe] versus [Fe/H] abundance trends, where X is the element indicated in each plot. Data for Hercules-stream stars are marked by gray circles. Data for the thin- and thick-disk samples from Bensby *et al.* (2003, 2005) are marked by open and filled circles, respectively.

iron-peak-element trends are all remarkably tight, but add no more than a consistency check for the Hercules-stream abundance trends.

Even though the Hercules stars do not split up as nicely in the abundance plots as they did in the age–metallicity plot, they do indeed mainly reproduce the trends outlined by the thin and thick disks, in fact, predominantly the thick disk. So what does this suggest about the origin of stars of the Hercules stream?

3.1 A unique population?

The Hercules-stream stars are certainly unique in the sense that they form a well-defined enhancement in the local velocity distribution. Its age and abundance spreads, however, tend to support the hypothesis that it is a mixture of populations.

3.2 A mixture of thin and thick disks?

For [Fe/H] $\gtrsim -0.2$ there are several stars with thin-disk chemical compositions. The stars that follow the trends outlined by the thick disk are, however, more numerous. At supersolar metallicities ([Fe/H] > 0) there is really no way to tell whether we are looking at the thin or the thick disk.

If the Hercules stream has a dynamical origin, consisting of stars originating at smaller Galactocentric radii, are we tracing the inner thin disk or the inner thick disk? Since no detailed abundance data for the inner disk(s) are currently available one can only speculate. Either it could be that there is a substantial radial metallicity gradient in the thick disk (and that is why we see so many metal-rich thick-disk stars in the Hercules stream), or the chemical properties of the inner thin disk could differ considerably from what we observe in the Solar neighborhood. It is clear, however, that the Hercules stream divides into the distinct abundance trends of the nearby thin and thick disks.

3.3 Bulge stars?

This is not likely because authors of recent studies find bulge stars to have large α-enhancements at Solar and supersolar [Fe/H] Fulbright *et al.* 2007; Zoccali *et al.* 2006).

3.4 A recent merger?

It cannot be excluded that the stars of the Hercules stream could have originated from a recent merger event between the Milky Way and another system. This system must then have had properties very similar to the present properties of the Galactic thin and thick disks. Thus, such a hypothetical merging galaxy would have chemical characteristics that depart considerably from those of local dwarf galaxies (e.g. Venn *et al.* 2004) and would presumably be more similar to a major spiral galaxy.

4 Conclusion

We conclude that the stars in the Hercules stream seem to be a mixture of thin and thick-disk stars (especially the thick disk), supporting models that suggest that their kinematics are due to dynamical interactions with the Galactic bar. We note that these models suggest an inner-disk origin for the Hercules stream, so we will need to compare our abundances with inner-disk data, which we plan to obtain in the near future.

Acknowledgment

This work was supported by the National Science Foundation, grant AST-0448900. S.F. is a Royal Swedish Academy of Sciences Research Fellow supported by a grant from the Knut and Alice Wallenberg Foundation.

References

Arifyanto, M. I., & Fuchs, B. (2006), *A&A* **449**, 533–538
Asplund, M., Grevesse, N., Sauval, A. J., Allende Prieto, C., & Blomme, R. (2005), *A&A* **431**, 639–705
Asplund, M., Gustafsson, B., Kiselman, D., & Eriksson, K. (1997), *A&A* **318**, 521–534
Bensby, T., Feltzing, S., & Lundström, I. (2003), *A&A* **410**, 527–551
   (2004), *A&A* **421**, 969–976
Bensby, T., Feltzing, S., Lundström, I., & Ilyin, I. (2005), *A&A* **433**, 185–203
Bernstein, R., Shectman, S. A., Gunnels, S. M., Mochnacki, S., & Athey, A. E. (2003), *Proc. SPIE* **4841** 1694–1704
Dehnen, W. (1998), *AJ* **115**, 2384–2396
   (1999), *ApJ* **524**, L35–L38
   (2000), *AJ* **119**, 800–812
Demarque, P., Woo, J.-H., Kim, Y.-C., & Yi, S. K. (2004), *ApJS* **155**, 667–674
Ecuvillon, A., Israelian, G., Pont, F., Santos, N. C., & Mayor, M. (2007), *A&A* **461**, 171–182
Edvardsson, B., Andersen, J., Gustafsson, B., Lambert, D. L., Nissen, P. E., & Tomkin, J. (1993), *A&A* **275**, 101
ESA (1997), *The HIPPARCOS and TYCHO Catalogues*. Noordwijk: European Space Agency
Famaey, B., Jorissen, A., Luri, X. *et al.* (2005), *A&A* **430**, 165–186
Fulbright, J. P., McWilliam, A., & Rich, R. M. (2007), *AJ* **661**, 1152–1179
Fux, R. (2001), *A&A* **373**, 511–535
Gustafsson, B., Bell, R. A., Eriksson, K., & Nordlund, A. (1975), *A&A* **42**, 407–432
Haywood, M. (2006), *MNRAS* **371**, 1760–1776

Helmi, A., Navarro, J. F., Nordström, B., Holmberg, J., Abadi, M. G., & Steinmetz, M. (2006), *MNRAS* **365**, 1309–1323
Kim, Y., Demarque, P., Yi, S. K., & Alexander, D. R. (2002), *ApJS* **143**, 499–511
Raboud, D., Grenon, M., Martinet, L., Fux, R., & Udry, S. (1998), *A&A* **335**, L61–L64
Schuster, W. J., Moitinho, A., Márquez, A., Parrao, L., & Covarrubias, E. (2006), *A&A* **445**, 939–958
Skuljan, J., Hearnshaw, J. B., & Cottrell, P. L. (1999), *MNRAS* **308**, 731–740
Venn, K. A., Irwin, M., Shetrone, M. D., Tout, C. A., Hill, V., & Tolstoy, E. (2004), *AJ* **128**, 1177–1195
Zoccali, M., Lecureur, A., Barbuy, B. *et al.* (2006), *A&A* **457**, L1–L4

8
An abundance survey of the Galactic thick disk

Bacham E. Reddy,[1] David L. Lambert[2] & Carlos Allende Prieto[2]

[1]*Indian Institute of Astrophysics, Bangalore, India*
[2]*McDonald Observatory and Department of Astronomy, University of Texas, Austin, TX, USA*

We present the results of our recent abundance survey of the Galactic thick disk. We selected from the Hipparcos catalog 176 sample stars satisfying the following criteria: they are nearby ($d \leq 150\,\text{pc}$) subgiants and dwarfs, of spectral types F and G, and with thick-disk kinematics ($V_{\text{LSR}} \leq -40\,\text{km}\,\text{s}^{-1}$, $|W_{\text{LSR}}| \geq 30\,\text{km}\,\text{s}^{-1}$). Assuming that the velocity distribution of each stellar population is Gaussian, we assigned stars with a probability $P \geq 70\%$ to one of the three components. This resulted in 95 thick-disk stars, 17 thin-disk stars, and 24 halo stars. The remaining 40 objects cannot be unambiguously assigned to one of the three components.

We derived abundances for 23 elements from C to Eu. The thick-disk abundance patterns are compared with earlier results from the thin-disk survey of Reddy *et al.* (2003). The levels of α-elements (O, Mg, Si, Ca, and Ti), thought to be produced dominantly in Type-II supernovae, are enhanced in thick-disk stars relative to the values found for thin-disk members in the range $-0.3 > [\text{Fe/H}] > -1.2$. The scatter in the abundance ratios [X/Fe] at a given [Fe/H] for thick-disk stars is consistent with the predicted dispersion due to measurement errors, as is the case for the thin disk, suggesting a lack of "cosmic" scatter. The observed compositions seem consistent with a model of galaxy formation by mergers in a Λ CDM universe.

1 Introduction

Understanding the origin and evolution of galactic disks is a major endeavor in contemporary astrophysics. Observational scrutiny of our Galactic disk offers many obvious advantages and observational data not available when examining external

The Metal-rich Universe, eds. G. Israelian and G. Meynet. Published by Cambridge University Press.
© Cambridge University Press 2008.

spiral galaxies. In particular, it is possible to obtain chemical compositions for stars in the Solar neighborhood with different kinematical properties; the stars are sufficiently bright that detailed abundance analyses are possible and close enough for accurate proper motions (and radial velocities) to be obtained, and finally to derive the space velocities. Combining compositions and kinematics makes it possible to trace the chemical history of the stellar populations involved, notably, the thin disk, the thick disk, and the halo.

The composition of the local thin disk was investigated in previous studies of F–G dwarfs (e.g. Edvardsson *et al.* 1993; Reddy *et al.* 2003). Since the suggestion that there are two different stellar populations coexisting in the disk of the Galaxy, many studies have been devoted to characterizing them (Gilmore & Reid 1983). Fuhrmann's (1998) study of the Mg abundance in nearby stars suggested that the thick- and thin-disk stars have different chemical histories. The plausibility of this idea was later strengthened by the results for other elements (e.g. Bensby *et al.* 2003, 2005). Here, we present a much larger sample covering more elements, using the same techniques as we applied in our thin-disk survey (Reddy *et al.* 2003).

2 Sample selection

Stars were first selected from the Hipparcos catalogue according to the following criteria: the stars had a declination north of $-30°$ so that they were observable from the W. J. McDonald Observatory; a $B-V$ color corresponding to an effective temperature of 5,000–6,500 K; an absolute visual magnitude in the range $2.5 \leq M_V \leq 6.0$; and a distance of less than 150 pc. Our initial selection of thick-disk candidates was also based on space motions computed using radial velocities collected from several catalogs.

Stars were separated into one of the three components by assigning the probability of each star being a member of the thin disk, the thick disk, and the halo (see Figure 8.1). This procedure follows the approach used in earlier studies by Bensby *et al.* (2003, 2005) and Mishenina *et al.* (2004). We assume that the sample is a mixture of the three components, and that the velocity of each is represented by a Gaussian distribution. We also consider the local relative numbers of thin-disk, thick-disk, and halo stars in the statistics (Soubiran *et al.* 2003; Robin *et al.* 2003).

High-resolution spectra ($R \approx 60,000$) were obtained for all program stars with the Harlan J. Smith 2.7-m telescope at the W. J. McDonald Observatory, using the 2dcoudé echelle spectrometer (Tull *et al.* 1995). The reduced spectra have signal-to-noise ratios ≈ 100–200 covering 3,800–10,000 Å with gaps between orders in the red and near-infrared.

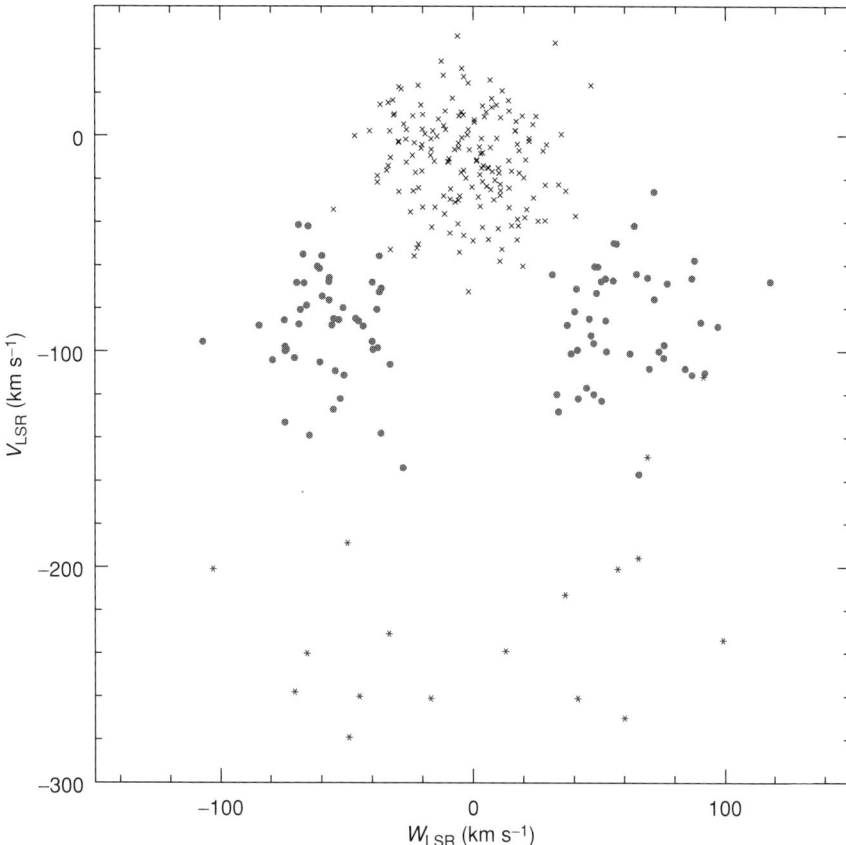

Figure 8.1. Sample stars in the *U–V* plane for the three components (cross, thin disk; circles, thick disk; and asterisks, halo). Note the increase in the *W*-velocity dispersion as one scans the diagram from top to bottom. The biases imposed by our selection criteria for thick-disk stars are apparent ($V < -40$ and $|W| > 30$).

3 Results

Our results on the thick-disk composition are based on 95 thick-disk stars satisfying the requirement that $P_{\text{thick}} > 70\%$. The abundance results are grouped into four categories: Mg-like and other Mg-like, Ni-like, and heavy elements. Figure 8.2 shows the run of the [X/Fe] ratio versus [Fe/H] for thick-disk, thin-disk, and halo stars. This figure conveys the impression that the abundances of Mg-like elements for the thin and thick disks are different, while the patterns for thick-disk and halo stars are similar.

The Mg-like elements are the so-called α-elements: C, O, Mg, Si, Ca, and Ti – the latter included in Figure 8.2. Other Mg-like elements are Al (also included in Figure 8.2), Sc, V, Co, and possibly Zn. The trend of [X/Fe] against [Fe/H] for Mg-like elements in the thick disk may be characterized by a shallow slope for

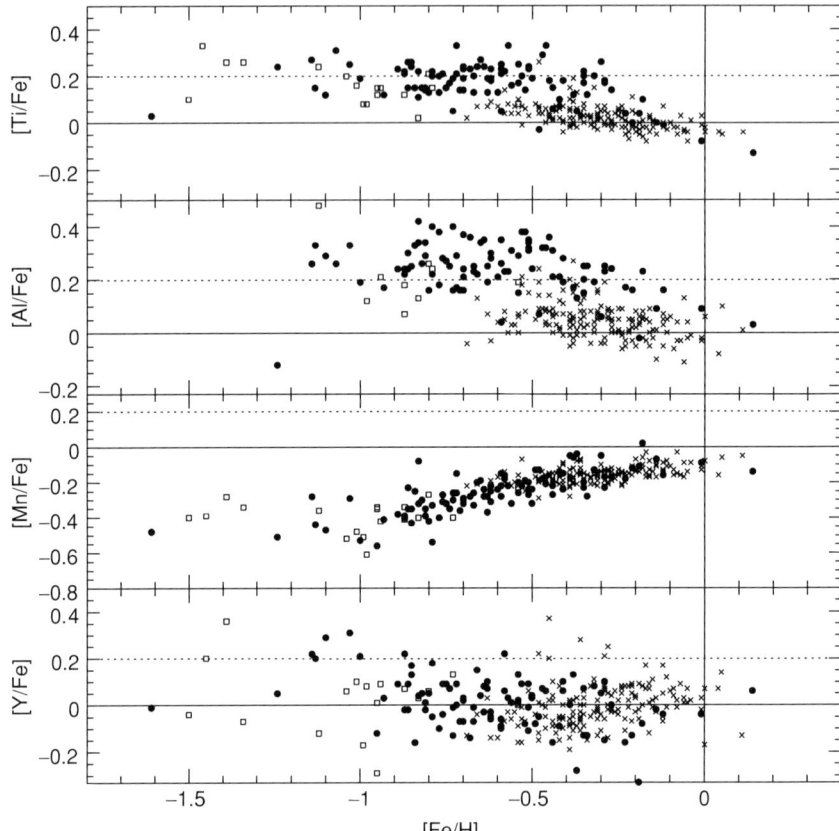

Figure 8.2. [X/Fe] versus [Fe/H] for four elements, Ti, Al, Ni, and Ba, representing four groups of elements (see the text): Mg-like, other Mg-like, Ni-like, and heavy elements, respectively. Results for three components: (thin disk, crosses; thick disk, filled circles; and halo, open squares) are shown.

[Fe/H] ≤ -0.3 with the mean value offset from the trend for thin-disk stars. Ratios [X/Fe] are enhanced compared with those for the thin-disk stars at overlapping [Fe/H] values below [Fe/H] ≈ -0.3. Ratios of elements like Al, V, and Zn are also enhanced in thick-disk stars compared with those for stars in the thin disk. These results were not expected, since the abundances of their counterparts, Na and Fe-peak elements, are the same for the thin- and thick-disk stars. However, the data from our large sample confirm earlier results (Prochaska *et al.* 2000; Bensby *et al.* 2005).

The Ni-like elements are Na, Mn, Cr, Ni, and Cu, for which we found similar abundance ratios for the thin- and thick-disk stars. Our results confirm earlier results based on a few stars (Prochaska *et al.* 2000; Bensby *et al.* 2005). Of special interest, perhaps, is the example of Mn (see Figure 8.2), where [Mn/Fe] is a quite steeply increasing function of Fe, yet data for thin- and thick-disk stars obey a single relation. The contrast odd–even between Na (Ni-like) and Al (Mg-like) is also interesting.

Synthesis of elements beyond the iron group occurs by two neutron-capture processes: the s-process occurring in AGB stars, and the r-process occurring (probably) in Type-II supernovae. Our analysis includes the s-process elements Y (see Figure 8.2), Ce, Nd, and Ba, and an r-process element, Eu. For the elements Y, Ba, Ce, and Nd we found similar abundances for the two components, and their contributions are those of Ni-like elements. Europium is a Mg-like element. The scatter in [Eu/Fe] at a given [Fe/H] is consistent with the measurement errors. The large scatter in Eu for the thick-disk stars compared with that for the thin-disk stars may be attributed to the larger number of cool stars in our thick-disk sample.

4 Discussion and conclusion

The origin of the thick disk is believed to lie in the formation of the early Galaxy through mergers of smaller (proto-)galaxies in the context of a ΛCDM universe. Here, we restrict our remarks largely to interpretations of the abundance differences between thick and thin disks and the similarities between the thick disk and the halo. Recent detections of accretion of dwarf galaxies show that merger is a continuing way of life for the Galaxy (Yanny et al. 2003). Theoretical ideas about mergers in the usual ΛCDM Universe predict that the rate of mergers was much higher in the past, with a marked decline in the rate at redshift $z \sim 1$ or about 8 Gyr ago, which explains why the Galactic thick-disk stars are old. One additional condition is required: the thick disk must run out of gas after a few Gyr, preventing the formation of young thick disk stars.

Dalcanton et al. (2005) note that there are three ways in which a high σ_W may be achieved through merging: (A) heating of a thin disk in a merger (which may, but need not, lead to the disruption of the thin disk); (B) direct accretion of stars from satellite galaxies; and (C) star formation in merging gas-rich systems. The three scenarios are not necessarily incompatible.

On the basis of the available abundance data on the thin and thick disks and of the published simulations of disk formation through mergers in a ΛCDM universe, scenario (C) appears to be a plausible leading explanation for the origin of the thick and thin disks. The realization that thick- and thin-disk stars of the same [Fe/H] differ in composition and the strong suggestion that thick- and thin-disk stars span overlapping but distinctly different ranges in [Fe/H] has consequences for constraining models of chemical evolution of the Galactic disk, especially for models of the Solar neighborhood.

References

Bensby, T., Feltzing, S., & Lundström, I. (2003), *A&A* **410**, 527
Bensby, T., Feltzing, S., Lundström, I., & Ilyin, I. (2005), *A&A* **433**, 185
Dalcanton, J. J., Seth, A. C., & Yoachim, P. (2005), ArXiv astro-ph/0509700
Edvardsson, B., Andersen, J., Gustafsson, B., Lambert, D. L., Nissen, P. E., & Tomkin, J. (1993), *A&A* **275**, 101
Gilmore, G., & Reid, N. (1983), *MNRAS* **202**, 1025.
Fuhrmann, K. (1998), *A&A* **338**, 161
Mishenina, T. V., Soubiran, C., Kovtyukh, V. V., & Korotin, S. A. (2004), *A&A* **418**, 551
Prochaska, J. X., Naumor, S. O., Carney, B. W., McWilliam, A., & Wolfe, A. M. (2000), *AJ* **120**, 2513
Reddy, B. E., Tomkin, J., Lambert, D. L., & Allende Prieto, C. (2003), *MNRAS* **340**, 304
Robin, A. C., Reylé, C., Derrière, S., & Picaud, S. (2003), *A&A* **409**, 523
Soubiran, C., Bienaymé, O., & Siebert, A. (2003), *A&A* **398**, 141
Tull, R. G., MacQueen, P. J., Sneden, C., & Lambert, D. L. (1995), *PASP* **107**, 251
Yanny, B., Newberg, H. J., Grebel, E. K., *et al*. (2003), *ApJ* **588**, 824

Part II

Abundances in the Galaxy: Galactic stars in clusters, bulges and the centre

9

Galactic open clusters with supersolar metallicities

Sofia Randich

INAF – Osservatorio di Arcetri, Largo E. Fermi 5, I-50125 Firenze, Italy

Galactic open clusters provide a key tool to address a variety of issues related to the formation and evolution of stars and the Galactic disk. In the last few years a metallicity higher than Solar has been derived/confirmed spectroscopically for a few clusters, the most famous example being the very old NGC 6791, for which a metallicity [Fe/H] ~ 0.4 has recently been reported. In this paper current knowledge of these supersolar-metallicity clusters is reviewed and their properties and abundance patterns are compared with those of non-metal-rich clusters and other Galactic populations. Possible implications for their origin and for the metallicity gradient in the disk are briefly discussed. A summary of recent surveys for planets in metal-rich clusters is also provided, together with new results on Li abundances for the 3-Gyr-old metal-rich cluster NGC 6253.

1 Introduction

Galactic open clusters (OCs) provide homogeneous samples of stars whose parameters (age, distance, reddening) can be determined much more easily than for field stars. The OCs span large intervals of ages, metallicities, and positions in the Galactic disk; furthermore, their brightest members are strong-line red giants that are very well suited for measurements of radial velocity and chemical composition. Therefore, OCs are widely and traditionally recognized as excellent tracers of the structure, kinematics, and chemistry of the Galactic disk, as well as of the evolution of stars and their properties.

In recent years spectroscopic studies of OCs have received increasing attention thanks to the advent of state-of-the-art high-resolution spectrographs and, in

The Metal-rich Universe, eds. G. Israelian and G. Meynet. Published by Cambridge University Press.
© Cambridge University Press 2008.

particular, multi-object facilities. Among the other results, these studies have allowed us to confirm the existence of a few OCs with metallicities higher than Solar and to derive their detailed chemical composition.

In the following sections we will address several questions related to these metal-rich OCs, namely the following. What is the fraction of metal-rich clusters? What are their properties? How do their abundance patterns compare with those of other Galactic populations? Can we put constraints on their origin? That is, were they born in the Galactic disk or, rather, in the bulge (or inner side of the Galaxy) and then scattered in the disk? How do they contribute to the radial metallicity gradient in the Galactic disk?

Independently of their origin, the very existence of metal-rich OCs provides unique samples with which to investigate various critical issues, such as the dependence on metallicity of the initial mass function (IMF) and mixing processes in stellar interiors. In addition, clusters with higher than Solar [Fe/H] are in principle ideal sites in which to search for extra-Solar planets. Regarding the issue of the IMF, we refer to Chapter 24 in these proceedings. We will present in Sections 5 and 6 below a few highlights concerning planet searches in OCs and the results of a lithium study concerning the metal-rich cluster NGC 6253.

2 On the existence of metal-rich clusters

The very old, very massive cluster NGC 6791 is probably the most famous example of a metal-rich OC. Interestingly, in the first photometric study of this cluster (Kinman, 1965) a possible metal deficiency was reported and only in later studies (in particular spectroscopic ones) was its metal-rich nature ascertained (Origlia *et al.* 2006 and references therein). Besides NGC 6791, other clusters with higher-than-Solar metallicity have been detected by photometric or spectroscopic means during the last few years. In order to estimate their fraction, we plot in Figure 9.1 the distribution of [Fe/H] values for all clusters for which metallicity is available (from photometry or spectroscopy) and that for OCs with spectroscopic [Fe/H] (from either low- or high-resolution measurements). The information was retrieved from the catalog of Dias *et al.* (2002) (see also www.astro.iag.usp.br/wilton/). Of the 1756 OCs contained in the catalog (at the time of writing), for only 143 (i.e., less than 10%) is an estimate of metallicity available and for only 68 has there been a determination of [Fe/H] from spectroscopy. In a few cases (like NGC 6791 itself) we updated the [Fe/H] values given in the catalog with more recent and higher-quality determinations.

Considering the whole sample of OCs for which [Fe/H] data are available, about 30% (39 of 143), 12% (17 of 143), and 1.5% (2 of 143) of the clusters have [Fe/H] greater than 0.05, 0.1, and 0.2, respectively. These percentages change to ~23%

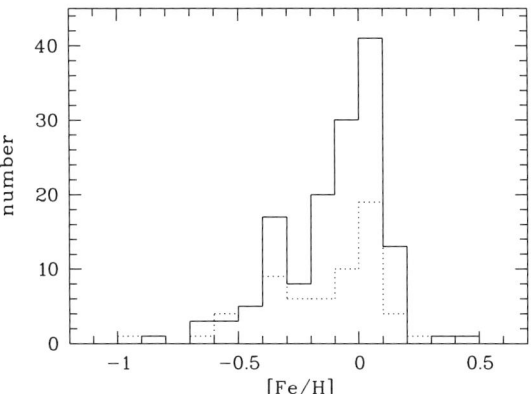

Figure 9.1. The metallicity distribution of Galactic open clusters. Solid line: all clusters from the catalog of Dias *et al.* (2002) for which [Fe/H] determinations are available. Dashed line: clusters with spectroscopic [Fe/H] available.

(16 of 68), 15% (10 of 68), and 4.4% (3 of 68) for clusters with [Fe/H] from spectroscopy. With the caveat that the sample of OCs with a metallicity determination is far from complete and certainly biased toward closer clusters, these percentages suggest that a non-negligible fraction of OCs has a metallicity moderately higher than Solar ([Fe/H] > 0.1), while only a very small fraction of the whole cluster population has a metallicity considerably higher than Solar ([Fe/H] > 0.2).

We choose [Fe/H] = 0.1 (i.e. an iron content ~25% higher than Solar) as a (somewhat arbitrary) threshold between metal-rich and non-metal-rich. Also, the discussion in the following sections will be based on the spectroscopic sample only, under the assumption that [Fe/H] values from spectroscopy are more reliable than estimates from photometry. Determinations of spectroscopic metallicities for a much larger number of clusters are indeed badly needed; we note, however, that there is an overall good agreement between the spectroscopic and photometric metallicity distributions shown in Figure 9.1.

3 Properties of metal-rich clusters

The ten clusters with spectroscopic [Fe/H] ≥ 0.1 currently known are listed in Table 9.1, together with their properties and references. In Figure 9.2 we show the distribution of ages of OCs with [Fe/H] below and above 0.1. Since the sample of clusters for which spectroscopic metallicity data are available is not complete, we cannot carry out a comparison on a statistical basis. Nevertheless, Figure 9.2 suggests that the sample with [Fe/H] ≥ 0.1 has a flatter age distribution than the other one, which peaks at ages older than ~1 Gyr. Metal-rich clusters are not necessarily young, in contrast to what one would naively expect, and they are evenly distributed

Table 9.1. *Properties of OCs with [Fe/H] ≥ 0.1*

Cluster	[Fe/H]	Age (Gyr)	R_{gc} (kpc)	z (pc)	Reference
IC 4725	+0.15	0.09	7.9	−47.9	Luck (1994)
NGC 6705	+0.10	0.2	6.9	−91	González & Wallerstein (2000)
NGC 6475	+0.14	0.25	8.2	−24	Sestito et al. (2003)
Hyades	+0.13	0.6	8.5	−17.1	Paulson et al. (2003)
NGC 6134	+0.15	0.69	7.6	−3.2	Carretta et al. (2004)
NGC 5822	+0.10	1.2	7.8	57.4	Luck (1994)
IC 4651	+0.10	1.7	7.7	−122.2	Pasquini et al. (2004)
NGC 6253	+0.36	3.0	7.0	−164	Sestito et al. (2007)
NGC 6791	+0.40	8.0	8.2	800	Carraro et al. (2006)
Praesepe	+0.27	0.6	8.6	85	Pace et al. (2008)

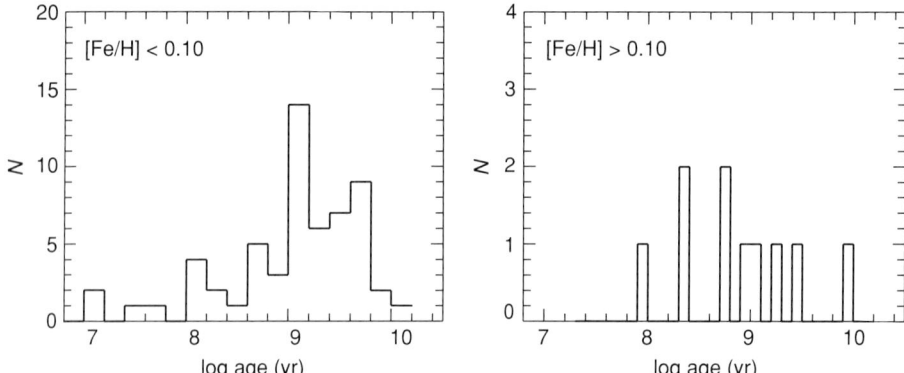

Figure 9.2. Age distributions of clusters with metallicities [Fe/H] < 0.1 and [Fe/H] ≥ 0.1 (left- and right-hand panels). Ages have been taken from the catalog of Dias et al. (2002).

in the age interval between ∼0.1 and 8 Gyr: NGC 6791, the most metal-rich cluster, is indeed among the oldest Galactic OCs. Conversely, and surprisingly, we note the lack of metal-rich OCs younger than 100 Myr. Figures 9.3 and 9.4 are similar to Figure 9.2, but the distribution of Galactocentric distances and heights above the Galactic plane are shown. Again, there seems to be a difference between metal-rich clusters and those with [Fe/H] < 0.1. Namely, OCs with [Fe/H] ≥ 0.1 are all located at R_{gc} < 9 kpc, with more than half of them being closer than 8.5 kpc (R_{gc} of the Sun) to the Galactic Center; the low-metallicity sample instead covers a much larger range of R_{gc} values and most of them are located at R_{gc} larger than 10 kpc. In other words, no known metal-rich clusters exist in the outer regions of the Galactic disk, at least when the spectroscopic sample is considered. Regarding z values, with the exception of NGC 6791, all metal-rich clusters are located within 200 pc of the

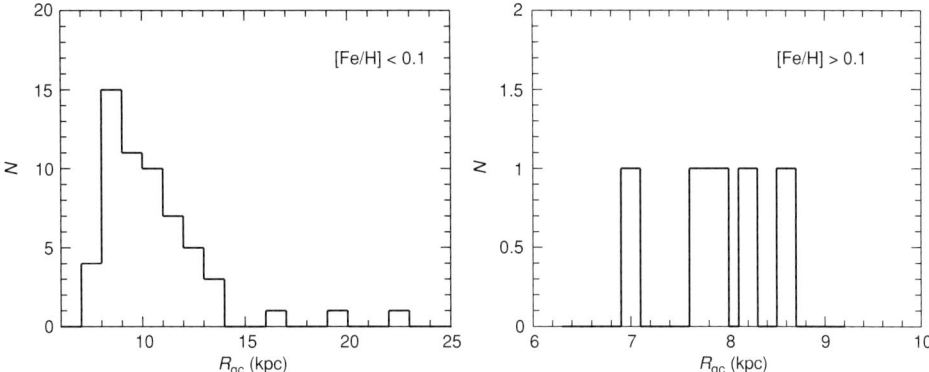

Figure 9.3. The same samples in as Figure 9.2, but showing the distributions of Galactocentric distances.

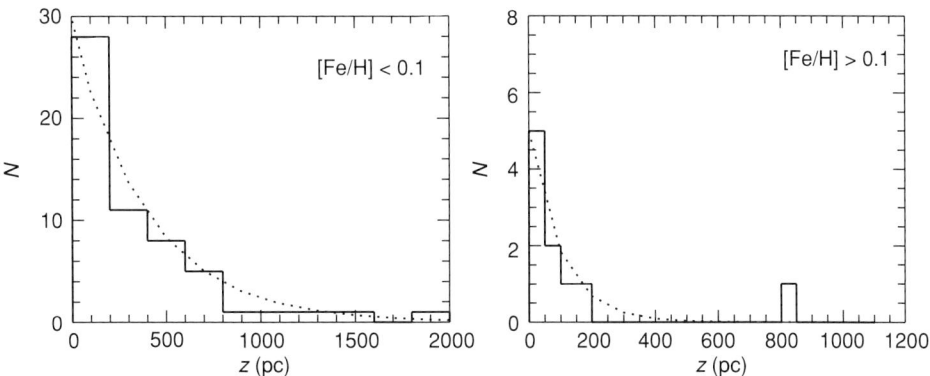

Figure 9.4. The same samples as in Figures 9.2 and 9.3, but showing the distributions of absolute distances from the Galactic plane. In each panel the solid line represents the empirical distribution, while the dashed curve denotes the best fit to it.

Galactic plane: their scale height is 100 pc, compared with the value of 450 pc for the non-metal-rich sample

4 Comparison with other populations

In order to gather clues on the origin of metal-rich clusters, in Figure 9.5 we compare their [X/Fe] (for the most abundant α-elements apart from oxygen) versus [Fe/H] distribution with those of other Galactic populations, namely bulge stars, thin- and thick-disk stars, bulge-like dwarf stars (stars with metallicities and kinematics characteristics of a probable inner-disk or bulge origin – e.g. Castro *et al.* (1997)). Note that α-element measurements are not available for all of the ten

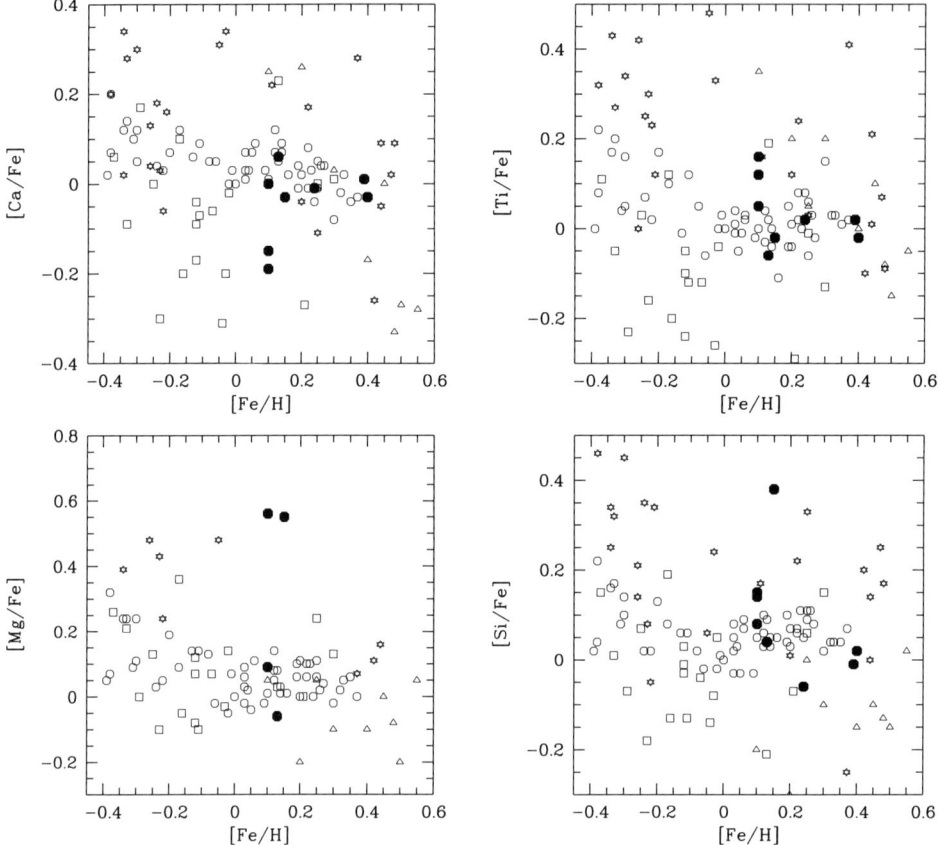

Figure 9.5. Plots of [X/Fe] versus [Fe/H] for metal-rich OCs (filled circles) and other Galactic populations, namely thin- and thick-disk stars, open circles (Bensby *et al.* 2003); bulge stars, open stars (McWilliam & Rich 1994; Fulbright *et al.* 2007); and bulge-like stars, open triangles and squares (Pompeia *et al.* 2003; Castro *et al.* 1997).

clusters listed in Table 9.1. As demonstrated by several other authors, Figure 9.5 shows a large spread in [α/Fe] ratios for any given population/metallicity; it is therefore difficult to discern whether OCs fit well into one or other of the Galactic components. With the exception of a few outliers, however, metal-rich clusters are characterized by a rather homogeneous abundance pattern. They do not exhibit significant α-enhancement (most [α/Fe] ratios are indeed close to zero), suggesting that there is a better agreement with the trend of disk stars rather than with that of the bulge or bulge-like populations. The latter, bulge stars in particular, may be characterized by α-enhancement at supersolar metallicities; see also Zoccali *et al.* (2006). Therefore, we tentatively conclude that the majority of metal-rich clusters

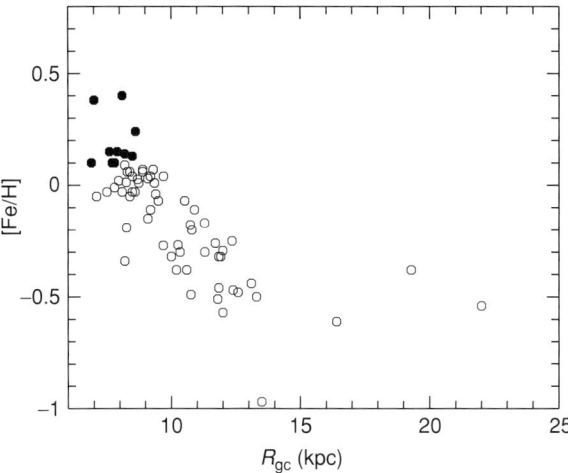

Figure 9.6. The distribution of [Fe/H] versus Galactocentric distance (R_{gc}). Filled symbols denote the 10 metal-rich clusters listed in Table 9.1, while open circles indicate the 58 non-metal-rich clusters with a spectroscopic determination of metallicity. The [Fe/H] values come from various sources and are not on the same scale. The figure is not meant to provide a quantitative estimate of the slope of the gradient, but merely to illustrate qualitatively the influence of metal-rich clusters on the gradient.

most likely originated in the local disk; a different origin might hold for NGC 6791 (Carraro *et al.* 2006; Bedin *et al.* 2006).

Regarding the discrepant clusters in Figure 9.5, namely NGC 5822 and IC 4725, their abundances come from Luck (1994) and are based on only one and two stars, respectively. Additional observations are needed before conclusions concerning their anomalous abundance trends can be drawn.

Finally, under the hypothesis that most metal-rich clusters were born in the disk, we can use them to trace the radial metallicity gradient. Since, as mentioned, all of these clusters are located at small Galactocentric radii, their inclusion in the determination of the gradient would make it steeper (see Figure 9.6). Note, however, that at any given R_{gc} there is a dispersion in metallicity; the reason for this dispersion and the possible dependence of metallicity on other parameters must be ascertained before the exact slope of the gradient can be derived.

5 Planet searches

It is now well established on observational grounds that stars hosting giant gas planets are more metal-rich than stars not harboring a planetary system (see e.g. Chapter 2 in these proceedings). The average metallicity of stars hosting a planet is [Fe/H] = +0.14 (Santos *et al.* 2001), compared with a mean value of [Fe/H] ∼ −0.1 for the Solar neighborhood. Also, the hypothesis that supersolar metallicity could correspond to a higher probability of hosting a planet has been proved by examination of statistically significant samples (e.g. Santos *et al.* 2004). Metal-rich clusters thus offer, in principle, ideal samples in which to look for planets. Several groups have started planet searches in metal-rich clusters, by means of both photometric (to detect transits) and radial-velocity surveys. At the time of writing (November 2006), to the best of our knowledge, only one candidate transiting planet has been detected, but the feasibility and/or the limits of the various methods have been demonstrated. We mention in particular the PISCES (Planets in Stellar Clusters Extensive Search) survey, a long-term project aimed at searching for transiting planets in clusters (Mochejska *et al.* 2005 and references therein): no good candidate has been detected in the very metal-rich NGC 6791, while the detection of a possible transiting-planet candidate in the metal-poor NGC 2158 ([Fe/H] = −0.46) has been reported (Mochejska *et al.* 2006). Planet searches in NGC 6791 have also been carried out by Bruntt *et al.* (2003); although they found three transit-like events, these were shown not to be candidate planets by Mochejska *et al.* (2006).

Regarding radial-velocity searches, most efforts have focused on the close-by, metal-rich Hyades (Paulson *et al.* 2004a, 2000b). They observed a sample of 94 stars and did not detect any close-in giant planets. They discussed the possibility of detecting planets in the Hyades and the effects of rotational modulation of stellar active regions. They concluded that, for young stars with relatively high levels of activity, planet detection is feasible only in the case of very-high-mass companions or with extremely good sampling (several observations per month). We suggest that NGC 6253, which is older, and thus considerably less active than the Hyades, but is close enough to allow radial-velocity surveys, is indeed a very good target for planet searches.

Finally, we mention that as a by-product of all the planet surveys in clusters several new binary systems have been detected.

6 Lithium and mixing in Solar-type stars

[7]Lithium (Li) is destroyed by proton capture at the relatively low temperature of 2.5 MK and it is depleted from stellar atmospheres when there operates a mechanism able to transport surface material down to the deeper stellar interiors where the temperature is high enough for Li burning. Thus, measurements of Li abundance in stars are unique tracers of internal mixing mechanisms. With the exception of the case of very-low-mass, fully convective stars, the physics driving Li depletion in stars is not well understood.

Focusing on stars similar to our Sun, extensive Li measurements in OCs of various ages have convincingly shown that they undergo Li depletion during the main-sequence phases (e.g. Sestito & Randich 2005 and references therein). In order to explain the observed depletion several extra (or non-standard) mixing mechanisms have been proposed; most of them predict that Li depletion should depend on metallicity, since a higher [Fe/H] would imply a deeper convective zone and thus a smaller amount of extra mixing needed at a given mass.

So far, the comparison of the Li distributions of clusters with similar ages but different iron contents has shown that both for young and for old clusters the metallicity does not seem to affect Li depletion greatly (e.g. Randich *et al.* 2000; Sestito *et al.* 2003; Jeffries 2006 and references therein). Those comparisons, however, were limited to clusters with differences in metallicities below ∼0.15 dex. The question then is "what happens when considering significantly higher [Fe/H] values?"

In the context of a big VLT/FLAMES project aimed at obtaining high-resolution spectra of a large sample of stars in a variety of old OCs (Randich *et al.* 2005) in order to put constraints on the evolution of the Galactic disk and light elements, we observed the extremely metal-rich cluster NGC 6253. We used the fiber link to UVES to acquire spectra of evolved cluster stars in order to derive the chemical composition of the cluster. We found [Fe/H] $= 0.36 \pm 0.07$ and about Solar [X/Fe] ratios for most elements (Sestito *et al.* 2007). Simultaneously, we took Giraffe ($R = 19{,}000$) spectra of 196 main-sequence cluster candidates to derive radial velocities and confirm membership, and to determine their Li abundances.

Of the 196 candidates, 105 were confirmed as members on the basis of their radial velocities (a contamination of seven stars is expected). In Figure 9.7 we compare the n(Li) versus $T_{\rm eff}$ distribution of NGC 6253 members with those of NGC 752 and IC 4651. These two clusters are slightly younger than NGC 6253 and have lower metallicities ([Fe/H] − ∼0 and +0.1, respectively). Both clusters should in principle exhibit a smaller amount of Li depletion.

Figure 9.7 clearly shows that, although similarly to the Solar-age, Solar-metallicity cluster M 67, NGC 6253 is characterized by a larger amount of dispersion than IC 4651 and NGC 752, the majority of NGC 6253 members are not more Li-depleted than stars in the other two clusters. This in turn implies that, at variance with model predictions, even a rather large difference in the overall metal content does not affect the rate of Li depletion, at least in the temperature range considered here.

7 Concluding remarks

Spectroscopic determinations of metallicity in OCs have confirmed the existence of a population of clusters with [Fe/H] from moderately to significantly above

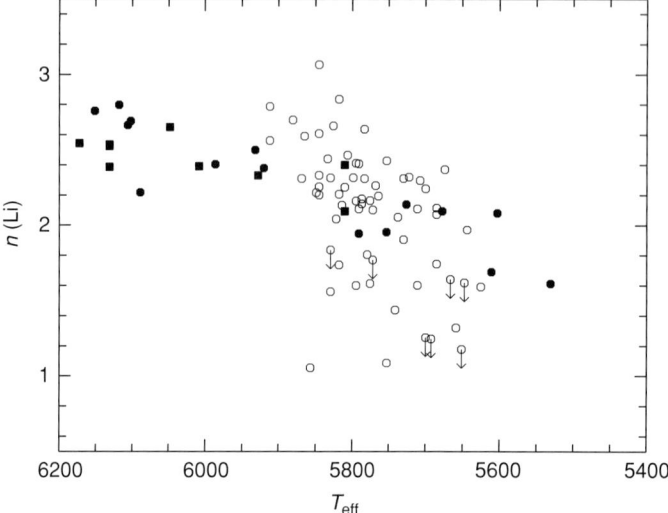

Figure 9.7. Lithium abundance (n(Li) = log(N(Li)/N(H)) + 12) versus effective temperature for NGC 6253 members (open circles) and members of the similar-age, but different-metallicity clusters NGC 752 ([Fe/H] $\sim$ 0, filled circles) and IC 4651 ([Fe/H] = 0.1, filled squares). The lithium abundances for these clusters were taken from Sestito & Randich (2005). Those for NGC 6253 were derived in a similar way. Lithium abundances were derived only for those of the 105 stars that were confirmed as cluster members for which secure photometry data were available.

Solar. At present, 10 out of 68 OCs with a spectroscopic [Fe/H] determination have [Fe/H] > 0.1. The properties of these clusters are different from those of non-metal-rich ones.

- They have a flatter age distribution; no metal-rich clusters younger than $\sim$100 Myr have been detected so far. This suggests that metal enrichment is due to the position in the disk, rather than age.
- All of them are located within 8.6 kpc of the Galactic Center, at variance with non-metal-rich clusters; also, their scale height above the Galactic plane is about a third that of clusters with (under-)Solar metallicity.
- The available [α/Fe] abundance pattern suggests that the majority of these clusters originated in the local disk. However, additional measurements both of abundances and, most importantly, of kinematic properties of the clusters are necessary in order to put this conclusion on firmer grounds and to constrain the radial metallicity gradient in a more secure way.
- Several projects aimed at searching for planets in metal-rich clusters are being carried out. So far only one candidate transiting planet has been detected.

- Finally, the comparison of the lithium pattern of the very-metal-rich NGC 6253 with those of clusters with different metallicities allows us to confirm definitively that the iron content of a star does not affect Li depletion in Solar-type stars.

References

Bedin, L. R., Piotto, G., Carraro, G., King, I. R., & Anderson, J. (2006), *A&A* **160**, L27–L30
Bensby, T., Feltzing, S., & Lundström, I. (2003), *A&A* **384**, 879–883
Bruntt, T. M., Grundhal, F., Tingley, B. *et al.* (2003), *A&A* **410**, 323–335
Carraro, G., Villanova, S., Demarque, P. *et al.* (2006), *ApJ* **643**, 1151–1159
Carretta, E., Bragaglia, A., Gratton, R. G., & Tosi, M. (2004), *A&A* **422**, 951–963
Castro, S., Rich, R. M., Barbuy, B., & McCarthy, J. K. (1997), *AJ* **114**, 376–387
Dias, W. S., Alessi, B. S., Moitinho, A., & Lèpine, J. R. D. (2002), *A&A* **389**, 871–873
Fulbright, J. P., McWilliam, A., & Rich, R. M. (2007), *ApJ*, **661**, 1152–1179
González, G., & Wallerstein, G. (2000), *PASP* **112**, 1081–1088
Jeffries, R. D. (2006), in *Chemical Abundances and Mixing in Stars in the Milky Way and its Satellites*, S. Randich and L. Pasquini (eds), Berlin: Springer, pp. 163–170
Luck, R. E. (1994), *ApJS* **91**, 309–346
Kinman, T. D. (1965), *ApJ* **142**, 655–680
McWilliam, A., & Rich, M. (1994), *ApJS* **91**, 749–791
Mochejska, B. J., Stanek, K. Z., Sasselov, D. D. *et al.* (2005), *ApJ* **129**, 2856–2868 (2006), *AJ* **131**, 1090–1105
Origlia, L., Valenti, E., Rich, M. R., & Ferraro, F. R. (2006), *ApJ* **646**, 499–504
Pace, G., Pasquini, L., Randich, S., & François, P. (2008), *A&A* submitted
Pasquini, L., Randich, S., Zoccali, M., Hill, V., Charbonnel, C., & Nordström, B. (2004), *A&A* **424**, 951–963
Paulson, D. B., Sneden, C., & Cochran, W. D. (2003), *AJ* **125**, 3185–3195
Paulson, D. B., Saar, S. H., Cochran, W. D., & Henry, G. W. (2004a), *AJ* **127**, 1644–1652
Paulson, D. B., Cochran, W. D., & Hatzes, A. P. (2004b), *AJ* **127**, 3579–3586
Pompeia, L., Barbuy, B., & Grenon, M. (2003), *ApJ* **592**, 1173–1185
Randich, S., Pasquini, L., & Pallavicini, R. (2000), *A&A* **356**, L25–L29
Randich, S., Bragaglia, A., Pastori, L. *et al.* (2005), *ESO Messenger* **121**, 18–22
Santos, N. C., Israelian, G., & Mayor, M. (2001), *A&A* **373**, 1019–1031 (2004), *A&A* **415**, 1153–1166.
Sestito, P., Randich, S., Mermilliod, J.-C., & Pallavicini, R. (2003), *A&A* **407**, 289–301
Sestito, P., & Randich, S. (2005), *A&A* **442**, 615–627
Sestito, P., Randich, S., & Bragaglia, A. (2007), *A&A*, **465**, 185–196
Zoccali, M., Lecureur, A., Barbuy, B. *et al.* (2006), *A&A* **457**, L1–L4

10

Old and very-metal-rich open clusters in the BOCCE project

Angela Bragaglia,[1] Eugenio Carretta,[1] Raffaele Gratton[2] & Monica Tosi[2]

[1] INAF – Osservatorio Astronomico di Bologna, Via Ranzani 1, I-40127 Bologna, Italy
[2] INAF – Osservatorio Astronomico di Padova, Vicolo dell'Osservatorio 5, I-35122 Padova, Italy

The Bologna Open Cluster Chemical Evolution (BOCCE) project is intended to study the disk of our Galaxy using open clusters as tracers of its properties. We are building a large sample of clusters, deriving homogeneously their distance, age, reddening, and detailed chemical composition. Among our sample we have several objects more metal-rich than the Sun and we present here first results of the analysis for NGC 6819, IC 4651, NGC 6134, NGC 6791, and NGC 6253, the last two being the most metal-rich open clusters known.

1 Introduction: the BOCCE project

Open clusters (OCs) are good tracers of the properties of the Galactic disk, since they are seen over the whole disk, cover the entire age interval of the disk, and trace its chemical abundances both at present and in the past (e.g. Friel 1995). To study the Galactic disk we are building a large sample of OCs studied in the most homogeneous way possible. In particular, we are interested in defining the radial abundance distribution in the disk and understanding whether it has changed during the disk lifetime, since this is important for chemical-evolution models. This metallicity distribution is generally described as a radial gradient (e.g. Friel *et al.* 2002), but it has also been found to be step-like (Twarog *et al.* 1997), and new data for the outermost part of the disk seem to imply a flattening of the gradient (e.g. Yong *et al.* 2005).

We named our program BOCCE, which stands for Bologna Open Cluster Chemical Evolution project. We employ (i) deep, precise photometry to derive ages, distances, and reddenings by comparison with synthetic color–magnitude diagrams

The Metal-rich Universe, eds. G. Israelian and G. Meynet. Published by Cambridge University Press.
© Cambridge University Press 2008.

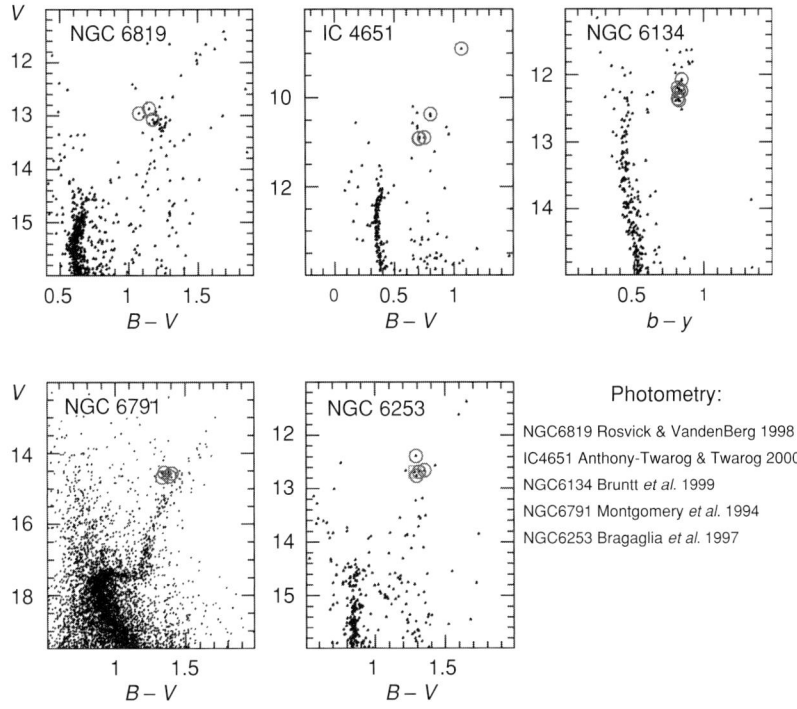

Figure 10.1. Color–magnitude diagrams for the five metal-rich BOCCE open clusters already examined (at least partially); sources of photometry data are indicated. For each cluster the stars for which we collected spectra – mostly red clump ones – are shown by larger symbols

(CMDs), see Bragaglia *et al.* (1997) and Bragaglia & Tosi (2006) for an update of the photometric part of our work; (ii) medium-resolution spectra to derive membership in terms of radial velocities (e.g. D'Orazi *et al.* 2006); and (iii) high-resolution spectra to derive detailed abundances (e.g. Carretta *et al.* 2004).

We try to cover the full extent of the metallicity range of OCs, and we present here our results on five clusters richer in metals than the Sun: NGC 6819, IC 4651, NGC 6134, NGC 6253, and NGC 6791 in order of ascending metallicity.

2 Data and results

Our strategy for high-resolution spectroscopy is to obtain spectra of a few stars per cluster, choosing stars known to be cluster members (see Figure 10.1). We concentrate on clump stars, as an optimal compromise between high luminosity (for a high signal-to-noise ratio) and temperatures that are not too cold (to minimize problems of line crowding and uncertainties in the model atmospheres). Furthermore, our choice of a single evolutionary status increases the homogeneity of the analysis.

Table 10.1. *Metal-rich open clusters examined so far in the BOCCE sample*

Cluster	Age (Gyr)	Number of stars	Radial velocity (km s^{-1})	Spectrograph	[Fe/H]	Reference
NGC 6819	2	3	4.2 ± 1.1	SARG	+0.07	Bragaglia et al. (2001)
IC 4651	1.7	5	−29.7 ± 0.5	FEROS	+0.11	Carretta et al. (2004)
NGC 6134	0.7	6	−24.5 ± 0.2	FEROS/UVES	+0.15	Carretta et al. (2004)
NGC 6791	9	4	−47.2 ± 0.8	SARG	+0.47	Gratton et al. (2006)
NGC 6253	3	4	−28.3 ± 0.3	FEROS/UVES	+0.46	Carretta et al. (2007)

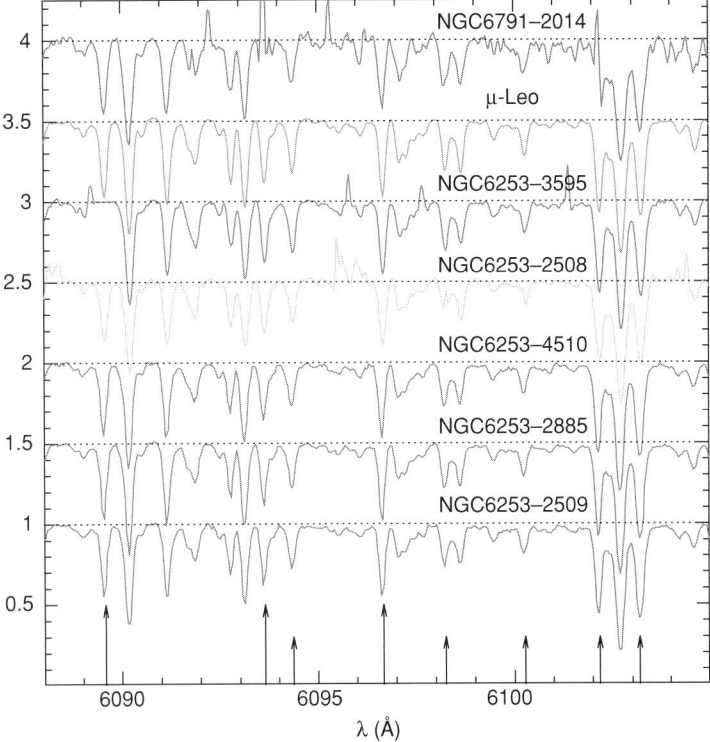

Figure 10.2. Spectra for stars in the two most metal-rich BOCCE open clusters, NGC 6791 and NGC 6253, shown together with the well-studied, metal-rich field giant μ-Leo (e.g. Gratton & Sneden 1990). The arrows indicate Fe lines. Star NGC6253-2508 is most probably a binary, since the radial velocity measured on the present spectrum differs by about 10σ from the cluster average and from that measured on an old, lower-quality spectrum (Carretta *et al.* 2000), while its membership probability based on proper motion is high (M. Montalto and S. Desidera, private communication).

We have already collected data for 15 clusters, using various telescopes and spectrographs; in particular, we employed SARG at the TNG, FEROS at the 1.5-m ESO, and UVES at the VLT for the clusters presented here (see Table 10.1, where information on the five clusters can be found, and Figure 10.2 for examples of spectra). We will later enlarge our sample using archive data and the spectra collected in a companion survey with FLAMES at the VLT; see Randich *et al.* (2005) and Chapter 9. With the latter we will also be able to understand possible systematic differences resulting from the various analysis procedures and assumptions.

We make every effort to keep our analysis procedure as homogeneous as possible: the model grids used to derive abundances, the line lists, the oscillator strengths, the Solar reference abundances, and the method of measurement of equivalent

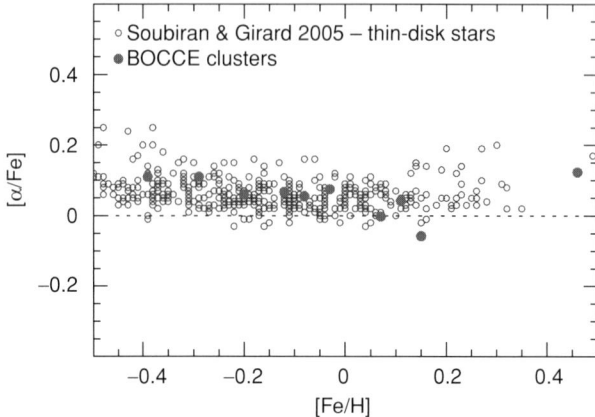

Figure 10.3. A comparison between the run of α-elements with [Fe/H] of field thin-disk stars (Soubiran & Girard 2005) and BOCCE clusters (some values are still provisional).

widths are always the same. Abundances are usually derived using equivalent widths (EWs). Spectrum synthesis is used as a check, see Carretta *et al.* (2004) for a discussion, except for the most metal-rich OCs, for which lines are too blended to allow us to obtain reliable EWs at the resolution we have ($R = 30{,}000$–$48{,}000$), in which cases we use synthesis of selected lines.

We have obtained the metallicity (see Table 10.1) and have measured (or are working on) the elemental ratios of light, α, Fe-group, and heavier elements. Our results are generally in agreement with existing determinations, e.g. Pasquini *et al.* (2004) for IC 4651; and Peterson & Green (1998), Carraro *et al.* (2006), and Origlia *et al.* (2006) for NGC 6791.

One final consideration: are OCs really good tracers of the disk abundances? To check this issue, we made a comparison of abundance ratios obtained for OCs in the BOCCE sample and thin-disk field stars by Soubiran & Girard (2005). The run with [Fe/H] of α-elements and other elements considered is encouraging (Figure 10.3): the two populations seem to follow the same trends. The only notable exception is Na, which seems to be overabundant in OCs with respect to field stars. The same result has been found by others, see e.g. Yong *et al.* (2005). The difference is worth investigating and it could have something to do with the different evolutionary status of stars in the two samples (dwarfs for the field one and giants for the OC one), which could be either real or due to some systematic difference in the analysis (e.g. Mishenina *et al.* 2006).

Further considerations are postponed until a larger number of clusters has been examined (both the photometric and the spectroscopic data), forming a truly homogeneous sample on which to draw reliable conclusions.

References

Anthony-Twarog, B. J., & Twarog, B. A. (2000), *AJ* **110**, 2282–2295
Bragaglia, A., & Tosi, M. (2006), *AJ* **131**, 1544–1558
Bragaglia, A., Tessicini, G., Tosi, M., Marconi, G., & Munari, U. (1997), *MNRAS* **284**, 477–488
Bragaglia, A., Carretta, E., Gratton, R. G. *et al.* (2001), *AJ* **121**, 327–336
Bruntt, H., Frandsen, S., Kjeldsen, H., & Andersen, A. I. (1999), *A&A* **140**, 135–143
Carraro, G., Villanova, S., Demarque, P., McSwain, M. V., Piotto, G., & Bedin, L. R. (2006), *ApJ* **643**, 1151–1159
Carretta, E., Bragaglia, A., Tosi, M., & Marconi, G. (2000), in *Stellar Clusters and Associations: Convection, Rotation, and Dynamos*, eds. R. Pallavicine, G. Micela, & S. Sciortiono (San Francisco, CA, Astronomical Society of the Pacific), pp. 273–276
Carretta, E., Bragaglia, A., Gratton, R. G., & Tosi, M. (2004), *A&A* **422**, 951–962
Carretta, E., Bragaglia, A., & Gratton, R. G. (2007), *A&A* **473**, 129
D'Orazi, V., Bragaglia, A., Tosi, M., Di Fabrizio, L., & Held, E. V. (2006), *MNRAS* **368**, 471–478
Friel, E. D. (1995), *ARA&A* **33**, 381–414
Friel, E. D., Janes, K. A., Tavarez, M. *et al.* (2002), *AJ* **124**, 2693–2720
Gratton, R. G., & Sneden C. (1990), *A&A* **234**, 366–386
Gratton, R. G., Bragaglia, A., Carretta, E., & Tosi, M. (2006), *ApJ* **642**, 462–469
Mishenina, T. V., Bienaymé, O., Gorbaneva, T. I. *et al.* (2006), *A&A* **456**, 1109–1120
Montgomery, K. A., Janes, K. A., & Phelps, R. L. (1994), *AJ* **108**, 585–593
Origlia, L., Valenti, E., Rich, M. R., & Ferraro, F. R. (2006), *ApJ* **646**, 499–504
Pasquini, L., Randich, S., Zoccali, M., Hill, V., Charbonnel, C., & Nordström, B. (2004), *A&A* **424**, 951–963
Peterson, R. C. & Green, E. M. (1998), *ApJ* **502**, L39
Randich, S., Bragaglia, A., Pastori, L. *et al.* (2005), *The Messenger* **121**, 18–22
Rosvick, J. M., & VandenBerg, D. A. (1998), *AJ* **115**, 1516–1523
Soubiran, C., & Girard, P. (2005), *A&A* **438**, 139–151
Twarog, B. A., Ashman, K. M., & Anthony-Twarog, B. J. (1997), *AJ* **114**, 2556–2585
Yong, D., Carney, B. W., & Teixera de Almeida, M. L. (2005), *AJ* **130**, 597–625

11

Massive-star versus nebular abundances in the Orion nebula

S. Simón-Díaz[1,2]

[1]*Instituto de Astrofísica de Canarias, E-38200 La Laguna, Tenerife, Spain*
[2]*LUTh, Observatoire de Meudon, 92195 Meudon Cedex, France*

The search for consistency between nebular and massive-star abundances has been a longstanding problem. I briefly review what has been done regarding this topic, also presenting a recent study focused on the Orion nebula: the O and Si stellar abundances resulting from a detailed and fully consistent spectroscopic analysis of the group of B stars associated with the Orion nebula are compared with the most recent nebular gas-phase results.

1 Introduction

Photospheres of OB stars are representative of the interstellar material from which they were born due to their relative youth. The evolutionary characteristics of blue massive stars imply that these objects and the associated ionized nebulae – H II regions – must share the same chemical composition.[1]

Traditionally, chemical abundance studies in spiral and irregular galaxies have been based on the emission line spectra of H II regions. This is logical, since H II regions are luminous and have high surface brightness (in the emission lines) relative to individual stars in galaxies. Therefore, it is relatively easy to obtain high-quality spectroscopic observational data, even with small and medium-sized

---

[1] There are certain cases in which this is not completely fulfilled. (1) Strong stellar mass-loss may expose already-contaminated underlying layers on the surface of the star; (2) Authors of several studies of OBA-type stars have found observational evidence of stellar-surface contamination by products from the CNO bicycle; rotating models by Maeder & Meynet (2000) predict that mixing of nuclear processed material at the surface will increase with stellar mass, age, initial rotational velocity and decreasing metallicity. (3) Certain elements in the nebular material can be depleted onto dust grains; in this case the gas-phase abundances derived through a spectroscopic study of the H II region could be somewhat lower than the stellar ones.

The Metal-rich Universe, eds. G. Israelian and G. Meynet. Published by Cambridge University Press.
© Cambridge University Press 2008.

telescopes. This has made possible both the detailed study of individual nebulae (Esteban *et al.* 2004) and the determination of radial gradients in the Milky Way (Shaver *et al.* 1983; Afflerbach *et al.* 1997; Esteban *et al.* 2005) and other spiral galaxies (see Chapter 17 in these proceedings), imposing observational constraints on the chemical-evolution models of these galaxies (see Chapter 43 in these proceedings).

Although this is a commonly used methodology, it is not without some difficulties and problems. For example, it is known that the optical recombination lines (ORLs) of ionized nebulae indicate higher abundances than do collisionally excited lines (CELs); temperature fluctuations, density condensations, and abundance inhomogeneities have been proposed as explanations to solve the ORL/CEL problem; however, none of these explanations is completely satisfactory (Esteban 2002). I refer the reader to Chapter 17 for a review of the use and limitations of strong-line methods in the determination of nebular abundances. Stasińska (2005) has recently shown that, for metal-rich nebulae, the derived abundances based on direct measurements of the electron temperature (T_e) may deviate systematically from the real ones. Finally, one must keep in mind other sources of uncertainties in the nebular abundance determination such as the atomic data, reddening corrections, and the possible depletion of elements by their accretion onto dust.

Among the stellar objects, blue massive stars can easily be identified at large distances due to their high luminosities. Therefore, massive stars offer a unique opportunity to study present-day chemical abundances in spiral and irregular galaxies as an alternative method to classical H II-region studies. However, the amount of energy released by these stars is so large that it produces dramatic effects on the star itself: these stellar objects undergo mass outflows throughout their lifetimes (so-called stellar winds) and their atmospheres depart from LTE conditions, two facts that make their modeling quite complex. It has not been until very recently that the development of massive-star model atmospheres and the growth of computational efficiency have allowed a reliable abundance analysis of these objects. Nowadays, it is feasible; however, one must take into account that there are some effects that can affect the final results and must be treated carefully: (1) the hypothesis governing the stellar-atmosphere modeling (LTE versus NLTE, plane-parallel versus spherical models, inclusion of line-blanketing effects), (2) atomic models and atomic data; and (3) establishment of the stellar parameters and microturbulence.

Nebular and stellar methodologies are now working in tandem to advance our knowledge of the metallicity content of irregular and spiral galaxies (from the Milky Way to far beyond the Local Group). Since they sample similar spatial and temporal distributions, these objects offer us a unique framework in which the reliability of the abundances derived using both methodologies can be tested.

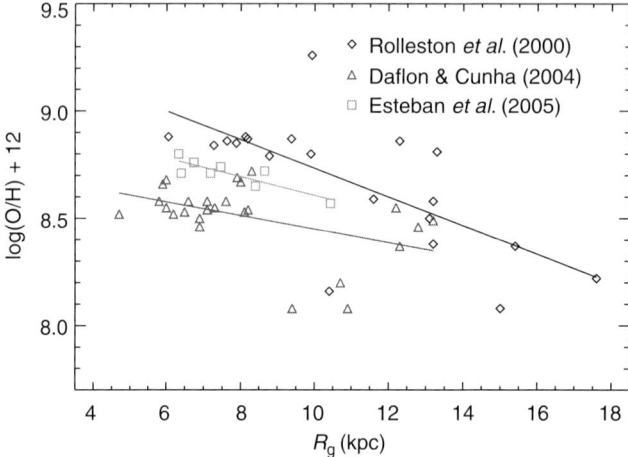

Figure 11.1. A comparison of three recent determinations of the O abundance gradient in the Milky Way. Although the comparison of values found in the literature for the gradient obtained by means of massive-star and H II-region studies seems to be in agreement within the intrinsic uncertainties (see the text), this might not be the case when comparing results from individual studies. Note also the offset in absolute abundances resulting from the three studies.

2 Do stellar and nebular abundances agree?

Two types of studies can be addressed to investigate whether there is a consistency of abundance results from stellar and nebular studies: the comparison of radial abundance gradients in spiral galaxies or mean abundances in irregular galaxies (the *global approach*) and the comparison of absolute abundances in the same Galactic region (the *local approach*). Values found in the literature for the oxygen-abundance gradient in the Galaxy range from -0.04 to -0.08 in the nebular case (Esteban *et al.* 2005 and references therein) and from -0.03 to -0.07 in the stellar case (Gummersbach *et al.* 1998; Rolleston *et al.* 2000; Daflon & Cunha 2004). One could conclude that results from the two methodologies are in agreement within the intrinsic uncertainties; however, this might not be the case when comparing results from individual studies (see e.g. Figure 11.1).

In the extragalactic case, Urbaneja *et al.* (2005b) found good agreement in the comparison of stellar and H II-region O abundances (based on direct determinations of the T_e of the nebulae) in M33; however, other workers studying spiral galaxies have found that the stellar–nebular agreement tends to be very dependent on the calibrations used in the strong nebular methods (Trundle *et al.* 2002; Urbaneja *et al.* 2005a).

Examples of the comparison between stellar and nebular abundances in the same galactic region or mean abundances in the Magellanic Clouds can be found in Cunha

& Lambert (1994), Korn *et al.* (2002), and Trundle *et al.* (2004). Although, within the typical errors of the stellar and nebular analyses, the abundances determined by the two methodologies are fairly consistent, more work remains necessary. For example, until very recently, there has not been any consistent detailed comparison of stellar and nebular abundances in the same star-forming region.

With the aim of rectifying this omission, Simón-Díaz *et al.* (2006) selected the Orion nebula, a well-studied and spatially resolved H II region with a cluster of a few massive stars inside it (the Trapezium cluster). Below, I will present the results of a detailed and fully consistent spectroscopic analysis of three B0.5V stars associated with the Orion nebula for deriving their O and Si abundances. The resulting abundances are compared with the most recent nebular gas-phase results.

3 Massive stars and nebular abundances in Orion nebula

The stellar spectroscopic observational dataset consists of INT at the IDS 2.5-m spectra of θ^1-Ori A, θ^1-Ori D, and θ^2-Ori B, three B0.5V stars located inside the Orion nebula. Details on the spectroscopic analysis for the stellar parameters and oxygen-abundance determination can be found in Simón-Díaz *et al.* (2006). The analysis of silicon abundances was presented in Simón-Díaz (2005), and will be published elsewhere soon. Both stellar parameters and abundances were determined by means of the stellar-atmosphere code FASTWIND (Santolaya-Rey *et al.* 1997, Puls *et al.* 2005), a NLTE code taking into account line-blanketing and stellar-wind effects. A detailed O and Si abundance analysis by multiplets had been performed beforehand, using the slow-rotator B0.2V star τ-Sco, to estimate the abundance uncertainties related to the stellar parameters, microturbulence, and line-to-line abundance dispersion.

Table 11.1 summarizes the results from the stellar-abundance determination, together with the values found in the literature for the nebular gas-phase abundances. The most recent and complete analysis of the chemical composition of the Orion nebula is that presented by Esteban *et al.* (2004); they used a wide variety of collisionally excited and recombination lines from very deep UVES at the VLT 8.2-m spectra for the abundance determination. Table 11.1 shows the oxygen gas-phase abundance proposed by those authors, together with their estimated gas-plus-dust oxygen abundance, which takes into account a dust depletion factor of 0.08 dex obtained by Esteban *et al.* (1998). For the nebular silicon-abundance determination I refer the reader to the works by Rubin *et al.* (1993) and Garnett *et al.* (1995); those authors estimated the Si abundances in the Orion nebula by using IUE high-dispersion spectra.

Table 11.1. *Results from the stellar and nebular abundance analyses in the Orion nebula*

Reference	Subject	Abundance (dex)	Comments
Simón-Díaz et al. (2006)	Orion nebula, B0.5V stars	$\epsilon(O) = 8.63 \pm 0.10$	θ^1-Ori A, D, θ^2-Ori B
Esteban et al. (2004)	Gas phase	$\epsilon(O) = 8.65 \pm 0.03$	ORLs, CELs, $t^2 = 0.022$
Esteban et al. (2004)	Gas + dust	$\epsilon(O) = 8.73 \pm 0.03$	Dust depletion factor 0.08
Simón-Díaz (2005)	Orion nebula, B0.5V stars	$\epsilon(Si) = 7.55 \pm 0.20$	θ^1-Ori A, D, θ^2-Ori B
Rubin et al. (1993)	Gas phase	$\epsilon(Si) = 6.65$	
Garnett et al. (1995)	Gas phase	$\epsilon(Si) = 6.60$–7.15	

4 Results and discussion in the context of the metal-rich Universe

The mean value of the stellar oxygen abundances derived by Simón-Díaz et al. (2006) is in agreement with the nebular gas-phase oxygen abundance proposed by Esteban et al. (2004), however the dust-plus-gas nebular oxygen abundance is somewhat larger than the stellar value. These results may imply that the dust depletion factor for oxygen is lower than had previously been considered for the Orion nebula.

In the case of silicon abundances, the stellar results are systematically larger than the nebular values. This suggests that a certain amount of nebular silicon is depleted onto dust grains; however, the exact determination of the silicon-depletion factor is complicated due to uncertainties associated with the derived silicon abundances, both stellar and nebular.

I would like finally to note the importance of this kind of comparison for the study of the metal-rich Universe. In high-metallicity environments, nebular T_e diagnostic lines are faint and hence hardly measurable, so strong-line methods (which depend on calibrations) must be used. Moreover, even if very deep spectra are obtained, derived nebular abundances may be affected by biases (Stasińska 2005). These deficiencies in the nebular methodology can be solved by using massive stars, since the strength of metal lines in the stellar spectra increases with metallicity, and hence stellar abundance diagnostic lines are (even!) more clearly seen in those high-metallicity environments.

References

Afflerbach, A., Churchwell, E., & Werner, M. W. (1997), *ApJ* **478**, 190
Cunha, K., & Lambert, D. L. (1994), *ApJ* **426**, 170
Daflon, S., & Cunha, K. (2004), *ApJ* **617**, 1115
Esteban, C. (2002), *Revista Méxicana de Astronomía y Astrofísica (Serie de Conferencias)* **12**, 56
Esteban, C., Peimbert, M., Torres-Peimbert, S., & Escalante, V. (1998), *MNRAS* **295**, 401
Esteban, C., Peimbert, M., García-Rojas, J. *et al.* (2004), *MNRAS* **355**, 229
Esteban, C., García-Rojas, J., Peimbert, M. *et al.* (2005), *ApJ* **618**, 95
Garnett, D. R., Dufour, R. J., Peimbert, M. *et al.* (1995), *ApJ* **449L**, 77
Gummersbach, C. A., Kaufer, A., Schaefer, D. R. *et al.* (1998), *A&A* **338**, 881
Korn, A. J., Keller, S. C., Kaufer, A. *et al.* (2002), *A&A* **385**, 143
Maeder, A., & Meynet, G. (2000), *ARA&A* **38**, 143
Puls, J., Urbaneja, M. A., Venero, R. *et al.* (2005), *A&A* **435**, 669
Rolleston, W. R. J., Smartt, S. J., Dufton, P. L., & Ryans, R. S. I. (2000), *A&A* **363**, 537
Rubin, R. H., Dufour, R. J., & Walter, D. K. (1993), *ApJ* **413**, 242
Santolaya-Rey, A. E., Puls, J., & Herrero, A. (1997), *A&A* **323**, 488
Shaver, P. A., McGee, R. X., Newton, L. M. *et al.* (1983), *MNRAS* **204**, 53
Simón-Díaz, S. (2005), unpublished PhD thesis, Universidad de la Laguna
Simón-Díaz, S., Herrero, A., Esteban, C., & Najarro, F. (2006), *A&A* **448**, 351
Stasińska, G. (2005), *A&A* **434**, 507
Trundle, C., Dufton, P. L., Lennon, D. J. *et al.* (2002), *A&A* **395**, 519
Trundle, C., Lennon, D. J., Puls, J., & Dufton, P. L. (2004), *A&A* **417**, 217
Urbaneja, M. A., Herrero, A., Bresolin, F. *et al.* (2005a), *ApJ* **622**, 862
Urbaneja, M. A., Herrero, A., Kudritzki, R.-P. *et al.* (2005b), *ApJ* **635**, 311

12

Abundance surveys of metal-rich bulge stars

Jon P. Fulbright,[1] R. M. Rich[2] & A. McWilliam[3]

[1]*Department of Physics & Astronomy, The Johns Hopkins University,*
3400 N. Charles Street, Baltimore, MD 21218, USA
[2]*Division of Astronomy, Department of Physics & Astronomy, UCLA, Los Angeles,*
CA 90095-1562, USA
[3]*Observatories of the Carnegie Institution of Washington, 813 Santa Barbara Street,*
Pasadena, CA 91101, USA

We present the results from optical high-resolution spectroscopic surveys of the Milky Way bulge. The bulge is observed to have stars with [Fe/H] values up to at least +0.5 and [Mg/H] values up to at least +0.8. Age information from color–magnitude diagrams suggests these stars formed at nearly the same time as old metal-rich globular clusters, and the abundance ratios imply that the chemical evolution of the bulge was dominated by Type-II supernovae, including progenitors at least as metal-rich as those seen in the local disk today.

1 Introduction

The elemental-abundance properties of the bulge can help provide some answers to some of the must fundamental questions we have about the Galactic bulge. How did the bulge form? Did it involve merging processes? What fraction of the stars in the bulge formed in external galaxies? What is the relation of the bulge to the bar and the inner disk?

We discuss how these questions have been and will be addressed by abundance analyses in more detail in Fulbright *et al.* (2006a, 2006b), and Chapter 14 in this volume should provide additional insights. For this contribution, our focus will be on the metal-rich stars in the bulge and what they tell us about the formation of the bulge at early times.

In particular, the so-called alpha-elements (here we mean O, Mg, Si, Ca, and Ti) provide a great deal of information about star-formation timescales. The longstanding paradigm is that alpha-elements are produced in Type-II supernovae resulting from short-lived massive stars, whereas large amounts of iron are produced in Type-Ia supernovae that occur on timescales 1–2 orders of magnitude longer (Tinsley 1979, Weaver *et al.* 1989, Timmes *et al.* 1995, McWilliam 1997). The yields of numerical models of SNe, together with corroborating abundance determinations for very metal-poor stars, are the theoretical underpinning that supports this paradigm.

McWilliam & Rich (1994) (hereafter MR94) was the pioneering work on the detailed abundance properties of the bulge. The authors of MR94 found that stars in Baade's Window have enhancements in [Mg/Fe] and [Ti/Fe] to well above the Solar ratios at all metallicities. This result provides the fundamental empirical evidence for a rapid formation of the bulge (within <1 Gyr) in star-formation models of Matteucci & Brocato (1990), Matteucci *et al.* (1999), and Ballero *et al.* (2006). Additionally, we know that cluster and field stars in the bulge are equally old from color–magnitude studies such as Ortolani *et al.* (1995), Feltzing & Gilmore (2000), Kuijken & Rich (2002), and Zoccali *et al.* (2003). A synthesis of results from the studies of our bulge and extragalactic (proto-)bulges can be found in both Rich (2006) and Chapter 1 in this volume.

Despite the value of MR94, the work was done at relatively low resolution ($R \sim 17{,}000$) and signal-to-noise ratio (~ 50). Improvements in equipment (large telescopes, better echelle spectrographs, and high-resolution spectrographs for the near-IR) and abundance-analysis techniques, and the availability of supplemental data such as the OGLE reddening maps of Stanek (1996) and the 2MASS point-source catalog of Cutri *et al.* (2003) have made it possible for observers to re-address the question of abundances in the bulge with greater precision, more elements, more stars, etc. For the rest of this contribution, we will review just a fraction of these new results.

2 Data sources

The data for field stars in the bulge will come from our optical studies (Fulbright *et al.* 2006a, 2006b) and the near-IR studies of Rich & Origlia (2005) and Cunha & Smith (2006). Zoccali *et al.* (2006) and Lecureur *et al.* (2006) (and Chapter 14 in this volume) are also studying a large sample of stars in the bulge with the VLT.

We also omit from this discussion the results of Pompeia *et al.* (2003), who analyzed a sample of nearby old metal-rich stars with peculiar orbits that may associate them with the bulge. We will limit ourselves to *in situ* studies, but, briefly, their alpha-element abundances found for this population are unlike those seen both for bulge field and for cluster stars.

We also discuss the results from several studies of stars within globular clusters that lie within a few degrees of the Galactic Center. The clusters discussed here and the data sources are given in Table 12.1. We do not include the study of

Table 12.1. *Bulge globular-cluster abundance data used in this paper (the value $[\alpha_{Ex}/Fe]$ is the mean of the $[Si/Fe]$, $[Ca/Fe]$, and $[Ti/Fe]$ abundances)*

Cluster	(l, b)	[Fe/H]	[O/Fe]	[Mg/Fe]	[α_{Ex}/Fe]	Reference
HP-1	−3, +2	−0.99	+0.40	+0.10	+0.12	Barbuy et al. (2006)
NGC 6342	+5, −7	−0.60	+0.30	+0.38	+0.33	Origlia et al. (2005a)
NGC 6528	+1, −4	+0.07	+0.07	+0.14	+0.21	Carretta et al. (2001)
NGC 6528	+1, −4	−0.11	+0.15	+0.07	−0.09	Zoccali et al. (2004)
NGC 6528	+1, −4	−0.17	+0.32	+0.35	+0.32	Origlia et al. (2005a)
NGC 6553	+5, −3	−0.55	...	+0.33	+0.39	Barbuy et al. (1999)
NGC 6553	+5, −3	−0.16	+0.50	+0.41	+0.20	Cohen et al. (1999)
NGC 6553	+5, −3	−0.3	...	...	+0.30	Origlia et al. (2002)
NGC 6553	+5, −3	−0.21	+0.20	...	...	Melendez et al. (2003)
Terzan 4	−4, +1	−1.60	+0.54	+0.41	+0.51	Origlia et al. (2004)
Terzan 5	+4, +2	−0.24	+0.28	+0.30	+0.32	Origlia et al. (2004)
UKS-1	+5, +1	−0.78	+0.27	+0.30	+0.31	Origlia et al. (2005b)

Lee & Carney (2002) of three very metal-poor ([Fe/H] < −2) clusters that lie within 2 kpc of the Galactic Center. At this time, there are no known bulge field stars with detailed abundance results in that metallicity range. The Lee & Carney (2002) clusters may be members of the inner halo.

The globular-cluster results come from three main research groups. The symbols we use in the figures link the results from various clusters to one of the three groups. One group, led by J. Cohen and E. Carretta, studied NGC 6528 and NGC 6553 with Keck/HIRES in the optical. We mark their results with triangles. Another group, using optical data from ESO telescopes, includes B. Barbuy, M. Zoccali, and J. Meléndez. Their cluster results are marked by squares or four-pointed stars. R. Origlia and R. Rich and collaborators studied several clusters using near-IR spectra taken with the Keck telescope. These points are marked by pentagons and five-pointed stars.

One question when comparing results from abundance studies is whether or not the results have systematic offsets. Fortunately, the authors of all the used field-stars studies Arcturus as their differential abundance standard. Arcturus is a full dex more metal-poor than the most metal-rich bulge stars, so there may be resulting problems with the analysis.

Fortunately, there are nearby metal-rich disk stars to use as testbeds for the analysis. Fulbright *et al.* (2006a, 2006b) have included stars like µ-Leo, Hyades giants, and other bright, nearby giants in the analysis. The results for these local disk giants can be compared with results from local disk dwarfs to look for systematic offsets. In practice, we have found that it is very important to use echelle data with the same resolution for these local giants as for the bulge giants, but we believe our abundance zero-point problems to be small ($\leq$0.1 dex). Analyzing of old open clusters such as NGC 188 and NGC 6791 may be a way to compare the results from optical studies using K-giants with those from the near-IR studies using M-giants.

3 The hydrostatic alpha-elements: oxygen and magnesium

We first look at the two alpha-elements primarily made in hydrostatic equilibrium in massive stars: oxygen and magnesium. Oxygen is primarily produced during the He-burning phases, whereas magnesium is made during the C-burning phases. The yields of both are proportional to the mass of the respective burning shell (Woosley & Weaver 1995, Timmes *et al.* 1995). This means that, if there is no mass loss, more massive stars will have larger shells, and therefore produce more O and Mg than do lower-mass stars. Models of Type-Ia SNe do not produce significant amounts of either O or Mg.

We plot the observed bulge [O/Fe] ratios in Figure 12.1. The top panel shows only the bulge field stars and in the lower panel we add in the bulge globular clusters.

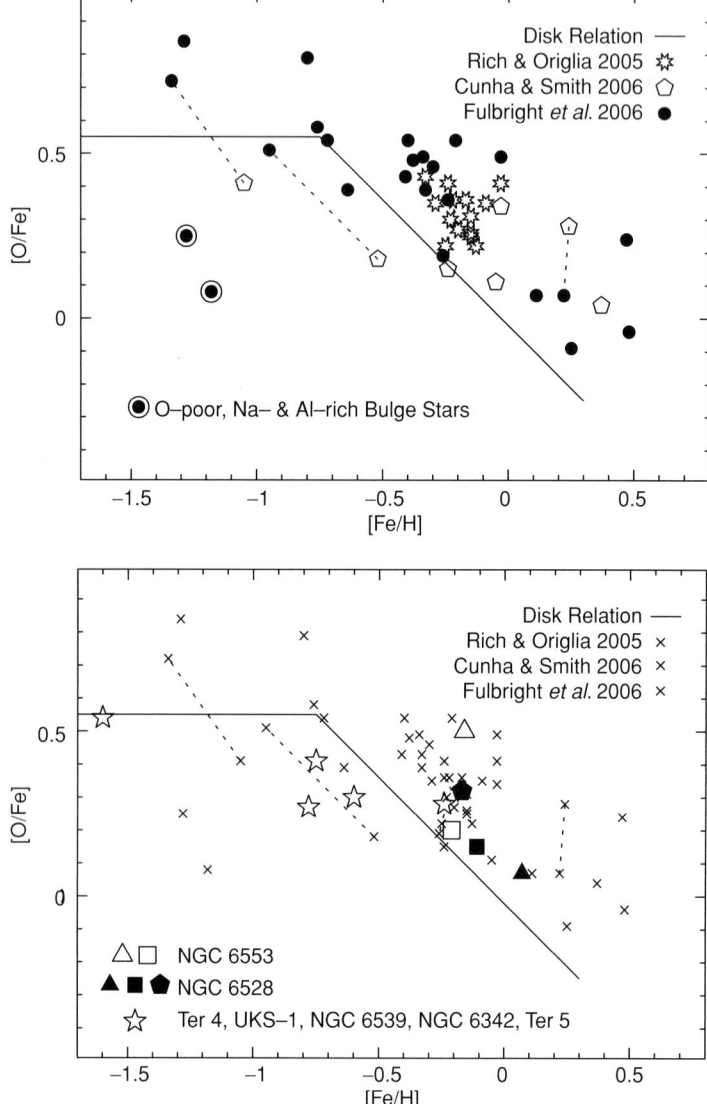

Figure 12.1. Top panel: [O/Fe] versus [Fe/H] for bulge field stars from three surveys in Baade's Window. The solid line gives a rough indication of where disk and halo field stars lie. Field stars in the bulge lie along, or slightly above, the disk–halo relation at the metal-poor end, but have higher [O/Fe] values for the most metal-rich stars. Dashed lines connect multiple observations of the same star, as discussed in the text. Two metal-poor Baade's Window stars have lower [O/Fe] and higher [Na/Fe] and [Al/Fe] ratios than other metal-poor bulge stars. They lie along the Na–O anti-correlation, as if they were globular-cluster stars like those in M4. Bottom panel: the same plot as the top panel, but here all the field stars are marked by crosses and the large symbols reflect the mean values from studies of bulge globular clusters. The triangles mark data from Cohen et al. and Carretta et al. (see Table 12.1), the square points and HP-1 data from Barbuy et al. and Zoccali et al., and the pentagons and stars data from Origlia, Rich, and collaborators (the order of clusters given on the star-symbol label is that of increasing metallicity). The bulge cluster stars lie in the same region as the field stars.

Oxygen has been observed in all three of our bulge field-star surveys. Fulbright et al. (2006b) used the [O I] 6,300 Å line, whereas the near-IR groups used the OH molecular absorption lines available in that wavelength region.

There are a few cases in which the same star has been analyzed by two different groups. The multiple observations of these stars (three in common between Fulbright et al. (2006b) and Cunha & Smith (2006), two in common between Fulbright et al. (2006b) and Rich & Origlia (2005)) are connected by dotted lines. In the former case the disagreement can be large – but in two of the three cases the oxygen abundances are in good agreement, and it is the Fe abundances that differ by sizable amounts. In the latter case, the agreement is very good and the points are very close to each other in Figure 12.1.

Also plotted is a solid line representing the rough trend of values seen in disk and halo field stars. The bulge [O/Fe] ratios lie at or above the disk/halo relations at the highest metallicities both for the field and for the cluster stars. At the highest metallicities, the [O/Fe] ratio is about +0.3 dex higher. The basic interpretation is the ratio of Type-II to Type-Ia supernovae contributions in the bulge has to be higher than that in the local disk.

In the panels of Figure 12.2, we see a similar result – the [Mg/Fe] values of the metal-rich bulge stars and clusters lie well above that of the metal-rich disk. However, the [Mg/Fe] ratios undergo only a mild decline at the highest metallicities: [Mg/Fe] $\approx$ +0.3 at [Fe/H] = +0.5. The [Mg/O] ratio rises from roughly Solar at [Fe/H] = −0.5 to +0.3 at [Fe/H] = +0.5. This effect cannot be reproduced by the inclusion of Type-Ia ejecta because these supernovae do not produce significant amounts of either element. If O and Mg are produced in the shells of massive stars, then how can the yields change with increasing metallicity?

One possible answer is that mass loss by the Type-II progenitor could possibly decrease the oxygen yield by lowering the mass of the He-burning shell without significantly affecting the mass of the C-burning shell. The details of this scenario are given in Fulbright (2006b), but there is only mixed observational and theoretical support for this type of Wolf–Rayet mass loss altering oxygen yields at high metallicity.

The abundances of the light odd-Z elements sodium and aluminum lend more support for the scenario that the high [Mg/Fe] values at high metallicity are the true measure of the Type-II contribution of the bulge. Both [Na/Fe] and [Al/Fe] ratios for bulge stars stay far above Solar at all metallicities, like [Mg/Fe] ratios. Neither Na nor Al is predicted to be made in Type-Ia supernovae, so, again, the only option is that the chemical evolution of the bulge is dominated by Type-II ejecta.

Exceptions to the high [Mg/Fe] ratios in the bulge are the cluster HP-1 and possibly NGC 6528. The [Mg/Fe] ratio for the metal-poor cluster HP-1 lies well below what is expected for metal-poor systems. One solution is that this cluster

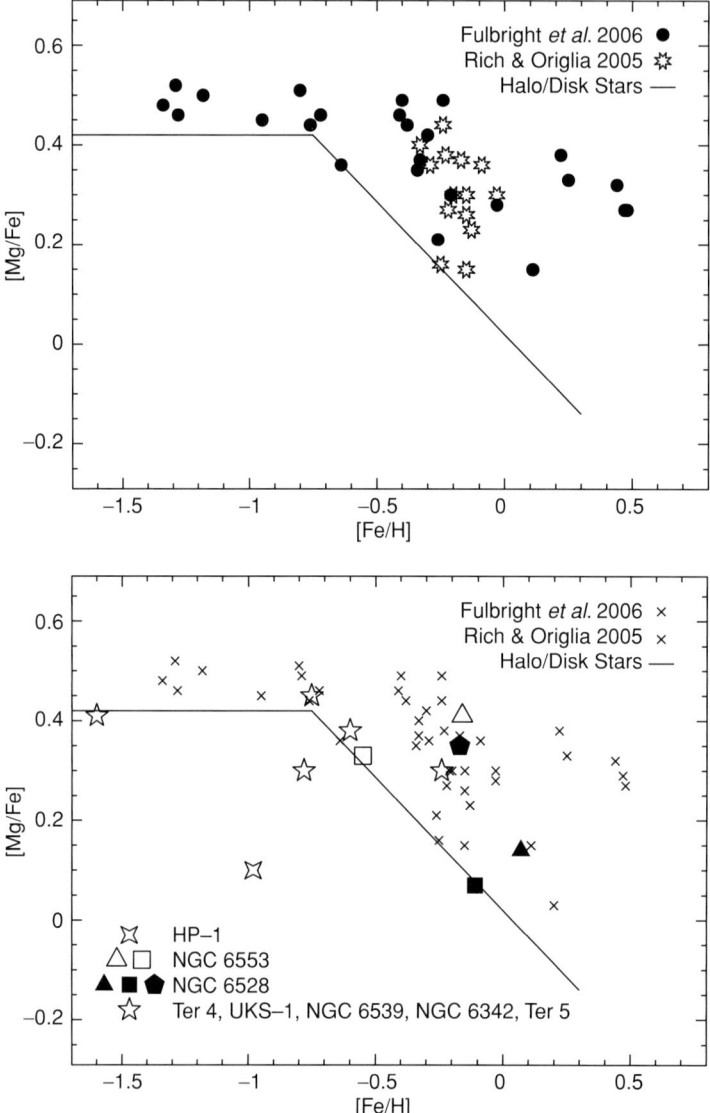

Figure 12.2. Top panel: [Mg/Fe] versus [Fe/H] for bulge field stars from two surveys in Baade's Window. The solid line gives a rough indication of where disk and halo field stars lie. Field stars in the bulge lie above the disk–halo relation at all metallicities, with only slightly lower [Mg/Fe] values for the most metal-rich stars. This is among the strongest evidence that the chemical evolution of the bulge was dominated by Type-II supernovae. Bottom panel: the same plot as the top panel, but here all the field stars are marked by crosses and the large symbols (the same ones as in Figure 12.1) reflect the mean values from studies of bulge globular clusters. As with [O/Fe], the bulge globular clusters lie in similar regions, although two of the three observations of NGC 6528 lie on the lower envelope of the field-star distribution. Data for the cluster HP-1 lie well below both the bulge and the disk–halo relations.

is truly alpha-poor, like the relatively young clusters Ruprecht 106 and Pal 12 (Brown *et al.* 1997). The younger age allows for more Type-Ia enrichment to dilute the [α/Fe] ratios. The other solution is mentioned in Barbuy *et al.* (2006). They found that the spectroscopic [Fe/H] solution was about 0.5 dex higher than what one would assume from the shape of the red giant branch and the cluster's blue horizontal branch. Lowering the [Fe/H] value by itself would raise the various [element/Fe] ratios.

Finally, two of the Baade's Window giants examined by Fulbright *et al.* (2006b) have low [O/Fe] ratios yet high [Na/Fe] and [Al/Fe] values. This is reminiscent of what is seen in many globular clusters. In fact, these two stars have abundances that place them on the same region of the [Na/Fe] versus [O/Fe] and [Al/Fe] versus [O/Fe] diagrams as the stars of the similar-metallicity globular cluster M4 (Ivans *et al.* 1999). These two stars are the first non-cluster stars in the Milky Way found to exhibit this effect. The origin of the so-called Na–O and Al–O anti-correlations in globular cluster stars is unknown, but two possible theories for the existence of bulge field stars exhibiting this pattern are that these stars were stripped from metal-poor globular clusters like M4 or that the conditions that caused the abundance trends in globular-cluster stars were present in the bulge when these stars formed.

4 The explosive alpha-elements: silicon, calcium, and titanium

The elements silicon, calcium, and titanium are all believed to be created in the explosive nucleosynthesis phase of Type-II supernovae (Woosley & Weaver 1995). Owing to our belief in their common origin, we have averaged the three [X/Fe] ratios of these elements to reduce the effects of random observational scatter. We define $[\alpha_{Ex}/Fe] = [(Si + Ca + Ti)/Fe] = ([Si/Fe] + [Ca/Fe] + [Ti/Fe])/3$.

We plot the results for this combination in Figure 12.3. As we have seen in Figures 12.1 and 12.2, the bulge field stars and cluster stars have $[\alpha_{Ex}/Fe]$ values as high as or higher than those of the other populations of the galaxy (again with the exception of HP-1 and the Zoccali *et al.* result for NGC 6528). As before, we conclude that this indicates a higher ratio of Type-II to Type-Ia supernovae contributions than elsewhere in the Galaxy at a given metallicity. Note that MR94 found high [Ti/Fe] values for their metal-rich stars. Owing to the superior spectra of the Fulbright *et al.* (2006b) analysis, we favor the more recent result.

As in the case of the decrease in bulge [O/Fe] ratios at high metallicity, the drop in $[\alpha_{Ex}/Fe]$ values cannot arise purely from Type-Ia contributions. Type-Ia supernovae make large amounts of Fe-group elements, but they do contribute some lighter elements as well. The yields decrease with decreasing atomic number – a given Type Ia makes more Ti than Ca, but more Ca than Si, and so on (Thielemann *et al.* 1996). These heavy alpha-element contributions from Type-Ia SNe can

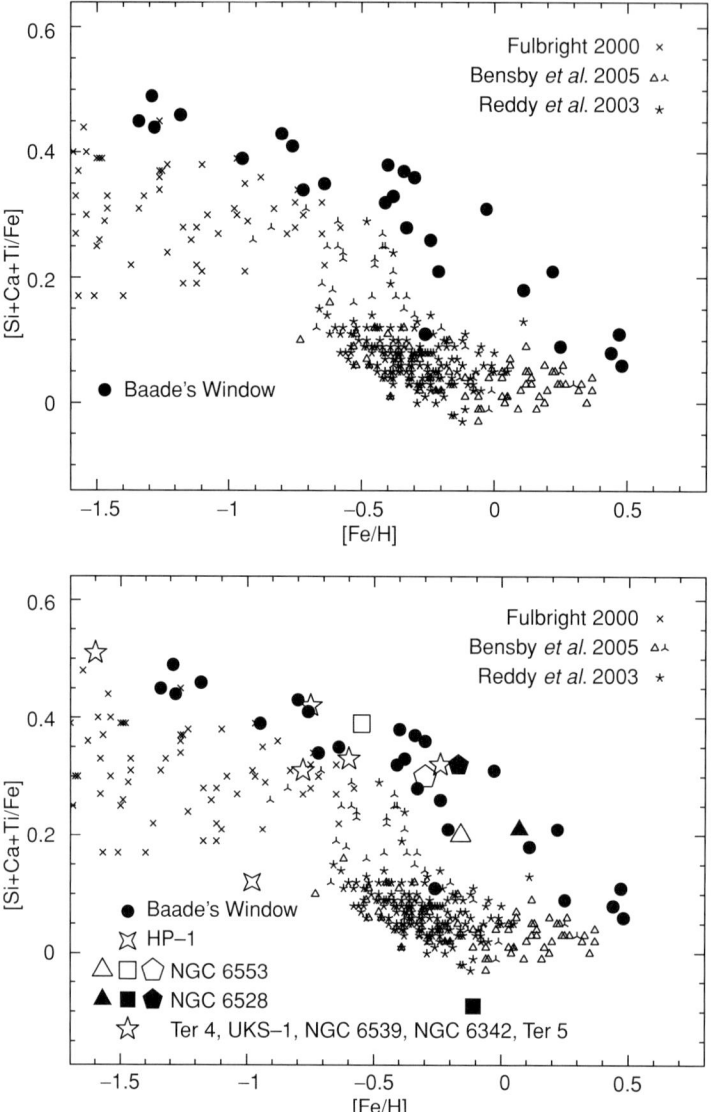

Figure 12.3. Top panel: [(Si + Ca + Ti)/Fe] (i.e. [α_{Ex}/Fe]) versus [Fe/H] ratios for bulge, disk, and halo stars from several surveys. The bulge field-star data are from Fulbright *et al.* 2006, while the thin-disk data are from Reddy *et al.* and Bensby *et al.* (small triangles), and the three-point crosses indicate the region's thick-disk stars. The Baade's Window data lie at or above those for all of the other populations in the Galaxy at all metallicities. Bottom panel: the same as above, but with the bulge globular-cluster data added (using the same symbols as in Figures 12.1 and 12.2). As with the oxygen and magnesium data, data for the bulge clusters lie along the range of the bulge field-star data. The cluster HP-1 and the Zoccali *et al.* data point for NGC 6528 lie well below all other points. The HP-1 point here agrees with what was seen with its [Mg/Fe] ratio. The NGC 6528 point is well below that of the other two studies of that cluster (the Zoccali *et al.* [Mg/Fe] ratio for NGC 6528 in Figure 12.2 also lies below that of the other two studies, whereas the [O/Fe] ratios are in good agreement).

be seen as a lower decline in [X/Fe] ratios from the "plateau" values in the metal-poor halo to Solar values in the disk. For example, the "plateau" value for [Mg/Fe] in the halo is about +0.5 dex, compared with +0.3 dex for [Ti/Fe] (Fulbright 2000). The same amount of Fe is added to the system in both cases, but more Ti was added at the same time (presumably by Type-Ia SNe) to lessen the extent of the "dilution."

This means that if Type-Ia SNe were the reason for the drop in the [α_{Ex}/Fe] ratios at high metallicities then the [Mg/Fe] ratio should have dropped more. Yet the opposite is seen in the bulge. We cannot exclude all Type-Ia contributions, of course – if the previous estimates of a 500-Myr timescale for star formation in the bulge still hold then we expect a reasonable amount of Type-Ia ejecta to be included in the stars.

In the top panel of Figure 12.3 we have included the data for individual field halo and disk stars from various surveys to show the internal scatter within the various populations. The rms scatter between bulge points (about 0.04 dex) from Fulbright *et al.* (2006b) is similar to that of the thin disk (Reddy *et al.* and Bensby *et al.* points) and smaller than that of the halo (the metal-poor Fulbright (2000) points).

The lesser scatter seen for the bulge stars indicates that the bulge composition evolved much more homogeneously than that of the halo. The high [α_{Ex}/Fe] values in the bulge suggest that either Type-II SNe in the bulge had more massive progenitor masses than did that in the halo (equal to the highest mass function in the halo), or the halo experienced more nucleosynthesis contributions from Type-Ia SNe than did the bulge. The range of [α/Fe] ratios seen in the halo certainly indicates that it experienced a very inhomogeneous enrichment history. More extreme evidence of alpha/Fe dispersion in the halo is already well established (Nissen & Schuster 1997, Brown *et al.* 1997, Fulbright 2002).

Wyse & Gilmore (1992) proposed that the bulge formed from Galactic-spheroid (halo) gas, because of the similarity of the specific angular momenta of these two systems, and because the low mean metallicity indicates that 90% of the spheroid gas was lost. These abundances provide an interesting test of this bulge-formation idea: since the [α_{Ex}/Fe] ratios in the most metal-poor bulge stars are higher than those of stars in the halo, and because the most metal-poor bulge stars, at [Fe/H] ~ -1.3 dex, are of metallicity similar to the mean metallicity for the halo, the metal-poor bulge stars could not have been made from halo gas with the average metallicity and composition seen today. However, the metal-poor bulge stars could have been produced from halo gas, provided that the halo composition at that time was similar to the top $\sim$20% of the halo [α_{Ex}/Fe] ratios seen today, i.e. provided that the mean halo composition changed with time. Thus, this result suggests that, if the bulge formed out of halo gas, then the onset of bulge formation occurred before $\sim$80% of the chemical enrichment of today's stellar halo had occurred.

5 Abundance gradients in the bulge?

At this point in time, the published bulge field-star data are focused on stars in Baade's Window at $(l, b) = (+1, -4)$. However, the good agreement between the field and globular-cluster star abundances suggests that there might not be large abundance gradients across the bulge. However, the line of sight through Baade's Window passes about 500 pc from the Galactic Center, and the scale length of the bulge is about 300 pc (Wyse *et al.* 1997). It may be possible that gradients exist within that range. Both our group and the group of Zoccalli, Lecureur *et al.* have high-resolution data for multiple fields both closer to and further from the Galactic Center than Baade's Window, so the question of abundance gradients in the bulge may be answered shortly.

6 Concluding remarks

Some of the main results from recent high-resolution abundance studies of metal-rich stars in the bulge are the following:

- The abundance properties of both bulge field and globular-cluster stars generally agree with each other, with the exception of the unusual cluster HP-1.
- The high-resolution near-IR techniques being used on bulge giants yield results roughly consistent with the results from more traditional optical studies. There is still some disagreement in the results for individual stars, but the overall conclusions for the chemical evolution of the bulge as a whole are identical.
- For all of the alpha elements, the [X/Fe] ratios of bulge stars are as high as or higher than those in any other Galaxy population studied to date. We conclude that this indicates that the ratio of contributions from Type-II SNe to those from Type-Ia SNe reaches its highest value in the bulge.
- Color–magnitude information both for the bulge field stars and for globular-cluster stars indicates that the bulge is very old. This combined with the high [α/Fe] ratios in the bulge forces us to conclude that the bulge formed very quickly, possibly before the bulge of halo stars had formed.
- The lower [O/Fe] and [α_{Ex}/Fe] in the bulge cannot be due purely to Type-Ia contributions.
- The narrow scatter of the [α_{Ex}/Fe] values in the bulge indicates that either the bulge was well mixed when it formed or the various star-forming regions that eventually mixed into the bulge had identical chemical evolution histories – more so than the stellar halo, for example.
- Metal-poor giants in the bulge field have the same light-element (O, Na, and Al) star-to-star abundance variations as those seen for globular-cluster stars. The origin of this phenomenon in globular-cluster stars may hold important clues on star formation in the early bulge.

References

Ballero, S. K., Matteucci, F., Origlia, L., & Rich, R. M. (2006), *A&A*, submitted.
Barbuy, B. *et al.* (1999), *A&A* **341**, 539
Barbuy, B. *et al.* (2006), *A&A* **449**, 349
Bensby T., Feltzing S., & Lundström, I. (2003), *A&A* **410**, 527
Bensby T., Feltzing S., Lundström I., & Ilyin I. (2005), *A&A* **433**, 185
Brown, J. A., Wallerstein, G., & Zucker, D. (1997), *AJ* **114**, 180
Carretta, E., Cohen, J. G., Gratton, R. G., & Behr, B. B. (2001), *AJ* **112**, 1496
Cohen, J. G., Gratton, R. G., Behr, B. B., & Carretta, E. (1999), *ApJ* **523**, 739
Cunha, K., & Smith, V. V. (2006), *ApJ*, at press
Cutri, R. M., *et al.* (2003), 2MASS All-Sky Catalog of Point Sources, *VizieR* **II/246**
Feltzing, S., & Gilmore, G. (2000), *A&A* **355**, 949
Fulbright, J. P. (2000), *AJ* **120**, 1841
  (2002), *AJ* **123**, 404
Fulbright, J. P., McWilliam, A. & Rich, R. M. (2006), *ApJ* **636**, 821
Fulbright, J. P., McWilliam, A. & Rich, R. M. (2006), astro-ph/0609087
Kuijken, K., & Rich, R. M. (2002), *ApJ* **124**, 2054
Lecureur *et al.* (2006), astro-ph/0610346
Lee, J.-W., & Carney, B. W. (2002), *AJ* **124**, 151
Matteucci, F., & Brocato, E. (1990), *ApJ* **365**, 539
Matteucci, F., Romano, D., & Molaro, P. (1999), *A&A* **341**, 458
McWilliam, A. (1997), *ARA&A* **35**, 503
McWilliam, A., & Rich, R. M. (1994), *ApJS* **91**, 749 (MR94)
Melendez *et al.* (2003), *A&A* **411**, 417
Nissen, P. E., & Schuster, W. J. (1997), *A&A* **326**, 751
Origlia, L., & Rich, R. M. (2004), *AJ* **127**, 3422
Origlia, L., Rich, R. M., & Castro, S. (2002), *AJ* **123**, 1559
Origlia, L., Valenti, E., & Rich, R. M. (2005a), *MNRAS* **356**, 1276
Origlia, L., Valenti, E., Rich, R. M., & Ferraro, F. R. (2005b), *MNRAS* **363**, 897
Pompeia, L., Barbuy, B., & Grenon, M. (2003), *ApJ* **592**, 1173
Ortolani, S. *et al.* (1995), *Nature*, **377**, 701
Reddy, B. E., Tomkin, J., Lambert, D. L., & Allende Prieto, C. (2003), *MNRAS* **340**, 304
Rich, R. M., & Origlia, L. (2005), *ApJ* **634**, (1293)
Rich, R. M., Fulbright, J., McWilliam, A., & Origlia, L. (2006), in *Stellar Populations Cozumel meeting* ASP conf series, at press
Stanek, K. Z. (1996), *ApJ* **460**, L37
Thielemann, F.-K., Nomoto, K., & Hashimoto, M. (1996), *ApJ* **460**, 408
Timmes, F. X., Woosley, S. E., & Weaver, T. A. (1995), *ApJS* **98**, 617
Tinsley, B. M (1979), *ApJ* **229**, 1046
Wheeler, J. C., Sneden, C., & Truran, J. W. Jr. (1989), *ARA&A* **27**, 279
Woosley, S. E., & Weaver, T. A. (1995), *ApJS* **101**, 181
Wyse, R. F. G., & Gilmore, G. (1992), *AJ* **104**, 144
Wyse, R. F. G., Gilmore, G., & Franx, M. (1997), *ARA&A* **35**, 637
Zoccali, M., *et al.* (2003), *A&A* **399**, 931
Zoccali, M., *et al.* (2004), *A&A* **423**, 507
Zoccali, M., *et al.* (2006), astro-ph/0609052

13

Metal abundances in the Galactic Center

Francisco Najarro

Instituto de Estructura de la Materia, CSIC, Serrano 121, 29006 Madrid, Spain

Thanks to the impressive evolution of IR detectors and the new generation of line-blanketed models for the extended atmospheres of hot stars we are able to derive accurately the physical properties and metallicity estimates of massive stars. Here, we review quantitative spectroscopic studies of massive stars in the three Galactic Center clusters: the Quintuplet, Arches, and Central clusters. Our analysis of the LBVs for the Quintuplet cluster provides a direct estimate of chemical abundances of α-elements and Fe in these objects. For the Arches cluster, we introduce a method based on the N abundance of WNL stars and the theory of evolution of massive stars. For the Central cluster, new observations reveal IRS8 to be an outsider with respect to the rest of the massive stars in the cluster in terms of both age and location. Using the derived properties of IRS8, we present a new method by which to derive metallicity from the O III feature at 2.115 µm. Our results indicate that the three clusters have Solar metallicity.

1 Introduction

The detection of a He I emission-line cluster (Krabbe *et al.* 1991) in the central parsec raised the question of the physical nature of its members and their role in the energetics of this region and triggered a substantial improvement in the atmospheric models for hot stars in the near-infrared. Thus, reliable values for luminosities, temperatures, mass-loss rates, helium abundances, and ionizing photons were obtained for the members of the cluster with the brightest He I lines (Najarro *et al.* 1994, 1997), which solved the energy puzzle of the Central cluster and placed important constraints on the theory of the evolution of massive stars.

The Metal-rich Universe, eds. G. Israelian and G. Meynet. Published by Cambridge University Press.
© Cambridge University Press 2008.

The Galactic Center (GC) is clearly a unique environment in the Galaxy insofar as massive-star formation is concerned, since it contributes 10% of the present Galactic star-formation activity. Hence, studying this region is an ideal way to address crucial questions such as those concerning the universality of the initial mass function (IMF) and the maximum mass a star can possess (Figer 2004). In fact, the GC hosts three dense and massive star clusters that have formed in the inner 30 pc within the past 5 Myr. The Central cluster and the Quintuplet and Arches clusters host more stars with initial masses above $100 M_\odot$ than anywhere else in the Galaxy. Each cluster has a mass of around $\sim 10^4 M_\odot$. The extreme youth (2–2.5 Myr) (Figer *et al.* 2002; Najarro *et al.* 2004) of the Arches cluster allows to address the above questions using photometry alone (Figer 2004). The Central and Quintuplet clusters, being twice as old, may have lost their most massive members to the supernova stage.

On the other hand, such a fundamental aspect as metallicity remains to be addressed. Indeed, the issue of metallicity in the Galactic Center is still a source of controversy. From gas-phase measurements Shields & Ferland (1994) obtained twice Solar metallicity from argon and nitrogen emission lines while a Solar abundance was derived for neon. For the cool stars, using LTE-differential analysis with other cool supergiants, Carr *et al.* (2000) and Ramírez *et al.* (2000) obtained strong indications for a Solar Fe abundance. Further, Maeda *et al.* (2000) obtained four times Solar abundance by fitting the X-ray local emission around Sgr. A East, and Koyama *et al.* (2006) recently derived 3.5 times Solar abundance from the diffuse GC X-ray emission. It is therefore crucial to obtain metallicity estimates from direct analysis of hot stars and confront them with those from the cool-star and gas-phase analyses. Spectroscopic studies of photospheres and winds of massive hot stars are ideal tracers of metal abundances because they provide the most recent information about the natal clouds and environments where these objects formed.

In this paper, we review progress both in infrared observations and in quantitative infrared spectroscopy of massive stars in the Quintuplet, Arches, and Central clusters, which allows us to obtain direct abundance estimates of N, C, O, Si, Mg, and Fe in the GC.

2 Improved observations and models

In the field of hot stars, high-quality IR-spectra have been obtained during the last decade (Morris *et al.* 1996; Hanson *et al.* 1996; Figer *et al.* 1997; Blum *et al.* 1997). Most of these spectra were obtained with low–mid ($R \sim 500$–2,000) resolution, which is enough to classify the stars but insufficient in most cases to perform accurate quantitative spectroscopic studies. An example of

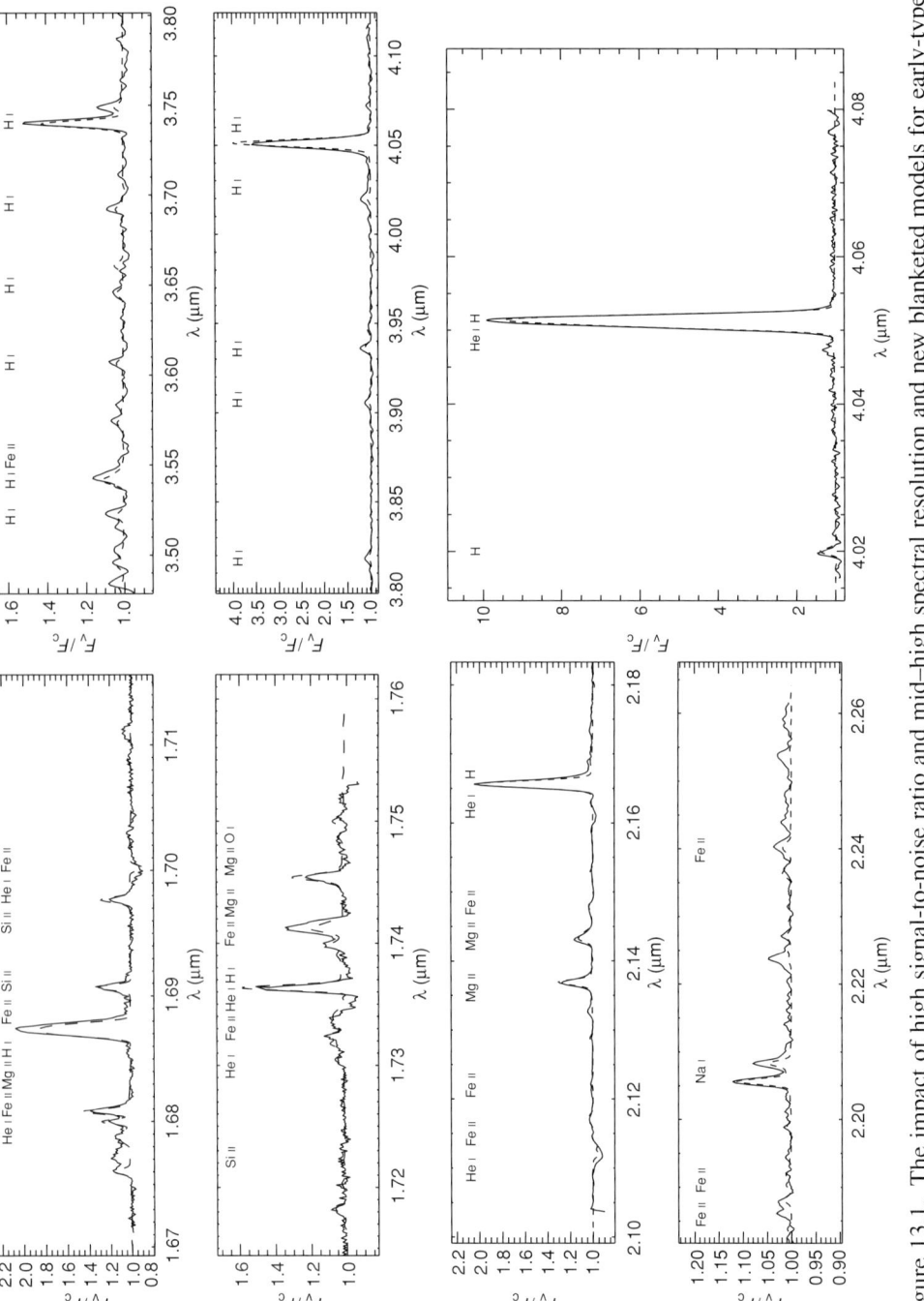

Figure 13.1. The impact of high signal-to-noise ratio and mid-high spectral resolution and new blanketed models for early-type stars in the infrared, showing UKIRT-CGS4 ($R \sim 5{,}000$) H-, K-, and L-band and Brα observations of the "Pistol Star" in the Quintuplet cluster and new model fits.

mid–high-resolution observations of early-type stars with strong winds is shown in Figure 13.1, which displays the spectra for several IR bands of the "Pistol Star," an LBV in the Quintuplet cluster, which were obtained with UKIRT-CGS4 with a resolution of $R \sim 5,000$. The number of new observational constraints provided by the new spectroscopic data is striking in comparison with the limited quality of previous low–mid-resolution observations, from which we could gather information only from some H and He lines. On inspecting Figure 13.1 we immediately note that the key diagnostic lines would be fully blurred at $R \sim 1,000$. The availability of high-quality IR spectroscopic data has been improved substantially with a new generation of IR-spectrographs on 8-m class telescopes (ISAAC, NIRSPEC, SINFONI, etc.).

The updated model (Hillier & Miller 1998a) is a line-blanketing method based on the standard iterative, non-LTE method to solve the radiative-transfer equation for the expanding atmospheres of early-type stars. New species, O, Mg, Ca, Si, Na, Al, Fe, etc., are included and the blanketing ensures that the effect of continua on lines and of lines on the continua as well as overlapping lines are handled automatically; see Hillier & Miller (1998a, 1999) for a detailed discussion of the method. The new model is then prescribed by the stellar radius, R_*, the stellar luminosity, L_*, the mass-loss rate $\dot{M}$, the velocity field, $v(r)$, the volume-filling factor f, and the abundances of the element considered.

In this review, we assume the Solar abundances derived by Anders & Grevesse (1989) because they are the ones entering evolutionary-model calculations.

3 The Quintuplet cluster

The Quintuplet cluster (Glass et al. 1987; Figer et al. 1999a, 1999b) contains a variety of massive stars, including WN, WC, WN9/Ofpe, LBV, and less-evolved blue supergiants. The presence of such stars constrains the cluster age to be about 4 Myr, assuming coeval formation. The cluster provides enough ionizing flux ($\sim 10^{51}$ photons s^{-1}) to ionize the nearby "Sickle" H II region and enough luminosity ($\sim 10^{7.5} L_\odot$) to heat the nearby molecular cloud, M0.20-0.033. Its total mass is estimated to be $\sim 10^4 M_\odot$. The presence of two LBVs, the "Pistol Star" and #362 (Geballe et al. 2000), with IR spectra rich in metal lines (see Figure 13.1), allows us to obtain a direct estimate of the metallicity of the objects and hence constrain the metal-enrichment history of the region. Furthermore, using the Fe II, Si II, and Mg II lines (Najarro et al. in preparation) we can measure the α-elements versus Fe ratio and infer whether the initial mass function (IMF) is dominated by massive stars or is like for other clusters in the Galaxy with a steeper slope. If the IMF is dominated by massive stars, we should expect enhanced yields of α-elements relative to Fe through a higher than average ratio of SN II versus SN Ia events.

3.1 The "Pistol Star" and star #362

To model the "Pistol Star" and star #362 we have assumed the atmosphere to be composed of H, He, C, N, O, Si, and Fe. The new blanketed models provide a significant improvement in our knowledge of the physical properties of the "Pistol Star" compared with the results we obtained in Figer *et al.* (1998) using non-blanketed models. The new model solves the dichotomy between the "high"- and "low"-luminosity (T_{eff}) solutions in Figer *et al.* (1998) through the analysis of the metal lines (see the excellent fits in Figure 13.1). The Si II, Mg II, and Fe II lines "choose" the low-luminosity model. We find a luminosity of around $1.75 \times 10^6 L_\odot$ and an effective temperature of $T_{\text{eff}} \sim 11{,}000$ K. This result, which reduces the previous estimate of the star luminosity by a factor of two, shows the importance of the new generation of models. Given the T_{eff} and high wind density of the object, we do find a degeneracy in the H/He ratio; see also Hillier *et al.* (1998b). In principle, fits of virtually equal quality may be obtained with H/He number ratios varying from 10 to 0.05. The only line that may help to break this degeneracy is the He I 2.112-µm absorption line, which seems to favor H/He number ratios between 3 and 0.05. An important consequence of this degeneracy is the mass fractions derived for Fe, Si, and Mg. Our models show that, once the H/He ratio falls below unity, the resulting metal abundances have to be scaled down. In other words, if H/He ≤ 1 then we will obtain only an upper limit on the metal abundances. We obtain Solar iron abundance as an upper limit for the "Pistol Star" (see the line fits in Figure 13.1). This estimate is rather robust since there is a large number of Fe II diagnostic lines. Further, if we assume that the object displays He enrichment consistent with an LBV evolutionary phase, H/He ≥ 1, we may conclude that the "Pistol Star" has *Solar* Fe abundance, in agreement with previous estimates from differential analysis of cool stars in the GC (Carr *et al.* 2000; Ramírez *et al.* 2000). The silicon abundance is obtained from the three Si II lines in the H band. The lines at 1.69 µm are extremely sensitive to the effective temperature of the star as well as to the transition zone between the star's photosphere and wind, while the line at 1.718 µm constitutes a more robust abundance diagnostic. In any case, our Si $\approx 1.4 \text{Si}_\odot$ result should be regarded with some caution. Magnesium, on the other hand, provides more diagnostic lines through Mg II both in the H band and in the K band. We regard Mg $\approx 1.6 \text{Mg}_\odot$ as our current best estimate. This slight enrichment in α-elements versus Fe seems to favor the situation of an IMF dominated by massive stars in the GC.

For star #362 (Geballe *et al.* 2000), nearly a twin of the "Pistol Star," we also obtain a luminosity of around $1.7 \times 10^6 L_\odot$ and an effective temperature of $T_{\text{eff}} \sim 105{,}00$ K (see the line fits in Figure 13.2). We derive a *Solar* iron abundance as upper limit as well and obtain Si $\approx 1.8 \text{Si}_\odot$ and Mg $\approx 2.2 \text{Mg}_\odot$, in agreement with the slight enrichment in α-elements versus Fe found for the "Pistol Star."

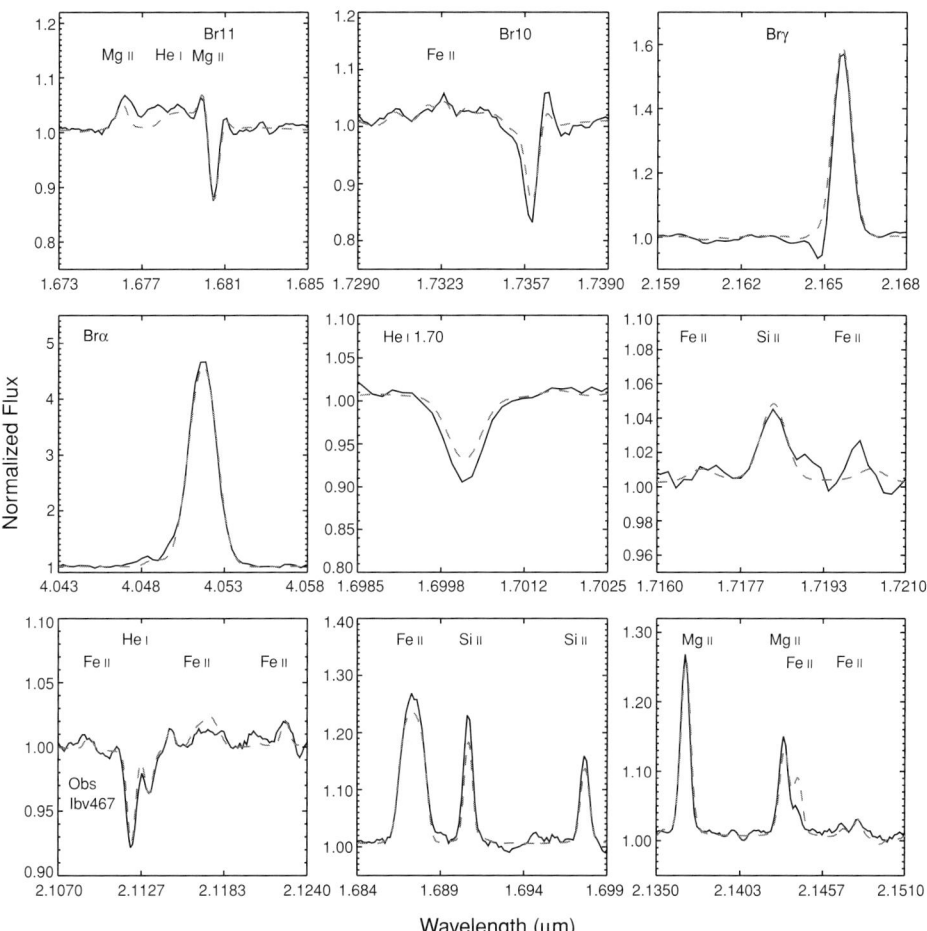

Figure 13.2. Some UKIRT-CGS4 ($R \sim 5,000$) H-, K-, and L-band and Brα observations of the LBVc "#362" in the Quintuplet cluster and model fits.

4 The Arches cluster

The Arches cluster (Figer *et al.* 1999b, 2002) is the youngest and densest cluster at the GC, containing thousands of stars, including at least 160 O stars and around 10 WNLs, namely WN stars still showing H at the surface (Chiosi & Maeder 1986). The cluster is very young ($\leq$2.5 Myr), and the only emission-line stars present are WNLs and OIf$^+$ stars having infrared spectra dominated by H, He I, He II, and N III lines. Some have weak C III/IV lines. The cluster satisfies all requirements for studying the high-mass IMF slope and estimating an upper mass cutoff: individual members can be resolved, there is a large amount of mass in stars, and it is young enough that its most massive members are not pre-supernovae and old enough for its stars to have emerged from their natal coccoons.

118 *Metal abundances in the Galactic Center*

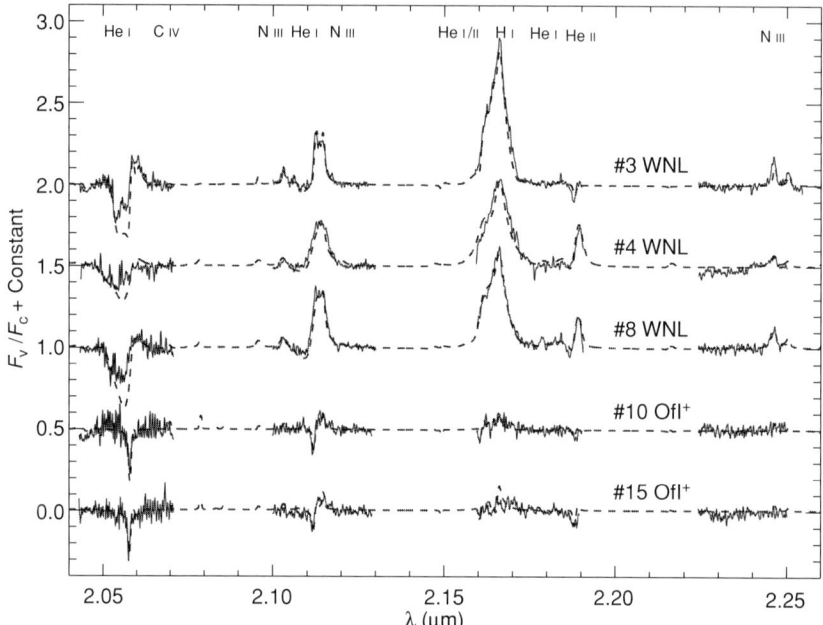

Figure 13.3. Observed spectra (solid) and model fits (dashed) for the three WNL and two OfI+ Arches stars.

4.1 Metallicity studies

The absence of late B supergiants and LBVs prevents one from obtaining direct estimates of the important α-elements versus Fe metallicity ratio as in the Quintuplet cluster and the Central cluster. Since it is the youngest cluster at the GC, any hint about its metallicity would constitute our "last-minute" picture of the chemical enrichment of the central region in the Milky Way. To analyze the stars in the Arches cluster we have assumed the atmosphere to be composed of H, He, C, N, O, Si, and Fe. Observational constraints are provided by the K-band spectra of the stars (see Figure 13.3) and the narrow-band HST/NICMOS photometry (filters F_{F110W}, F_{F160W}, and F_{F205W}) and Pα equivalent width (filters F_{F187N} and F_{F190N}). Object identifications are given according to Figer *et al.* (2002). Below we present the results of our analysis (Najarro *et al.* 2004).

The reduced spectra and model fits are shown in Figure 13.3. The top three spectra correspond to some of the most luminous stars in the cluster. As described in Figer *et al.* (2002), these are nitrogen-rich Wolf–Rayet stars with thick and fast winds. The bottom two spectra in Figure 13.3 correspond to slightly less-evolved stars with the characteristic morphology of OIf+ stars. Of concern are the N III 8–7 lines at 2.103 μm and 2.115 μm as well as the N III 5p ^{2}P–5s ^{2}S doublet at 2.247 μm and 2.251 μm. Figer *et al.* (1997) showed that these N III lines appear only for a narrow range of temperatures and wind densities, which occur in the WN9h

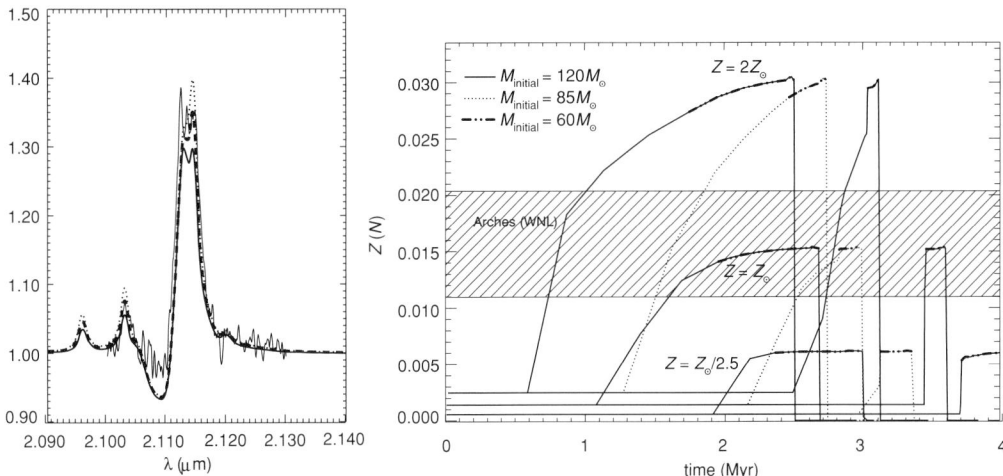

Figure 13.4. Left: the leverage of error estimates on N abundance. The observed 2.10–2.13-μm region (solid) is shown together with current best-fit (dashed) and 30% enhanced (dotted) and 30% depleted (dashed–dotted) nitrogen mass fractions. Right: nitrogen mass abundance versus time calculated using Geneva models. The measurements require Solar metallicity and an age of 2–2.5 Myr.

(WNL) stage. The fairly distinct nature and energies of the multiplets involved in each of these two N III line sets provide strong constraints for the determination of the nitrogen abundance. Thus, at the signal-to-noise ratio of our spectra, our models show that the WNLs N III lines can easily track relative changes as small as 20% in the nitrogen abundance, and a 30% error should be regarded as a safe estimate, as shown in Figure 13.4 on the left.

Of particular importance is that we obtained roughly the same surface abundance fraction of N, $Z(N)$, in our analysis for all three WNL objects (∼1.6%), which is well above the upper limit found for the OIf$^+$ stars (∼0.6%). Although WNL stars do not exhibit any primary diagnostic line in their K-band spectra from which to estimate metallicity, the crucial role of $Z(N)$ in determining their metallicity is immediately apparent if we make use of the stellar-evolution models for massive stars.

According to the evolutionary models of Schaller *et al.* (1992) and Charbonnel *et al.* (1993), a star entering the WNL phase still shows H at its surface, together with strong enhancement in levels of helium and nitrogen and strong depletion in levels of carbon and oxygen, as expected from processed CNO material. During this phase, the star maintains a nearly constant $Z(N)$ value, which essentially depends only linearly on the original metallicity (see the right-hand panel of Figure 13.4), being basically unaffected by the mass-loss rate assumed and the occurrence of stellar rotation during evolution (Meynet & Maeder 2004). Since we expect the CNO abundance in the natal cloud to scale with that of the rest of metals, the nitrogen surface abundance must trace the metallicity of the cluster. The parameters derived for these stars (Najarro *et al.* 2004) indicate that this is indeed the case for objects #3, #4, and #8. The derived $Z(N)$ (~1.6%) is that expected for *Solar* metallicity from the evolutionary models. The reliability of our method is demonstrated in the right-hand panel of Figure 13.4, where we display the nitrogen mass fraction as a function of time for stars with initial masses of 60, 85, and 120 times $M_\odot$, and metallicities equivalent to 2, 1, and 0.4 times Solar, assuming the canonical mass-loss rates (Schaller *et al.* 1992). Our results for the WNL and O stars (the cross-hatched region) require Solar metallicity and an age of 2–2.5 Myr; see also Najarro *et al.* (2004).

5 The Central cluster revisited

The Central cluster hosts a large number of massive stars that formed during in the past 10 Myr (Becklin *et al.* 1978; Krabbe *et al.* 1991, 1995; Najarro *et al.* 1994, 1997). The last census by Eisenhauer *et al.* (2005) and Paumard *et al.* (2006) includes at least 80 massive stars, with ~50 OB stars close to the main sequence or in their early supergiant phase and 30 more-evolved massive stars which appear to be confined to two disks (Paumard *et al.* 2006). There is also a group of about a dozen B stars within the central arcsecond (the "s" stars). Interestingly, Paumard *et al.* (2006) do not detect any OB star outside the central 0.5 pc and find that the stellar contents of the two disks indicate a common age of 6 ± 2 Myr, with O8–9I as earliest spectral type detected.

To test whether the new generation of models and observations could alter considerably our current picture of the evolved massive stars in the Central cluster (Najarro *et al.* 1994, 1997), as has been the case for the "Pistol Star" (see above), we have started a re-analysis campaign using high-resolution observations obtained with NIRSPEC (Keck) and our up-to-date line-blanketed code. Figure 13.5 displays K-band observations and new model fits to the AF star and IRS16NE. Our new fits indicate a slightly higher temperature and luminosity for the AF star but are still consistent with the values and error estimates obtained by Najarro *et al.* (1994, 1997). On the other hand, for IRS16NE, the presence of metal lines (Mg II) implies a downward revision by ~6,000 K of the temperature estimate from Najarro *et al.* (1997), and therefore a much lower luminosity for this object.

5.1 IRS 8: an outsider in the GC

The nature of the GC source IRS 8 (Becklin & Neugebauer 1975), one of the brightest compact mid-infrared sources in the Central cluster, was unknown until

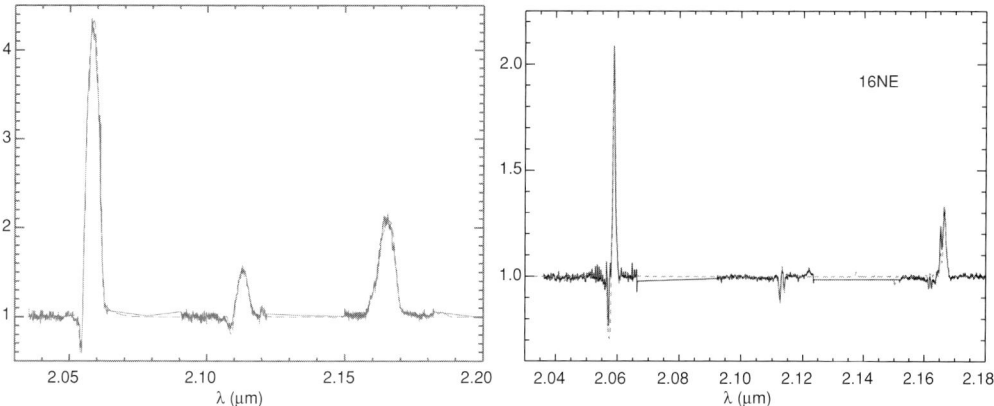

Figure 13.5. New K-band high-resolution observations and model fits to the AF star (left) and 16NE (right). For the AF star our models indicate a slightly higher but still consistent temperature and luminosity with the values obtained by Najarro et al. (1994, 1997). For 16NE, the presence of metal lines implies a downward correction of up to 6,000 K to the temperature estimate from Najarro et al. (1997).

adaptive-optics H- and K-band imaging revealed that the bulk of its infrared emission originates from a classic bowshock (Rigaut et al. 2003; Geballe et al. 2004). Geballe et al. (2004) showed that the IRS 8 bowshock is a straightforward consequence of the interaction of a dense and high-velocity wind from a hot star (hereafter IRS 8*) that is traversing moderately dense interstellar gas. To investigate the nature of the central source, we obtained mid-resolution ($R \sim 900$) K-band spectra using the Gemini adaptive-optics module ALTAIR to feed the near-infrared spectrograph NIRI; see also Geballe et al. (2006).

Figure 13.6 shows the resulting normalized K-band spectrum of IRS 8* compared with online-available K-band spectra from the Hanson et al. (1996) catalog for O stars ranging from O4 to O6.5 and various luminosity classes. The resolving powers for all template spectra have been degraded to 800 for direct comparison with the observed spectrum. From Figure 13.6 we judge that IRS 8* falls within the O5–O6.5 and III–If ranges, with likely O5–O6 spectral type and if luminosity class. Given the strong spectral similarities of IRS 8* to the O5–O6 supergiants in Cyg OB2 (see Figure 13.6), we computed model fits covering that parameter domain, drawing on our analysis of the Cyg OB2 stars for which UV, optical, and IR spectra are available (Najarro et al. in preparation). Figure 13.6 displays our best-fitting model (dashed line), which we obtained using the line-blanketing method presented in previous sections. See Geballe et al. (2006) for a thorough discussion of the analysis.

Of concern is the re-identification of the strong emission feature at 2.116 μm in IRS 8* which has been attributed in the past to C III and N III $n = 8$–7 transitions and is present over a very wide range of O spectral types and luminosities (Hanson

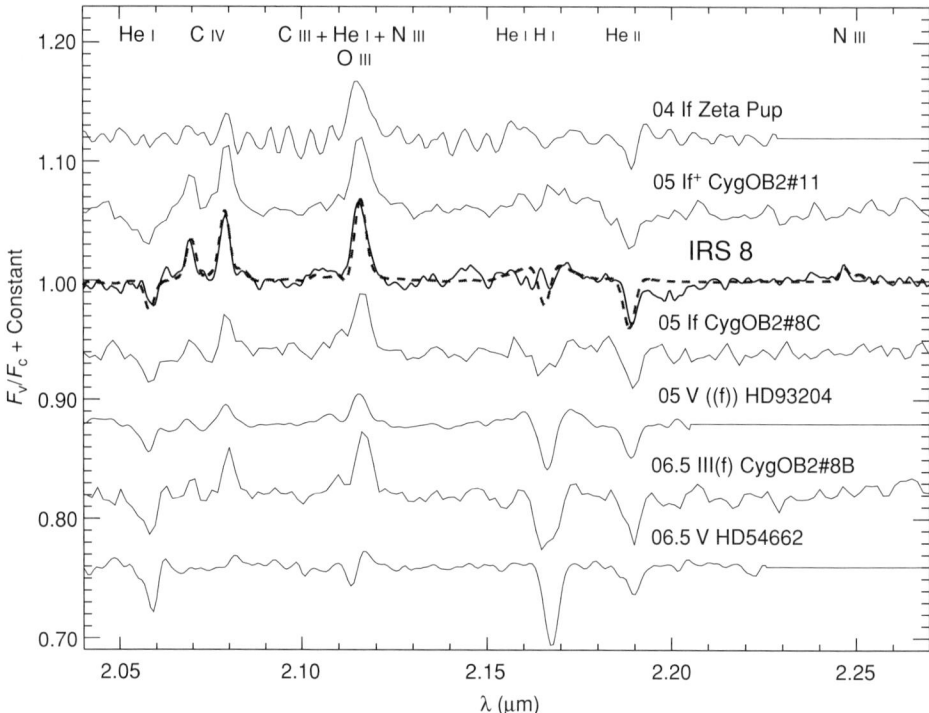

Figure 13.6. Spectral type determination of IRS 8* by comparison of the resulting normalized spectrum with K-band spectra from Hanson *et al.* (1996) degraded to a resolution of $R = 800$. Also displayed (dashed) is a model fit with stellar parameters corresponding to an O5.5 If star (see the text).

et al. 1996). Our investigation (Geballe *et al.* 2006) indicates that the 2.116-μm feature in IRS 8* is dominated by O III $n = 8$–7 transitions. Further, the O III component of the 2.116-μm feature depends largely on the oxygen abundance and only slightly on the gravity, effective temperature, wind density, and velocity field. Thus, this feature may be a powerful diagnostic of oxygen abundance, and therefore an important metal-abundance determiner, over a wide range of O spectral types (Najarro *et al.* in preparation). Using it we obtain an oxygen abundance of 0.8–1.1 times Solar in IRS 8*, which indicates that the cloud in which IRS 8* formed was of *Solar* metallicity.

Our analysis suggests that IRS 8*, although only 1 pc from the center, does not fit into the current picture of the Central cluster of hot stars. It is of much earlier spectral type than any of the stars classified by Paumard *et al.* (2006). Currently it is the only known OB star outside the central 0.5 pc region of the cluster. Figure 13.7 shows the position of IRS 8* (solid cross) as estimated from our model fits on the HR diagram compared with various evolutionary scenarios. The age of 2.8 Myr and absence of surface enrichment obtained with tracks of stars without rotation as used

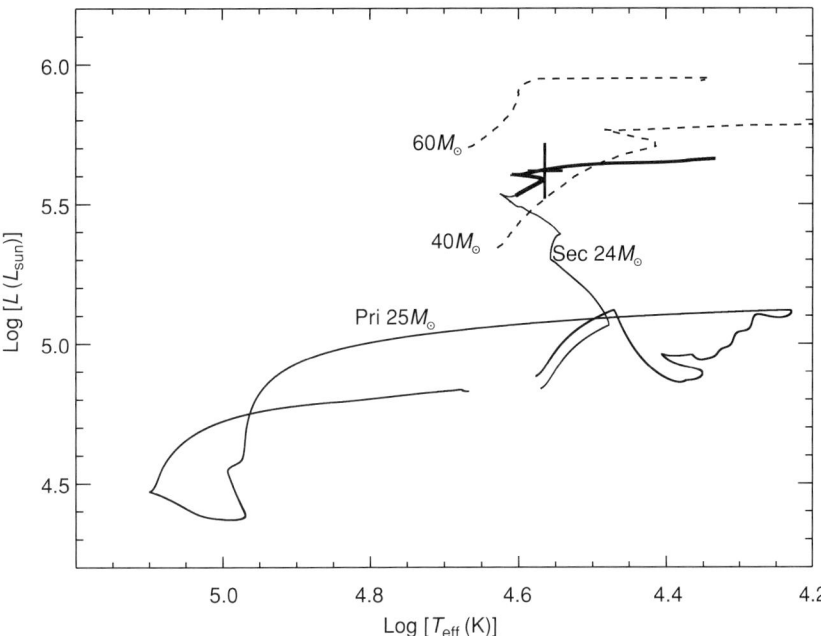

Figure 13.7. The position of IRS 8* (solid cross) in the HR diagram as estimated from model fits, compared with various evolutionary scenarios. The current location of IRS 8* is reached after 3.6 and 7.1 million years for the evolutionary cases of a single star (dashed) and massive close binary (solid), respectively. See Geballe et al. (2006) for a thorough discussion.

by Paumard et al. (2006) (Maeder & Meynet 2003) is clearly at odds both with the current estimate for the age of the GC cluster and with the abundance pattern derived from our models. The situation improves when evolutionary models accounting for rotation (dashed lines in Figure 13.7) are considered (3.5 Myr and CNO-processed material on the stellar surface). Except for the age, which is still well below the estimate obtained by Paumard et al. (2006), on using rotating models for a single-burst scenario we obtain stellar parameters fully consistent with those derived from our modeling.

The crucial question thus is whether this star is really much younger than the cluster and probes the existence of ongoing (or at least much more recent) star formation, or, on the contrary, the star is either an impostor or a cluster member that underwent a rejuvenation cure. A possible way out is provided if the star was originally a member of a massive close binary system. In such a case, we could be looking now at the secondary star, with the primary either having exploded as a supernova or being in an evolutionary phase during which it is much dimmer in the K band than the secondary. Using the models of Wellstein & Langer (1999) we found that for a massive close binary system with initial masses of $25M_\odot$ and

$24M_\odot$ (their model 10a) the current position of IRS 8* may be elegantly explained (see Geballe *et al.* (2006) for a thorough discussion) without contradiction with the age of the GC cluster (solid lines in Figure 13.7). Similar scenarios are a possible explanation for some of the overluminous He I objects in the central parsec.

6 Conclusions

Our result of Solar metallicity for the Central cluster, the Arches cluster, and the Quintuplet cluster runs counter to the trend in the disk (Rolleston *et al.* 2000; Smartt *et al.* 2001) but is consistent with the findings from cool-star studies (Carr *et al.* 2000; Ramírez *et al.* 2000). This may imply that the ISM in the disk does not extend inward to the GC, or that the GC stars are forming out of an ISM that has an enrichment history that is distinctly different from that in the disk. Our result is more consistent with the values found for the bulge (Frogel *et al.* 1999; Felzing & Gilmore 2000).

Acknowledgments

I would like to thank Don Figer, John Hillier, Rolf Kudritzki and, Tom Geballe for invaluable discussions, and acknowledge receipt of grants PNAYA-2003-02785-E and AYA2004-08271-C02-02.

References

Anders, E., & Grevesse, N. (1989), *Geochimi. Cosmochimi. Acta* **53**, 197
Becklin, E. E., & Neugebauer, G. (1975), *ApJ* **200**, L71
Becklin, E. E., Matthews, K., Neugebauer, G., & Willner, S. P. (1978), *ApJ* **219**, 121
Blum, R. D., Ramond, T. M., Conti, P. S., Figer, D. F., & Sellgren, K. (1997), *AJ* **113**, 1855
Carr, J. S., Sellgren, K., & Balachandran, S. C. (2000), *ApJ* **530**, 307
Charbonnel, C., Meynet, G., Maeder, A., Schaller, G., & Schaerer, D. (1993), *A&AS* **101**, 415
Chiosi, C., & Maeder, A. (1986), *AR A&A* **24**, 329
Eisenhauer, F., *et al.* (2005), *ApJ* **628**, 246
Feltzing, S., & Gilmore, G. (2000), *A&A* **355**, 949
Figer, D. F., McLean, I. S., & Najarro, F. (1997), *ApJ* **486**, 420
Figer, D. F., Najarro, F., Morris, M. *et al.* (1998), *ApJ* **506**, 384
Figer, D. F., McLean, I. S., & Morris, M. (1999a), *ApJ* **514**, 202
Figer, D. F., Kim, S. S., Morris, M., Serabyn, E., Rich, R. M., & McLean, I. S. (1999b), *ApJ* **525**, 750
Figer, D. F., Najarro, F., Gilmore, D., Morris, M., Kim, S. S. (2002), *ApJ* **581**, 258
Figer, D. F. (2004), in proceedings of "IMF50"
Frogel, J. A., Tiede, G. P., & Kuchinski, L. E. (1999), *AJ* **117**, 2296
Glass, I. S., Catchpole, R. M., & Whitelock, P. A. (1987), *MNRAS* **227**, 373
Geballe, T. R., Najarro, F., & Figer, D. F. (2000), *ApJ* **530**, L97
Geballe, T. R., Rigaut, F., Roy, J.-R., & Draine, B. T. (2004), *ApJ* **602**, 770
Geballe, T. R., Najarro, F., Rigaut, F., & Roy, J. (2006), arXiv:astro-ph/0607550
Hanson, M. M., Conti, P. S., & Rieke, M. J. (1996), *ApJS* **107**, 281

Hanson, M. M., Kudritzki, R.-P., Kenworthy, M. A., Puls, J., & Tokunaga, A. T. (2005), *ApJS* **161**, 154
Hillier, D. J., & Miller, D. L. (1998), *ApJ* **496**, 407
(1999), *ApJ* **519**, 354
Hillier, D. J., Crowther, P. A., Najarro, F., Fullerton, A. W. (1998), *A&A* **340**, 483
Krabbe, A., Genzel, R., Drapatz, S., & Rotaciuc, V. (1991), *ApJ* **382**, L19
Krabbe, A., *et al. ApJ* **447**, L95
Koyama, L., *et al.* (2006), *PASJ*, arXiv:astro-ph/0609215
Maeda, Y., *et al.* (2002), *ApJ* **570**, 671
Maeder, A., & Meynet, G. (2003), *A&A* **404**, 975
Meynet, G., Maeder, A., Schaller, G., Schaerer, D., & Charbonnel, C. (1994), *A&AS* **103**, 97
Meynet, G., & Maeder, A. (2004), *A&A* **404**, 975
Morris, P. W., Eenens, P. R. J., Hanson, M. M., Conti, P. S., & Blum, R. D. (1996), *ApJ* **470**, 597
Najarro, F., Hillier, D. J., Kudritzki, R. P. *et al.* (1994), *A&A* **285**, 573
Najarro, F., Krabbe, A., Genzel, R., Lutz, D., Kudritzki, R. P., & Hillier, D. J. (1997), *A&A*, **325**, 700
Najarro, F., Figer, D. F., Hillier, D. J., & Kudritzki, R. P. (2004), *ApJ* **611**, L108
Paumard, T., *et al.* (2006), *ApJ*, in press (astro-ph 0601268)
Ramírez, S. V., Sellgren, K., Carr, J. S. *et al.* (2000), *ApJ* **537**, 205
Rigaut, F., Geballe, T. R., Roy, J.-R., & Draine, B. T. (2003), in *Galactic Center Workshop 2002: The Central 300 Parsecs of the Milky Way*, ed. A. Cotera *et al.*, *Astron. Nachr.*, **324**, 551
Rolleston, W. R., Smartt, S. J., Dufton, P. L., & Ryans, R. S. I. (2000), *A&A* **363**, 537
Schaller, G., Schaerer, D., Meynet, G., & Maeder, A. (1992), *A&A* **96**, 269
Shields, J. C., & Ferland, G. J. (1994), *ApJ* **430**, 236
Smartt, S. J., Venn, K. A., Dufton, P. L., Lennon, D. J., Rolleston, W. R., & Keenan, F. P. (2001), *A&A* **367**, 86
Wellstein, S., & Langer, N. (1999), *A&A* **350**, 148

14

Light elements in the Galactic bulge

A. Lecureur[1], V. Hill[1], M. Zoccali[2] & B. Barbuy[3]

[1]*Observatoire de Paris-Meudon, GEPI and CNRS UMR 8111, 92125 Meudon Cedex, France*
[2]*Popular Universidad Católica de Chile, Departamento de Astronomía y Astrofísica, Casilla 306, Santiago 22, Chile*
[3]*Universidade de São Paulo, IAG, Rua do Matão 1226, São Paulo 05508-900, Brazil*

We present abundance results for 53 bulge giant stars using high-resolution spectra obtained with FLAMES/UVES at the ESO/VLT for various regions of the Bulge ($-12 < b < -4$). The trend of the four light elements O, Na, Mg and Al indicates a chemical enrichment of the bulge dominated by massive stars at all metallicities. For [Fe/H] > -0.5, [O/Fe], [Na/Fe], [Mg/Fe] and [Al/Fe] are enhanced relative to both the thin- and the thick-disc trend. This suggests that the bulge formed on a shorter timescale than did the Galactic discs.

Using Mg as a proxy for metallicity (instead of Fe) we further show the following (i) The [O/Mg] ratio for bulge stars follows and extends to higher metallicities the decreasing trend of [O/Mg] found in the galactic discs. (ii) The [Na/Mg] ratio trend with increasing [Mg/H] is found to increase in three distinct sequences in the thin disc, the thick disc, and the bulge. The bulge trend is well represented by the predicted metallicity-dependent yields of massive stars, whereas the galactic discs have Na/Mg ratios that are too high at low metallicities, indicating an additional source of Na from AGB stars. (iii) In contrast to the case with Na, there appears to be no systematic difference in the [Al/Mg] ratio between bulge and disc stars, and the theoretical yields for massive stars agree with the observed ratios, leaving no space for an AGB contribution to Al.

1 Introduction

Although some physical characteristics of the Galactic bulge (a wide metallicity distribution, old stellar populations . . .) are now commonly admitted, the nature of its formation is still being debated. Is our bulge a *classical bulge* in which most stars originated from a short phase of star formation when the Universe was only a

The Metal-rich Universe, eds. G. Israelian and G. Meynet. Published by Cambridge University Press.
© Cambridge University Press 2008.

few Gyr old or a *pseudobulge* resulting from the secular evolution of the disc driven by the presence of the bar (Kormendy & Kennicutt, 2004)?

To distinguish between these two scenarios and understand the chemical evolution of the bulge, we selected four fields towards the Galactic bulge in order to obtain a good sampling of the bulge along its Galactic latitude: a low-reddening window at $(l, b) = (0, -6)$, Baade's Window at $(l, b) = (1, -4)$, the Blanco field at $(l, b) = (0, -12)$ and a field in the vicinity of the globular cluster NGC 6553 at $(l, b) = (5, -3)$.

2 Data

The observational data were collected at the VLT-UT2 with the FLAMES fibre spectrograph. In the four fields, spectra of ∼1000 K giants were obtained with the GIRAFFE arm of the instrument (R ∼ 20 000). This work describes the analysis of data for the 53 stars also observed with the UVES arm of the instrument, with a resolution of ∼48 000, in the range 4800–6800 Å. The final sample contains 13 red clump stars and 40 RGB stars from four separate regions of the bulge. Among them, 11 stars are located in the $b = -6$ window, 26 stars in Baade's Window, 5 stars in the Blanco field and 13 stars in the NGC 6553 field. The signal-to-noise ratio of the spectra is typically between 20 and 50 per pixel, allowing the derivation of accurate abundances.

3 Analysis

The analysis of our bulge clump and RGB stars was performed differentially relative to the well-known giant μ-Leo. The more metal-poor giant Arcturus ([Fe/H] ∼ −0.5 dex) was also the used as a comparison star. The stellar-atmosphere models that we used were interpolated in a grid of the most recent OSMARCS models available. The spectrum synthesis was done using the LTE spectral analysis code "turbospectrum" (Alvarez & Plez, 1998) as well as the spectrum-synthesis code of Barbuy *et al.* (2003), whereas abundances from EQWs (measured with DAOSPEC (Stetson & Pancino, 2006)) were derived using the Spite programs (Spite 1967, and subsequent improvements over the years).

3.1 Determination of the stellar parameters

With the Fe I linelist relative to μ-Leo, the effective temperature $T_{\rm eff}$ and the microturbulence velocity $V_{\rm t}$ were determined with the spectroscopic analysis: $T_{\rm eff}$ by imposing the condition of excitation equilibrium on Fe I lines, and $V_{\rm t}$ in order that lines of different strengths give the same abundance. Using the photometric data coming from the OGLE and 2MASS catalogues, the surface gravity was computed from the luminosity of the star assuming that all stars lie at the same distance ($R =$ 8 kpc).

3.2 Abundances of O, Na, Mg and Al

The linelists (molecules and atoms) around each of the O, Na, Al and Mg lines studied were adjusted to reproduce the observed spectra of the Sun, Arcturus and μ-Leo. The final abundances were computed by minimizing the ξ^2 value between observed and synthetic spectra. We refer the reader to Lecureur *et al.* (2006) for a detailed description of the analysis.

4 Results and discussion

4.1 Mixing

The mean [C/Fe] and [N/Fe] values (-0.04 and $+0.43$, respectively) confirm that all the stars studied here have passed the first dredge-up. Moreover, the [C + N/Fe] ratio is constant (within uncertainties) for the whole sample and close to the Solar [C + N/Fe] ratio, ruling out the possibility of mixing in deep ON-cycle layers which could have explained the O–Na anticorrelation that we observed within the bulge sample. Therefore the O, Na, Mg and Al abundances reflect the initial abundances in the gas from which the star formed and can be used to describe the chemical evolution of the bulge. However, part of the sodium can be explained by invoking the first dredge-up. This can amount to 0.15 dex according to models and is observed to be up to 0.30 dex in Solar-neighbourhood giants, as shown by Mishenina *et al.* (2006). The excess of Na observed (see Figure 14.1) is much larger than this and we therefore think that a large part of it is initial.

4.2 Abundances of O, Na, Mg and Al

These four elements were chosen because they are produced mainly by massive stars that explode as Type-II supernovae. Therefore their ratios to iron (which is produced mainly by Type-I supernovae) reflect the formation timescale of the parent population. As shown in Figure 14.1, wherever the thin and thick discs are well separated, the bulge manifests a third distinct trend. For Al and Mg, the separation is clearly visible for [Fe/H] < 0, whereas for O the distinction among the three populations was confirmed by statistical tests. Concerning Na, at subsolar metallicities a distinction can be made neither between bulge and disc stars nor

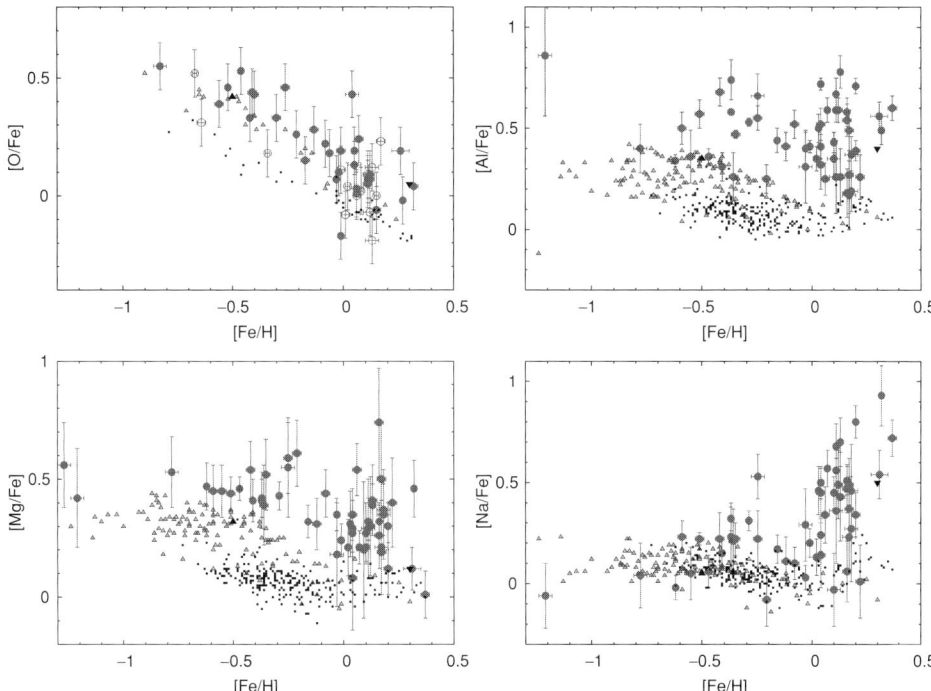

Figure 14.1. Plots of [O/Fe], [Mg/Fe], [Na/Fe] and [Al/Fe] ratios against [Fe/H] for our sample of bulge giants (filled grey circles) compared with the thin-disc (open triangles) and thick-disc samples (black squares) from Reddy *et al* (2006) and Bensby *et al.* (2004, 2005). The black upward- and downward-pointing triangles are for Arcturus and μ-Leo, respectively. In the [O/Fe] panel, only the Bensby *et al.* (2004) [O I] measurements are shown. Notice the clear separations between the thin-disc, thick-disc and bulge stars.

between the two discs. At high metallicities, the bulge stars have higher sodium abundances than than those of the two discs. The [Al/Fe], [Mg/Fe] and [O/Fe] ratios for the bulge stars are higher than those for the thick-disc ones, which are themselves more enhanced than those for the thin-disc stars. These results support a scenario in which the formation timescale of the bulge was shorter than that of the thick disc, which was in its turn shorter than that of the thin disc and the three populations have experienced different chemical trajectories.

4.3 Nucleosynthesis of the bulge

To investigate the enrichment by massive stars we compared our measurements with the predicted yields of Chieffi and Limongi (2004), using Mg instead of iron as an indicator of metallicity. Figure 14.2 shows the following.

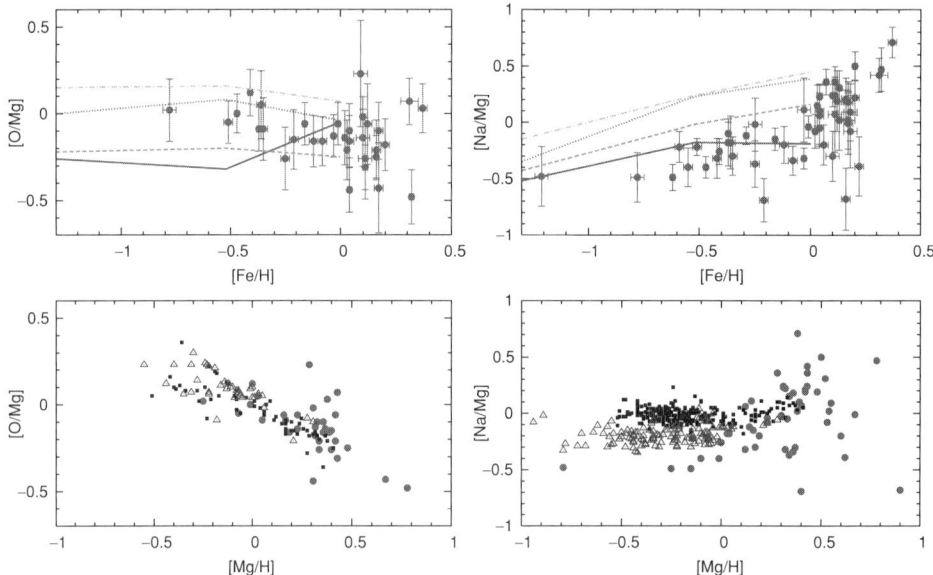

Figure 14.2. The run of the [O/Mg] (left) and [Na/Fe] (right) ratios with metallicity for the bulge stars of our sample (circles). In the top panels, the predicted yields for SN II of $15M_\odot$ (full line), $20M_\odot$ (dashed line), $25M_\odot$ (dotted line) and $35M_\odot$ (dash–dotted line) from Chieffi & Limongi (2004) are overlain. In the bottom panels, the bulge is compared with the Galactic thin (squares) and thick (triangles) discs, as in Bensby *et al.* (2004), using [Mg/H] as a proxy for metallicity.

- The [O/Mg] ratio decreases with [Fe/H], whereas the yields are essentially metallicity-independent. In the [O/Mg] versus [Mg/H] plot, thin disc, thick disc and bulge lie on the same sequence: the bulge merely extends the sequence to higher [Mg/H] values. The possibility of a decrease in production of O arising from an increase in winds with increasing metallicity has been invoked to explain the decrease in [O/Mg], which is not predicted by the yields. This point has still to be investigated from a theoretical point of view.
- The [Na/Mg] ratio increases with [Fe/H] with values lower than those predicted by Chieffi and Limongi, especially at low [Fe/H]. In the [Na/Mg] versus [Mg/H] plot, for [Mg/H] < 0 the three populations exhibit different trends according to their different formation timescales: the thin disc is above the thick disc, which itself is above the bulge. This tendency suggests an additional contribution of Na by longer-lived progenitors (such as SN I or intermediate-mass AGB). This additional source starts contributing first in the thin disc, at lower metallicities then in the thick disc and the bulge.

The [Al/Mg] ratio, which is plotted in Lecureur *et al.* (2006), undergoes an increase with metallicity that is well predicted by the metallicity dependence of the yields. In contrary to the case with Na, the three populations seem to be merged, despite the occurrence of a large dispersion of Al at high metallicity. This suggests that the contribution by AGB stars may be important for Na but not for Al, possibly because of the higher temperature required to sustain the Mg–Al cycle than for the Ne–Na cycle.

5 Future work

The analysis of the UVES spectra will be extended to the abundance measurements of other α-elements (Ca, Si, Ti), the iron peak and neutron-capture elements (Ba, Eu). We also plan to investigate for the whole GIRAFFE sample ($\sim$1000 stars) the possible correlations between the chemical data and the kinematic data inside one of the four fields or between the fields.

References

Alvarez, R., & Plez, B. (1998), *A&A* **330**, 1109
Barbuy, B., Perrin, M.-N., Katz, D. *et al.* (2003), *A&A* **404**, 661
Bensby, T., Feltzing, S., & Lundström, I. (2004), *A&A* **415**, 155
Bensby, T., Feltzing, S., Lundström, I., & Ilyin, I. (2005), *A&A* **433**, 185
Chieffi, A., & Limongi, M. (2004), *ApJ* **608**, 405
Kormendy, J., & Kennicutt, R. C. Jr. (2004), *MNRAS* **42**, 603
Lecureur, A., Hill, V., Zoccali, M. *et al.* (2006), *A&A* at press, astro-ph/0610346
Mishenina, L., Bienaymé, O., Gorbaneva, T. *et al.* (2006), astro-ph/0605615
Stetson, P. B., & Pancino, E. (2006), in preparation
Zoccali, M., Lecureur, A., Barbuy, B. *et al.* (2006), *A&AL* **433**, 185

15

Metallicity and age of selected G–K giants

Luca Pasquini,[1] M. Döllinger,[1] J. Setiawan,[2] A. Hatzes,[3] L. Girardi,[4]
L. da Silva,[5] J. R. de Medeiros,[6] A. Weiss[7] & O. Von Der Lühe[8]

[1] *European Southern Observatory, Garching bei München, Germany*
[2] *Max-Planck-Institut für Astronomie, Heidelberg, Germany*
[3] *Tautemburg Observatory, Germany*
[4] *INAF – Trieste, Italy*
[5] *Observatorio Nacional, Rio de Janeiro, Brazil*
[6] *UFRN, Natal, Brazil*
[7] *Max-Planck-Institut für Astronomie, Garching bei München, Germany*
[8] *Kipenheuer Institut für Sonnenphysik, Freiburg, Germany*

We have derived metallicity, masses, and ages for two samples of nearby giant stars, which have been observed with the aim of understanding their nature of the radial-velocity (RV) variability and to search for planetary companions. Our stars have reliable Hipparcos parallaxes, and for several we also have measured angular diameters; the parameters we retrieve from our inversion process are in very good agreement with the observed ones. Among our results, we find that the stars regarded as candidates to host planetary companions are not preferencially metal-rich, which is at odds with what is found for main-sequence stars. We also find that stars younger than ~ 1 Gyr can be described by a single metallicity and that an age–metallicity relationship applies to our samples.

1 Introduction

About six years ago we started an observational program with the FEROS spectrograph at the ESO La Silla, to investigate the radial-velocity (RV) variability in evolved late-type stars (Setiawan *et al.* 2003a). The main goals of this program were to determine whether G–K giants are RV variable, and why. Setiawan *et al.* (2004) showed that a large proportion of our stars are indeed RV variables, and two giants hosting extra-Solar planets were found (Setiawan *et al.* 2003b, 2005). Unfortunately, the accuracy obtained with FEROS at the ESO 1.5-m telescope was limited to about $20\,\mathrm{m\,s^{-1}}$; and this implies that we were missing a number of RV variables due to our limited accuracy. We therefore started a similar program in the northern sky with the 2.2-m telescope at Tautemburg Observatory (TLS) equipped

The Metal-rich Universe, eds. G. Israelian and G. Meynet. Published by Cambridge University Press.
© Cambridge University Press 2008.

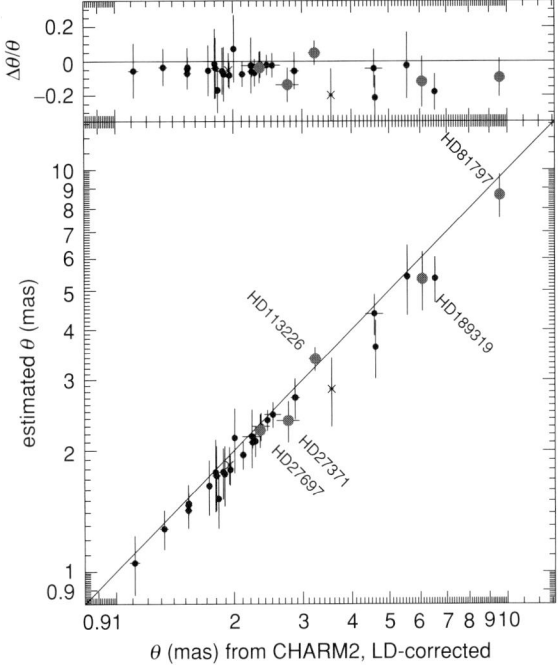

Figure 15.1. A comparison between estimated and measured stellar radii (in milli-arcseconds) for the giants of our S sample. The entries in the CHARM2 catalog were corrected for limb-darkening. The solid line represents the one-to-one relation. The large dots indicate stars with lunar occultation or LBI measurements. The fractional difference between the two samples is 6%.

with an iodine cell that guarantees accuracy to within a few m s^{-1}. This northern sample is composed of 62 stars. The first results (Döllinger et al. 2007b, A&A in preparation) show that 91% of the stars are indeed RV variables at the improved accuracy reached in our observations. Among them, we have a handful of exoplanet-host candidates, of which one has already been confirmed (Döllinger et al. 2007a). In order to understand how RV variability of giant stars evolves with the stellar parameters, we must be able to derive firm values of stellar parameters, such as age and mass. In particular, a reliable estimate of the mass is crucial for determining the companion mass. Unlike for main-sequence stars, it is difficult to calculate stellar masses from effective temperatures of colors. We present here the first results from this analysis.

2 Sample and data analysis

All our sample stars have parallaxes measured by Hipparcos to better than 10%, and therefore accurate absolute magnitudes. On the other hand, in order to break the age–metallicity degeneracy present along the giant branch, an independent

Table 15.1. *Statistics about RV variability from our N and S samples. The larger percentage of RV variables in the N sample is attributed to the RV accuracy obtained at TLS (3–$5\,m\,s^{-1}$) with being higher than that obtained for the S sample ($23\,m\,s^{-1}$); see Setiawan et al. (2004).*

Type	S sample	N sample
Binaries	13 (17%)	11 (16%)
Variable	43 (56%)	45 (75%)
Long-period	–	9 (15%)
Short-period	–	36 (60%)
Constant	21 (27%)	6 (9%)
Confirmed planets	3	1
Accuracy ($m\,s^{-1}$)	22	5

estimate of the stellar metallicity was required. By using our high-resolution, high signal-to-noise-ratio spectra, we were able to determine the stellar metallicity and effective temperature, by using an LTE analysis and many (typically 50–60) Fe I and Fe II lines. We performed a classical analysis, which resulted in very robust results; the zero points of metallicity were obtained by analyzing a Solar spectrum and by comparing our results with the parameters obtained in the literature for two well-studied giants: ε-Vir and the Hyades giant HD27371. The metallicity and effective temperatures retrieved spectroscopically were used in conjunction with a modified Jørgensen and Lindegren (2005) method and the Padova tracks (Girardi *et al.* 2000) to create probability distribution functions for mass and age for each star. The presence of Hyades stars allowed us to check the good agreement with parameters retrieved with other methods (e.g. main-sequence fitting) for this cluster. Further checks have been performed, such as comparison between observed and retrieved colors, spectroscopic and evolutionary temperatures, and observed and computed stellar radii. The result of this comparison is shown in Figure 15.1 for the southern sample, and the results are extremely good, with a fractional difference between our estimate and the CHARM2 (Richichi *et al.* 2005) values of 0.06 and an rms scatter of 0.06 (da Silva *et al.* 2006). In the following the two samples are kept separate and labeled N and S.

3 Results and discussion

The first results concern the general trends of RV variability. As we can see from Table 15.1, which summarizes our results, most of the stars in the N sample are RV variables, at an accuracy of $5\,m\,s^{-1}$. The lower percentage of RV variables in the S sample is clearly due to the lower accuracy obtained with FEROS.

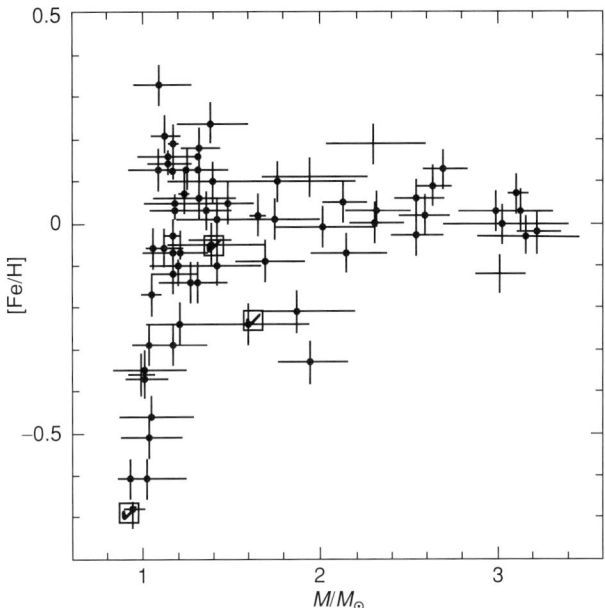

Figure 15.2. The metallicity distribution versus mass for the S stars. The three stars hosting exoplanets are marked with box-and-tick symbols. Our candidates are not preferentially metal-rich.

Another interesting aspect of our statistics is that we have a large number of planet-host candidates (15% in the N sample and, a similar number in the S sample). This percentage is very high, and, if most of these stars are confirmed to host exoplanets, it is much higher than what is usually found in comparable surveys of main-sequence stars. This discrepancy is even larger when considering that among our evolved stars there are no short-period, hot Jupiters (Mayor and Queloz 1995), since they would be engulfed by their extended stellar atmospheres. Our stars have, on average, masses much larger than the typical masses of the main-sequence stars surveyed in RV studies. This might be one possible key for the interpretation of the results; it would indicate a very steeply increasing probability of formation of exoplanets with increasing stellar mass.

Remarkably, of the four stars which have been confirmed to host planets so far, none is metal-rich. Admittedly, our metallicity distribution for giant stars does not peak towards metal-rich stars, but the planet-hosting stars do not belong to the metal-rich tail of our distribution. Conversely, one of them (HD 47536) is the most metal-poor star of both samples. This can be clearly seen from Figure 15.2, which shows the metallicity–mass distribution of the S sample, in which the planet-host candidates are marked.

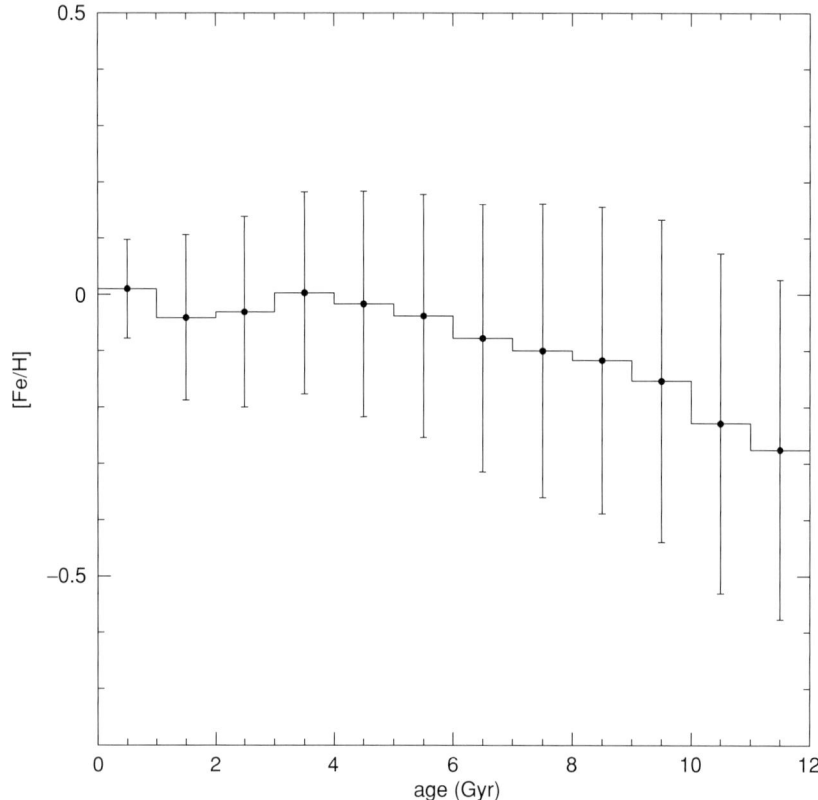

Figure 15.3. The metallicity–age distribution for our S stars. The sample has been re-sampled to give a similar number of stars for each age bin. In spite of a large metallicity spread for old stars, a clear age–metallicity relationship is present.

Finally, thanks to our accurate age reconstruction, we are able to investigate (for the first time using evolved stars) the age–metallicity relationship in the Solar neighborhood.

The first striking point is that the young (i.e. aged <1 Gyr) stars have a small spread in [Fe/H] ($\sigma = 0.09$), comparable to the measurement errors. In spite of a larger spread in the metallicity distribution of older stars, a clear age–metallicty relationship is present in our sample, as shown in Figure 15.3, which shows the age–metallicity relationship for the S stars. The relationship would indicate a very modest increase in [Fe/H] over the last 3–4 Gyr, together with an increase of ~0.3–0.5 dex in the previous 10 Gyr. This result seems at odds with the results obtained by Nordström *et al.* (2004) for a much larger sample of main-sequence stars. It is beyond the scope of this paper to discuss this discrepancy; however, we note that determination of the age of main-sequence stars might be very critical, as pointed out e.g. by Pont and Eyeer (2004), who found a clear age–metallicity relationship from their analysis of main-sequence stars, once ages had been reviewed critically.

References

Girardi, L., Bressan, A., Bertelli, G., & Chiosi, C. (2000), *A&AS* **141**, 371
da Silva, L., Girardi, L., Pasquini, L. *et al.* (2006), *A&A* **458**, 609
Döllinger, M. P., Hatzes, A. P., Pasquini, L. *et al.* (2007a), *A&A* **472**, 649
Döllinger, M. P., Pasquini, L., Hatzes, A. P. *et al.* (2007b), *A&A* submitted
Jørgensen, B. R., & Lindegren, L. (2005), *A&A* **436**, 127
Mayor, M., & Queloz, D. (1995), *Nature* **378**, 355
Nordström, B., Mayor, M., Andersen, J. *et. al.* (2004), *A&A* **418**, 989
Pont, F., & Eyer, L. (2004), *MNRAS* **351**, 487
Richichi, A., & Percheron, I. (2005), *A&A* **431**, 773
Setiawan, J., Pasquini, L., da Silva, L. *et al.* (2003a), *A&A* **397**, 1151
Setiawan, J., Hatzes, A. P., von der Lühe, O. *et al.* (2003b), *A&A* **398**, L19
Setiawan, J., Pasquini, L., da Silva, L. *et al.* (2004), *A&A* **421**, 241
Setiawan, J., Rodmann, J., da Silva, L. *et al.* (2005), *A&A* **437**, L19

Part III

Observations – abundances in extragalactic contexts

16

Stellar abundances of early-type galaxies

S. C. Trager

Kapteyn Astronomical Institute, Rijksuniversiteit Groningen, Postbus 800, NL-9700 AV Groningen, the Netherlands

It is currently impossible to determine the abundances of the stellar populations star by star in dense stellar systems more distant than a few megaparsecs. Therefore, methods to analyse the composite light of stellar systems are required. I review recent progress in determining the abundances and abundance ratios of early-type galaxies. I begin with 'direct' abundance measurements: colour–magnitude diagrams of stars and planetary nebulae in nearby early-type galaxies. I then give an overview of 'indirect' abundance measurements: inferences from stellar-population models, with an emphasis on cross-checks with 'direct' methods. I consider the variations of early-type galaxy abundances as a function of mass, age and environment in the local Universe. I conclude with a list of continuing difficulties in the modelling that complicate the interpretation of integrated spectra and I look ahead to new methods and new observations.

1 Introduction

The spheroidal components of galaxies – elliptical and lenticular (S0) galaxies and the bulges of spirals – contain at least half of the present-day stellar mass in the local Universe (Schechter & Dressler 1987; Fukugita *et al.* 1998). An understanding of the formation and evolution of these objects is therefore necessary in order to understand the dominant galaxy types in the local Universe.

The abundances of stars in these objects give us direct handles on their formation and evolution. The gross metallicity of a galaxy tells us about its overall chemical evolution. Abundances of specific elements are much more useful, since these give direct evidence regarding the nucleosynthetic processes that occurred during the formation of that star (and, by extension, of the stellar population and even galaxy).

The Metal-rich Universe, eds. G. Israelian and G. Meynet. Published by Cambridge University Press.
© Cambridge University Press 2008.

As an example, the ratio [α/Fe] is crudely an indicator of the ratio of SNe II, resulting from the explosion of massive stars and producing the α-elements, to SNe Ia, resulting from the explosion of low- to intermediate-mass stars and producing most of the Fe-peak elements. This leads to [α/Fe] being commonly used as a tracer of the timescale of star formation, since it will decrease with increasing duration of star formation as SNe Ia become more important in the nucleosynthesis in a stellar population (e.g. Worthey *et al.* 1992; Greggio 1997; Trager *et al.* 2000b; Thomas *et al.* 2005). This is an oversimplification, of course, as can be seen in Chapters 28 and 44 in this volume.

Unfortunately, the Universe has made measuring stellar abundances in these objects difficult for us. Spheroids of galaxies are very dense and the crowding makes it impossible to make accurate photometric measurements of individual stars closer than one effective radius away from their centres in all spheroidal objects but our own Galactic bulge (and in that case dust presents a formidable barrier), even with the Hubble Space Telescope (HST). Attempting to determine abundances of individual stars from high-resolution spectra, as we do in the Milky Way, is accordingly impossible beyond the Local Group with the current generation of 8–10-m ground-based telescopes and the 2.5-m HST. The best we can hope for is to determine the colour–magnitude diagrams (CMDs) of the outer regions of these galaxies and infer their abundances from the distribution of stars on the giant branches, attempt to determine abundances from planetary nebulae – which stand out from the crowd as emission-line objects – and finally infer abundances from the integrated light of the galaxies. Each of these methods will be discussed below.

In the following, I refer to as 'direct' methods those techniques that determine abundances from resolved stellar populations. Colour–magnitude diagrams fall into this category, even though abundances of *individual* stars may be poorly determined, as do abundances from analyses of planetary-nebula spectra. I refer to as 'indirect' methods those techniques that determine abundances from unresolved stellar populations, in particular analysis of absorption-line strengths. I discuss direct abundance measurements in Section 2 and indirect abundance measurements in Section 3. I summarise the results in Section 4.

To conclude this section, I mention that there are other methods for determining the abundances of early-type galaxies, including the analysis of their globular-cluster systems and determinations of the abundance of X-ray-emitting gas. However, for reasons of space and to do with complications in relating these determinations to the stellar population of the host galaxy, I will not discuss them here. For the former, the interested reader can refer to Trager (2004) and Brodie & Strader (2006); for the latter, see e.g. Humphrey & Buote (2006). Finally, Jablonka has covered the subject of bulges of spiral galaxies in Chapter 27 in this volume.

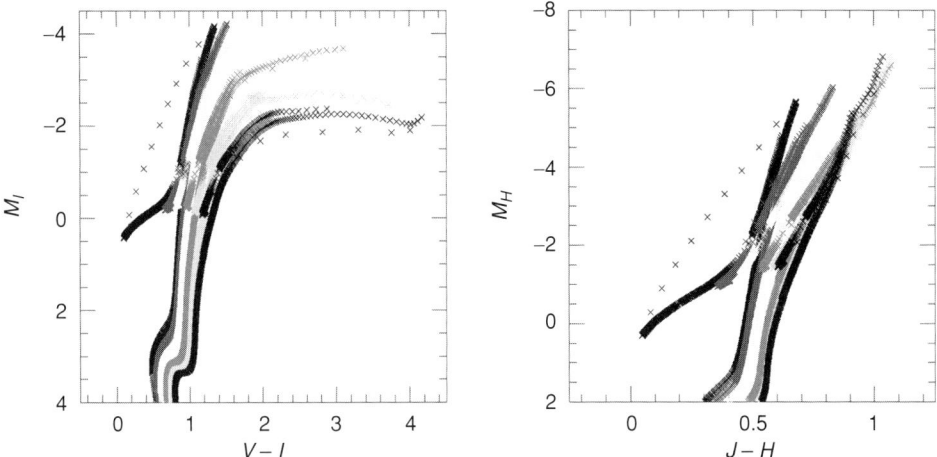

Figure 16.1. The giant branches of old (12-Gyr) stellar populations from the BaSTI (Pietrinferni *et al.* 2004, 2006) isochrone collection, in the optical (left) and near-infrared (right). These isochrones include the horizontal-branch and AGB phases. Metallicities, from left to right in both panels, are [Fe/H] = −2.27, −1.49, −0.96, −0.35, 0.06 and 0.40.

2 Direct abundance measurements

I begin with an overview of determinations of stellar population abundances of nearby galaxies using direct abundance measurements: CMDs and planetary nebulae.

2.1 Abundances from colour–magnitude diagrams

Currently, CMDs have been determined for at least seven (non-dwarf) elliptical and S0 galaxies: M32 (Grillmair *et al.* 1996), NGC 5128 (Rejkuba *et al.* 2005 and references therein), NGC 3379 (Gregg *et al.* 2004), NGC 3115, NGC 5102, NGC 404 (Schulte-Ladbeck *et al.* 2003), and Maffei 1 (Davidge & van den Bergh 2001). For all of these galaxies, CMDs have required HST imaging in the optical or near-infrared (NIR), except for Maffei 1, which was observed in the NIR with the CFHT adaptive-optics system. However, in only three cases – M32, NGC 3379 and NGC 5128 – have the CMDs reached *significantly below* the tip of the red giant branch (TRGB), as is required for reasonable abundance measurements (Figure 16.1, left). In the other cases, only the asymptotic giant branch (RGB) was detected (Maffei 1) or only stars less than one magnitude below the tip of the red giant branch (RGB) were detected. In NGC 3779, although the NICMOS F110W (J) and F160W (H) measurements extend to three magnitudes below the tip of the asymptotic giant branch (AGB), the tight spacing of isochrones of different metallicities and the

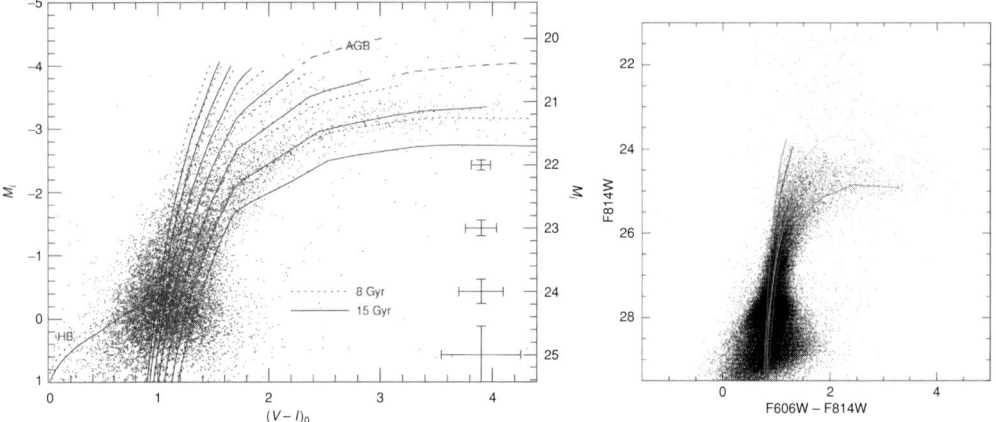

Figure 16.2. Colour–magnitude diagrams of M32 (left) from Grillmair et al. (1996) and NGC 5128 (right) from Rejkuba et al. (2005). In the left panel, isochrones from Worthey (1994) have been over-plotted with metallicities in the range [Fe/H] = −1.2 to 0 dex in 0.2-dex steps for ages of 10 Gyr (dotted lines) and 15 Gyr (solid lines). Fiducial globular-cluster colour–magnitude sequences for NGC 6341, NGC 6752, 47 Tuc, NGC 5927, and NGC 6528 (from left to right) have been over-plotted in the right panel.

co-mingling of the AGB and RGB over this magnitude range make precision estimates of stellar abundances very difficult, and only ranges of acceptable metallicities can be inferred, rather than accurate metallicity *distributions* (Figure 16.1, right) (Gregg et al. 2003).

In M32, Grillmair et al. (1996) obtained a CMD at a distance of $r \sim 3r_e$ with WFPC2 on the HST. Because M31 is superimposed on M32, the resulting CMD was statistically cleaned of M31 stars by using a nearby M31 disc field. The resulting CMD reaches a depth just below the 'red clump', the pile-up of red horizontal-branch (helium-burning) stars about four magnitudes below the TRGB (Figure 16.2, left). There was no evidence of a blue horizontal branch (BHB) in this CMD, although the limiting magnitude in V prevented ruling out its presence completely. This absence has been confirmed in deeper WFPC2 (Worthey et al., in preparation) and ACS/HRC (Lauer et al., in preparation) observations. This is a crucial result, since BHB stars trace the most metal-poor stars in our Galaxy: at metallicities [Fe/H] < 1.5, helium-burning stars are dominantly BHB stars in Galactic globular clusters. Their lack in M32 suggests that this galaxy is deficient in metal-poor stars, which is confirmed by its metallicity distribution inferred from the distribution of RGB stars (Figure 16.3, left). Therefore M32 has a 'G-dwarf problem' (Worthey et al. 1996), for which the usual suggested solution is pre-enrichment of the galaxy (Tinsley 1980).

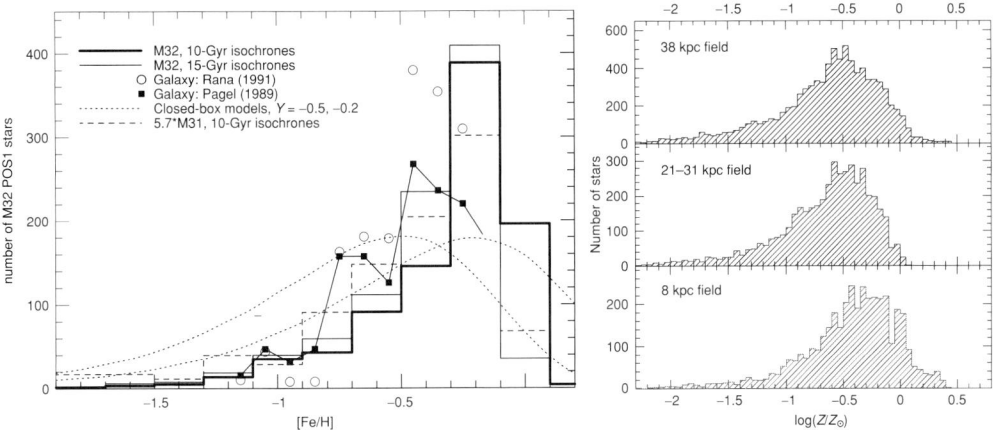

Figure 16.3. Metallicity distribution functions for M32 (left) from Grillmair *et al.* (1996) and NGC 5128 (right) from Rejkuba *et al.* (2005) inferred from comparing the distribution of RGB stars with isochrones. For M32, the isochrones used were 10 or 15 Gyr old (see legend); for NGC 5128, the isochrones used were 12 Gyr old. For M32, closed-box chemical-evolution ('simple') models have been over-plotted with two different choices of the yield.

For NGC 5128, Rejkuba *et al.* (2005), following earlier studies of inner fields by Harris & Harris (2000, 2002), obtained a CMD at a distance of $r \sim 7r_e$ with the ACS/WFC on the HST. As for M32, Rejkuba *et al.* obtained a CMD down to the red clump (Figure 16.2, right) and found no evidence of a BHB. Again, it is difficult to rule out its presence completely, but the metallicity distribution inferred from the distribution of RGB stars (Figure 16.3, right) suggests that there are few (but some) very-metal-poor stars in NGC 5128.

By counting stars on the RGB between isochrones of different metallicities (but typically constant age), a metallicity distribution function (MDF) can be determined. The MDFs from M32 and NGC 5128 are shown in Figure 16.3. Both galaxies have a deficiency of metal-poor stars relative to a closed-box ('simple') chemical-evolution model, as discussed above. Galaxy M32 has a sharp peak in its MDF at [Fe/H] $= -0.25$, using isochrones from Worthey (1994). In Section 3.2 below I will compare this inferred MDF with results from stellar-population analyses. The MDF of NGC 5128 has a significant radial gradient and peaks at [Fe/H] ≈ -0.6 to ≈ -0.3, from the outermost to the innermost fields, using isochrones from VandenBerg *et al.* (2000), re-calibrated to match Galactic globular clusters; see Rejkuba *et al.* (2005) for more details. Note that the lack of high-metallicity stars in the 21–31-kpc field of NGC 5128 is a selection effect due to insufficiently deep *V*-band observations, so the shape of this MDF cannot be fairly compared with the inner and outer fields. Even so, in both galaxies, there are stars of *at least*

Solar metallicity, and there is evidence for super-metal-rich stars, particularly in the innermost field.

There are certain cautions to be kept in mind when considering CMD-inferred MDFs. First, old, single-age stellar populations are *assumed* in order to compute these distributions. This is due to the lack of age information in CMDs that are insufficiently deep: without CMDs that reach down below the main-sequence turn-off(s), the age is unknown (although the presence of bright AGB stars can indicate intermediate-aged populations). A shift, or worse, a distribution, in stellar-population age can change both the *mean* and *the shape* of the inferred MDF. This can be seen by comparing the MDFs of M32 inferred from the 10- and 15-Gyr-old isochrones in Figure 16.3 and imaging a mixture of these populations. Second, the assumed abundance *pattern* (as parametrised by, say, [α/Fe]) can also alter both the mean and the shape of the inferred MDF. This is because changes in the abundance pattern can alter the temperature of the stars on the giant branch (e.g. Salaris *et al.* 1993). On the other hand, this might not be a significant problem for these two galaxies: for M32, because its abundance pattern is very close to that of the Sun (Worthey 2004), and the isochrones used to infer its MDF have scaled-Solar composition (Worthey 1994); and for NGC 5128, because, although the galaxy probably has [α/Fe] > 0 because it is a giant elliptical (see below), the isochrones used to infer its MDFs were α-enriched (Rejkuba *et al.* 2005).

2.2 Abundances from planetary nebulae

Planetary nebulae (PNe) are the second-to-last phase of the life of all low- and intermediate-mass stars. Although this phase is quite short, the large flux of stars through it means that many PNe are visible at any given time. Because this phase is so short, their density is much lower than that of the field stars from which they evolved, and therefore they are much less crowded on the sky. They are emission-line objects and are fairly easy to distinguish from the absorption-line background of their hosts. Using standard emission-line techniques, it is possible to determine elemental abundances directly for each PNe.

However, PNe are quite faint. Special wide-field spectroscopy techniques have been developed to determine the kinematics of PNe systems around galaxies (e.g. Douglas *et al.* 2002), which utilise a narrow-band region around the [O III]λ5007 line. However, these techniques do not (yet) allow abundance determinations, since these require a larger wavelength region: oxygen abundances require measurements of the [O III]λ4363, 5007 and [S II]λ6716, 6731 lines. Moreover, the [O III]λ4363 line is very weak. Unfortunately, without this line, the electron temperature T_e cannot be measured directly, and strong-line techniques (such as those used for H II regions) cannot be used for PNe because the temperature of the central stars varies significantly from object to object, and photoionisation models do not give unique determinations (Stasińska *et al.* 2006). On the other hand, unlike in H II regions, the [O III]λ4363 line is still visible for metal-rich PNe because of the high density of the nebulae and the high temperatures of the central stars (Stasińska *et al.* 2005). With large telescopes ($\geq$4 m) and sensitive spectrographs, bright PNe in early-type galaxies become reasonable tracer objects for the abundance patterns of their hosts.

Many early-type galaxies host well-known PNe systems, although have abundance measurements been attempted for only a handful of them. Here I will concentrate on the three cases that I know of for which independent abundance determinations have been obtained by other methods: M32 (Stasińska *et al.* 1998), NGC 5128 (Walsh *et al.* 1999, 2006) and NGC 4697 (Méndez *et al.* 2005). Of these, only in M32 and NGC 5128 have PNe with [O III]λ4363 lines been detected: nine PNe in M32 (Stasińska *et al.* 1998) and two (so far) in NGC 5128 (Walsh *et al.* 2006). None of the PNe in NGC 4697 show [O III]λ4363 individually, although it is detected in an average spectrum (Méndez *et al.* 2005). I will therefore (reluctantly) no longer consider it here, since the T_e of the individual PNe cannot be determined and therefore their abundances are likely to be very uncertain. In M32, Stasińska *et al.* (1998) found $\langle[O/H]\rangle = -0.5$, in reasonable agreement with, and at projected galactocentric distances similar to those of, the Grillmair *et al.* (1996) CMD study. Walsh *et al.* (1999) find a similar average [O/H] abundance ratio for NGC 5128 in a field close to the centre, although for none of these PNe has [O III]λ4363 been detected. Walsh *et al.* (2006) appear to confirm this result using two PNe with T_e estimates and also find evidence for two distinct populations of PNe in NGC 5128, a 'bulge' and a 'halo' population.

Abundances from PNe appear to be a rich area for further study, but the results of Méndez *et al.* (2005) and Walsh *et al.* (2006) suggest that perhaps telescopes of the current generation are at their limits even for PNe in galaxies as close as NGC 5128 (3.8 Mpc) (Rejkuba *et al.* 2005).

3 Indirect abundance measurements

Given the difficulties of determining the abundances and abundance distributions star by star even in nearby early-type galaxies, another method is required. Analysis of the integrated light of these galaxies allows us to determine abundances and abundance patterns in entire stellar populations over a significant fraction of the history of the Universe, currently out to $z \sim 0.8$ (Jørgensen *et al.* 2005).

3.1 Stellar-population models

Here I give a short overview of the method, which involves applying stellar populations to strengths of absorption lines tuned to decouple age and metallicity (and abundance ratios) in the spectra. A more complete description can be found in Trager (2004); for details the reader is referred to Trager *et al.* (2000a), with updates in Trager *et al.* (2008). An unfortunate drawback of this method at present is its inability to determine MDFs; only mean quantities (weighted in a peculiar way described below) can be determined.

At first glance, it might seem natural to use optical broad-band colours to determine rough abundances of galaxies: metal-poor stellar systems are blue and metal-rich ones are red. This is, of course, because the increasing opacity from metals removes light from the blue and moves it into the red, particularly on the RGB, which provides at least half of the optical light of galaxies. Unfortunately, age has the same effect: older populations are cooler because they have lower mass on the RGB and are therefore redder. One might expect to improve the situation by using metallic absorption lines like the Mg b feature, the MgH triplet (Mg_2), or the Fe lines at 5270 Å and 5335 Å, which would allow a direct measurement of the metallicity. Unfortunately, these lines have the same problem as broad-band colours: they are formed in the atmospheres of the cool RGB stars and are sensitive to their temperatures – and are therefore also subject to variations in stellar-population age. This age–metallicity degeneracy (e.g. O'Connell 1980) was finally broken by Worthey (1994) and Buzzoni *et al.* (1994), who independently showed that a plot of Balmer-line strength (such as Hβ) as a function of metal-line strength could allow an independent measurement of stellar-population age and metallicity, a result first demonstrated by Rabin (1982). This is because the temperature of the main-sequence turn-off (MSTO) of a stellar population is more sensitive to age than to metallicity, and the Balmer lines of hydrogen are non-linearly sensitive to the temperature of the MSTO.

Stellar-population models that predict the metal- and Balmer-line strengths of stellar populations can then be built (e.g. Worthey 1994). Stellar-interior calculations, in the form of isochrones, are combined with stellar fluxes, determined either empirically or theoretically, and stellar absorption-line strengths, which are almost always determined empirically (but interpolation methods vary), to produce predicted line strengths as a function of stellar-population age and composition. Modern models (e.g. Trager *et al.* 2000a; Thomas *et al.* 2003) now allow for variations in abundance ratios like [α/Fe], since models based on line strengths of Solar-neighbourhood stars produce (for example) Mg b-line strengths too weak for a given $\langle$Fe$\rangle$ line strength relative to giant elliptical galaxies (e.g. Peterson 1976; O'Connell 1980; Peletier 1989; Worthey *et al.* 1992) (Figure 16.4).

3.2 Stellar-population analyses

Once stellar-population models are available, stellar-population ages, metallicities and abundance ratios can be read off diagrams like Figure 16.4. (Note, however,

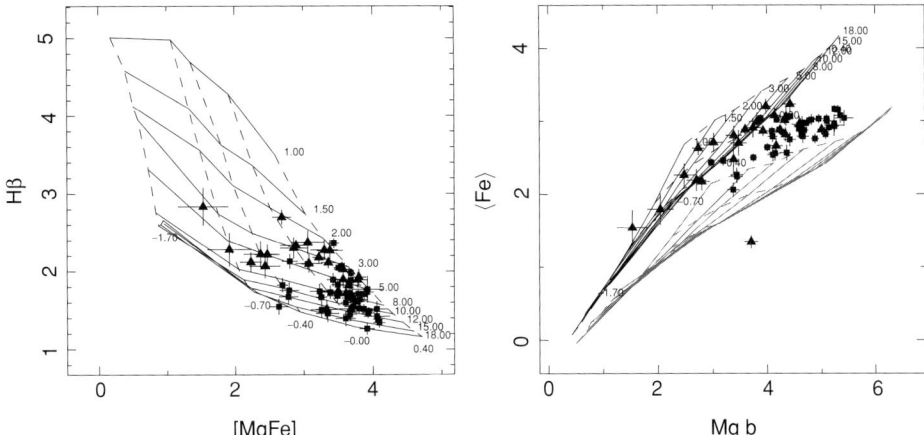

Figure 16.4. Stellar populations of nearby early-type (elliptical and S0) galaxies. Data for elliptical galaxies in the field are taken from González (1993), data for S0s in the field are from Fisher et al. (1996), and data for elliptical galaxies and S0s in the Fornax Cluster are from Kuntschner (2000). Data for ellipticals are squares and data for S0s are triangles. Model grids from Bruzual & Charlot (2003) as modified by Trager et al. (2008), see Serra & Trager (2007), for $[\alpha/\text{Fe}] \neq 0$ are over-plotted. Solid lines are for constant age; dashed lines are for constant metallicity. In the left panel, $[\text{MgFe}] = \sqrt{\text{Mg b} \times \langle \text{Fe} \rangle}$ is a metallicity indicator reasonably free of non-Solar-abundance-ratio effects (Thomas et al. 2003). This means that SSP-equivalent age and metallicity can be read from this panel. In the right panel, model grids have $[\alpha/\text{Fe}] = 0$ and $+0.3$ from left to right. This panel is used to estimate $[\alpha/\text{Fe}]$.

that stellar-population parameters are actually inferred from χ^2 minimisation of the model line strengths against the observed line strengths.) All early-type galaxies span a range in metallicity and age (Trager et al. 2000b), with the mean age being older in clusters than in the field (Thomas et al. 2005). Early-type galaxies in all environments typically have $-0.5 < [\text{Z}/\text{H}]_{\text{SSP}} < +0.5$ and $0 < [\alpha/\text{Fe}]_{\text{SSP}} < +0.4$ (depending exactly on the stellar-population model used). There appears to be no significant difference in metallicity between different environments (Thomas et al. 2005; Sánchez-Blázquez et al. 2006), although small differences in [C/Fe] or [N/Fe] are possible (Sánchez-Blázquez et al. 2006).

An important caveat must be understood when reading diagrams like Figure 16.4. The ages and compositions inferred from these diagrams are those of the equivalent single stellar population (SSP) (we call them the 'SSP-equivalent' parameters). That is, these are the ages and compositions that the objects would have *if* they were composed solely of a single population formed in a single burst at the SSP-equivalent age with a chemical composition given by the SSP-equivalent metallicity and abundance ratio(s). This means that the composite populations of real galaxies are treated in terms of their *line-strength-weighted mean* population parameters. In the case of

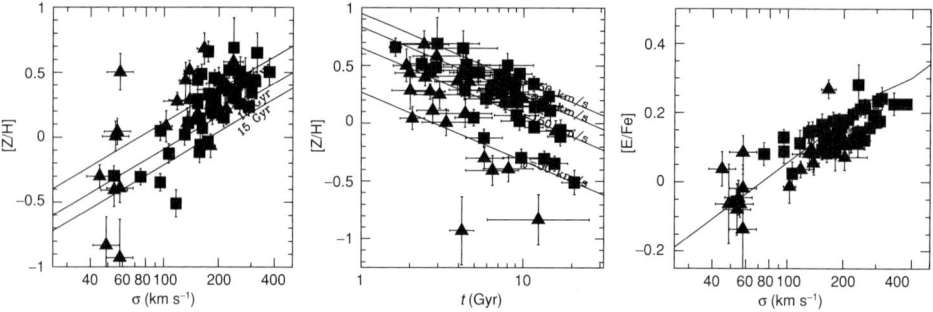

Figure 16.5. The variation of stellar-population parameters with age and velocity dispersion (σ). The plots are for, from left to right, the velocity-dispersion–metallicity plane, the age–metallicity plane, and the velocity-dispersion–abundance-ratio plane. Points are as in Figure 16.4. Stellar-population parameters were inferred from models of Worthey (1994) altered as described in Trager et al. (2006). Lines in the left and middle panels are projections of the Z-plane (Trager et al. 2000b). In the left panel, the lines correspond to ages of 5, 10 and 15 Gyr from top to bottom; in the middle panel, the lines correspond to $\sigma = 50, 150, 250$ and 350 km s^{-1}. The line in the right panel is the [E/Fe]–σ relation of Trager et al. (2000b).

M32, we can study what effect the MDF has on the inferred SSP-equivalent metallicity. The light-weighted mean metallicity from the MDF is [Fe/H] = -0.25, as can be inferred from Figure 16.3 (left). Using absorption-line measurement extrapolated (slightly) to precisely the same position in the galaxy, [Fe/H]$_{\rm SSP}$ = -0.32 (Trager et al. 2000b). This excellent agreement between the metallicities inferred from the integrated light and the CMD is possible only in the absence of a metal-poor tail in the MDF, since hot stars corrupt the abundances inferred from line strengths (Trager et al. 2005). More extensive analyses using synthetic star-formation histories and chemical-evolution models show that the inferred SSP-equivalent metallicities and [α/Fe] ratios are very close to their light-weighted or even mass-weighted mean values (Serra & Trager 2007; Trager & Somerville, in preparation). This is certainly *not* the case for the SSP-equivalent ages, which can easily be skewed towards young ages by the presence of very small amounts of young stars.

Using long-slit spectra of M32 and sophisticated stellar-population models that include in a limited way the variation of individual elements as a function of Fe abundance, Worthey (2004) finds [Fe/H] = $+0.02$, [C/Fe] = $+0.08$, [N/Fe] = -0.13 and [Mg/Fe] = -0.18 near the nucleus. This confirms the generally held notion that M32 has an abundance pattern very much like that of the Sun (O'Connell 1980; Grillmair et al. 1996; Trager et al. 2000a). The inferred nitrogen abundance appears to be too low compared with the nitrogen abundances of the PNe (Stasińska et al. 1998), but this could be due to self-enrichment of nitrogen in the PNe.

With this caveat in mind, the variations of stellar-population parameters with velocity dispersion (as a tracer of the mass) can be examined. The left panel of Figure 16.5 shows the mass–metallicity relation for field galaxies. That a narrower mass–metallicity relation is not apparent in this plot is due to the *real* anticorrelation between SSP-equivalent age and metallicity of galaxies in this sample (middle panel): older field galaxies have lower metallicities at fixed velocity dispersion. The dispersion in ages causes a broadening of the mass–metallicity relation. A narrow mass–metallicity relation is, however, seen clearly in cluster samples (e.g. Nelan *et al.* 2005; Trager *et al.* 2008), because these samples have small age spreads. At the same time, there is a clear mass–[α/Fe] relation in all environments, which is only weakly dependent on (Thomas *et al.* 2005; Nelan *et al.* 2005) or independent of (Trager *et al.* 2000b) SSP-equivalent age. The physical mechanisms driving these relations are likely to be a combination of rapid star formation, so that SNe II products dominate over SNe Ia products, and metal-enriched outflows, in which high-mass galaxies retain their SNe II products – which dominate the overall metallicity of the systems – more easily than low-mass galaxies do (Trager *et al.* 2000b); see also Chapters 28 and 44 in this volume.

3.3 Continuing annoyances

Nothing is easy, of course. Besides the effects of composite populations on the inferred stellar-population parameters and the fact that MDFs cannot (currently) be determined from the integrated light, two major annoyances persist when trying to infer abundances from absorption-line measurements. The first is the unknown oxygen abundances of early-type galaxies. The oxygen abundance controls the temperature of the MSTO (Salaris & Weiss 1998), and so the inferred age – and therefore metallicity – is dependent on the oxygen abundance. It is not yet clear whether a good tracer of oxygen abundance is available in the optical (but see below). The second is the hot-star content of these galaxies. Trager *et al.* (2005) have shown that these stars, such as BHB and blue straggler stars, not only corrupt age measurements for the oldest objects (e.g. Maraston & Thomas 2000) but also corrupt abundance measurements. Correcting for this effect requires excellent models and data with sensitivity in the blue, but this has been done for very few galaxies.

4 Summary

Early-type galaxies are dominantly metal-rich, of typically Solar metallicity, and usually enriched in SNe II products. Their compositions – both bulk metallicity and abundance ratios – depend on their mass and possibly on their environment. Their MDFs appear to violate 'simple' chemical-evolution models and therefore these galaxies suffer, like the Milky Way, from the G-dwarf problem. They appear to have complex star-formation histories: galactic winds and rapid star formation have apparently played crucial roles in their chemical evolution; and there is mounting evidence for recent star formation, suggesting that perhaps recent accretion or mergers are playing a role. Better understanding of their nucleosynthetic properties would be very helpful to disentangle these histories.

The future for understanding the nucleosynthetic histories of early-type galaxies is bright. Serven *et al.* (2005) have developed 54 new indices over a wide wavelength range in the optical that will help to determine C, N, O, Na, Mg, Al, Si, Ca, Sc, Ti, V, Cr, Mn, Fe, Co, Ni, Sr and Ba abundances. A significant computational effort is under way by Worthey and collaborators (including the author), the intention being to produce fully self-consistent stellar-interior and -atmosphere models (and emergent spectra) to take advantage of these indices and their predecessors, the Lick/IDS indices. At the same time, line-strength measurements are being pursued both with two-dimensional spectrographs like SAURON to study local galaxies (e.g. Kuntschner *et al.* 2006) and with multi-slit spectrographs on large telescopes to study very distant ones (e.g. Jørgensen *et al.* 2005). New observations of nearby early-type galaxies with the ACS on the HST (such as the Lauer *et al.* and Worthey *et al.* studies of M32) will greatly improve our knowledge of their MDFs (and ages). It is hoped that the next generation of large telescopes will enable star-by-star chemical studies of the nearest elliptical galaxies!

Acknowledgments

I thank the SOC and LOC, particularly Garik Israelian, for the kind invitation to review this subject and for their financial support that enabled my participation. I wish also to thank my collaborators, Alan Dressler, Sandy Faber, Wendy Freedman, Carl Grillmair, Tod Lauer, Ken Mighell, Paolo Serra, Rachel Somerville and Guy Worthey for allowing me to present results from ongoing studies. Thanks are due to Carl Grillmair and Marina Rejkuba for allowing me to reprint two figures each. Finally, I would like to thank Grażyna Stasińska for a very helpful discussion about the use of planetary nebulae as abundance indicators, Patricia Sánchez-Blázquez for several helpful conversations and Marina Rejkuba for reading a draft of the manuscript.

References

Bender, R., Burstein, D., & Faber, S. M. (1992), *ApJ* **399**, 462
Brodie, J. P., & Strader, J. (2006), *ARA&A* **44**, 193
Bruzual, G., & Charlot, S. (2003), *MNRAS* **344**, 1000
Buzzoni, A., Mantegazza, L., & Gariboldi, G. (1994), *AJ* **107**, 513
Davidge, T. J., & van den Bergh, S. (2001), *ApJ* **553** L133
Douglas, N. G., Arnaboldi, M., Freeman, K. C. *et al.* (2002), *PASP* **114**, 1234

Fisher, D., Franx, M., & Illingworth, G. (1996), *ApJ* **459**, 110
Fukugita, M., Hogan, C. J., & Peebles, P. J. E. (1998), *ApJ* **503**, 518
González, J. J. (1993), Ph.D. Thesis, University of California, Santa Cruz, CA
Gregg, M. D., Ferguson, H. C., Minniti, D., Tanvir, N., & Catchpole, R. (2004), *AJ* **127**, 1441
Greggio, L. (1997), *MNRAS* **285**, 151
Grillmair, C. J., Lauer, T. R., Worthey, G. *et al.* (1996), *AJ* **112**, 1975
Harris, G. L. H., & Harris, W. E. (2000), *AJ* **120**, 2423
Harris, W. E., & Harris, G. L. H. (2002), *AJ* **123**, 3108
Humphrey, P. J., & Buote, D. A. (2006), *ApJ* **639**, 136
Jørgensen, I., Bergmann, M., Davies, R., Barr, J., Takamiya, M., and Crampton, D. (2005), *AJ* **129**, 1249
Kuntschner, H. (2000), *MNRAS* **315**, 184
Kuntschner, H., Emsellem, E., Bacon, R. *et al.* (2006), *MNRAS* **369**, 497
Maraston, C., & Thomas, D. (2000), *ApJ* **541**, 126
Méndez, R. H., Thomas, D., Saglia, R. P., Maraston, C., Kudritzki, R. P. & Bender, R. (2005), *ApJ* **627**, 727
Nelan, J. E. Smith, R. J. Hudson, M. J., *et al.* (2005), *ApJ* **632**, 137
O'Connell, R. W. (1980), *ApJ* **236**, 430
Pagel, B. E. J. (1989), in *Evolutionary Phenomena in Galaxies*, eds. J. E. Beckman & B. E. J. Pagel (Cambridge, Cambridge University Press), p. 201
Peletier, R. F. (1989), Ph.D. Thesis, Rijksuniversiteit Groningen
Peterson, R. C. (1976), *ApJ* **210**, L123
Pietrinferni, A., Cassisi, S., Salaris, M., & Castelli, F. (2004), *ApJ* **612**, 168 (2006), *ApJ* **642**, 797
Rabin, D. (1982), *ApJ* **261**, 85
Rana, N. C. (1991), *AR&A* **29**, 129
Rejkuba, M., Greggio, L., Harris, W. E., Harris, G. L. H., & Peng, E. W. (2005), *ApJ* **631**, 262
Salaris, M., & Weiss, A. (1998), *A&A* **335**, 943
Salaris, M., Chieffi, A., & Straniero, O. (1993), *ApJ* **414**, 580
Sánchez-Blázquez, P., Gorgas, J., Cardiel, N., & González, J. J. (2006), *A&A* **457**, 809
Schechter, P. L., & Dressler, A. (1987), *AJ* **94**, 563
Schulte-Ladbeck, R. E., Drozdovsky, I. O., Belfort, M., & Hopp, U. (2003), *Ap&SS* **284**, 909
Serra, P., & Trager, S. C. (2007), *MNRAS* **374**, 769
Serven, J., Worthey, G., & Briley, M. M. (2005), *ApJ* **627**, 754
Stasińska, G., Richer, M. G., & McCall, M. L. (1998), *A&A* **336**, 667
Stasińska, G., Vílchez, J. M., Pérez, E., Corradi, R. L. M., Mampaso, A., & Magrini, L. (2005), *AIPC* **804**, 262
Stasińska, G., Vílchez, J. M., Pérez, E. *et al.* (2006), in *Planetary Nebulae Beyond the Milky Way*, ed. L. Stanghellini, J. R. Walsh & N. G. Douglas (Berlin, Springer-Verlag), pp. 234ff.
Thomas, D., Maraston, C., & Bender, R. (2003), *MNRAS* **339**, 897
Thomas, D., Maraston, C., Bender, R., & Mendes de Oliveira, C. (2005), *ApJ* **621**, 673
Tinsley, B. M. (1980), *FCPh* **5**, 287
Trager, S. C. (2004), in *Origin and Evolution of the Elements*, ed. A. McWilliam & M. Rauch (Cambridge, Cambridge University Press), pp. 388ff.
Trager, S. C., Faber, S. M., & Dressler, A. (2008), *MNRAS* in press
Trager, S. C., Faber, S. M., Worthey, G., & González, J. J. (2000a), *AJ* **119**, 1645

Trager, S. C., Faber, S. M., Worthey, G., & González, J. J. (2000b), *AJ* **120**, 165
Trager, S. C., Worthey, G., Faber, S. M., & Dressler, A. (2005), *MNRAS* **362**, 2
VandenBerg, D. A., Swenson, F. J., Rogers, F. J., Iglesias, C. A., & Alexander, D. R. (2000), *ApJ* **532**, 430
Walsh, J. R., Walton, N. A., Jacoby, G. H., & Peletier, R. F. (1999), *A&A* **346**, 753
Walsh, J. R., Jacoby, G., Peletier, R., & Walton, N. A. (2006), in *Planetary Nebulae Beyond the Milky Way*, ed. L. Stanghellini, J. R. Walsh & N. G. Douglas (Berlin, Springer-Verlag), pp. 262ff.
Worthey, G. (1994), *ApJS* **95**, 107
  (2004), *AJ* **128**, 2826
Worthey, G., Dorman, B., & Jones, L. A. (1996), *AJ* **112**, 948
Worthey, G., Faber, S. M., & González, J. J. (1992), *ApJ* **398**, 69

17

Measuring chemical abundances in extragalactic metal-rich H II regions

Fabio Bresolin
Institute for Astronomy, University of Hawaii, 2680 Woodlawn Drive, HI 96822, USA

1 Chemical abundances of metal-rich H II regions: why?

Ionized nebulae (H II regions) trace the sites of massive-star formation in spiral and irregular galaxies. The rapid evolution of these stars, ending in supernova explosions, and the subsequent recycling of nucleosynthesis products into the interstellar medium, make H II regions essential probes of the present-day chemical composition of star-forming galaxies across the Universe. The study of nebular abundances is therefore crucial for understanding the chemical evolution of galaxies. In the following pages I will provide an optical astronomer's perspective on some of the issues concerning the measurement of abundances in metal-rich H II regions, by focusing on the observational difficulties that are peculiar to the high-metallicity regime, discussing some of the most recent abundance determinations from H II regions in the metal-rich zones of spiral galaxies, and indicating some possibilities for further progress. Throughout this paper I will use the oxygen abundance as a proxy for the metallicity (oxygen makes up roughly half of the metal content of the interstellar medium), and assume the Solar value from Asplund *et al.* (2004), $12 + \log(O/H)_\odot = 8.66$. Elements besides oxygen will not be discussed in great detail.

1.1 Motivations

Why measure abundances of metal-rich H II regions? After all, as we will see in Section 2, metal-rich H II regions pose difficulties to the observer that are not present at lower metallicities, i.e. roughly below half the Solar O/H value. However, high abundances are encountered in a variety of astrophysical contexts, and the study of ionized nebulae often provides the only way to measure these abundances. Here

The Metal-rich Universe, eds. G. Israelian and G. Meynet. Published by Cambridge University Press.
© Cambridge University Press 2008.

are a few examples that motivate detailed studies of metal-rich H II regions, aimed at improving our abundance diagnostics.

(a) Galactic abundance gradients have been known in spiral galaxies since the pioneering work carried out in the 1970s; see the compilations by Vila-Costas & Edmunds (1992) and Zaritsky *et al.* (1994). Even after the recent downward revision of the metallicity in metal-rich H II regions from empirical methods (Bresolin *et al.* 2004), the central parts of spiral galaxies reach or exceed the Solar metallicity (Pilyugin *et al.* 2004, 2006a).

(b) Chemical-evolution models of galaxies aim at explaining the observed abundance gradients and abundance ratios on the basis of assumptions about the stellar yields and initial mass function, the star-formation rate and efficiency, the presence of gas inflows and outflows, etc. (Henry *et al.* 2000; Chiappini *et al.* 2003; Carigi *et al.* 2005). There is a major uncertainty in the input data regarding the metallicity of the inner regions of spiral galaxies.

(c) The luminosity–metallicity relation of local ($z < 0.1$) star-forming galaxies depends critically on the methods adopted for measuring abundances in nearby H II regions, in particular at the high-metallicity end (Salzer *et al.* 2005). A similar uncertainty exists for more remote galaxies ($z < 0.9$) (Kobulnicky & Kewley 2004; Maier *et al.* 2005). The finding that the brightest (and more massive) galaxies exceed the Solar metallicity by a factor of a few, namely up to $12 + \log(O/H) = 9.2$ is dependent on the calibration of strong-line methods, such as those discussed in Section 3.

(d) The same conclusion can be drawn when considering the mass–metallicity relation for low- and high-redshift ($z \leq 3$) star-forming galaxies (Tremonti *et al.* 2004; Savaglio *et al.* 2005; Erb *et al.* 2006).

(e) Solar-like metallicities are measured (via strong-line methods) at the largest redshifts so far investigated (Shapley *et al.* 2004); see Chapter 20 in this volume. Once more, the resulting abundances and the implications for the chemical evolution of early galaxies depend on the accuracy of local calibrations of strong-line methods.

(f) The number ratio of Wolf–Rayet stars of type WC to those of type WN is predicted to be an increasing function of the stellar metallicity (Meynet & Maeder 2005); see Chapter 29 in this volume. We have observational confirmations of this prediction (Massey 2003; Crockett *et al.* 2006), although the results in the high-metallicity regime are somewhat fuzzy due to uncertainties in measuring abundances for the Wolf–Rayet parent galaxies.

(g) Although there have been claims that the upper limit of the stellar initial mass function of the ionizing clusters of H II regions might decrease with increasing abundance (Thornley *et al.* 2000), other authors suggest that this is not the case (Pindao *et al.* 2002). Obtaining accurate abundances at the metal-rich end is critical for the reliability of this conclusion.

(h) A metallicity dependence of the Cepheid period–luminosity relation is possible, although the issue is far from being settled. Sakai *et al.* (2004) showed how the distances derived to metal-rich galaxies are sensitive on the method adopted to measure the metallicity from H II regions.

2 Difficulties at high metallicity

Obtaining a *direct* measurement of chemical abundances in H II regions requires that we have a good idea of the value of the electron temperature, T_e. The strengths of the forbidden emission lines (collisionally excited) from various metal ions commonly detected in nebular spectra are strongly sensitive to T_e, besides being sensitive to the abundance of the originating ionic species, $N(X^{+i})$:

$$I_\lambda \propto N(X^{+i})\epsilon_\lambda \tag{17.1}$$

since the line emissivity ϵ_λ has essentially an exponential dependence on T_e:

$$\epsilon_\lambda \propto \Omega(T) T_e^{-0.5} e^{-\chi/(k T_e)}. \tag{17.2}$$

Taking the ratio of forbidden lines corresponding to atomic transitions that originate from widely separated energy levels, such that their relative population is sensitive to the electron temperature, provides a measure of T_e. One uses the ratio of the *auroral* lines (transitions from the second-lowest excited level to the lowest excited level) to the corresponding *nebular* lines (transitions from the first excited level to the ground level). The auroral lines are generally quite faint, and relatively high temperatures are required to excite enough atoms via collisions to the energy levels from which these lines originate. The case of oxygen is probably the most well known; specifically, for [O III] one uses the ratio of $\lambda 4363$ (auroral) to $\lambda\lambda 4959$, 5007 (nebular; the ground level is actually split by spin–spin interactions, hence the two nebular lines).

The heart of the problem in the high-metallicity nebular business is illustrated in Figure 17.1. An increase in the metallicity has the effect of increasing the cooling in the nebula, which occurs mostly via line emission from metals, mostly oxygen, first through the [O III] optical forbidden lines and, at higher abundances (lower temperatures), through the far-IR [O III] hyperfine transitions at $\lambda\lambda 52, 88\,\mu\mathrm{m}$. Figure 17.1 shows that as one moves to low temperatures (say $T_e < 8{,}000\,\mathrm{K}$) the [O III] $4363/(4959 + 5007)$ line ratio soon becomes very small, due to its exponential dependence. In fact, for extragalactic work the ratio hits the typical limit imposed by the capabilities of current 8–10-m telescopes (represented by the horizontally streched rectangle in Figure 17.1) already for metallicities well below the Solar value (notice the position of the Solar symbol on the dotted curve, around $6{,}000\,\mathrm{K}$).

Qualitatively the picture remains the same when considering auroral-to-nebular line ratios for other ions; however, we can consider cases in which these ratios are larger, at a given temperature, than in the case of [O III]. This happens for ions for which the upper level (from which the auroral transition originates) lies at a lower energy above the ground level with respect to the [O III] $\lambda 4363$ case (5.3 eV), for example for [N II] $\lambda 5755$ (4.0 eV) and [S III] $\lambda 6312$ (3.4 eV). Observationally

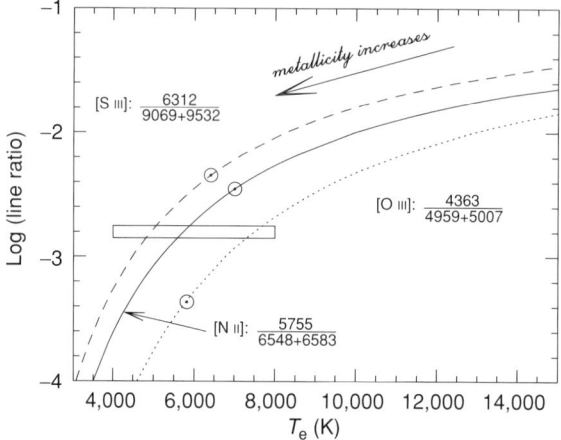

Figure 17.1. Dependences of commonly used auroral-to-nebular line ratios on electron temperature T_e. The arrow points in the direction of increasing metallicity (increased cooling results in lower T_e). In the high-metallicity regime, T_e is typically below 8,000 K. The Solar symbols on each of the three curves indicate the line ratios observed for extragalactic H II regions of approximately Solar metallicity. The open rectangle represents the current observational limit reached with an 8-m-class telescope. Although there is no hope of being able to measure electron temperatures of metal-rich nebulae from observations of the [O III] λ4363 auroral line, it is possible to derive T_e (and therefore abundances) using the auroral lines [S III] λ6312 and [N II] λ5755 even at metallicities above Solar.

this situation creates a big advantage: even at supersolar metallicities ratios such as [S III] 6312/(9069 + 9532) and [N II] 5755/(6548 + 6583) are measurable in bright extragalactic H II regions with current instrumentation. The Solar symbols on the [S III] and [N II] curves, representing the typical line ratios measured for Solar-metallicity extragalactic H II regions (e.g. Bresolin *et al.* 2005) lie well above the typical observational limit of 8–10-m telescopes. Note that some of the coolest extragalactic nebulae for which a direct T_e determination exists have temperatures in the range 5,000–6,000 K.

2.1 Caveat 1: assumptions on the temperature structure of ionized nebulae

The analysis of the chemical composition of extragalactic H II regions on the basis of collisionally excited lines commonly simplifies the structure of the nebulae by regarding them as composed of spherically symmetric, concentric shells, centered on the ionizing source. Different ionic species dominate the emission in different shells. Typically the [O III] and [Ne III] lines are assumed to originate from a central, high-excitation zone, whereas the [O II], [N II], and [S II] lines are emitted from the low-excitation outer region. An intermediate-excitation zone is sometimes introduced for [S III] and [Ar III]. Each of these emitting zones is characterized by particular ionic temperatures, e.g. $T(O^{++})$, $T(O^+)$, and so forth, which can differ from the line temperatures measured from the auroral-to-nebular line ratios, such as $T[\text{O{\sc iii}}]$, $T[\text{O{\sc ii}}]$, etc.

Photoionization models provide useful relationships between the temperatures in the various zones (Stasińska 1978). For example, Garnett (1992) found

$$T[\text{S{\sc iii}}] = 0.83\, T[\text{O{\sc iii}}] + 1{,}700\,\text{K} \tag{17.3}$$

and

$$T[\text{O{\sc ii}}] = T[\text{N{\sc ii}}] = T[\text{S{\sc ii}}] = 0.70\, T[\text{O{\sc iii}}] + 3{,}000\,\text{K}. \tag{17.4}$$

Similar relations have been found by other authors (Campbell *et al.* 1986; Pilyugin *et al.* 2006b). Since at high metallicity, namely approximately $12 + \log(\text{O/H}) > 8.5$, the [O{\sc iii}] temperature cannot be measured directly using the $\lambda 4363$ auroral line, one can then calculate T[O{\sc iii}] from either T[S{\sc iii}] or T[N{\sc ii}], or some clever combination of the available temperatures. It is important to remember that virtually all [O{\sc iii}] ionic abundances measured in high-metallicity extragalactic H{\sc ii} regions are derived using this technique. Real ionized nebulae have naturally a much more complicated structure than is assumed in this stratification model; see the study of NGC 588 in M33 by Jamet *et al.* (2005). Some reassurance on the validity of the assumed temperature relations for real nebulae comes from the works of Kennicutt *et al.* (2003) on M101 and Bresolin *et al.* (2004) on M51.

3 Strong-line methods

Often we cannot rely on the direct method described in Section 2 to measure nebular abundances. This occurs whenever auroral lines cannot be detected, for example in the case of the spectrum of a cool H{\sc ii} region of very high metallicity, or in the case of the spectrum of a faint, high-redshift galaxy, where only the brightest emission lines can be seen. Among the various strong-line methods, the R_{23} indicator originally proposed by Pagel *et al.* (1979) stands out as arguably the most popular. For in-depth discussions on this and additional methods, which will be discussed only marginally here, the reader should consult the papers by Kewley & Dopita (2002) and Pérez-Montero & Díaz (2005).

In the Pagel *et al.* method, one considers the two most important stages of ionization of oxygen, both emitting collisionally excited lines in the visually accessible range of the spectrum:

$$R_{23} = ([\text{O{\sc ii}}]\lambda 3727 + [\text{O{\sc iii}}]\lambda\lambda 4959, 5007)/\text{H}\beta.$$

Like all strong-line diagnostics, R_{23} as an abundance indicator has a statistical value based on the fact that the hardness of the ionizing radiation correlates with metallicity. Considering H{\sc ii} regions of increasing metallicity (lower temperature), the optical [O III] lines become progressively fainter as the emission shifts to the far-IR fine-structure lines.

Calibrations of R_{23} in terms of the nebular chemical composition abound in the literature. Many are based on grids of photoionization models (Edmunds & Pagel 1984; McGaugh 1991; Kewley & Dopita 2002). Empirical calibrations are based on abundances derived from auroral lines (Pilyugin 2001). The end-user of any of these calibrations should always keep in mind the systematic abundance differences of up to 0.5 dex that one obtains from the same input emission-line fluxes using different calibrations.

3.1 Empirical calibrations from deep spectroscopy of extragalactic H II regions

Necessary input data for empirical calibrations of strong-line methods are obviously the direct abundances obtained from detailed studies of high-metallicity H II regions focusing on the inner zones of giant spiral galaxies. Early works on the abundances of high-metallicity H II regions from auroral lines include those of Kinkel & Rosa (1994) on the region Searle 5 in M101, the anchor point at high metallicity of the early Pagel et al. (1979) calibration, Castellanos et al. (2002) on CDT1 in NGC 1232, and Kennicutt et al. (2003) on Searle 5 and H1013 in M101. The availability of 8-m-class telescopes in recent years has expanded the field considerably. There are now direct abundance determinations for about a dozen H II regions in M51 (Bresolin et al. 2004; Garnett et al. 2004b), and a larger sample in five southern spirals has been studied by Bresolin et al. (2005).

Figure 17.2 shows the relation between R_{23} and the nebular abundance derived from the availability of auroral lines, essentially [S III] $\lambda 6312$ and [N II] $\lambda 5755$, as described earlier. The data points have been taken from the literature, emphasizing with larger symbols the extension to low R_{23} values (higher abundance) derived from observations with 8-m-class telescopes. The double-valued nature of the R_{23} indicator is apparent; the degeneracy can be removed through the availability of emission-line diagnostics that are monotonic with the oxygen abundance (e.g. [N II] $\lambda 6583/H\alpha$). The scatter in the data shows the sensitivity of R_{23} to additional parameters besides the chemical abundance, such as the ionization parameter and the effective temperature of the ionizing stars (Pérez-Montero & Díaz 2005). Note that nebular abundances still hover around the Solar value even at the smallest observed R_{23}, and only a handful of points appear to be more metal-rich than the Solar value. In addition, the curves in Figure 17.2 clearly illustrate the well-known discrepancy between R_{23} calibrations obtained empirically, here taken from the P-method of Pilyugin & Thuan (2005), from auroral line observations and those obtained from photoionization-model grids; here Kewley & Dopita (2002) is taken

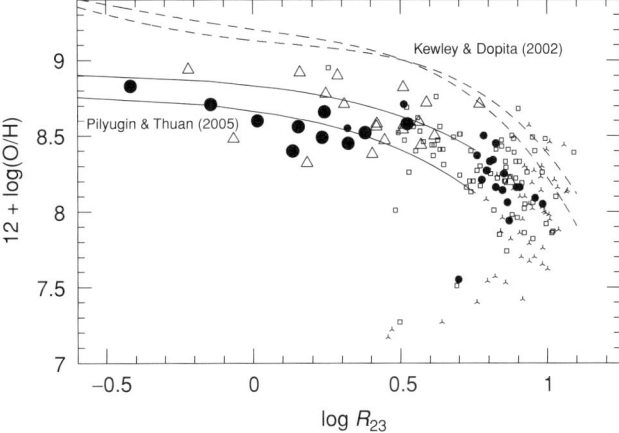

Figure 17.2. This abundance versus R_{23} diagram shows observational data for extragalactic H II regions, for which the O/H abundances have been derived from the detection of auroral lines (the data points constitute an incomplete collection drawn from various sources in the literature). The high-metallicity samples of Bresolin *et al.* (2004, 2005) are data shown with enlarged symbols (full circles and triangles). The small full circles are data from the M101 sample studied by Kennicutt *et al.* (2003). The curves show the calibrations of the R_{23} indicator derived empirically by Pilyugin & Thuan (2005) and from photoionization-model grids by Kewley & Dopita (2002). In both cases two curves are drawn, corresponding to two different values of the ionization parameter.

as a representative example. This discrepancy amounts to approximately 0.3 dex, and its complete explanation still eludes us.

3.2 Implications of the recent empirical results

The direct abundances obtained recently for M51 and other metal-rich galactic environments have important consequences for the derivation of chemical abundances from strong-line methods. These results in fact concern most of the investigations mentioned in the introduction. For example, when looking at the radial abundance gradients in spiral galaxies we see a decrease in the central O/H values, by up to a factor of three, relative to the results based on older R_{23} calibrations, or in general those based on photoionization-model grids (Kennicutt *et al.* 2003). The abundance gradients consequently become flatter. The central abundances of spirals regarded as being among the most metal-rich in our neighborhood (such as M51, NGC 3351, and M83) are basically Solar, or exceed the Solar value by just a small amount, lying in the range $12 + \log(\text{O/H}) = 8.60-8.75$ (Pilyugin *et al.* 2006a). Values up to three times Solar had been claimed earlier for some of these galaxies. It must be pointed out that when considering the luminosity–metallicity and mass–metallicity relationships for star-forming galaxies the adoption of different calibrations has the effect of modifying the zero points, but the general trends remain essentially unaltered (Ellison & Kewley 2005).

3.3 Other strong-line methods

It should be at least briefly mentioned here that there are other strong-line methods besides R_{23} that can be used to determine abundances of high-metallicity star-forming regions. They rely on emission lines that can generally be measured even in low signal-to-noise-ratio nebular spectra, such as [N II] $\lambda 6583$ and [S II] $\lambda\lambda 6717, 6731$. The accuracy of the derived O/H measurements should always be regarded as being at the 0.2–0.3-dex level. Among the most useful methods for the analysis of high-redshift star-forming galaxies is N2 = [N II] $\lambda 6583/\text{H}\alpha$ (Denicoló *et al.* 2002; Pettini & Pagel 2004), because the emission lines involved are very close in wavelength, and can be observed with the same setup when moving for example to the infrared K band, where the rest-frame spectral lines end up at redshift $z \sim 2$. Among the drawbacks affecting N2 are the saturation above the Solar abundance and the sensitivity to the chemical history of nitrogen.

The sulfur-line-based

$$S_{23} = ([\text{S II}]\, \lambda\lambda 6717, 6731 + [\text{S III}]\, \lambda\lambda 9069, 9532)/\text{H}\beta$$

(Vílchez & Esteban 1996) has been calibrated in terms of both oxygen and sulfur abundances (Pérez-Montero *et al.* 2006). The advantage of this indicator derives from its small sensitivity to the ionization parameter compared with that of R_{23}, but its usefulness is limited to redshifts $z < 0.1$. Stasińska (2006) has recently introduced abundance indicators based on high-excitation lines only, [Ar III] $\lambda 7135$/[O III] $\lambda 5007$ and [S III] $\lambda 9069$/[O III] $\lambda 5007$, that alleviate one potential problem of indicators involving low-excitation lines, such as N2, namely the fact that the latter lines can be produced both in the diffuse galactic interstellar medium and in the star-forming regions.

3.4 Caveat 2: temperature gradients

Metal-rich H II regions are not isothermal clouds of ionized gas: both large-scale temperature gradients and small-scale fluctuations are predicted to exist. The presence of large-scale T_e gradients is a consequence of the efficient cooling of the inner regions via far-IR [O III] fine-structure lines (there are no corresponding fine-structure [O II] lines). Strong gradients are expected to develop above the Solar metallicity, with the inner region cooler by several thousands of degrees than the outer zones (Stasińska 1980; Garnett 1992). This has important consequences for the chemical analysis. Owing to the exponential dependence of the line emissivities on T_e, the observed line strengths are weighted strongly toward warmer regions. At high metallicity the optical [O III] collisionally excited lines originate therefore not in the cooler, high-excitation zone, but instead in the warmer O^+ zone. As a consequence, using the auroral-to-nebular line-ratio scheme to derive T_e leads one to over-estimate the ionic temperature: T[O III] $> T(O^{++})$. Under these conditions, the oxygen abundance is under-estimated, possibly by large factors (Stasińska 2005).

4 Seeking independence from the temperature dependence: metal recombination lines

How reliable are the chemical abundances derived from the auroral-line method at high metallicity? Are the effects of temperature gradients measurable? There are several ways to carry out these tests. Using objects other than H II regions to trace galactic abundance gradients is becoming feasible with high-signal-to-noise-ratio spectroscopy of planetary nebulae and luminous stars in nearby galaxies. As an example, the blue supergiant metallicities derived by Urbaneja *et al.* (2005) in M33 are in good agreement with the H II-region abundances calculated by Vílchez *et al.* (1988) and Crockett *et al.* (2006). Detailed photoionization modeling also offers a possible route; however, as mentioned earlier, current discrepancies between empirical and model-based abundances are not fully understood. A further approach envisions the analysis of nebular lines that depend far less than the collisionally excited lines on the electron temperature, such as optical metal recombination lines and IR fine-structure lines. In both cases the line emissivity is only moderately dependent on T_e, even though it can be a strong function of the gas density. For example, considering Equation (17.2) for fine-structure transitions, in which the excitation potentials are very small (in the range 0.01–0.04 eV, versus 2–3 eV for optical transitions), the exponential term becomes virtually unity, and only a moderate T_e dependence is left.

The use of [O III] $\lambda\lambda 52, 88$ μm to derive abundances in extragalactic H II regions is limited to only a few cases in the literature. Garnett *et al.* (2004) have used ISO spectra of region CCM10 in M51 to derive $12 + \log(O/H) = 8.8$, which compares well to $12 + \log(O/H) = 8.6$ obtained by Bresolin *et al.* (2004) on the basis of auroral-line analysis. Additional far-IR studies include those of H II regions in the Milky Way by Peeters *et al.* (2002), Martín-Hernández *et al.* (2002), and Rudolph *et al.* (2006). Alternatively, it is also possible to use the mid-IR, accessible to Spitzer, to study the [Ne II] $\lambda 12.8$ μm and [Ne III] $\lambda 15.6$ μm lines, as done for M33 by Willner & Nelson-Patel (2002).

The emissivity of recombination lines from metals is also only weakly dependent on T_e: $\epsilon_\lambda \sim \alpha_{\text{eff}}(\lambda) \sim T_e^{-1}$. When measuring abundances relative to hydrogen, the T_e dependence becomes negligible, since H recombination lines have a similar sensitivity to T_e. The main drawback in their use for abundance studies is their weakness. Among the strongest metal recombination lines are the O II multiplet at 4651 Å, and C II λ4267. These lines are beautifully detected in the high-resolution spectra of some well-known H II regions in the Milky Way, such as the Orion nebula (Esteban *et al.* 2004) and the Triffid nebula (García-Rojas *et al.* 2006). Detection in extragalactic nebulae can still be challenging, and results for only one relatively high-metallicity, namely $12 + \log(\text{O/H}) = 8.5$, H II region, NGC5461 in M101 (Esteban *et al.* 2002), have been published so far.

In general, it is found that oxygen abundances obtained from recombination lines are 0.2–0.3 dex larger than those derived from collisionally excited lines under the assumption (tacitly implied in the discussion so far) that temperature fluctuations are absent in H II regions. A similar result is obtained when comparing O/H derived from a combination of optical and far-IR [O III] lines or from optical data alone (Jamet *et al.* 2005). This systematic difference can be explained by invoking the existence of temperature fluctuations in H II regions, as argued by M. Peimbert and collaborators in a series of papers; see Peimbert *et al.* (2007) for a recent review. With mean-square temperature fluctuations t^2 in the 0.02–0.06 range, one can reconcile the abundances from collisionally excited lines with those from recombination lines for the H II regions for which both sets of emission lines have been analyzed.

Figure 17.3 shows the same diagram as Figure 17.2, with the addition of Galactic and extragalactic H II regions for which a comparison between abundances obtained from collisionally excited lines and from metal recombination lines has been carried out so far. The sources can be found in Peimbert & Peimbert (2005) and Peimbert *et al.* (2007). These objects populate the diagram in its upper branch only down to $\log R_{23} \simeq 0.5$. Efforts to extend this sample to higher metallicities with high-resolution spectroscopy at 8–10-m telescopes are under way. Bresolin (2007) recently measured the C II λ4267 line, as well as the mean-square temperature fluctuation $t^2 = 0.06$, in H1013, an H II region in the inner, metal-rich parts of M101. The oxygen abundance derived within the scheme developed originally by Peimbert (1967) and Peimbert & Costero (1969) is $12 + \log(\text{O/H}) = 8.9$ (1.7 × Solar), about 0.3 dex larger than the value obtained without accounting for temperature fluctuations. These two values are represented by the open and solid square symbols in Figure 17.3.

The results obtained from recombination lines are in good agreement with the R_{23} calibration derived from photoionization models, as Figure 17.3 shows. This is a consequence of the fact that, even though photoionization models fail at reproducing the intensity of an auroral line like [O III] λ4363, calling for an as-yet-unidentified

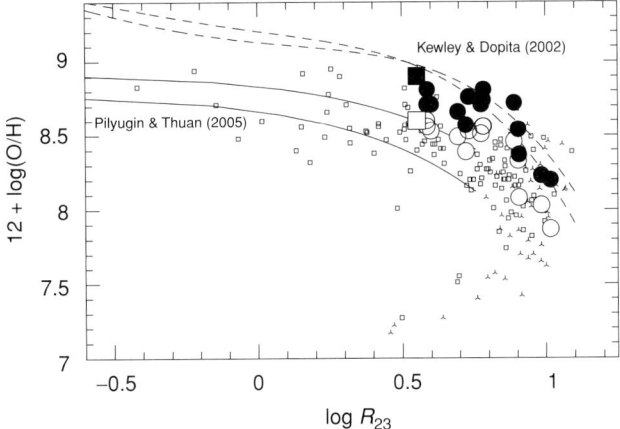

Figure 17.3. The same as Figure 17.2, with the observational data points shown with small symbols. The larger symbols represent Galactic and extragalactic H II regions for which abundances have been derived both from collisionally excited lines (open symbols) and from metal recombination lines (full symbols); the sources are specified in Peimbert *et al.* (2007). Data for the H II region H1013 in M101 studied by Bresolin (2007) are shown by squares.

heating source (Stasińska & Schaerer 1999), the predictions for the strong lines [O II] λ3727 and [O III] λλ4959, 5007 are less sensitive to the temperature structure of the nebulae. The virtual lack of dependence of the abundances obtained from metal recombination lines on T_e makes them more suitable to provide an accurate calibration of strong-line methods, such as R_{23}, than collisionally excited lines. The extension of the recombination-line measurements to the high-metallicity regime then becomes imperative.

References

Asplund, M., Grevesse, N., Sauval, A. J., Allende Prieto, C., & Kiselman, D. (2004), *A&A* **417**, 751
Bresolin, F. (2007), *ApJ* **656**, 186
Bresolin, F., Garnett, D. R., & Kennicutt, R. C. Jr. (2004), *ApJ* **615**, 228
Bresolin, F., Schaerer, D., González Delgado, R. M., & Stasińska, G. (2005), *A&A* **441**, 981
Campbell, A., Terlevich, R., & Melnick, J. (1986), *MNRAS* **223**, 811
Carigi, L., Peimbert, M., Esteban, C., & García-Rojas, J. (2005), *ApJ* **623**, 213
Castellanos, M., Díaz, A. I., & Terlevich, E. (2002), *MNRAS* **329**, 315
Chiappini, C., Romano, D., & Matteucci, F. (2003), *MNRAS* **339**, 63
Crockett, N. R., Garnett, D. R., Massey, P., & Jacoby, G. (2006), *ApJ* **637**, 741
Denicoló, G., Terlevich, R., & Terlevich, E. (2002), *MNRAS* **330**, 69
Edmunds, M. G., & Pagel, B. E. J. (1984), *MNRAS* **211**, 507
Ellison, S. L., & Kewley, L. J. (2005), astro-ph/0508627
Erb, D. K., Shapley, A. E., Pettini, M., Steidel, C. C., Reddy, N. A., & Adelberger, K. L. (2006), *ApJ* **644**, 813

Esteban, C., Peimbert, M., García-Rojas, J., Ruiz, M. T., Peimbert, A., & Rodríguez, M. (2004), *MNRAS* **355**, 229
Esteban, C., Peimbert, M., Torres-Peimbert, S., & Rodríguez, M. (2002), *ApJ* **581**, 241
García-Rojas, J., Esteban, C., Peimbert, M. *et al.* (2006), *MNRAS* **368**, 253
Garnett, D. R. (1992), *AJ* **103**, 1330
Garnett, D. R., Edmunds, M. G., Henry, R. B. C., Pagel, B. E. J., & Skillman, E. D. (2004a), *AJ* **128**, 2772
Garnett, D. R., Kennicutt, R. C. Jr., & Bresolin, F. (2004b), *ApJL* **607**, L21
Henry, R. B. C., Edmunds, M. G., & Koppen, J. (2000), *ApJ* **541**, 660
Jamet, L., Stasińska, G., Pérez, E., González Delgado, R. M., & Vílchez, J. M. (2005), *A&A* **444**, 723
Kennicutt, R. C. Jr., Bresolin, F., & Garnett, D. R. (2003), *ApJ* **591**, 801
Kewley, L. J., & Dopita, M. A. (2002), *ApJS* **142**, 35
Kinkel, U., & Rosa, M. R. (1994), *A&A* **282**, L37
Kobulnicky, H. A., & Kewley, L. J. (2004), *ApJ* **617**, 240
Maier, C., Lilly, S. J., Carollo, C. M., Stockton, A., & Brodwin, M. (2005), *ApJ* **634**, 849
Martín-Hernández, N. L., Peeters, E., Damour, F. *et al.* (2002), *Revista Méxicana de Astronomía y Astrofísica*, **12**, 41
Massey, P. (2003), *ARAA* **41**, 15
McGaugh, S. S. (1991), *ApJ* **380**, 140
Meynet, G., & Maeder, A. (2005), *A&A* **429**, 581
Pagel, B. E. J., Edmunds, M. G., Blackwell, D. E., Chun, M. S., & Smith, G. (1979), *MNRAS* **189**, 95
Peeters, E., Martín-Hernández, N. L., Damour, F. *et al.* (2002), *A&A* **381**, 571
Peimbert, M. (1967), *ApJ* **150**, 825
Peimbert, M., & Costero, R. (1969), *Boletin de los Observatorios Tonantzintla y Tacubaya*, **5**, 3
Peimbert, A., & Peimbert, M. (2005), *Revista Méxicana de Astronomía y Astrofísica*, **23**, 9
Peimbert, M., Peimbert, A., Esteban, C. *et al.* (2007), *Revista Méxicana de Astronomía y Astrofísica*, **29**, 72
Pérez-Montero, E., & Díaz, A. I. (2005), *MNRAS* **361**, 1063
Pérez-Montero, E., Díaz, A. I., Vílchez, J. M., & Kehrig, C. (2006), *A&A* **449**, 193
Pettini, M., & Pagel, B. E. J. (2004), *MNRAS* **348**, L59
Pilyugin, L. S. (2001), *A&A* **369**, 594
Pilyugin, L. S., & Thuan, T. X. (2005), *ApJ* **631**, 231
Pilyugin, L. S., Vílchez, J. M., & Contini, T. (2004), *A&A* **425**, 849
Pilyugin, L. S., Thuan, T. X., & Vílchez, J. M. (2006a), *MNRAS* **367**, 1139
Pilyugin, L. S., Vílchez, J. M., & Thuan, T. X. (2006b), *MNRAS* **370**, 1928
Pindao, M., Schaerer, D., González Delgado, R. M., & Stasińska, G. (2002), *A&A* **394**, 443
Rudolph, A. L., Fich, M., Bell, G. R. *et al.* (2006), *ApJS* **162**, 346
Sakai, S., Ferrarese, L., Kennicutt, R. C. Jr., & Saha, A. (2004), *ApJ* **608**, 42
Salzer, J. J., Lee, J. C., Melbourne, J., Hinz, J. L., Alonso-Herrero, A., & Jangren, A. (2005), *ApJ* **624**, 661
Savaglio, S., Glazebrook, K., Le Borgne, D. *et al.* (2005), *ApJ* **635**, 260
Shapley, A. E., Erb, D. K., Pettini, M., Steidel, C. C., & Adelberger, K. L. (2004), *ApJ* **612**, 108
Stasińska, G. (1978), *A&A* **66**, 257
  (1980), *A&A* **85**, 359
  (2005), *A&A* **434**, 507
  (2006), *A&A* **454**, L127

Stasińska, G., & Schaerer, D. (1999), *A&A* **351**, 72
Thornley, M. D., Schreiber, N. M., Förster, L. D. *et al.* (2000), ApJ **539**, 641
Tremonti, C. A., Heckman, T. M., Kauffmann, G. *et al.* (2004), *ApJ* **613**, 898
Urbaneja, M. A., Herrero, A., Kudritzki, R.-P., *et al.* (2005), *ApJ* **635**, 311
Vila-Costas, M. B., & Edmunds, M. G. (1992), *MNRAS* **259**, 121
Vílchez, J. M., & Esteban, C. (1996), *MNRAS* **280**, 720
Vílchez, J. M., Pagel, B. E. J., Díaz, A. I., Terlevich, E., & Edmunds, M. G. (1988), *MNRAS* **235**, 633
Willner, S. P., & Nelson-Patel, K. (2002), *ApJ* **568**, 679
Zaritsky, D., Kennicutt, R. C. Jr., & Huchra, J. P. (1994), *ApJ* **420**, 87

18

On the maximum oxygen abundance in metal-rich spiral galaxies

Jose M. Vílchez,[1] Leonid Pilyugin[2] & Trinh X. Thuan[3]

[1] *Instituto de Astrofísica de Andalucía (CSIC), Apartado Postal 3004, 18080 Granada, Spain*
[2] *Main Astronomical Observatory of the National Academy of Sciences of Ukraine, 03680 Kiev, Ukraine*
[3] *Astronomy Department, University of Virginia, P.O. Box 400325, Charlottesville, VA 22904, USA*

We discuss recent results based on our ongoing work on the study of the chemical abundances in the central part of spiral galaxies. A robust technique has been used to extrapolate the derived radial abundance gradients of oxygen to the center of the respective galaxies, taking into account the recent ff relation of Pilyugin (2005) and the new model-independent correction for electron-temperature structure within the H II regions as well as the contribution of a possible oxygen depletion by dust grains. In this way, a typical value for the expected maximum O/H abundance in spiral galaxies is derived. Implications of this result for the metallicity–luminosity relation and for the chemical abundances derived for high-redshift-galaxy samples are briefly discussed.

1 Motivation

Metallicity is one of the fundamental parameters governing the evolution of galaxies. The main aim of this conference has been the study of the metal-rich Universe; but what exactly is meant by "metal-rich" in our Universe of observed galaxies remains a key question. In the case of spiral galaxies it is well known that they exhibit radial abundance gradients (e.g. Pagel *et al.* 1980; McCall *et al.* 1985; Vílchez *et al.* 1988; Garnett 1989; Pilyugin *et al.* 2004 and references therein). Though there is a collection of results dealing with the subject of gradients of chemical abundances in galaxies, the abundance of the interstellar medium (ISM) at their centers is still not well known, in particular for their H II regions. In fact this is not a minor issue, since this parameter plays a key role in constraining models of galaxy evolution and the values of elemental yields. A corollary of this issue is the now-fashionable study of the mass–metallicity relation of galaxies, or of

The Metal-rich Universe, eds. G. Israelian and G. Meynet. Published by Cambridge University Press.
© Cambridge University Press 2008.

its metallicity–luminosity version: how does the maximum metallicity depend on galaxy mass or luminosity? In order to answer this question, we first have to sort out a mostly technical problem: is it possible to derive a reliable value for the maximum ISM oxygen abundance of a disk galaxy?

Oxygen abundances of the ISM in the Milky Way (MW) have been derived by making use, for example, of measurements of the interstellar O I λ1356 Å absorption line (e.g. Sofia & Meyer 2001). Also, massive-stars have been used to derive the abundance gradient in the MW and in nearby galaxies (e.g. Smartt & Rolleston 1997; Smartt et al. 2001; Urbaneja et al. 2005 and references therein) and also for massive-star clusters near the center of our Galaxy such as the Arches cluster (see Chapter 13 in these proceedings). Planetary nebulae are also very useful for investigating abundance gradients (e.g. Stasińska et al. 2006). A powerful tool for deriving the chemical properties of the gas in nearby and distant galaxies is provided by giant H II regions: these objects are luminous enough to be observed in distant galaxies and for their chemical abundances to be derived.

2 Strategies for deriving the maximum value of O/H

In our recent work we have followed a direct strategy to derive the maximum oxygen abundance of a spiral galaxy. Galaxy disks typically have negative O/H abundance gradients – a well-known result since the seventies (e.g. Smith 1975; Shields & Searle 1978; Pagel et al 1979); the idea has been to obtain reliable O/H gradients across the disks of the galaxies and extrapolate them to their respective centers. In doing that, the oxygen abundance at a given radius r, O/H(r), will reach the maximum value (O/H)$_{max}$ when the galactocentric radius $r = 0$. This derivation will have strong implications for e.g. the derivation of the oxygen yield in galaxies, or for the understanding of the mass–metallicity (metallicity–luminosity) relation; also it could be useful for the interpretation of the observations of high-redshift luminous emission-line galaxies.

2.1 Some words of caution

It is important to bear in mind several caveats regarding the derivation of the maximum value of O/H. One is the shape of radial abundance gradients, which can be affected by dynamical mechanisms in the disks. This aspect cannot be overlooked: observers sometimes naively expect to derive well-behaved smooth radial gradients, which might not exist at all in galaxies subject to important gas flows; e.g. consider the effects produced by cyclonic/anticyclonic gas flows near co-rotation according to the models of Vorobyov (2006). Another important caveat comes from the fact that the standard direct abundance derivation for H II regions relies on accurate measurements of the electron temperature. However, these measurements are rare for the inner more metal-rich H II regions (see Chapter 17 in these proceedings). Also, the exact value of the O/H abundance of an H II region can be subject to several corrections related to possible electron-temperature fluctuations, to the presence of chemical inhomogeneities, or to depletion on dust (see Chapter 11 in these proceedings). The exact geometry of the H II regions may also play a role in the understanding of some observed discrepancies between the abundances derived from optical collisionally excited lines or recombination lines and those derived from forbidden lines in the infrared (see Chapter 17 in these proceedings). All in all, typical corrections proposed for the oxygen abundance of giant H II regions can reach values of up to 0.3 dex. This correction factor, though important, must be compared with the errors affecting the determination of abundances in distant H II regions, which are usually of the order of 0.1–0.2 dex. What is relevant for the subject of this paper is the fact that recent work predicts that a direct standard derivation of the O/H abundance for high-metallicity H II regions can give totally wrong values, due to the existence of strong internal temperature gradients that have previously been unaccounted for, and which are a consequence of their high metallicity (e.g. Stasińska 2005). Recent observations of H II regions in M101 (Bresolin, 2006) do not seem to suffer from this effect. Nonetheless, it seems that strong internal electron-temperature gradients in H II regions can be produced even in the case of near-primordial metallicities, as recent population-III-star H II-region models have shown (Abel *et al.* 2007).

2.2 Our approach

The approach we have followed in order to derive a robust estimation for the maximum oxygen abundance in spiral galaxies is to use new abundance-derivation tools. This approach is based on firm physical assumptions that are independent of the particular temperature structure in H II regions, as shown in Figure 18.1 (Pilyugin *et al.* 2006). In order to derive O/H abundances consistently for our extended sample of H II regions in galaxies, the ff relation (Pilyugin 2005) between nebular and auroral [O III] line fluxes has been used. The relation between $T_{[O III]}$ and $T_{[O II]}$ temperatures presented in Figure 18.1 (left) was derived by Pilyugin *et al.* (2006) by applying nebular boundary conditions to the H II-region ionization structure, giving $t_2 = 0.72 t_3 + 0.26$, where $t_2 = T_{[O II]}/10^4$ and $t_3 = T_{[O III]}/10^4$. In Figure 18.1 (right), the ff relation shows a tight correlation between [O III] $\lambda 4363/H\beta$ (R) and [O III] $\lambda\lambda 4959, 5007/H\beta$ (R_3), which appears to hold for high-abundance ($\log(O/H) \geq -3.75$) H II regions (Pilyugin 2005; Pilyugin *et al.* 2006). The expected value of the electron temperature of an H II region can be derived using this ff relation.

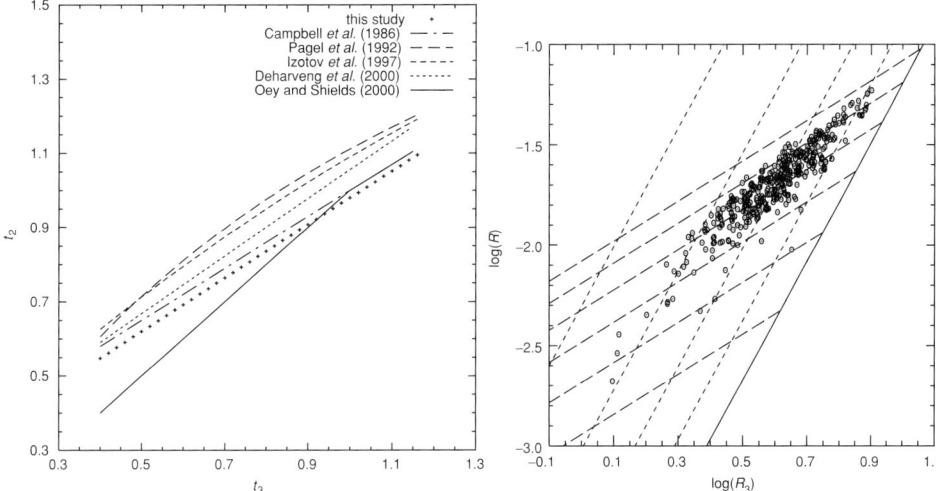

Figure 18.1. Left: a model-independent relation between the expected electron temperatures corresponding to [O III], t_3, and [O II], t_2 (in units of 10^4 K), compared with other derivations in the literature (Pilyugin et al. 2006). Right: the ff relation of Pilyugin (2005) between the flux ratios of auroral, [O III] $\lambda 4363$/Hβ (R), and nebular, [O III] $\lambda\lambda$ 4959, 5007/Hβ (R_3), lines of [O III]; see the text for details.

3 Results and discussion

The abundance of oxygen has been derived for an extended sample of giant H II regions using our compilation of high-quality spectroscopic data for spiral galaxies. The O/H gradient derived for the MW gives an average oxygen abundance at the radius of the Solar vicinity which agrees very well with the abundance of oxygen derived from measurements of the interstellar O I $\lambda 1356$ Å absorption line of 284–390 oxygen atoms per million hydrogen atoms (Meyer et al. 1998; Sofia & Meyer 2001; Cartledge et al. 2004), equivalent to $12 + \log(O/H) = 8.45$–8.59.

For the sample of galaxies, the extrapolation to their centers of the O/H abundance gradients obtained provides maximum gas-phase oxygen abundances that do not appear to exceed $12 + \log(O/H)_{max} \sim 8.87$ for the largest galaxies (Pilyugin et al. 2006). Assuming 0.08 dex as the maximum fraction of oxygen locked onto dust grains, the maximum ISM oxygen abundance can be up to $12 + \log(O/H)_{max} \sim 8.95$ or slightly larger, if temperature fluctuations are present. This abundance value would constitute, *de facto*, the upper limit of metallicity in the metallicity–luminosity relation. This relation, as has recently been shown (Pilyugin et al. 2004, 2006), in fact reveals a flattening of the O/H abundance in the luminosity range of the most massive galaxies. Recent works (Tremonti et al. 2004; Lamareille et al. 2006) have concerned this range of the metallicity–luminosity relation in extended samples of galaxies, although their O/H values appear to be larger. Similarly, the flattening of the metallicity–luminosity relation appears to be consistent with the findings of Erb *et al.* (2006) for high-redshift star-forming galaxies.

Acknowledgments

We thank the LOC/SOC for a well-organized and very productive meeting. This research was partially funded by the "Estallidos" project: grant AYA2004-08260-C03-02 of the Spanish PNAYA.

References

Abel, T., Wise, J. H., & Bryan, G. L. (2007), *ApJ* **659**, L87
Bresolin, F. (2006), astro-ph/0610690
Cartledge, S. I. B., Lauroesch, J. T., Meyer, D. M., & Sofia, U. J. (2004), *ApJ* **613**, 1037–1048
Erb, D., Shapley, A. E., Pettini, M., Steidel, C. C., Reddy, N. A., & Adelberger, K. L. (2006), *ApJ* **644**, 813–828
Garnett, D. (1989), *ApJ* **345**, 282–297
Lamareille, F., Contini, T., Brinchmann, J., Le Borgne, J.-F., Charlot, S., & Richard, J. (2006), *A&A* **448**, 907–919
McCall, M. L., Rybski, P. M., & Shields, G. (1985), *ApJS* **57**, 1–62
Meyer, D. M., Jura, M., & Cardelli, J. A. (1998), *ApJ* **493**, 222–229
Pagel, B. E. J., Edmunds, M. G., Blackwell, D. E., Chun, M. S., & Smith, G. (1979), *MNRAS* **189**, 95–113
Pagel, B. E. J., Edmunds, M. G., & Smith, G. (1980), *MNRAS* **193**, 219–230
Pilyugin, L. S. (2005), *A&A* **436**, L1–L4
Pilyugin, L. S., Vílchez, J. M., & Contini, T. (2004), *A&A* **425**, 849–869
Pilyugin, L. S., Vílchez, J. M., & Thuan, T. X. (2006), *MNRAS* **370**, 1928–1934
Shields, G. A., & Searle, L. (1978), *ApJ* **222**, 821–832
Smartt, S., Venn, K. A., Dufton, P. L., Lennon, D. J., Rolleston, W. R. J., & Keenan, F. P. (2001), *A&A* **367**, 86–105
Smartt, S., & Rolleston, W. R. J. (1997), *ApJ* **481**, L47–L50
Smith, H. E. (1975), *ApJ* **199**, 591–610
Sofia, U., & Meyer, D. M. (2001), *ApJ* **554**, L221–L224
Stasińska, G., Vílchez, J. M., Pérez, E., *et al.* (2006), in *PN beyond the Milky Way* (Berlin, Springer-Verlag), p. 234
Stasińska, G. (2005), *A&A* **434**, 507–520
Tremonti, C., Heckman, T. M., Kauffmann, G. *et al.* (2004), *ApJ* **613**, 898–913
Urbaneja, M. A., Herrero, A., Kudritzki, R. P., *et al.* (2005), *ApJ* **635**, 311–335
Vílchez, J. M., Pagel, B. E. J., Días, A. I., Terlevich, E., & Edmunds, M. G. (1988), *MNRAS* **235**, 633–653
Vorobyov, E. I. (2006), *MNRAS* **370**, 1046–1054

19

Starbursts and their contribution to metal enrichment

Daniel Kunth

Institut d'Astrophysique de Paris, F-75014 Paris, France

I review the properties of starburst galaxies, compare the properties of the local ones with those of more distant starbursts and examine their role in the metal enrichment of the interstellar medium and the intergalactic–intracluster medium. Metallicity is not an arrow of time and, contrary to current belief, metal-rich galaxies can also be found at high redshift.

1 Introduction

For the purpose of this conference, let me first stress that known starburst galaxies in the local Universe are not metal-rich ($Z \leq Z_\odot$). Having written this, I remark however that use of the term 'metal-rich' differs depending upon one's field of interest. Stellar astrophysicists deal with individual stars as metal-poor as [X] ≈ -4, quite similar to the metallicities of some DLAs, hence they would regard Solar abundances as being large! On the other hand, the most dramatic case known, so far, of metal overabundance comes from spectroscopy of QSOs with [N/H] ≈ 1 (Hamann *et al.* 2002). Most of my talk will review our knowledge about local starbursts. While starburst galaxies are not known to be metal-rich objects, they nevertheless fit within the scope of this meeting. They are particularly interesting in their own right, while bearing very strong similarities to their high-redshift counterparts, because they are very significant components of our present-day Universe and are mostly important sources of chemical enrichment. Moreover, their study brings the important clue that metallicity is not providing us with an arrow of time, as too often believed. Moreover, high-redshift galaxies that are obscured and with high star-formation rates can be even more metal-rich than their low-luminosity analogues of the local Universe.

The Metal-rich Universe, eds. G. Israelian and G. Meynet. Published by Cambridge University Press.
© Cambridge University Press 2008.

2 Phenomenology and definitions

The concept of 'starburst' sometimes leads to confusion. A starburst 'event' refers to the formation of thousands of massive stars ($M_\odot \geq 20$) within a very small volume (a few parsecs) and on a timescale of only a few million years (Hoyos 2006). Such an event may occur in a normal galaxy such as our Galaxy or 30 Dor in the LMC. At variance, a starburst galaxy displays violent star formation on such a scale that the whole system is out of balance with respect to available resources. Perhaps the most fundamental definition of a starburst would be a galaxy in which the star-formation rate approaches the upper limit set by causality (Heckman 2005, his Figure 19.1). For a given total mass and gas mass fraction the upper bound of the star-formation rate (SFR) is proportionnal to σ^3, where σ is the stellar velocity dispersion (Murray *et al.* 2005). Other physical quantities are often considered, such as the duration within which the current SFR consumes the remaining interstellar gas (the inverse of this is called the efficiency). In such a context, a starburst produces the current stellar mass at the observed SFR on a timescale much shorter than a Hubble time. As pointed out by Kennicutt (1998), a starburst must develop a large SFR per unit area. Extreme starbursts have SFRs per unit area larger than 1 $M_\odot \, \text{yr}^{-1} \, \text{kpc}^{-2}$. These figures correspond to $L(\text{bol}) \geq 10^{10} L_\odot \, \text{kpc}^{-2}$ and hence are thousands of times larger than values for normal disc galaxies and consumption times of only 10^8 years. It is important to realise that this definition refers to major starbursts and is somewhat arbitrary: starburst properties appear to be mostly continuous across the range of amplitudes observed.

3 The local star-forming galaxies

Local star-forming galaxies provide roughly 10% of the total radiated energy and roughly 20% of all the high-mass-star formation in the present day (Brinchmann *et al.* 2004). Among local star-forming galaxies, sometimes referred to as H II galaxies, most are dwarfs: dwarf irregulars (dIrrs) and blue compact dwarfs (BCDs). They remain, however, a minority among the general population of dwarf galaxies (Kunth & Östlin, 2000; hereafter KO2000). In term of absolute SFRs most figures are not very impressive. Dwarfs range from $M_B \approx -14.0$ to $M_B \approx -17.0$ and have SFRs in the range $\approx (0.02-1) M_\odot \, \text{yr}^{-1}$, while giants have M_B of up to -21 and SFRs of $(20-40) M_\odot \, \text{yr}^{-1}$ at most. These galaxies possess a gigantic reservoir of H I gas that sometimes extends into huge halos surrounding the optical region. The mechanism that ignites strong bursts of star formation across these systems is still not known. Star formation may be triggered during encounter events (merging or interaction of galaxies) or by stochastic processes within the galaxy itself.

At optical wavelengths, their spectra are dominated by young stars and ionised gas, closely resembling those of giant H II regions in nearby spiral and irregular galaxies. Analysis of their spectra shows that most of them are metal-deficient. Towards the lower end of the metallicity distribution (O/H $\sim (1/50)Z_\odot$) we find galaxies like IZw18 and SBS0335-052. Owing to their extreme properties, it was conjectured by Searle & Sargent (1970) that these chemically unevolved galaxies could be young systems still in the process of forming.

In star-forming dwarfs, because of their low mass, star formation occurs over a large part of the discs. Some star-forming regions are very compact and remain confined within the inner few hundred parsecs of their hosts, whereas others have one or several star-forming clumps in off-nucleus regions. These events have strong and sometimes devastating impacts on various phases of the interstellar medium (ISM). The hot gas, typically observed in X-rays, can, at least in some cases, blow out from these galaxies (see Section 8). A starburst event is likely to provide an important mechanism for the metal enrichment of the intergalactic medium (IGM) in the early Universe. On small scales, some dwarf galaxies have numerous H I holes, suggesting that star formation is propagating and allowing the investigation of early phases of what later may become a blow-out of the gas. Molecular gas, the site of actual star formation, is notoriously difficult to detect in dwarfs. The reasons for this are not necessarily related to a dearth of molecular material, but rather arise from the low-metallicity environment characteristic for dwarf galaxies and the low H I density that inhibit the rate of formation of the diffuse molecular gas. It is also likely that the diffuse molecular hydrogen is destroyed by the incident UV flux from the massive stars in the star-forming region. Most of the remaining molecules should be in dense clumps, which are opaque to far-UV radiation, and do not contribute to the observed spectra (Vidal-Madjar *et al.* 2000; Hoopes *et al.* 2004).

4 The more-distant starbursts

At $0.4 \leq z \leq 1.0$ luminous compact blue galaxies (LCBGs) dominate the number density of galaxies at intermediate redshifts. They are small starburst systems of ($R_e \leq 3.0$ kpc) that have evolved more than any other galaxy class during the last 8 Gyr (Phillips *et al.* 1997; Guzmán *et al.* 2003). They are a major contributor to the observed enhancement of the UV luminosity density of the Universe at $z < 1$, their decline in number density being in concert with the rapid drop in the global SFR since $z = 1$. They appear to form a bridge in terms of redshift, size and luminosity between Lyman-break galaxies and local H II galaxies today. On the other hand, the work of Steidel and collaborators (e.g. Steidel *et al.* 1996) has confirmed the existence of a substantial population of star-forming galaxies at $z \sim 3$, with a co-moving number density of roughly 10%–50% that of present-day luminous galaxies ($L \geq L^\star$). It is clear that observational biases play a role since at $z \geq 2$ galaxies with $L \leq L^*$ are more difficult to study.

5 Measuring heavy-element abundances from ionised gas

The spectral analysis of the H II galaxies shows that many H II galaxies are metal-poor objects (probably this is the reason why I was invited to speak at this conference!), some of them – IZw18 and SBS 0335-052 – being among the most metal-poor systems known. Heavy-element abundances are relatively easy to measure in star-forming galaxies because they contain ionised gas clouds in which large numbers of hot stars are embedded. What is observed in the optical is narrow emission lines superimposed on a blue stellar continuum. These are identified as helium and hydrogen recombination lines and several forbidden lines: O, N, S, Ne, Ar, H and He lines have been measured to date (Izotov *et al.* 2006). Methods used in determining abundances are well understood and generally more reliable than those based on stellar absorption-line data because transfer problems become less important.

Oxygen is the most reliably determined element, since the major ionisation stages can all be observed. Moreover, the [O III] $\lambda 4363$ line allows an accurate determination of the electron temperature. The intrinsic uncertainty in this method is of the order of ~ 0.1 dex. Furthermore, when the electron temperature cannot be determined, empirical relations between the oxygen abundance and the [O II] $\lambda 3727$ and [O III] $\lambda\lambda 4959, 5007$ line strengths relative to that of Hβ are used, though with lower accuracy (0.2 dex or worse) (Pagel, 1997). For other species, in general, not all of the ionisation stages are seen and an ionisation correction factor must be applied to derive their total abundances.

The ultraviolet (UV) region is dominated by the hot stellar continuum and contains relatively weak emission lines except for those that originate in stellar winds. This region has provided ways of measuring nebular carbon and silicon abundances.

We caution that many parameters control the observed metallicity in a given galaxy. They are incorporated into chemical evolutionary models and include stellar evolution and nucleosynthesis, inflows and outflows and the problem of the mixing and dispersion timescales of freshly released heavy elements (KO2000). Moreover, metallicity measurements may be relevant to only one particular component of a galaxy. Kunth & Sargent (1986) suggested that H II gas could enrich itself with metals expelled by CC SNe on timescales shorter than the lifetime of a starburst (self-pollution). Recent FUSE observations of some dwarf galaxies revealed a possible disconnect in metallicity between H I and H II regions, although possible saturation effects on the line of sight may alter the result of such a comparison (Lebouteiller *et al.* 2006).

One important aspect of H II-region abundances is that they can be obtained also at great distances. This makes them powerful tools also for studying high-redshift galaxies, with the price that our view will be biased towards actively star-forming systems. For a discussion on possible problems associated with deriving abundances in very distant galaxies, see Kobulnicky *et al.* (1999).

5.1 Do we expect metal-rich systems?

Knowledge of the chemical composition could be used to constrain the age of any given galaxy, provided that the heavy-element abundance of galaxies increases forever with age and that the gas-phase metallicity can be assumed to be equal to the metal abundance of the stellar population. The first assumption implies that galaxies evolve in a 'closed-box' manner. In this model, the metal content of a galaxy should be a function of its gas mass fraction only. The predictive power of this model would therefore be enormous, should this hypothesis be true for at least some fraction of galaxies (but see Section 7). One can easily show, under the hypothesis of a Salpeter initial mass function (IMF) and normal stellar yields, that a natural limit of the order of $\approx 2 Z_\odot$ would be reached. To go beyond this one would need to bias the IMF strongly towards high masses or/and use up all the gas (100% efficiency!). However, galaxies (and star-forming ones in particular) are not known to deviate significantly from a normal IMF, while they are known to exchange large amounts of gas with the intergalactic medium (IGM) (outflows, infall, merging etc.). Hence it can be understood that the requirements applying to the cosmic enrichment of the Universe as a whole are not always satisfied for individual galaxies. One can, of course, argue that observationally very-metal-rich galaxies might escape emission-line surveys since their usual tracers such as the [O III] $\lambda\lambda 4959, 5007$ line, for instance, become extremely faint beyond $12 + \log(\text{O/H}) = 9.0$ (Stasińska, 2002).

The second requirement is more complex to verify. First of all the gas-phase metallicity does not necessarily directly relate to the heavy-element abundance of a galaxy. Second, it is very important to keep in mind that metals ejected from dying stars do not necessarily mix instantly into the general ISM of galaxies (Roy & Kunth 1995). This delay can be as large as one billion years (Tenorio-Tagle 1996). This point is further discussed in Section 8.

6 The luminosity–metallicity relation

Star-forming galaxies are likely to produce superwinds that will export metal-enriched gas to the IGM. Merger events will also disrupt the 'closed-box' model paradigm. The strongest evidence for a departure from simple closed box models comes from the luminosity–metallicity relation. The luminosity–metallicity relationship, in such a context, is a useful relationship between evolutionary state and metallicity, hence the location of galaxies in the M_B–$(12 + \log(\text{O/H}))$ plane is of crucial importance for understanding the details of the inner workings of

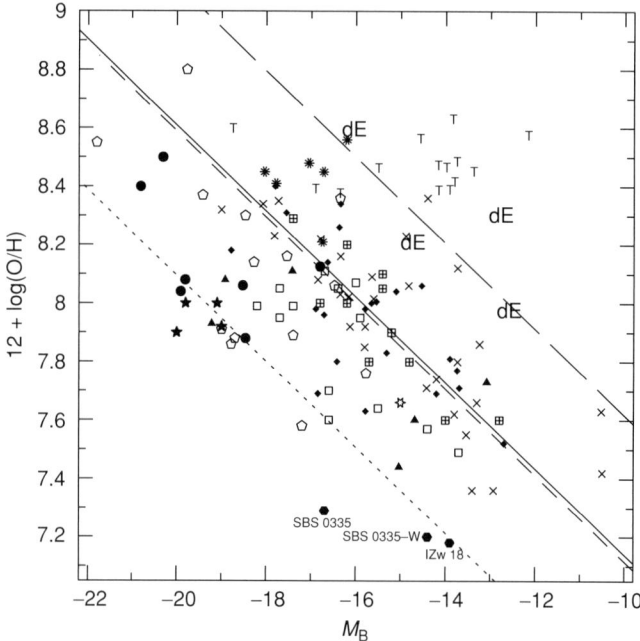

Figure 19.1. The luminosity versus metallicity diagram for dwarf galaxies. Crosses represent dIrrs. Filled symbols represent galaxies classified as BCGs or H II galaxies. Small filled diamonds show data for regular galaxies that can be classified as Type II or iE/nE, while filled circles are for galaxies that can be classified as Type I. Filled triangles are for BCGs for which no classification or images were available. Points for the three most metal-poor galaxies are labelled and shown as filled hexagons. Filled stars are data for luminous BCGs. Asterisks show the location of 'blue amorphous galaxies' (except for II Zw40 which is represented by the filled circle falling on the short-dashed line). Data for LSBGs are shown as open squares and open pentagons. The open star is for H I 1225 + 01, the H I cloud in Virgo. Boxes with plus signs inside are for quiescent (dIrr/LSBG) dwarfs. Candidate tidal dwarfs are shown as T and dEs are shown as dE. The solid line shows the M_B–O/H relation for dIrrs from Richer and McCall (1995), while the dotted line shows the same relation offset by 3.5 magnitudes, indicating the location of XBCGs. The short-dashed line shows the M_V–Z relation for dE/dSph from Caldwell (1998) assuming $(B-V) = 0.75$ and [O/H] = [Fe/H], while the long-dashed line shows the same relation assuming [O/H] = [Fe/H] + 0.5. Details and complete references for the object sources, their B magnitudes and their metallicities are listed in KO2000.

starbursts. One reason for the existence of this relationship – assuming that more-luminous galaxies are also more-massive – is that more-massive objects can lock up the metals created in their stars more easily than can less-massive galaxies because of their deeper gravitational wells. A metallicity–luminosity relation can also arise if smaller galaxies have larger gas fractions than do larger galaxies. This is seen statistically in the local Universe and can result from the fact that smaller galaxies evolve more slowly (lower SFR per unit gas mass) than larger ones. In Figure 19.1, we show the oxygen abundance versus luminosity diagram for dIrrs (crosses), BCGs (filled symbols) and LSBGs (open symbols), using data collected from the literature, with available abundances and integrated B magnitudes. At first sight many BCGs do not appear extreme in comparison with the normal dIrrs. Indeed, some BCGs are more metal-rich than dIrrs are at a given luminosity, while the opposite would be expected if BCGs were bursting dIrrs. These BCGs may be in a post-burst stage and the fresh metals may have become 'visible' already. Secondly, some 'extreme BCGs' appear much more metal-poor at a given luminosity. These extreme BCGs (XBCGs as listed in KO2000) are brighter by three magnitudes or more at a given metallicity, or equivalently 0.5 dex less metal-rich at a given luminosity relative to the dIrr relation. The intriguing XBCGs include IZw18 and SBS 0335-052 (Östlin *et al.* 1999, 2001).

Tidal dwarfs appear metal-rich for their metallicity because they were formed from the already-enriched gas of their parent galaxies (Duc & Mirabel 1998). Some arguments have ascribed the luminosity–metallicity correlation to the effects of selective losses of heavy elements from galaxies in supernova-driven outflows. A reduction in the effective yields (most probably due to galactic winds) can explain the observations: Garnett (2002) has shown that metal loss due to supernova-driven winds is a process at work in Irrs and some spiral galaxies. Indeed, galaxies with low rotational velocities (and hence low mass) lose a large fraction of metals, whereas galaxies with higher rotational velocities tend to retain metals. Not only is the metallicity–luminosity relation of galaxies inferred from the SDSS providing the best evidence for galactic winds (Tremonti *et al.* 2004) but also this relationship is expected to evolve with cosmic epoch.

Using emission-line ratios, Pettini *et al.* (2001 and references therein) found evidence that Lyman-break galaxies (LBGs) of under-Solar abundance at large redshifts ($z \geq 2.0$) are 5–40 times more luminous than local systems of similar metal content, confirming that the luminosity-to-metal ratio varies with time.

7 Cosmological context

It has been possible to sketch the star-forming history of the Universe at high redshifts (Chary & Elbaz 2001). It appears that the overall SFR of the galaxy population seems to increase on going from $z = 0$ to $z = 1$; this epoch is believed to be that of disc formation. The cosmic SFR becomes constant or declines slightly on going from $z = 1$ to $z = 4$ and this period is sometimes refered to as the spheroid epoch. In fact, at $z \geq 1$ more galaxies were involved in interactions and starbursts built on rapid timescales with high SFR per unit area, hence metallicity could be high. At high redshift, the highest SFRs occur in the most dusty galaxies (sub-millimetre sources), telling us that a young galaxy is not necessarily metal-poor! Numerical simulations indicate that metal production occurs at early times (due to spheroids consecutive to the strong starburst they experience), whereas at $z \leq 2$ the main sites of element production become the spiral discs (Calura & Matteucci 2004). Local irregular systems make a negligible contribution to the total element production.

Numerical simulations by Cen and Ostriker (1999) predict the evolution of the metal content of the Universe as a function of density by incorporating star formation and its feedback on the IGM. At a given gas density (corresponding e.g. to a rich cluster, a disc galaxy, dwarf etc.) their models predict an evolution with redshift, but more importantly show that metallicity is a stronger function of density than it is of age; and with a considerable scatter, moreover. At low redshift, one would expect a few per cent of the gas-rich dwarfs to have metallicities of the order of that of IZw18, without invoking their youth. This is also suggested by the Lyα absorption systems (Shull *et al.* 1998). Ferrara and Tolstoy (2000), discussing the feedback from star formation in dwarf galaxies, argue from this that low-mass galaxies could be a major contributor of metals in the IGM.

8 The fate of metals

8.1 The mixing timescales

The fraction of metals that mixes within the ISM crucially depends on the thermodynamical status of the ISM and on the star-formation history of the galaxy. The very first burst of star formation must occur in a very cold and dense medium; hence radiative losses of the expanding bubbles and superbubbles become significant (i.e. comparable to the thermal energy of the cavity) within a short timescale of the order of 10 Myr (MacLow & McCray 1988; Recchi *et al.* 2001). At later times, the ISM being warmer, more diluted and irregularly structured, radiative losses will be smaller; therefore the impact of late generation of stars into the ISM will be stronger. Metals freshly produced by these late generations of stars will be either directly channelled along the galactic outflow or released in a warm and tenuous medium on a much larger mixing timescale – as large as one billion years (Recchi *et al.* 2004; Tenorio-Tagle 1996).

In more general terms, Roy & Kunth (1995) concede that dwarf galaxies are expected to show kpc-scale abundance inhomogeneities. From an observational point of view, data for some galaxies support the fast-mixing scenario while data for others do not. This is true of NGC 5253, IIZw 40 and Mrk 996, where local N/H overabundances have been attributed to localised pollution from WR stars (Kobulnicky *et al.* 1997; Walsh & Roy 1993; Thuan *et al.* 1996). This indicates that rapid mixing can take place. However, this is not true for all young starbursts, even when WR stars are suspected or observed to be present (Oey & Shields 2000). On the other hand, IZw 18 appears to be rather homogeneous (Legrand *et al.* 2000), which does not advocate in favour of the concept of rapid self-enrichment. Arguments in favour of complete mixing over long timescales arise from the observation that disconnected H II regions within the same galaxies have nearly the same abundances. Six H II regions in the SMC have $\log(\text{O/H}) = 8.13$ (± 0.08) while four in the LMC give $\log(\text{O/H}) = 8.37$ (± 0.25) (Russell & Dopita 1990).

The possibility that metallicities in the neutral gas phase are orders of magnitude below the H II-region abundances would be an ultimate test of the hypothesis of large-scale inhomogeneities. Kunth and Sargent (1986) proposed a clean test for the self-enrichment hypothesis by using a QSO as background source to measure the abundance of metals in the neutral gas. Such a test was performed recently by Bowen *et al.* (2005), who measured the abundance of sulfur in the neutral gas of the dwarf SBS 1543 + 593 to be the same as that of oxygen in one of its neighbouring H II regions – a result at variance with expectations regarding self-enrichment pollution. A good benchmark for the enrichment process in the early evolution of a galaxy is our Milky Way. Metal-poor halo stars (Cayrel *et al.* 2004) exhibit a small spread in alpha-elements. This is in agreement with the hypothesis of instantaneous mixing that is commonly assumed in simple chemical-evolution models (e.g. Francois *et al.* 2004). Arnone *et al.* (2005) confirmed the need for a fast mixing process for the alpha-elements, although the large scatter in [Ba/Fe] shows that ejecta of stars in various mass ranges could have different mixing timescales.

8.2 The IGM enrichment

It is important to decide whether the newly produced metals leave the ISM and pollute the IGM. In principle, the energy released by stellar winds during the very early stage of a starburst (Leitherer *et al.* 1992) and by supernovae explosions after a burst of star formation often exceeds the binding energy of a dwarf galaxy, therefore the development of a galactic wind is likely . . . but . . .

8.2.1 Observations

Data concerning nearby far-infrared galaxies support the hypothesis of the occurrence of galactic winds. The nearest edge-on disc galaxies have clear kinematic signatures of an outflow along the minor axis with velocities of the order of 200–600 km s^{-1} (Heckman *et al.* 1990). As shown by HST, FUSE and Chandra observations, clear signatures of an outflow are present in many other nearby galaxies such as NGC1705 (Heckman *et al.* 2001), NGC1569 (Martin *et al.* 2002) and NGC3079 (Cecil *et al.* 2001), while that molecular disc–halo outflow is occurring in NGC3628

(Irwin & Sofue 1996). Kunth and colleagues (Mas-Hesse *et al.* 2003) have shown that the escape of the Lyα photons in star-forming galaxies strongly depends on the dynamical properties of their ISM. The Lyman-alpha profile in Haro 2 indicates a superwind of at least $200\,{\rm km\,s^{-1}}$, carrying a mass of $\sim 10^7 M_\odot$, which can be independently traced from the Hα component (Legrand *et al.* 1997). However, high-speed winds do not necessarily carry a lot of mass. Martin (1996) argues that a bubble seen in IZw 18 will ultimately blow out together with its hot-gas component. The fact that diffuse X-ray emissions around starburst galaxies correlate with the star formation per unit area (Strickland *et al.* 2004) corroborates the idea that these are the results of outflows driven by the starburst activity. The best examples of large-scale outflows driven by SNe feedback are, however, perhaps found at large redshifts (Pettini *et al.* 2001).

8.2.2 Interpretations

Even in spite of the ubiquity of outflow phenomena, there is no certainty that the outflowing gas will not recollapse towards the centre of a galaxy in the future. It is possible that an outflow takes the fresh metals with it and in some cases leaves a galaxy totally cleaned of gas, but the hot gas outside the H II regions may simply stay around in the halo. The presence of outflows has been used as an indication that supernova products and the whole of the ISM are easily ejected from the host dwarf systems, causing the contamination of the intra-cluster medium (Dekel & Silk 1986; De Young & Heckman 1994). This type of assumption is currently blindly used by cosmologists in their model calculations.

In a more refined approach, MacLow & Ferrara (1999) models with $\sim 10^6 M_\odot$ systems experience a complete blow-away, whereas with more-massive ones only a small fraction of the gas mass escapes the galactic potential well. In these blow-out phenomena a very large fraction of the metals is lost and for models with masses below or equal to $\sim 10^8 M_\odot$ almost all the metals are lost through the galactic wind. This result (that metals are ejected much more easily than pristine ISM) is pretty common (the same result was found by Strickland & Stevens (2000) and by D'Ercole & Brighenti (1999)), but is at variance with the vision of Silich & Tenorio-Tagle (2001). They argue that the indisputable presence of metals in galaxies implies that supernova products are not completely lost in all cases: for a dwarf galaxy, which has a weaker gravitational potential, these effects may result in gas loss from the galaxy *unless* the presence of a low-H I-density halo acts as a barrier. Model calculations developed by Silich & Tenorio-Tagle (1998) predict that superbubbles in amorphous dwarf galaxies must have already undergone blow-out and are at present evolving into an extended low-density halo. This should inhibit the loss of the swept-up and processed matter into the IGM. Recent Chandra X-ray observations are ambiguous: data for some young starbursts indicate that loss of metals from discs is occurring (Martin *et al.* 2002), but not in the case of NGC 4449 with an extended H I halo of around 40 kpc (Summers *et al.* 2003).

Legrand *et al.* (2001) have compared theoretical estimates by Mac Low & Ferrara (1999) and Silich & Tenorio-Tagle (2001) with data for some well-studied starburst galaxies. Derived values of the mechanical energy-injection rate were compared with the predictions of their hydrodynamical models. The net result is that all galaxies lie above the lower limit first derived by Mac Low & Ferrara (1999) for the ejection of metals from energised flattened disc-like ISM density distributions and most are *below* the limit for the low-density-halo picture.

8.3 Which galaxies enrich?

Which galaxies experience galactic winds and contribute to the chemical enrichment of the IGM? Gibson & Matteucci (1997) showed that dwarf galaxies can provide at most 15% of the intra-cluster medium (ICM), giant ellipticals being responsible for 20% of it and the remaining 65% being primordial. It has been proposed that only certain elements are lost (or that elements are lost in differing proportions), hence reducing the effective net yield of those metals compared with the predictions of a simple chemical-evolution model (Edmunds 1990). The SNe involved in such a wind are likely to be of Type II because Type-Ia SNe explode in isolation and will be less likely to trigger chimneys from which metals can be ejected out of the plane of a galaxy. In this framework O and part of Fe are lost, while He and N (largely produced by intermediate stars) are not. This would result in a cosmic dispersion in element ratios such as N/O between galaxies that have experienced mass loss and those that have not. In general, as we discussed above, the flatter a galaxy is, the easier is the development of a galactic wind (Strickland & Stevens 2000). Therefore, for elliptical galaxies the development of a galactic wind occurs only when the total thermal energy overcomes the binding energy of the galaxy, a condition that is not required in flattened systems. However, the initial SFR in an elliptical galaxy should be large enough to satisfy the condition for the onset of a galactic wind. Therefore IGM pollution due to galactic winds from giant ellipticals is likely.

Dwarf spheroidal galaxies with an even shallower potential well should favour the development of a galactic wind and their lack of gas might be explained by this phenomenon. However, this is not as easy at it looks, since Marcolini *et al.* (2006), simulating a spherically symmetric galaxy such as Draco, failed to produce a galactic wind. Hence the problem is difficult to solve from the hydrodynamics and I invite the reader to look at the paper of Skillman & Bender (1995) for a thorough discussion of these problems.

A very last point concerning iron was brought to my attention by S. Recchi. There is a very large amount of iron in the ICM, perhaps exceeding the amount of alpha-elements (e.g. M87) (Gastaldello & Molendi 2002). Since galactic winds should be enriched in alpha-elements because they are released by more numerous SNe II, there should be an overabundance of alpha-elements, which is at odds with observations. However, elements produced by SNe Ia are easily channelled along the wind (the hole in the ISM is already there, facilitating the efforts of elements to leave the parent galaxy), therefore simulations of starburst galaxies give larger ejection efficiencies for iron-peak elements. Models with continuous star formation have similar ejection efficiencies of alpha-elements and iron-peak elements.

Acknowledgment

I thank D. Garnett, C. Hoyos, G. Östlin, S. Recchi and D. Strickland for stimulating discussions.

References

Arnone, E., Ryan, S. G., Argast, D., Norris, J. E., & Beers, T. C. (2005), *A&A* **430**, 507
Bowen, D. V., Jenkins, E. B., Pettini, M., & Tripp, T. M. (2005), *ApJ* **635**, 880
Brinchmann J., *et al.* (2004), *MNRAS* **351**, 1151
Calura, F., & Matteucci, F. (2004), *MNRAS* **350**, 351
Cayrel, R., *et al.* (2004), *A&A* **416**, 1117
Cecil, G., Bland-Hawthorn, J., Veilleux, S., & Filippenko, A. V. (2001), *ApJ* **555**, 338
Cen R., & Ostriker, J. P. (1999), *ApJ* **519**, L109
Chary, R., & Elbaz, D. (2001), *ApJ* **556**, 562
Dekel, A., & Silk, J. (1986), *ApJ* **303**, 39
De Young, D., & Heckman, T. (1994), *ApJ* **431**, 598
D'Ercole, A., & Brighenti, F. (1999), *MNRAS* **309**, 941
Duc, P.-A., & Mirabel, I. F. (1998), *A&A* **333**, 813
Edmunds, M. G. (1990), *MNRAS* **246**, 678
Ferrara, A., & Tolstoy, E. (2000), *MNRAS* **313**, 291
François, P. *et al.* (2004), *A&A* **421**, 613
Garnett, D. R. (2002), *ApJ* **581**, 1019
Gastaldello, F., & Molendi, S. (2002), *ApJ* **572**, 160
Gibson, B. K., & Matteucci, F. (1997), *ApJ* **475**, 47
Guzmán, R., Östlin, G., Kunth, D. *et al.* (2003), *ApJ* **586**, L45
Hamann, F., Korista, K. T., Ferland, G. J., Warner, C., & Baldwin, J. (2002), *ApJ* **564**, 592
Heckman, T. (2005), *ASSL*: Vol 329; *Starbursts: From 30 Dor to Lyman Break Galaxies*, p. 3
Heckman, T. M., Armus, L., & Miley, G. K. (1990), *ApJS* **74**, 833
Heckman, T. M. *et al.* (2001), *ApJ* **554**, 1021
Hoopes, C. G., Sembach, K. R., Heckman, T. M. *et al.* (2004), *ApJ* **612**, 825
Hoyos C. (2006), On the Nature of Distant LCBs, unpublished thesis, Universidad Autonóma de Madrid
Irwin, J. A., & Sofue, Y. (1996), *ApJ* **464**, 738
Izotov, Y. I., Stasińska, G., Meynet, G., Guseva, N. G., & Thuan, T. X. (2006), *ApJ* **448**, 955
Kennicutt R. (1998), *ApJ* **498**, 541

Kobulnicky, H. A., Skillman, E. D., Roy, J.-R., Walsh, J. R., Rosa, M. R. (1997), *ApJ* **477**, 679
Kobulnicky, H. A., Kennicutt, R. C. Jr., & Pizagno, J. L. (1999), *ApJ* **514**, 544
Kunth, D., & Sargent, W. L. W. (1986), *ApJ* **300**, 496
Kunth, D., & Östlin, G. (2000), *A&ARev* **10**, 1
Lebouteiller, V., Kunth, D. *et al.* (2006), *A&A* **459**, 161
Legrand, F., Kunth, D., Mas-Hesse, J. M., & Lequeux, J. (1997), *A&A* **326**, 929
Legrand, F., Kunth, D., Roy, J.-R., Mas-Hesse, J. M., & Walsh, J. R. (2000), *A&A* **355**, 891
Legrand, F., Tenorio-Tagle, G., Silich, S., Kunth, D., & Cerviño, M. (2001), *ApJ* **560**, 630
Leitherer, C., Robert, C., & Drissen, L. (1992), *ApJ* **401**, 596
Mac Low, M., & Ferrara, A. (1999), *ApJ* **513**, 142
Mac Low, M.-M., & McCray, R. (1988), *ApJ* **324**, 776
Marcolini, A., D'Ercole, A., Brighenti, F., & Recchi, S. (2006), *MNRAS* **371**, 643
Martin, C. L. (1996), *ApJ* **465**, 680
Martin, C. L., Kobulnicky, H. A., & Heckman, T. M. (2002), *ApJ* **574**, 663
Mas-Hesse, J. M., Kunth, D., Tenorio-Tagle, G., Leitherer, C., Terlevich, R. J., & Terlevich, E. (2003), *ApJ* **598**, 858
Murray, N., Quataert, E., & Thompson, T. (2005), *ApJ* **618**, 569
Oey, M. S., & Shields, J. C. (2000), *ApJ* **539**, 687
Östlin, G., Amram, P., Masegosa, J., Bergvall, N., & Boulesteix, J. (1999), *A&AS* **137**, 419
Östlin, G., Amram, P., Bergvall, N., Masegosa, J., Boulesteix, J., & Marquez, I. (2001), *A&A* **373**, 800
Pagel, B. E. J. (1997), *Nucleosynthesis and Chemical Evolution of Galaxies* (Cambridge, Cambridge University Press)
Pettini, M., Shapley, A. E., Steidel, C. C. *et al.* (2001), *ApJ* **554**, 981
Phillips, A. C., Guzmán, R., Gallego, J. *et al.* (1997), *ApJ* **489**, 543
Recchi, S., Matteucci, F., & D'Ercole, A. (2001), *MNRAS* **322**, 800
Recchi, S., Matteucci, F., D'Ercole, A., & Tosi, M. (2004), *A&A* **426**, 37
Roy, J.-R., & Kunth, D. (1995), *A&A* **294**, 432
Russell, S. C., & Dopita, M. A. (1990), *ApJS* **74**, 93
Searle, L. & Sargent W. L. W. (1970), *ApJ* **162**, L155
Shull, J. M. *et al.* (1998), *AJ* **116**, 2094
Stasińska, G. (2002), *Revista Méxicana de Astronomía y Astrofísica Conference Series* **12**, 62
Silich, S. A., & Tenorio-Tagle, G. (1998), *MNRAS* **299**, 249
(2001), *ApJ* **552**, 91
Skillman, E. D., & Bender, R. (1995), *Revista Méxicana de Astronomía y Astrofísica Conference Series* **3**, 25
Steidel, C. C., Giavalisco, M., Pettini, M., Dickinson, M., & Adelberger, K. L. (1996), *ApJ* **462**, L17
Strickland, D. K., & Stevens, I. R. (2000), *MNRAS* **314**, 511
Strickland, D. K., Heckman, T. M., Colbert, E. J. M., Hoopes, C. G., & Weaver, K. A. (2004), *ApJ* **606**, 829
Summers, L. K., Stevens, I. R., Strickland, D. K., & Heckman, T. M. (2003), *MNRAS* **342**, 690
Tenorio-Tagle, G. (1996), *AJ* **111**, 1641
Thuan, T. X., Izotov, Y. I., & Lipovetsky, V. A. (1996), *ApJ* **463**, 120
Tremonti, C. A., *et al.* (2004), *ApJ* **613**, 898
Vidal-Madjar, A. *et al.* (2000), *ApJ* **538**, L77
Walsh J. R., & Roy J.-R. (1993), *MNRAS* **262**, 27

20

High metallicities at high redshifts

Max Pettini

Institute of Astronomy, Madingley Road, Cambridge CB3 0HA, UK

As well as being the realm of the first stars, the high-redshift regime is a window on some of the most metal-rich components in our Universe, the massive galaxies destined to become today's ellipticals and the black holes at their centres at a time of peak activity. While much has been learnt in recent years about these 'get-rich-quick' objects, progress is still hampered by the same limitations as apply to nearby metal-rich stars and H II regions: our methods for exploring the super solar-metallicity regime require considerable improvement before they can be considered to be reliable. I illustrate this conclusion with a few recent case studies of active galactic nuclei, star-forming galaxies and damped Lyman-alpha systems.

1 Introduction

There are two basic points I should like to convey in this brief review. The first is that the metal-rich Universe is not only a local affair: even at high redshifts (in the present context $z = 2$–3), we can easily locate astrophysical environments where chemical evolution has progressed to Solar and supersolar metallicities. Second, in interpreting observations of sources at high redshifts we rely on metallicity diagnostics developed for nearby stars, H II regions and interstellar gas (with the additional complication of the generally lower signal-to-noise ratio of the data). Consequently, the concerns discussed at this meeting on the reliability of abundance determinations in the super-metal-rich regime and on the consistency (or otherwise) of various metallicity calibrators also limit our ability to investigate the metal-rich Universe at high redshifts.

On a more positive note, the last few years have seen remarkable progress in constructing increasingly detailed descriptions of the abundances of chemical

The Metal-rich Universe, eds. G. Israelian and G. Meynet. Published by Cambridge University Press.
© Cambridge University Press 2008.

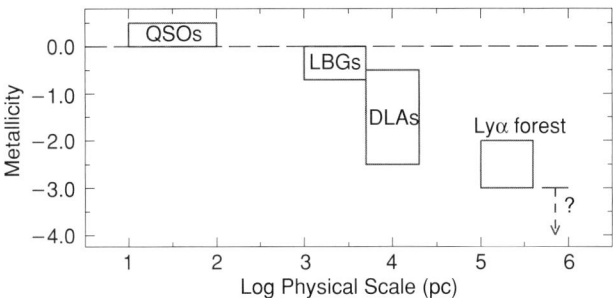

Figure 20.1. A snapshot of the metallicities of various components of the high-redshift ($z = 2.5$) Universe. The logarithm of the metallicity relative to Solar (indicated by the long-dashed line at 0.0) is plotted against the typical linear scale of the structures to which it refers. The term 'Lyman-break galaxies' (LBGs) is used as a shorthand here to refer to a more general class of actively star-forming galaxies.

elements in various environments and over most of the age of the Universe, from large surveys of present-day galaxies with the Sloan Digital Sky Survey to QSOs with the highest redshifts known at $z > 6$. The overall picture which has gradually been coming into focus is summarised in the sketch in Figure 20.1. When we plot the metallicities measured in various environments against the typical physical scale of the structures to which they refer, we find a clear trend of decreasing metal abundance with increasing scale. Another way to interpret the trend is to view the x-axis of Figure 20.1 as a scale of decreasing overdensity relative to the cosmic mean. Evidently, it is the depth of the potential well within which the baryons find themselves that drives the pace at which gas is processed through stellar nucleosynthesis – a point that was already clear to interested theoreticians several years ago (e.g. Cen & Ostriker 1999). Thus, at any one epoch, there can be a range of three to four orders of magnitude in the degree of metal enrichment which one would measure, depending on environment. On the other hand, metallicity depends weakly on cosmic time: the age–metallicity relation of the Universe as a whole appears to be shallow, just as is the case for the stellar disc of the Milky Way – see the discussion by Pettini (2006).

From Figure 20.1 we see that QSOs and actively star-forming galaxies are the two classes of objects which are of particular interest to the theme of the present meeting, although I shall also briefly mention damped Lyman-α systems (DLAs). Each of these deserves, and indeed has been the subject of, a full review in its own right, which is well beyond the scope of this brief article. Instead I shall focus on what I see as the critical issues (in 2006) in abundance studies of QSOs, LBGs and DLAs.

2 Active galactic nuclei

There appears to be a general consensus that the gas associated with active galactic nuclei (AGN) is of Solar to supersolar metallicity (Hamann & Ferland 1999). The realisation in recent years that the growth of supermassive black holes at the centres of galaxies is closely linked to nuclear star formation and to the growth of bulges makes these very high metallicities relevant not only to gas in the extreme environment close to an AGN but, presumably, also to the stellar populations in the bulges of the host galaxies. Thus, some $(10^8–10^9)M_\odot$ of the stellar bulge (comparable to the black-hole mass) may have supersolar metallicity (Baldwin *et al.* 2003).

Clearly, it would be of considerable interest to measure element abundances in this regime with a degree of accuracy sufficient to (a) provide empirical constraints on models of galactic chemical evolution in the supersolar regime and (b) allow us to compare the results with other metallicity measures such as those applicable to the stellar light of elliptical galaxies, for example. So, how are metallicities of AGN measured? This is accomplished by fitting the relative strengths of various emission lines with the best available photoionisation models (e.g. Hamann *et al.* 2002; Nagao *et al.* 2006), but fundamentally it comes down to measuring the ratios of N lines to lines of He, C and O and relying on the secondary nature of nitrogen to deduce the oxygen abundance and hence the 'metallicity'. In other words, the whole procedure rests on an extrapolation of the (N/O) versus (O/H) trend seen in local H II regions, where the abundance of N grows with the square of the O abundance (or possibly even more steeply) when $(O/H) \gtrsim \frac{1}{2}(O/H)_\odot$ (e.g. Henry *et al.* 2000). Thus, the determination of metallicities in AGN is closely tied to the current debate on the most appropriate nebular abundance indicators in the metal-rich regime, reviewed by Fabio Bresolin in Chapter 17 of this volume. The large scatter in the values of (N/O) which apparently results when (O/H) is deduced from temperature-sensitive auroral lines (see, for example, Figure 8 of Bresolin *et al.* 2004) certainly sounds a note of caution in the interpretation of the nitrogen emission lines in AGN.

2.1 Metallicities from absorption lines

As in other areas of astrophysics, a cross-check of the high metallicities deduced from the broad emission lines in AGN with an independent abundance diagnostic is highly desirable. The narrow absorption lines seen in QSO spectra at redshifts close to the emission redshift are generally considered to be different from those of most 'intervening' QSO absorbers (in the sense that they arise in the QSO environment) and there are claims in the literature that the metallicity of these 'associated' absorbers is indeed supersolar (e.g. Petitjean *et al.* 1994). The difficulty

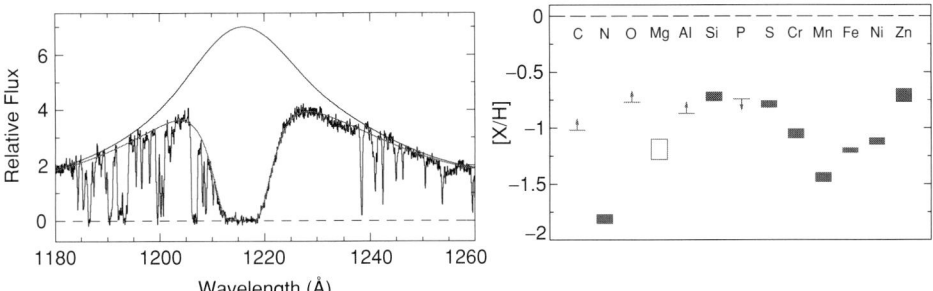

Figure 20.2. Left panel: a portion of the spectrum of Q2343–BX415 obtained with the Echelle Spectrograph and Imager (ESI) on the Keck II telescope, showing the region encompassing the Lyman-alpha (Lyα) line. A DLA with $N(\text{H\,\textsc{i}}) = 1 \times 10^{21}$ cm^{-2} absorbs away most of the QSO Lyα emission line since both are at redshift $z = 2.57393$. Right panel: the chemical-abundance pattern determined by Rix *et al.* (2007) for the proximate DLA in Q2343–BX415. Metallicities are shown relative to Solar (long-dashed line at [X/H] = 0.0); the height of each box reflects the uncertainty in the abundance of the corresponding element.

here is that these abundance determinations require photoionisation models to take into account unobserved ion stages, given that most proximate absorption systems exhibit a high degree of ionisation.

Among the various classes of QSO absorbers, those with the most secure and extensive abundance measures are the DLAs (Wolfe *et al.* 2005) – when the neutral-hydrogen column density exceeds $N(\text{H\,\textsc{i}}) = 2 \times 10^{20}$ cm^{-2}, the gas is predominantly neutral and one observes directly the dominant ion stages of a relatively large variety of elements in the rest-frame ultraviolet (UV) absorption spectrum. Proximate DLAs are rare, but Rix *et al.* (2007, in preparation) have recently studied in detail the chemical composition of one such system in Q2343–BX415. This faint ($\mathcal{R} = 20.2$) QSO was discovered in the course of the survey for UV-selected (or 'BX') galaxies at $z \simeq 2$ by Steidel *et al.* (2004).

As can be seen from Figure 20.2 (left panel), the QSO spectrum reveals a strong ($N(\text{H\,\textsc{i}}) = 1 \times 10^{21}$ cm^{-2}) DLA at a redshift that coincides with that of the QSO emission lines ($z_{\text{abs}} = z_{\text{em}} = 2.57393$). In their study, Rix *et al.* determined the abundances (or limits) for 13 elements, from C to Zn (illustrated in the right panel of Figure 20.2), and concluded that the properties of this DLA are broadly similar to those of more conventional damped systems at $z_{\text{abs}} < z_{\text{em}}$. In particular, its abundance pattern is not unusual, with a metallicity of $Z \simeq \frac{1}{5} Z_\odot$ (which evidently applies to α-capture elements and to Zn) and a mild depletion of refractory species similar to that seen in diffuse interstellar clouds in the Milky Way. The elements N and Mg are somewhat anomalous in exhibiting lower abundances than expected on the basis of the other eleven elements considered. We do not know the location of this DLA relative to BX415. The low degree of ionisation and excitation of the gas (the latter measured via the fine-structure line of C II at 1335 Å – see Wolfe *et al.* (2005)) make it unlikely that it is in the immediate vicinity of the AGN. However, the DLA may be formed in the interstellar medium (ISM) of the QSO host or of a nearby galaxy that is part of the same large-scale structure. In any case, these results show that absorption-line systems at the same redshifts as the background QSOs against which they are being viewed are not always metal-rich. In the case of BX415 there may in fact be little difference between the metallicity of the DLA and that of the QSO itself, which appears to be sub-solar given the weakness of its N emission lines (Rix *et al.* 2007 in preparation).

3 Star-forming galaxies

Among the various classes of objects considered in this review, star-forming galaxies at $z = 2$–3 are those for which in the last few years there has been most progress towards determining their physical properties, including metallicity. Here I concentrate in particular on recent results on galaxies in what used to be referred to as the 'redshift desert' ($z \simeq 2$), which somewhat ironically has turned out to be one of the most intensely studied cosmic epochs, partly because of its accessibility to ground-based observations and partly because we now understand it to be the epoch when mass assembly in galaxies was at its peak.

Observations of galaxies in the $z = 2$–3 redshift range now span almost the entire electromagnetic spectrum, from X-rays to radio, and spectroscopic surveys include thousands of such objects with measured redshifts. Among this rich complement of data, the near-infrared bands have played a pivotal role. Although it is still somewhat difficult to access efficiently from the ground, given the high and variable terrestrial background and the lack until now of multi-object spectrographs, the observed near-IR regime has the advantage of sampling rest-frame optical wavelengths at which metallicity and other physical diagnostics are better developed than in many other wavelength intervals.

In particular, the rest-frame optical colours measured with J, H and K photometry are sufficient, when combined with the more easily obtained rest-frame UV colours, to give an adequate description of the galaxies' spectral energy distributions (SEDs). As shown by Shapley *et al.* (2005), the SEDs can be interpreted with state-of-the-art stellar-population models in terms of the past history of star formation experienced by the galaxies and one can deduce, among other quantities, an estimate of the assembled stellar mass, as illustrated in Figure 20.3.

For the UV-selected (and hence actively star-forming) galaxies at $z \sim 2$ considered by Erb *et al.* (2006b), assembled stellar masses are in the range $\sim (10^9$–$10^{11.4}) M_\odot$, with a mean $\langle M_{\text{stars}} \rangle = (3$–$4) \times 10^{10} M_\odot$. At one end of this range we have galaxies where star formation has apparently just turned on (in the

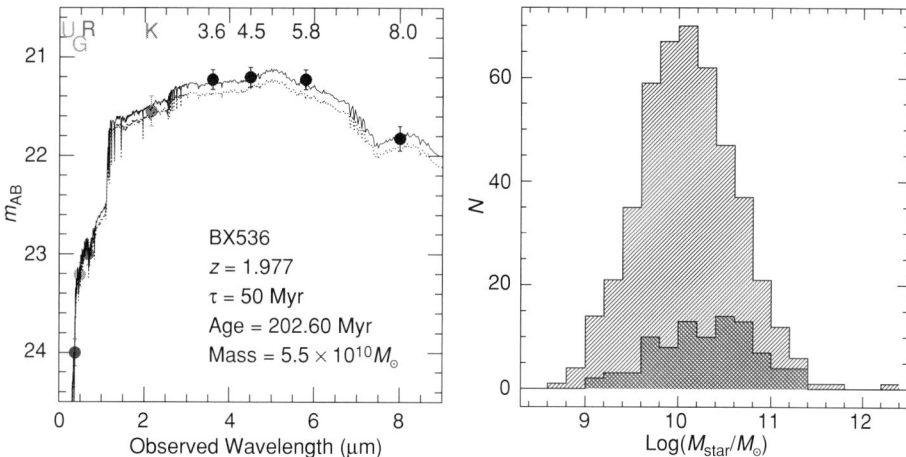

Figure 20.3. Left panel: an example of SED fitting, reproduced from Shapley *et al.* (2005). A wide grid of parameter space in the spectral synthesis models of Bruzual & Charlot (2003) is explored to determine the age, assembled stellar mass, past history of star formation (continuous or exponentially declining) and reddening that best fit the observed optical and IR photometry. The two fits shown here are respectively with and without the Spitzer IRAC data. Shapley *et al.* found that the addition of the 3.6–8-μm IRAC data, while reducing the uncertainties in the physical parameters deduced, does not lead to significantly different star-formation histories for most of the galaxies considered. Right panel: histograms of assembled stellar masses deduced from SED fitting of BX galaxies, reproduced from Erb *et al.* (2006b). The larger histogram is for a sample of 461 BX galaxies with near-IR photometry, while the smaller histogram pertains to a sub-set of these for which Erb *et al.* detected nebular Hα emission with near-IR spectroscopy.

last few 10^7 years) and most of the baryons are still in gaseous form (gas fractions $f > 0.8$). At the other end of the scale, some 15% of the BX galaxies in the Erb *et al.* (2006b) sample have ages comparable to the age of the Universe at the time when we observe them and have by then already turned most of their gas into stars ($f < 0.2$). When the gas fractions are taken into account, the total baryonic masses range from $M_{\text{stars+gas}} \simeq 10^{10.3} M_\odot$ to $10^{11.5} M_\odot$, in reasonable agreement with the dynamical masses indicated by the widths of the Hα emission lines and consistent with the masses $M_{\text{DM}} \simeq (10^{11.8}$–$10^{12.2}) M_\odot$ of the dark-matter halos in which these galaxies reside, as deduced from their clustering properties (Adelberger *et al.* 2005). As argued by Adelberger *et al.*, such masses indicate that today's ellipticals and massive bulges are the descendants of the galaxies we see undergoing vigorous star formation at $z = 2$.

3.1 Metallicities of star-forming galaxies at $z \simeq 2$

As reviewed by Pettini (2006), the problem of determining the metallicities in star-forming galaxies at $z = 2$–3 has been tackled with a variety of methods, using

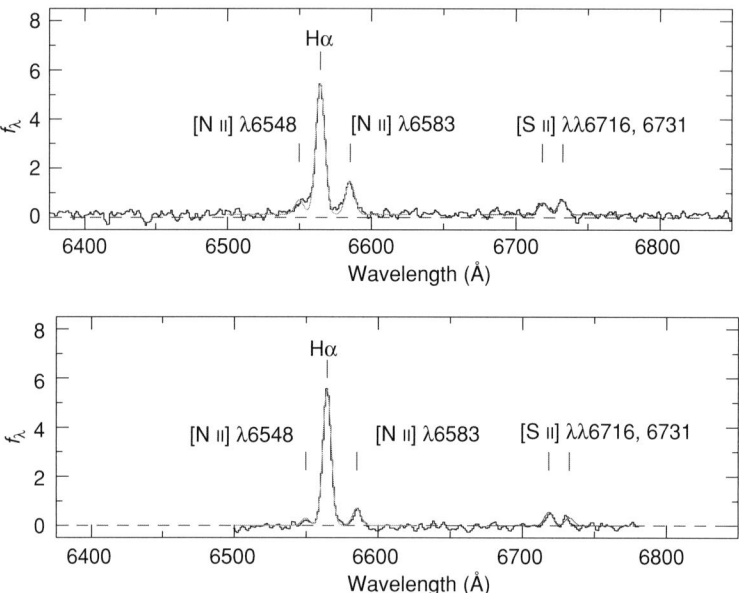

Figure 20.4. Composite spectra of BX galaxies at $z \simeq 2.2$ from the survey by Erb *et al.* (2006a). Upper panel: galaxies brighter than $K_s = 20$ have a mean ratio [N II]/H$\alpha = 0.25$, which indicates an oxygen abundance of $12 + \log$ (O/H) $= 8.56$, or about four fifths Solar, if the local calibration of the N2 index with (O/H) applies to these galaxies. Lower panel: galaxies fainter than $K_s = 20$ have [N II]/H$\alpha = 0.13$ and $12 + \log$ (O/H) $= 8.39$, or about half Solar.

nebular emission, interstellar absorption and stellar features in the UV, including both wind and photospheric lines. These different approaches have given roughly concordant answers (to within a factor of ~ 2) in the few cases so far when they could be cross-checked, but a realistic assessment of the current situation would be to say that we are still some way off from developing reliable metallicity indicators that are easily applicable to high-z galaxies. The diagnostic which has had the widest application up to now is the N2 index – the ratio of the intensities of the [N II] $\lambda 6583$ and Hα emission lines – which was most recently calibrated by Pettini & Pagel (2004) in terms of the oxygen abundance in nearby H II regions. One of the advantages of this index is that it relies on the ratio of two emission lines that are closely spaced in wavelength, thereby disposing of the need for accurate flux and reddening determinations (see Figure 20.4).

When Erb *et al.* (2006a) applied the local N2 calibration to their sample of 87 $z \sim 2$ galaxies grouped in six bins of stellar mass, a clear mass–metallicity relation emerged (see Figure 20.5). There are several points that can be made from consideration of Figure 20.5. First, these results certainly dispel any naive ideas that high-redshift galaxies may be generally metal-poor – on the contrary, 85%

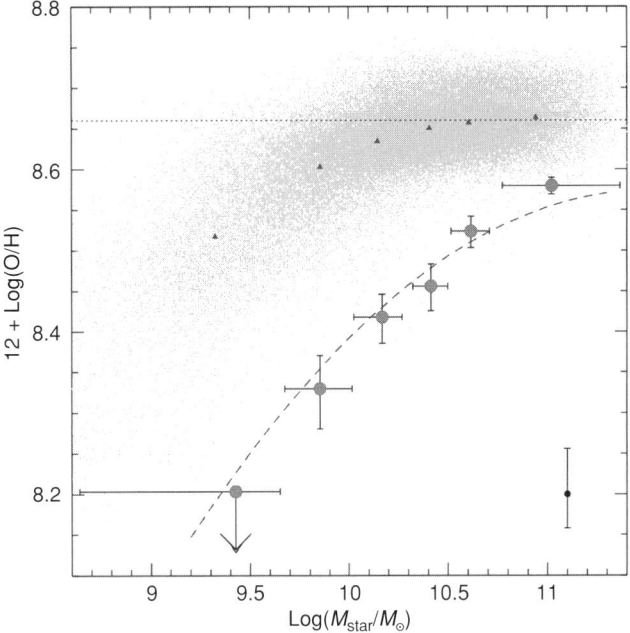

Figure 20.5. The stellar mass–metallicity relation for star-forming galaxies at $z \sim 2$ (large circular points with error bars) and in the nearby ($z \sim 0.1$) Universe from the SDSS survey (numerous small grey points). The large circular points correspond to the sample of 87 UV-selected galaxies at $z \simeq 2$ grouped by Erb *et al.* (2006a) into six bins according to their stellar mass. The triangles show the mean metallicities of the SDSS galaxies in the same mass bins as the $z \simeq 2$ galaxies. The vertical error bar in the bottom right-hand corner indicates the uncertainty in the calibration of the N2 index used to deduce the oxygen abundances of both sets of galaxies. Since the N2 index saturates near Solar metallicity (indicated by the horizontal dotted line), the offset between the present-day and high-redshift relations is best derived by consideration of the lower-metallicity bins. At a given assembled stellar mass, galaxies at $z - 2$ appear to be about a factor of two lower in metallicity than today. Figure reproduced from Erb *et al.* (2006a).

of the BX galaxies in the Erb *et al.* sample have $(O/H) \gtrsim \frac{2}{5}(O/H)_\odot$. Second, it is hard to assess how metal-rich the most massive galaxies actually are, for the reason discussed by Fabio Bresolin at this meeting: the inherent uncertainties in the nebular abundance scale at supersolar metallicities. Specifically, the N2 index saturates near Solar oxygen abundance, as is also shown by the cloud of SDSS points in Figure 20.5. Possibly, in this regime the C IV and Si IV P-Cygni lines from luminous OB stars, which are commonly seen in the UV spectra of these galaxies, would be a more reliable tell-tale sign of supersolar abundances (Rix *et al.* 2004), but no clear-cut such cases have been identified yet at high or low z (see Chapter 6 in these proceedings). In any case, the Solar or slightly subsolar metallicities of

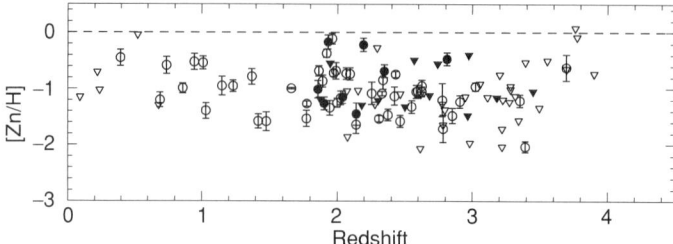

Figure 20.6. Abundances of Zn in 87 DLAs, from the compilation by Kulkarni *et al.* (2005) (open symbols), which brings together the results of several surveys for DLAs in optically selected QSO samples, and from the recent survey of 20 CORALS (radio-selected) QSOs by Akerman *et al.* (2005) (filled symbols). Triangles denote upper limits in DLAs whose Zn II $\lambda\lambda$2026, 2062 lines remain undetected.

even the most massive BX galaxies in Figure 20.5 are not necessarily inconsistent with the proposal that they are the progenitors of today's massive ellipticals, since the latter exhibit supersolar metallicities only in their inner regions, whereas the metallicity measures at high z refer to the integrated light of a whole galaxy. It will be of great interest to investigate the presence of abundance gradients in BX galaxies with spatially resolved observations, once integral field spectrographs with adaptive-optics correction are fully operative (Förster-Schreiber *et al.* 2006; Law *et al.* 2006).

A final point concerning the mass–metallicity relation is that, even within the uncertainties of the metallicity calibrations, there appears to be a real offset between the relation today and that at $z = 2$, in the sense that galaxies of a given stellar mass were a factor of ~ 2 less metal-enriched in the distant past (10 Gyr ago, or three quarters of the age of the Universe in today's 'consensus' cosmology) than they are now. This is weak evolution indeed, as argued earlier, and consistent with the shallow metallicity–redshift gradients deduced by observations of galaxies at intermediate redshifts (e.g. Savaglio *et al.* 2005, Maier *et al.* 2006).

4 Damped Lyman-alpha systems

Are DLAs relevant to a meeting on the metal-rich Universe? The answer would at first seem to be negative: it has been a source of some concern over the last decade that these absorption systems, which in principle should give us an unbiased measure of the progress of cosmic chemical evolution (e.g. Pei & Fall 1995), are generally metal-poor at all redshifts sampled (see Figure 20.6). Indeed, one could argue that DLAs offer us one of the best means to identify the most *metal-poor* gas at high redshifts and, with highly precise measures of its chemical composition, provide precious clues to the first cycles of nucleosynthesis, complementing those obtained from parallel stellar abundance studies in the Milky Way halo (e.g. Erni *et al.* 2006).

However, all this may be about to change. Thanks in part to the order-of-magnitude increase in the number of known QSOs – and therefore DLAs – brought about by the SDSS, there have been several reports in the last few months of discoveries of metal-rich DLAs, or sub-DLAs.[1] These systems may make a significant contribution to the census of metals, particularly at redshifts $z < 1$ (Kulkarni *et al.* 2006). Particularly intriguing are the claims of highly supersolar abundances, by factors of 4–5, in two cases (Péroux *et al.* 2006; Prochaska *et al.* 2006). In principle, metallicity measurements from interstellar absorption lines are the most straightforward to perform – the underlying physics is simple, and practical complications such as saturation of the metal lines and blending of the Lyα line with nearby unrelated absorption features would both work in the sense of *decreasing* the abundances deduced. Possibly, the gas is more highly ionised than can be judged from the available data. In any case, it is clear that these extreme cases are highly deserving of further study: at redshifts $z < 1$ imaging should presumably reveal the presence of massive, luminous galaxies close to the QSO sight-lines and, looking further ahead, observations of O I and N I absorption lines with a UV spectrograph on the refurbished Hubble Space Telescope should clarify the importance of ionisation corrections.

In a related development, Wild *et al.* (2006) recently suggested that absorption systems with strong Ca II $\lambda\lambda$3933, 3968 absorption lines (with an equivalent width greater than 0.5 Å for the stronger member of the doublet) are likely to be high-column-density DLAs. The Ca II doublet, which has been mapped extensively in the Galactic ISM since the discovery of its interstellar nature by Eddington (Adams 1941), has been relatively neglected at high redshifts, where it moves into a difficult part of the spectrum at $z > 1$. Now, with a database of QSO spectra as extensive as that available in the SDSS, it may provide an effective way to isolate DLAs specifically in that all-important redshift interval, $z < 1$, which accounts for more than half of the age of the Universe and yet is still relatively unexplored with 'conventional' DLAs, which require space observations to be identified in the first place. At these intermediate-to-low redshifts, it is possible with modern instrumentation to identify the galaxies responsible for the Ca II absorption and measure their physical properties. In a pilot study, admittedly aimed at the more luminous galaxies within

---

[1] The conventional definition of a damped Lyα system is one in which the column density of neutral hydrogen is above the threshold $N(\text{H I}) = 2 \times 10^{20}$ cm^{-2}, but absorption systems with column densities one order of magnitude lower than this value also exhibit (obviously weaker) damping wings in the Lyα line; they are usually referred to as sub-DLAs or 'super-Lyman-limit systems' – another example of the confusing power of labels in astronomy!

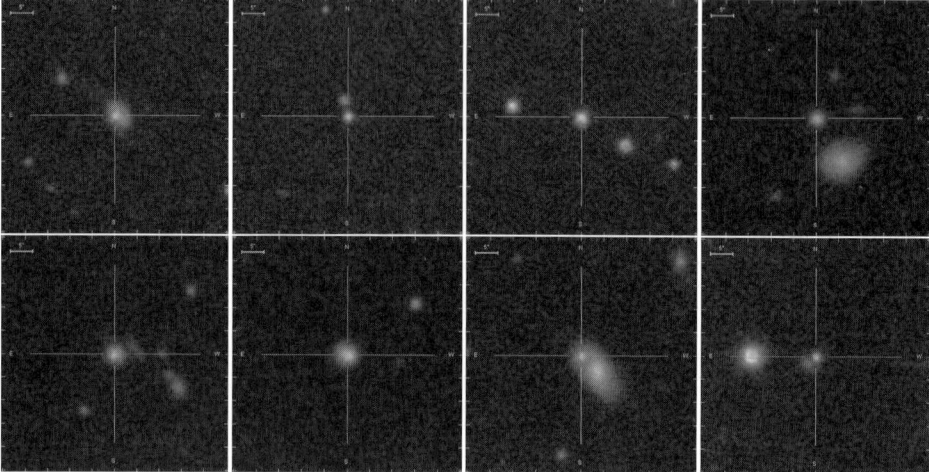

Figure 20.7. Colour composite (u, g, r, i, z) images of the eight SDSS QSOs with $z_{\rm abs} < 0.5$ Ca II absorption systems studied by Zych *et al.* (2007, in preparation). In each panel the galaxy closest to the QSO sight-line is at the same redshift as the Ca II lines in the QSO spectrum and is presumably the absorber. (The Ca II-bearing galaxy is directly beneath the QSO image in the third panel from the left in the top row and the second panel in the bottom row).

the SDSS sample of Ca II absorbers at $z_{\rm abs} < 0.5$, Zych *et al.* (2007, in preparation) found that all eight galaxies targeted (see Figure 20.7) are metal-rich, with nebular abundances (O/H) $\gtrsim$ (O/H)$_\odot$, and are sustaining vigorous rates of star formation in excess of a few Solar masses per year. These preliminary results are at odds with the conventional notion of metal-poor, quiescent, DLA galaxies and suggest that this picture may still be incomplete. They certainly stimulate efforts to clarify the relationship between Ca II- and H I-selected DLAs.

5 Concluding remarks

This timely and inspired meeting has highlighted the limitations of our knowledge of the metal-rich Universe. The tools at our disposal for measuring the final stages in the chemical evolution of galaxies are still rather blunt. One particular area of concern is that different metallicity indicators are used, of necessity, for different types of objects, with still a surprisingly little degree of overlap or cross-checking. Much remains to be done, and this is clearly the case at high, as well as low, redshifts. What I hope to have conveyed here is the key role that the high-redshift Universe can play in our understanding not only of the first structures to form and ignite stellar nucleosynthesis (an area where much attention is focused at present), but also of the most metal-rich objects we know, the quasars and massive elliptical galaxies, which at high redshifts we can observe directly during the epochs when they acquired most of their metals. The 'last', as well as the first, stars should provide the scientific motivation for the next generation of space and ground-based telescopes now being conceived.

Acknowledgments

I am grateful to Sam Rix and Berkeley Zych for permission to show results of their work in advance of publication.

References

Adams, W. S. (1941), *PASP* **53**, 73
Adelberger, K. L., Steidel, C. C., Pettini, M., Shapley, A. E., Reddy, N. A., & Erb, D. K. (2005), *ApJ* **619**, 697
Akerman, C. J., Ellison, S. L., Pettini, M., & Steidel, C. C. (2005), *A&A* **440**, 499
Baldwin, J. A., Ferland, G. J., Korista, K. T., Hamann, F., & Dietrich, M. (2003), *ApJ* **582**, 590
Bresolin, F., Garnett, D. R., & Kennicutt, R. C. Jr. (2004), *ApJ* **615**, 228
Bruzual, G., & Charlot, S. (2003), *MNRAS* **344**, 1000
Cen, R., & Ostriker, J. P. (1999), *ApJ* **519**, L109
Erb, D. K., Shapley, A. E., Pettini, M., Steidel, C. C., Reddy, N. A., & Adelberger, K. L. (2006a), *ApJ* **644**, 813
Erb, D. K., Steidel, C. C., Shapley, A. E., Pettini, M., Reddy, N. A., & Adelberger, K. L. (2006b), *ApJ* **646**, 107
Erni, P., Richter, P., Ledoux, C., & Petitjean, P. (2006), *A&A* **451**, 19
Förster-Schreiber, N. M., Genzel, R., Lehnert, M. D. *et al.* (2006), *ApJ* **645**, 1062
Hamann, F., & Ferland, G. (1999), *ARA&A* **37**, 487
Hamann, F., Korista, K. T., Ferland, G. J., Warner, C., & Baldwin, J. (2002), *ApJ* **564**, 592
Henry, R. B. C., Edmunds, M. G., & Köppen, J. (2000), *ApJ* **541**, 660
Kulkarni, V. P., Fall, S. M., Lauroesch, J. T. *et al.* (2005), *ApJ* **618**, 68
Kulkarni, V. P., Khare, P., Péroux, C., York, D. G., Laurocsch, J. T., & Meiring, J. D. (2007), *ApJ* **661**, 88
Law, D. R., Steidel, C. C., & Erb, D. K. (2006), *AJ* **131**, 70
Maier, C., Lilly, S. J., Carollo, C. M., Meisenheimer, K., Hippelein, H., & Stockton, A. (2006), *ApJ* **639**, 858
Nagao, T., Maiolino, R., & Marconi, A. (2006), *A&A* **447**, 863
Pei, Y. C., & Fall, S. M. (1995), *ApJ* **454**, 69
Péroux, C., Kulkarni, V. P., Meiring, J. *et al.* (2006), *A&A* **450**, 53
Petitjean, P., Rauch, M., & Carswell, R. F. (1994), *A&A* **291**, 29
Pettini, M. (2006), in *The Fabulous Destiny of Galaxies: Bridging Past and Present*, eds. V. Le Brun, A. Mazure, S. Arnouts, & D. Burgarella (Paris, Edition Frontières), pp. 319–333
Pettini, M. & Pagel, B. E. J. (2004), *MNRAS* **348**, L59
Prochaska, J. X., O'Meara, J. M., Herbert-Fort, S., Burles, S., Prochter, G. E., & Bernstein, R. A. (2006), *ApJ* **648**, L97

Rix, S. A., Pettini, M., Leitherer, C., Bresolin, F., Kudritzki, R.-P., & Steidel, C. C. (2004), *ApJ* **615**, 98
Savaglio, S., Glazebrook, K., Le Borgne, D. *et al.* (2005), *ApJ* **635**, 260
Shapley, A. E., Steidel, C. C., Erb, D. K. *et al.* (2005), *ApJ* **626**, 698
Steidel, C. C., Shapley, A. E., Pettini, M. *et al.* (2004), *ApJ* **604**, 534
Wild, V., Hewett, P. C., & Pettini, M. (2006), *MNRAS* **367**, 211
Wolfe, A. M., Gawiser, E., & Prochaska, J. X. (2005), *ARA&A* **43**, 861

21

Evolution of dust and elemental abundances in quasar DLAs and GRB afterglows as a function of cosmic time

Bryan E. Penprase,[1] Wallace Sargent[2] & Edo Berger[3]

[1] Pomona College Department of Physics and Astronomy, Claremont, CA, USA
[2] California Institute of Technology, Pasadena, CA, USA
[3] Observatories of the Carnegie Institution, Pasadena, CA, USA

We present results from comparisons of elemental abundances and dust content between damped Lyman-alpha (DLA) absorbers and gamma-ray-burst (GRB) afterglows, as determined by absorption-line spectroscopy. Our sample of DLA absorbers includes the results from 76 quasar spectra taken with the HIRES spectrograph of the Keck observatory, from which we obtain a sample of 38 DLA absorbers in the redshift range $2 < z < 4$. The GRB absorption lines were obtained in collaboration with the Caltech Carnegie NOAO GRB collaboration, in which rapid spectroscopy is obtained from newly discovered GRBs, to obtain high-quality optical spectra. We present results of O, N, C, Si, Zn, Cr, and C II*/C II ratios from a "core sample" of 15 of the best of the DLA absorbers, and detailed analysis of GRB 051111 and GRB 050505, which are at redshifts of $z = 1.549$ and 4.275, respectively. From our analysis we can see trends in the DLA dust content, and in [C/H], [N/H] and [O/H] values as a function of DLA redshift, as well as evidence for dust formation and highly excited dense gas within the disks of GRB host galaxies.

1 Introduction

Two of the best probes of the metallicity in the early Universe are quasar absorption lines, as revealed from damped Lyman-alpha (DLA) systems, and gamma-ray-burst (GRB) explosions in which the bright afterglow provides a transient probe of the high-redshift host galaxy. Many good estimates of trends in metallicity have been obtained from DLA systems, especially within the range $z < 3$ and for absorption lines such as Fe II, Mg II, Si II, Zn II, and Cr II, for which absorption from each species can be compared directly with H I column density, and in which the dominant

The Metal-rich Universe, eds. G. Israelian and G. Meynet. Published by Cambridge University Press.
© Cambridge University Press 2008.

ionization state coexists with the H I (Prochaska *et al.* 2003). From these studies it is clear that metallicity of DLA systems rises with decreasing redshift, correlating well with the prevailing scenario of star formation peaking with $z < 3$, and yet also with significant scatter, reflecting the varying impact parameter of the DLA sightline relative to the disks of the galaxy associated with the DLA. Additional information on the abundances of the lighter elements can be found from O I, N I, and C II absorption lines, which are difficult to measure due to the fact that most of these lines are saturated in typical DLA sightlines. One potential advantage of these absorption lines is the fine-structure excitation of O I* and C II*, which are sensitive to collisional excitation in galactic disks and to radiative excitation both from the cosmic background radiation and from star formation (Silva & Viegas 2002).

The rate of discovery of gamma-ray bursts (GRBs) has accelerated to nearly one per week, and networks of ground-based observatories are now able to follow up discoveries of GRBs with high-resolution spectroscopy, offering a unique probe of the early Universe within environments that are otherwise impossible to study. The brightness of a GRB within the first few hours is sufficient to enable detailed elemental analysis, by using the GRB afterglow to probe the disk of the "host galaxy." The key to succesful data acquisition is speed, and localization of the GRB host quickly to enable spectra to be acquired. As a probe of the interstellar medium (ISM) and elemental abundances, GRBs have some advantages over the DLA. Since the exploding star for a "long" GRB (the type typically studied with absorption lines) is thought to be a high-mass star, the GRB is physically located within disks of the host galaxy, often adjacent to star-forming complexes with enriched metallicity and some dust formation even at high redshifts. In addition, the GRB can probe highly excited regions in which lines of highly excited Fe II (in states like Fe II*, Fe II**, etc) and excited Si II are observed, to enable some constraints to be applied on the density, temperature, and radiation intensity within the environment of the GRB in the host galaxy.

In this work we present results for both DLA and GRB absorption-line spectra, and some of the key diagnostics of nucleosynthesis, excitation, and dust content within the high-redshift Universe.

2 Damped Lyman-alpha absorption-line abundances of Si II, O I, N I, C II, Zn II, and Cr II

Our sample of DLA absorption lines includes 38 excellent DLA spectra, from which we selected a subset for which minimal blending and saturation is seen for C II, O I, and Si II absorption lines. By using this subset of 15 of the best DLA systems, we are able to get better constraints on the Si II, O I, N I, and C II abundances, which usually are saturated in DLA systems. For each sightline, the absorption lines have been analyzed in multiple ways, including apparent-optical-depth (AOD) analysis, curve-of-growth (COG) analysis, and Voigt-profile analysis. Sample profiles and analysis steps were presented during the conference, and a basic summary of the trends in the results is provided here. Using H I-weighted abundances of [M/H], where M is each of the species of interest, we looked for trends in [M/H] against the redshift of the system. The trend of [O/H] is to remain flat in the aggregate from $2 < z < 4$, with median values of [O/H] $= -1.8 \pm 0.2$ within this redshift range, but with significant scatter. In the same redshift range, [C/H] increases slightly, with median values of [C/H] $= -2.0$ at $z = 4$ and [C/H] $= -1.6$ at $z = 2$. Abundances of N I are less than those of O I and C II, in accord with nucleosynthesis theory, which favors even-numbered nuclei, exhibit no significant trend with redshift, and have median values of [N/H] $= -2.3$.

One interesting result to look at is trends of abundance ratios of [C/O], [N/O], and [Si/O] against [O/H], which is common in stellar abundance studies, and which also can detect primary and secondary production of elements such as nitrogen (Pettini et al. 2002). Stellar observations of [C/O] and [O/H] show that there is a "valley" of [C/O] for [O/H] values of $-1.5 <$ [O/H] < 0.5 (Akerman et al. 2004), and for our DLA sample there is a consistent downward trend of [C/O] for our sampled values of [O/H], which are within the range $-2.7 <$ [O/H] < -1.5. The values of [N/O] are predicted to reach a floor at values of [O/H] < -1.5, reflecting the "primary" production of nitrogen from Population-III stars at very early epochs. Our data reveal a wide variation of [N/O] values within the observed range of [O/H], and are inconclusive in terms of detecting the floor of [N/O] $= -1.5$, since many of the DLAs have values of [N/O] < -1.5. Analysis of ratios of Si/O within the sample revealed no trend with [O/H], with median values close to Si/O $= -1.0$, suggesting that consistent enrichment of Si and O occurred in the early Universe through the same nucleosynthetic pathway. For those DLA systems for which we have observations of Zn II and Cr II absorption, we compared the abundances of [Zn/Cr] against redshift, and found that for redshifts of $1.5 < z < 3.2$ we have no significant trend in the values of [Zn/H] and [Cr/H], but in nearly all the DLA systems a significant overabundance of Zn over Cr is seen. The median level of [Zn/Cr] is approximately 0.2 dex, as expected for DLA systems with modest dust depletion, which would reduce the observed abundances of Cr as it is locked onto dust grains (Kulkarni et al. 2005).

Using the observed values of N(C II) and N(C II*) from our DLA sample, we computed also the cooling rate I_c in the DLA systems and the fine-structure excitation of the ISM within the DLAs. For our sample we see a signficant upward trend in the values of I_c from $2 < z < 3$, within which range the values of I_c increase from -27 to -26.4, consistent with an increase in star formation in the DLA galaxies, which correlates with C II* cooling (Wolfe et al. 2003). Likewise a computation of the excitation temperature from C II*/C II for the observed redshift range of our sample gives an upward trend that lies significantly above the expected excitation from the cosmic background radiation, suggesting the onset of significant excitation from the increase in star formation at redshifts $z < 3$.

3 Gamma-ray-burst afterglow scaling, dust, and element abundances

The growing number of detected GRBs is providing new sources and has begun to probe higher and higher redshifts, approaching quasars in providing the most distant probes of the early Universe, as shown by the recent discovery of a high-redshift ($z = 6.295$) GRB (Kawai et al. 2006). The GRB as a new probe of abundances and metallicity offers much promise, but first the context of the GRB within its host galaxy needs to be understood in order to interpret the spectroscopic results. A scale model of the GRB was presented at this conference, in which the host galaxy is scaled to the size of the Earth. At this scale, the GRB itself is a small clump of atoms, perhaps in a drink by the poolside, the photosphere producing the X-ray emission is the size of a grain of sand, and the afterglow which provides the optical light for our spectroscopic analysis is the size of the conference venue. Outside of this afterglow region a complex of star-forming clouds would extend, at this scale, to a region roughly the size of the Canary Islands. The unique aspect of the GRB afterglow is that it is imbedded within disks of high-redshift galaxies, in contrast to the DLA sightlines, which offer varying impact parameters with respect to the absorbing galaxies. For this reason the GRB can offer a better probe for detecting regions of higher metallicity within high-redshift regions, since the GRB is imbedded within gas recently enriched in metals from star formation.

Our analysis of GRB absorption lines has been limited by the small but increasing number of GRB observations at high resolution and high signal level. Recent observations of GRB 051111 (Penprase et al. 2006) and GRB 050505 (Berger et al. 2006) provide some illustrative results that indicate the presence of dust within the host galaxy, and offer interesting comparisons with DLA absorption-line results. For GRB 051111 at the redshift $z = 1.5$, we have obtained a high-quality Keck HIRES spectrum, and observed absorption lines of the species Al III, Cr II, Zn II, Fe II, Fe II*, Fe II**, Fe II***, Fe II****, Ni II, Si II, Si II*, S I, and Mg I (see Figure 21.1). By combining the column densities of all the absorption lines, we have been able to determine that the GRB host galaxy has a significant dust content, with a "warm-disk" depletion pattern, and a dust content 2.5 times greater than those of comparable DLA sightlines. The resulting amount of dust per unit metallicity from the GRB is comparable to that of the interstellar medium (ISM) in the Milky Way, suggesting that dust formation in the galaxy is coeval with metal

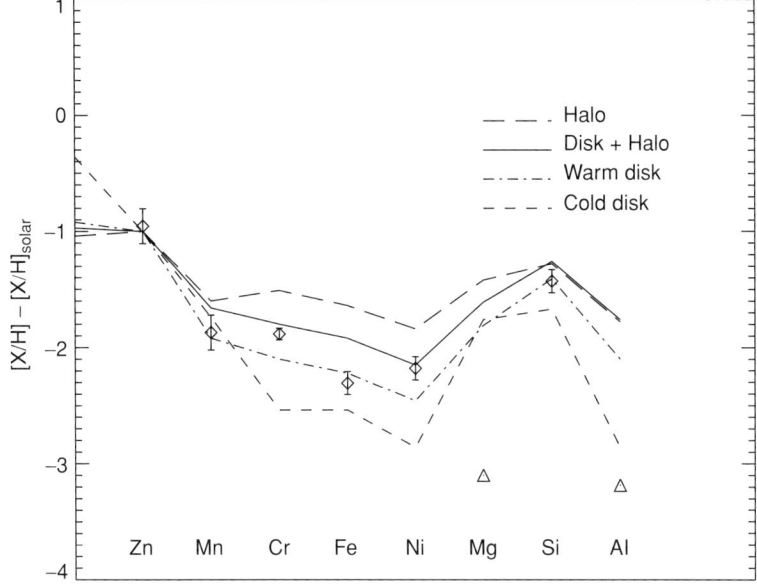

Figure 21.1. Abundance pattern observed for elements detected toward GRB 051111.

enrichment. The overall dust content is likely to be significantly less than that of Milky Way gas, however, and we estimate the value of [Fe/H] to be approximately −1.0 by comparisons with other GRB and DLA sightlines.

The GRB fine-structure excitation is seen to be significantly higher than those of comparable DLA sightlines, and excited fine-structure lines of C II, Si II, and Fe II indicate the presence of warm ($T > 3,000$ K) and dense ($n > 1,000$ cm^{-3}) gas within the GRB host galaxy (Penprase et al. 2006). From the three GRB sightlines mentioned above we observe excitation of N(Si II*)/N(Si II) = 0.174 for GRB 050505 and N(Fe II*)/N(Fe II) = 0.025 for GRB 051111. These values of excitation are rare for DLA samples, and an ongoing debate concerns whether they arise from collisionally excited dense gas within star-formation regions, or from radiatively excited gas near the GRB explosion itself (Prochaska et al. 2006). For the GRB 051111 sightline we have analyzed five different levels of Fe II excitation, which exhibits similar absorption profiles at all excitation levels, which is consistent the presence of with other species such as Mg I, and Mg II. The highly excited gas could arise from dense ISM, heated by the newly formed stars, and, if so, the GRB afterglow spectra can provide excellent diagnostics of conditions within galactic disks. As additional GRB sightlines are observed, it will become easier to separate the radiative and collisional excitation contributions, and it is likely that the GRB will be one of the best probes of metal enrichment and physical conditions at high redshifts in the coming years.

References

Akerman, C. J., Carigi, L., Nissen, P. E., Pettini, M., & Asplund, M. (2004), *A&A* **414**, 931–942
Berger, E., Penprase, B. E., Cenko, S. B. *et al.* (2006), *ApJ* **642**, 979–988
Kawai, N., *et al.* (2006), *Nature* **440**, 184–186
Kulkarni, V. P., Fall, S. M., Lauroesch, J. T. *et al.* (2005), *ApJ* **618**, 68–90
Penprase, B. E., Berger, E., Fox, D. B. *et al.* (2006), *ApJ* **646**, 358–368
Pettini, M., Ellison, S. L., Bergeron, J., & Petitjean, P. (2002), *A&A* **391**, 21–34
Prochaska, J. X., Gawiser, E., Wolfe, A. M., Castro, S., & Djorgovski, S. G. (2003), *ApJ* **595**, L9–L12
Prochaska, J. X., Chen, H. W., & Bloom, J. S. (2006), *A&A* **648**, 95–110
Silva, A. I., & Viegas, S. M. (2002), *MNRAS* **329**, 135–148
Wolfe, A. M., Gawiser, E., & Prochaska, J. X. (2003), *ApJ* **593**, 235–257

22

Dust, metals and diffuse interstellar bands in damped Lyman-alpha systems

Sara L. Ellison

Department of Physics & Astronomy, University of Victoria, 3800 Finnerty Road, Victoria, BC, Canada

Although damped Lyman-alpha (DLA) systems are usually considered metal-poor, it has been suggested that this could be due to observational bias against metal-enriched absorbers. I review recent surveys to quantify the particular issue of dust obscuration bias and demonstrate that there is currently no compelling observational evidence to support the hypothesis of a widespread effect due to extinction. On the other hand, a small subset of DLAs may be metal-rich and I review some recent observations of these metal-rich absorbers and the detection of diffuse interstellar bands in one DLA at $z \sim 0.5$.

1 Introduction

At this conference on the metal-rich Universe, a talk on damped Lyman-alpha (DLA) systems may seem misplaced, since there is now a considerable literature on these absorption-selected galaxies that describes them as metal poor (e.g. Pettini 2004 and references therein). Faced with the almost universally sub-solar abundances of DLAs over all redshifts (the average metallicity of DLAs is $\sim (1/30)\, Z_\odot$ at $z \sim 3$ and $\sim (1/10)\, Z_\odot$ at $z \sim 1$), questions have been raised concerning DLA-selection techniques. For example, are we missing a large fraction of metal-rich DLAs due to dust obscuration bias (Ostriker & Heisler 1984)? Or are the bulk of the metals in absorbers below the canonical DLA column-density threshold (Péroux et al. 2006)? In this contribution, I review the latest observations on the possibility of selection bias and discuss whether DLAs can ever be considered as metal-rich.

The Metal-rich Universe, eds. G. Israelian and G. Meynet. Published by Cambridge University Press.
© Cambridge University Press 2008.

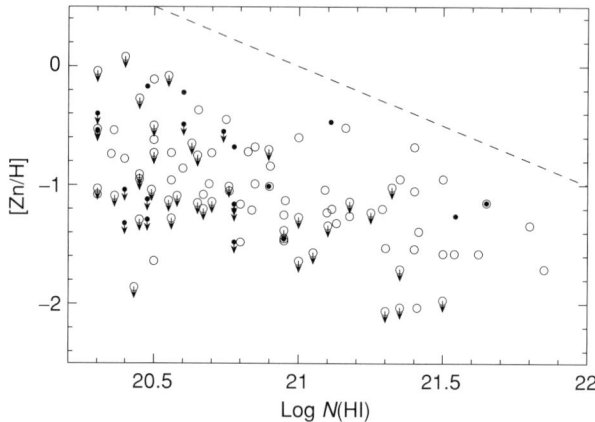

Figure 22.1. Metallicities from optically selected DLA surveys (open points) and from the radio-selected CORALS survey (filled points). The dashed line shows the 'dust filter' of Prantzos & Boissier (2000), although this is inconsistent with limits on DLA reddening (Ellison *et al.* 2005a). Figure adapted from Akerman *et al.* (2005).

2 Are DLA abundances biased due to dust obscuration?

Identification of DLAs relies on a bright background QSO on whose continuum the strong Lyman-α and metal-line species can imprint their absorption signature. If an intervening galaxy is rich in dust and metals, the very background QSO on which we rely will appear fainter and redder and may 'drop out' of traditional quasar surveys. This idea has been around for more than 20 years (Ostriker & Heisler 1984; Fall & Pei 1993) and it has been suggested that extinction bias may 'hide' more than 70% of gas at high redshift. Four pieces of observational support are often cited for the dust-bias scenario: (1) there is only mild redshift evolution in DLA metallicities, (2) metallicities are low compared with those of emission-line galaxies at similar redshifts, (3) continuum slopes for QSOs with DLAs steeper than those for QSOs without (Pei, Fall & Bechtold 1991) and (4) there is an anticorrelation between $N(H\ \textsc{i})$ and metallicity (Prantzos & Boissier 2000), see Figure 22.1.

In order to quantify the impact of dust bias, two surveys based on radio-selected QSOs have now been completed. Together, the Complete Optical and Radio Absorption Line System (CORALS) (Ellison *et al.* 2001) and the UCSD survey (Jørgenson *et al.* 2006) cover a redshift path $\Delta z \sim 100$ for 119 radio-selected QSOs with very deep (and, in the case of CORALS, complete) optical follow-up. In neither survey was an excess of absorbers found either at high or at low (Ellison *et al.* 2004) redshift and the neutral-gas content is in good agreement (within a factor of 2) with optical samples, see Figure 22.2. More importantly for the present discussion,

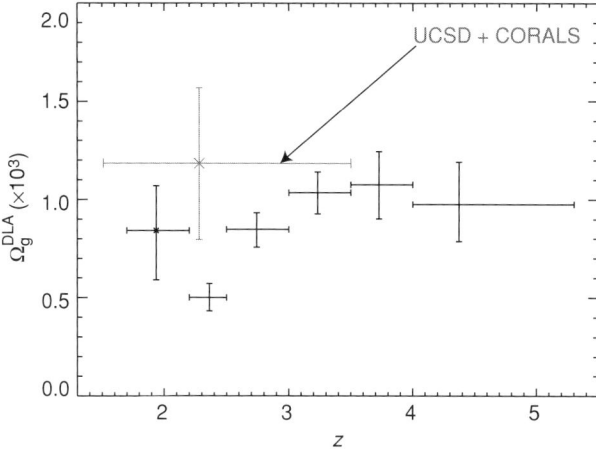

Figure 22.2. Mass density of neutral gas in DLAs from SDSS (points shown as a function of redshift) and radio-selected surveys (cross). Figure adapted from Jørgenson et al. (2006).

Akerman et al. (2005) have shown that the metallicities of DLAs in the CORALS sample are not significantly higher than those of optically selected samples and do not populate the parameter space at high $N(H\ I)$ and high metallicity (Figure 22.1). However, since metallicities are usually weighted by the rare high-$N(H\ I)$ absorbers, a larger sample is required in order to make this result robust.

So, how do we explain the observational 'evidence' in support of dust bias? Authors of a number of recent papers have proposed the idea that, although DLAs may be reservoirs rich in atomic gas, they do not generally flag the location of the bulk of star formation and, therefore, metals (e.g. Wolfe & Chen 2006). Indeed, emission-line spectroscopy of $z \sim 0.5$ galaxies causing absorption-line systems typically observes Solar abundances (Ellison et al. 2005b). Such observations indicate that high abundances *can* be found in absorption-selected galaxies, although they may be confined to smaller regions than the cross-section of DLA-producing gas. The cause of the anti-correlation between $N(H\ I)$ and metallicity is still being debated. However, Ellison et al. (2005a) have argued that the very low values of reddening that are now being determined for DLAs may soon make dust obscuration an unviable explanation. An alternative explanation may be that sight-lines that pass through high-column-density, high-metallicity gas are simply rare. Simulations support this idea (e.g. Johansson & Efstathiou 2006), showing that the cross-section for such gas at $z \sim 3$ is small. Finally, concerning the reddening of QSO continua, this may simply have been a case of a difficult measurement combined with small-number statistics. By fitting continua to $\sim$1500 SDSS spectra,

Murphy & Liske (2004) have found reddening to be very low, $E(B-V) < 0.02$, and a similarly low value has been found by Ellison *et al.* (2005a) in CORALS QSOs, $E(B-V) < 0.04$ based on optical-to-IR colours. Therefore, regardless of concerns about the modest sample size of radio-selected QSO surveys such as UCSD and CORALS, there is no compelling observational evidence to invoke a dust bias. Combined with this revised observational view of dust obscuration, it is interesting to note that theory is also re-assessing the effect of extinction. For example, Trenti & Stiavelli (2006) estimate the total gas density in DLAs to be underestimated by only $\sim 15\%$ in optical surveys.

3 Are all DLAs of low metallicity?

Although most DLAs are metal- and dust-poor, this is not to say that there is not a small fraction of absorbers exhibiting more extreme properties. These extreme cases open the door to some innovative analyses, permitting detection of unusual species and even shedding light on the Galaxy's physical properties.

3.1 Diffuse interstellar bands

Diffuse interstellar bands (DIBs) are common in the spectra of reddened stars in the Milky Way. Although they are numerous (there are now over 100 different known bands), the identification of the DIB carriers is one of the oldest outstanding mysteries in astronomical spectroscopy. Amongst the potential candidates are polycyclic aromatic hydrocarbons and long carbon chains. The DIBs have been detected in only a handful of extra-Galactic sight-lines, including the LMC and SMC (Snow 2001), and in one case, the broad 4428-Å feature in a $z \sim 0.5$ DLA (Junkkarinen *et al.* 2004). Since the strengths of some DIB lines, such as the 5780-Å feature, correlate with N(H I) in the Galaxy (Herbig 1993), selecting high-column-density DLAs may allow us to detect DIBs in high-redshift galaxies. We have been undertaking such a search (Lawton *et al.* (2008)) and summarise our findings in Figure 22.3. In only one case do we detect DIB absorption: in the $z \sim 0.5$ DLA towards AO 0235 + 164 (York *et al.* 2006), where we detect both the 5780-Å and the 5705-Å DIB. From this detection and upper limits from four other DLAs, we find that the 5780 Å line strengths in DLAs are weaker, by up to a factor of 10, for their N(H I) than Galactic sight-lines. Similar deficiencies are seen in Magellanic sight-lines, suggesting that DIB strength probably depends on metallicity and local physical properties as well as on N(H I). On the other hand, the 5780-Å and 5705-Å DIBs have similar equivalent width ratios in the $z \sim 0.5$ DLA and the Galaxy, possible evidence that these two bands originate from a similar carrier. Surprisingly, we do not detect the 6284-Å DIB, which is usually much stronger than the 5780-Å line

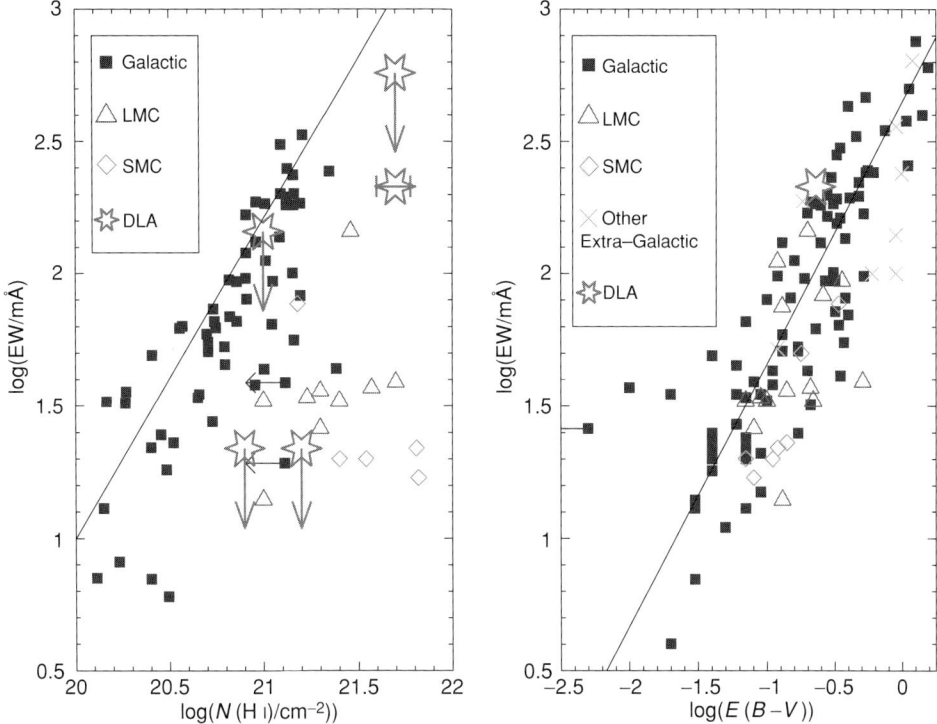

Figure 22.3. A search for DIBs in DLAs; only one detection of the 5780-Å DIB has been made (York *et al.* 2006). In general the 5780-Å DIB is weaker for its N(H I) by up to a factor of 10 compared with Galactic sight-lines, but is consistent with strengths in the Magellanic Clouds. In contrast, all sight-lines have similar dependences on $E(B-V)$. Figure adapted from York *et al.* (2006).

which we do detect. This indicates that the physical conditions in the DLA are different from those along the bulk of Galactic and Magellanic sight-lines that have been studied. The only known sight-line with similarly weak 6284-Å absorption is towards the sight-line Sk 143 located in the SMC wing. Finally, in contrast with the trends with N(H I), Galactic, Magellanic and DLA sight-lines exhibit a similar trend of $E(B-V)$ with DIB strength. Since DLAs generally have low $E(B-V)$ (Ellison *et al.* 2005a), this implies that DIBs are unlikely to be commonly detected in these absorbers.

3.2 Metal-strong DLAs in the SDSS

Herbert-Fort *et al.* (2006) have recently identified a sample of 'metal-strong' absorbers from the SDSS, characterised by strong heavy-element absorption that is clearly detected even in low-resolution SDSS spectra. These metal-strong DLAs comprise ~5% of the total population. In some cases, the very strong metal lines are a symptom of high $N(\text{H\,\textsc{i}})$. In other cases, the metallicity is truly high compared with those of known DLAs. Although they are relatively rare, the statistical power of Sloan is expected to reveal several hundred metal-strong DLAs. Follow-up observations with high-resolution spectrographs will yield data for a myriad of applications, including the search for rarely detected atomic transitions, work on molecular species and the study of isotopic ratios.

In closing, it is interesting to note that, although the definition of 'metal-rich' at this conference has been somewhat subjective, with the exception of AGN (see e.g. Chapter 20 in these proceedings) we rarely find metallicities above $2Z_\odot$ in individual stars, H\,\textsc{ii} regions or galaxies. However, a handful of very supersolar abundances (up to $5Z_\odot$) have recently been reported for QSO absorbers just below the traditional DLA column-density criterion of $N(\text{H\,\textsc{i}}) \geq 2 \times 10^{20}$ cm^{-2} (Péroux et al. 2006; Prochaska et al. 2006). The surprisingly high metallicities in these absorbers lead to a volley of new questions. How have these absorbers become so metal-rich at early times? What are their low-redshift analogues? What is the implication for the cosmic metals budget?

Acknowledgments

I am fortunate to enjoy stimulating collaborations with a large number of colleagues. In particular for the work described here, Chris Akerman, Chris Churchill, Pat Hall, Stéphane Herbert-Fort, Lisa Kewley, Brandon Lawton, Paulina Lira, Gabriela Mallen-Ornelas, Max Pettini, Jason X. Prochaska, Ted Snow and Brian York.

References

Akerman, C. J., Ellison, S. L., Pettini, M., & Steidel, C. C. (2005), *A&A*, **440**, 499
Ellison, S. L., Churchill, C. W., Rix, S. A., & Pettini, M. (2004), *ApJ* **615**, 118
Ellison, S. L., Hall, P. B., & Lira, P. (2005a), *AJ* **130**, 1345
Ellison, S. L., Kewley, L. J., & Mallen-Ornelas, G. (2005b), *MNRAS* **357**, 354
Ellison, S. L., et al. (2001), *A&A* **379**, 393
Fall, S. M., & Pei, Y. C. (1993), *ApJ* **402**, 479
Herbert-Fort, S., et al. (2006), *PASP* **118**, 1077
Herbig, G. (1993), *ApJ* **407**, 142
Johansson, P. H., & Efstathiou, G. (2006), *MNRAS* **371**, 1519
Jørgenson, R. A. et al. (2006), *ApJ* **646**, 730
Junkkarinen, V. T. et al. (2004), *ApJ* **614**, 658
Lawton, B., et al. (2008). *AJ*, submitted, arXiv: 0801.0447
Murphy, M. T., & Liske, J. (2004), *MNRAS* **354**, L31
Ostriker, J. P., & Heisler, J. (1984), *ApJ* **278**, 10
Pei, Y. C., Fall, S. M., & Bechtold, J. (1991), *ApJ* **378**, 6
Péroux, C., et al. (2006), *A&A* **450**, 53

Pettini, M. (2004), in *Proceedings of 'Cosmochemistry. The Melting Pot of the Elements'*, edited by C. Esteban, R. J. García López, A. Herrero, F. Sánchez, p. 257ff
Prantzos, N., & Boissier, S. (2000), *MNRAS* **315**, 82
Prochaska, J. X., *et al.* (2006), *ApJ* **648**, L97
Snow, T. P. (2001), in *Proceedings of 'Gaseous Matter in Galaxies and Intergalactic Space'*, edited by R. Ferlet, M. Lemoine, J.-M. Desert, B. Raban, pp. 63ff
Trenti, M., & Stiavelli, M. (2006), *ApJ* **651**, 51
Wolfe, A. M., & Chen, H.-W. (2006), *ApJ* **652**, 981
York, B. A., *et al.* (2006), *ApJ* **647**, L29

23

Tracing metallicities in the Universe with the James Webb Space Telescope

R. Maiolino,[1] S. Arribas,[2] T. Böker,[3] A. Bunker,[4] S. Charlot,[5]
G. de Marchi,[3] P. Ferruit,[6] M. Franx,[7] P. Jakobsen,[3] H. Moseley,[8]
T. Nagao,[9] L. Origlia,[10] B. Rauscher,[8] M. Regan,[11] H. W. Rix[12] &
C. J. Willott[13]

[1] *INAF – Astronomical Observatory of Rome, Italy*
[2] *CSIC – Departamento de Astrofísica Molecular e Infrarroja, Madrid, Spain*
[3] *European Space Agency – ESTEC, Noordwijk, the Netherlands*
[4] *School of Physics, University of Exeter, Exeter, UK*
[5] *Institute d'Astrophysique de Paris, Paris, France*
[6] *CRAL – Observatoire de Lyon, 9 Avenue Charles André, Saint-Genis Laval, France*
[7] *Leiden Observatory, Leiden, the Netherlands*
[8] *NASA – Goddard Space Flight Center, MD, USA*
[9] *National Astronomical Observatory of Japan, Osawa, Japan*
[10] *INAF – Astronomical Observatory of Bologna, Bologna, Italy*
[11] *Space Telescope Science Institute, Baltimore, MD, USA*
[12] *Max-Planck-Institut für Astronomie, Heidelberg, Germany*
[13] *Herzberg Institute of Astrophysics, Victoria, Canada*

The James Webb Space Telescope is a 6.6-m-aperture, passively cooled space observatory optimized for near-IR observations. It will be one of the most important observing facilities in the next decade, and it is designed to address numerous outstanding issues in astronomy. In this article we focus specifically on its capabilities to investigate the chemical abundances of various classes of astronomical objects and their metallicity evolution through the cosmic epochs.

1 An overview of JWST

The James Webb Space Telescope (JWST, formerly known as the Next Generation Space Telescope), which is being developed in a collaborative effort involving NASA, ESA, and the Canadian Space Agency (CSA), will be one of the prime astronomical facilities in the next decade. In this article we provide a short summary of the mission specifications. A more extensive overview can be found in Gardner *et al.* (2006), or at the Space Telescope Science Institute web site (http://www.stsci.edu/jwst/).

The Metal-rich Universe, eds. G. Israelian and G. Meynet. Published by Cambridge University Press.
© Cambridge University Press 2008.

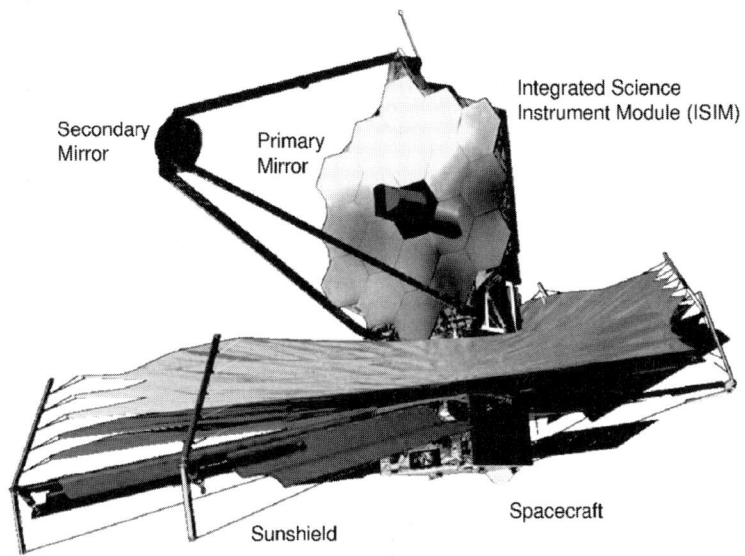

Figure 23.1. A schematic view of JWST, illustrating the main components of the observatory (credit: Northrop Grumman Space Technology).

JWST will be a 6.6-m-aperture deployable telescope, passively cooled to below 50 K, and optimized for near- and mid-IR observations. More specifically, JWST will observe in the wavelength range 0.6–28 μm. The telescope is designed to deliver diffraction-limited image quality at 2 μm (i.e. angular resolution $\sim 0''.07$) and, in terms of sensitivity, to be limited by the zodiacal light at $\lambda < 10$ μm.

JWST will be launched into orbit around the second Earth–Sun Lagrange point (L2), which provides a more stable environment and, most importantly, much less thermal emission from the Earth and the Moon, than any orbits around Earth. The launch is scheduled for 2013, on an Ariane 5 rocket. The minimum mission lifetime is 5 years, with the goal of lasting up to 10 years.

The primary mirror uses 18 semi-rigid segments made of beryllium with a thin gold coating (Figure 23.1), which will be phased periodically to maintain the required image quality. A sunshield is responsible for continuously keeping the telescope in the Sun's shade, thus allowing the required passive cooling. Six of the external mirror segments, the secondary mirror support, and the sunshield will be folded into the Ariane 5 rocket, and will be deployed during transfer to L2.

The Integrated Science Instrument Module (ISIM), located in the backplane of the primary mirror, will host four scientific instruments, which are described in the following.

NIRCam

The Near-Infrared Camera (NIRCam) (Horner & Rieke 2004) will be the key instrument of JWST in terms of sensitivity. It will perform the deepest observations ever obtained in the near-IR band, reaching a limiting flux of ∼1 nJy ($m_{AB} \approx 31.4$, signal-to-noise ratio 10), in deep integrations of 10^5 s. NIRCam will obtain images in the wavelength range 0.6–5 μm. A dichroic filter is used to observe simultaneously in two bands with two different cameras, optimized in the spectral regions 0.6–2.3 μm and 2.3–5 μm, respectively. The total field of view is $2'.2 \times 4'.4$. Both cameras are equipped with a wide set of broad-, intermediate-, and narrow-band filters. NIRCam has also coronographic capabilities.

NIRSpec

The Near-Infrared Spectrometer (NIRSpec) (te Plate *et al.* 2005) will be the first multi-object spectrograph in space, and will be the key instrument to trace metallicities, which is the main focus of this paper. Thanks to four arrays of 365×171 micro-shutters located on the focal plane, it will be possible to obtain spectra of more than 100 objects simultaneously over a field of view of $3'.5 \times 3'.5$. NIRSpec is also equipped with an integral field unit (IFU), using an image slicer, that will allow the observer to obtain two-dimensional spectra of a small field of $3'' \times 3''$. Six gratings will cover the 1–5-μm band, yielding spectral resolutions of $R \sim 1,000$ and $R \sim 2,700$. A single prism will cover the entire 0.6–5-μm band with a spectral resolution of ∼100. In the $R \sim 1,000$ mode NIRSpec will reach a line-detection sensitivity at $\lambda \sim 3$ μm of $\sim 4 \times 10^{-19}$ erg^{-1} cm^{-2} (signal-to-noise ratio 10), with deep exposures of 10^5 s. For comparison, this is the expected Hα flux of a galaxy at $z \sim 6$ forming stars at a rate of $\sim M_\odot$ yr^{-1}.

MIRI

The Mid-Infrared Instrument (MIRI) (Wright *et al.* 2004) operates in the spectral region 5–29 μm. It has an imaging module providing diffraction-limited, broad-band imaging over a field of view of $83'' \times 113''$, and also low-resolution ($R \sim 100$) slit spectroscopy in the range 5–10 μm. The imaging module also hosts masks for coronographic observations. MIRI also has an IFU, using four image slicers, which delivers two-dimensional spectra at $R \sim 3,000$ over the full 5–29-μm spectral range, with a field of view ranging from $3''.0 \times 3''.9$ to $6''.7 \times 7''.7$, depending on the wavelength. The sensitivity of MIRI will be limited by the thermal radiation emitted by the back side of the sunshield and scattered into the optical path of the instrument, which increases strongly with wavelength. At 10 μm the expected sensitivity is about 100 nJy in broad-band imaging (10^5 s, signal-to-noise ratio 10).

Such a sensitivity is about a factor of 50 better than that of Spitzer, and the angular resolution is a factor of 8 higher.

TFI

The Tunable Filter Imager (TFI) (Rowlands *et al.* 2004) uses a Fabry–Pérot interferometer to provide narrow-band imaging over the spectral range 1.6–4.9 μm (with a gap at 2.6–3.1 μm), with a resolution of $R \sim 100$, and over a field of view of $2'.2 \times 2'.2$. TFI also incorporates four coronographic occulting spots. The TFI is mounted on the same plate as the Fine Guidance Sensor (FGS), which uses two fields of view to provide fine guidance for JWST.

2 Tracing metallicities and chemical abundances with JWST

JWST will be a versatile astronomical facility designed to tackle numerous outstanding issues spanning most of the research fields in astronomy. The key scientific goals for JWST have been grouped into four main areas: (1) the investigation of planetary systems (including our own) and the potential for life; (2) the investigation of the birth of stars, their early evolutionary stages, and their proto-planetary systems; (3) the study of the assembly of galaxies, and their evolution in terms of dark matter, morphology, metallicity, stellar content, and gas content through the cosmic epochs; and (4) the identification of the first luminous sources in the Universe (Population-III stars) and the epoch of re-ionization. An extensive overview of these scientific cases is given in Gardner *et al.* (2006). In this short contribution we focus on the use of JWST to investigate the metallicity and chemical abundances in various classes of astronomical sources.

Individual stars in the Local Group

The stellar chemical abundances of the various galactic components (disk, bulge, halo, globular clusters) in galaxies of diverse masses and morphologies provide crucial information against which to test theories of galaxy formation and evolution (e.g. Helmi *et al.* 2006). Chemical abundances of evolved and metal-rich stars are extremely difficult, if not impossible, to investigate in the optical bands, due to the severe molecular blanketing and blending of the absorption features. Instead, the near-IR spectra of evolved stars (red giants and supergiants) display a wealth of prominent absorption features that are sensitive indicators of the abundances of key elements such as iron, carbon, and oxygen, as well as of various other α-elements such as calcium, magnesium, silicon, and titanium (Larsen *et al.* 2006; Rich *et al.* 2005; Origlia *et al.* 1997, 1999, 2006). At the maximum spectral resolution achieved by JWST with the NIRSpec ($R \sim 2,700$), many of these absorption features

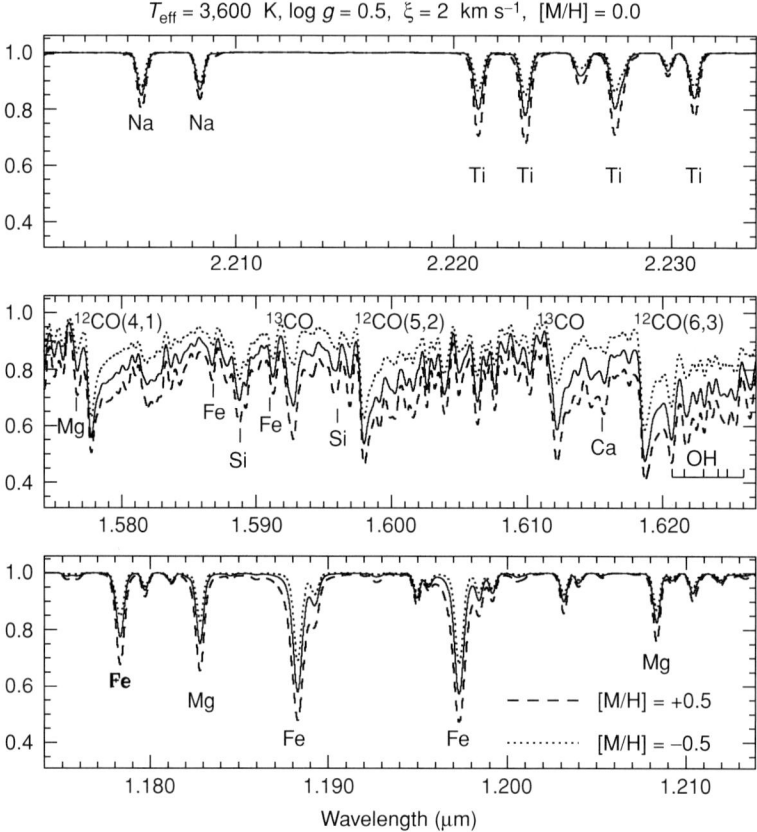

Figure 23.2. A model of the near-IR spectrum for a red giant star with $T_{\rm eff} = 3600$ K and Solar abundances, smoothed to the maximum spectral resolution of NIRSpec. The effect of varying the abundances by ± 0.5 dex is also shown.

are clearly separated and can be used to determine metallicities and relative elemental abundances. This is illustrated in Figure 23.2, where the model of the near-IR spectrum of a typical red giant star is convolved with the maximum spectral resolution achieved by NIRSpec. The same figure also shows the depth variation of the various stellar features on changing the abundances of all elements by ± 0.5 dex. With such a spectral resolution, the elemental abundances can be inferred with an accuracy of ± 0.2 dex, provided that the signal-to-noise ratio is high enough (>20). Thanks to the high sensitivity achieved by JWST with NIRSpec, it will be possible to derive detailed metallicities and abundance patterns for individual stars in all galaxies within the Local Group. As an example, a giant star near the tip of the red giant branch (RGB) at the distance of M31 has an apparent magnitude of $K_{\rm AB} = 22$; in this case NIRSpec will achieve a signal-to-noise ratio of 20 per resolution element in only two hours of integration at $R \sim 2{,}700$. With the micro-shutter array (MSA) it will be possible to obtain simultaneously spectra of about 100 stars in the bulge/disk/halo of M31; with the IFU it will be possible to obtain spectra of all giant stars in any globular cluster (GC) of M31, except possibly for their innermost region, where stellar crowding will deliver only integrated information of the GC cores (note that the IFU field of view is nicely matched to the size of GCs at the distance of M31).

NIRCam will be able to trace global metallicities of large numbers of stars by determining the location (and in particular the slope) of the RGB on the magnitude–color diagram M_K versus $J-K$. Indeed, the slope of the RGB ridge on this diagram correlates tightly with the stellar metallicity (Valenti et al. 2004, 2005). An accurate determination of its slope requires the RGB to be sampled at or below, the horizontal branch (i.e. stars with $M_K \approx -3$). In principle the sensitivity achieved by the NIRCam would allow us to measure the RGB slope even at the distance of the Virgo cluster. However, the angular resolution of JWST will allow us to resolve individual stars only in galaxies within the Local Group. At the distance of M31, only a few minutes of integration will be required to measure the RGB slope in a single exposure, with an unprecedented photometric accuracy. Therefore, NIRCam will be able to survey large stellar fields in several galaxies within the Local Group, identifying RGB slopes for several thousand stars, thus identifying metallicity variations from object to object and gradients within individual systems.

Galaxies at intermediate redshift

With JWST it will be possible to extend the near- and mid-IR metallicity diagnostics used for nearby galaxies to objects at $z \sim 1$–2, with important implications for our understanding of the evolution of the stellar and interstellar-medium (ISM) metallicity in the Universe.

The near-IR spectroscopic diagnostics discussed in the previous section are clearly detected in the spectra of nearby galaxies (e.g. Boisson et al. 2004; Dasyra et al. 2006), and have been widely used to determine stellar metallicities, especially in dusty systems. However, atmospheric absorption bands and lack of sensitivity have limited the use of these diagnostic tools to only up to $z \sim 0.1$. Instead, with NIRSpec it will be possible to observe the several H-band metal absorption features (Figure 23.2) up to $z \sim 2$. For a typical ULIRG at $z \sim 2$, $m(4.5\,\mu\text{m})_{AB} \sim 20.7$ (Egami et al. 2004; Frayer et al. 2004), NIRSpec will detect the continuum with a signal-to-noise ratio of ~ 20 with an integration time of about three hours at a spectral resolution of $R = 2{,}700$. The stellar features in the J-band, which is rich in Fe lines, will be observable by NIRSpec up to $z \sim 3$.

The mid-IR band of star-forming galaxies displays a number of emission lines that can be used to constrain the gas-phase metallicity. In particular, the strength of the fine-structure transitions of Ne^+ and Ne^{2+} at 12.8 μm and 15.6 μm, relative to the hydrogen recombination lines (Brα and Brβ), is a good indicator of the Ne

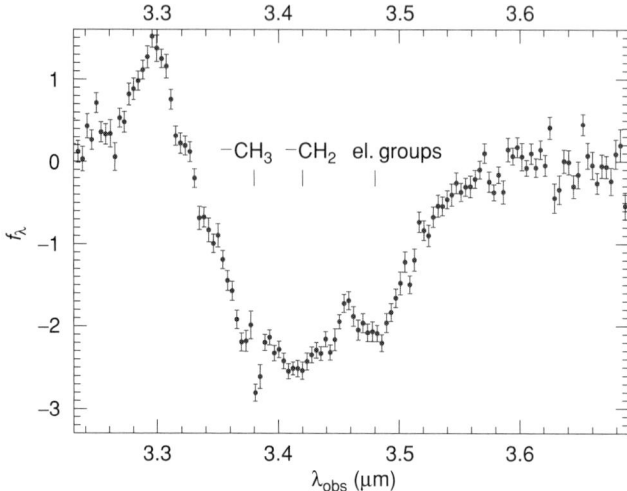

Figure 23.3. Dust absorption due to C—H stretching vibration in hydrocarbon dust grains observed in the ULIRG galaxy IRAS 19254-7245 at $z = 0.06$. From Risaliti et al. (2003). Individual dips due to the —CH_3 and —CH_2 vibrational transitions in hydrocarbon grains are marked.

abundance, which in turn is a good tracer of heavy elements produced by Type-II supernovae (Willner & Nelson 2002). In contrast to the more commonly used optical gaseous diagnostic, these mid-IR tracers have the advantage of being much less affected by dust extinction and also less sensitive to effects of temperature gradients within the H II regions. Spitzer-IRS has detected these lines in several objects, but still within the local Universe or in exceptionally luminous sources. MIRI will allow us to extend these metallicity mid-IR diagnostic tools easily up to $z \sim 1$. It will then be most important to compare the metallicity evolution inferred from MIRI observations with the results obtained from optical surveys, which are certainly biased against dusty galaxies and have provided some contradictory results (e.g. Savaglio et al. 2005; Liang et al. 2006).

The near-to-mid-IR spectral range also includes several dusty features that can be used to constrain the chemical composition of the ISM solid phase in galaxies. In particular, the 3–25-μm spectrum of galaxies is rich in polyaromatic hydrocarbon (PAH) emission features (Genzel & Cesarsky 2000; Brandl et al. 2004; Smith et al. 2004), and it often presents absorption (but also emission) by silicates at 10–20 μm (Spoon et al. 2004; Shi et al. 2006; Bressan et al. 2006) and absorption by ice-covered grains at 3.4 μm and by hydrocarbon grains at 3.3 μm (Imanishi et al. 2006; Risaliti et al. 2003). As an example, Figure 23.3 shows the hydrocarbon grains absorption observed in an infrared luminous galaxy at $z = 0.06$, together with the identification of the individual chemical dust components responsible for the individual absorption features. With NIRSpec and MIRI it will be possible to extend to high redshift such studies on the dust composition. Since the dusty absorption and emission features are generally broad, a moderate spectral resolution ($R \sim 100$) is enough for their detection and identification. For instance, MIRI will be able to detect the 3.3-μm dusty feature with a signal-to-noise ratio of ~ 20 through $R \sim 100$ spectroscopy in a luminous infrared galaxy at $z \sim 2$ with three hours of integration. Therefore, JWST will offer the unique opportunity to investigate the evolution of the chemical composition of dust with redshift, which will provide important information for understanding the origin of dust through the cosmic ages.

Galaxies at high redshift

The investigation of the metallicity evolution of galaxies is currently limited to $z < 4$, through the detection of optical emission lines in star-forming galaxies (e.g. Erb *et al.* 2006; Pettini *et al.* 2001) or of absorption lines in DLA (Prochaska *et al.* 2003; Kulkarni *et al.* 2005). At higher redshift ($z \sim 6$) the metallicity has been inferred only for a few, exceptionally luminous QSOs (Pentericci *et al.* 2002), while for the general population of (normal) galaxies the metallicity evolution between $z \sim 4$ and the epoch of re-ionization ($z \sim 7$–12) is largely unknown. This crucial redshift range is essentially unexplored because of both low sensitivity and limited spectral coverage of ground-based instrumentation. For instance, the classical R_{23} parameter (defined as $R_{23} = ([\text{O II}]3727 + [\text{O III}]4959 + [\text{O III}]5007)/\text{H}\beta$), which is widely used to trace gaseous metallicities, is observable from the ground only up to $z \sim 3.5$ (i.e. as long as [O III]5007 is observable in the K band). JWST, and NIRSpec in particular, will be able to extend these diagnostics to higher redshift and to fainter galaxy populations than is feasible from the ground. As an example, Figure 23.4 shows the simulated NIRSpec spectrum of a low-metallicity galaxy ($Z \sim 0.1 Z_\odot$) at $z = 8$, with $m_{\text{AB}} = 27$, observed with the MSA configuration at $R = 1,000$ and with an integration time of 10^5 s: all of the lines required to determine the R_{23} parameter are clearly detected, therefore providing an estimate of the metallicity. At $z > 9$ the [O III]5007 line leaves the NIRSpec range and will be observable only with MIRI, but only for brighter targets, due to the lower sensitivity of the latter. An alternative diagnostic for metallicity that is usable at high redshift (especially for metal-poor galaxies) is the ratio [Ne III]3869/[O II]3727. Indeed, the latter line ratio is seen to anti-correlate with metallicity, and at $Z < 0.1 Z_\odot$ the strength of [Ne III]3869 is comparable to that of [O II]3727 (Nagao *et al.* 2006a). These lines are observable in the NIRSpec band up to $z \sim 12$. The availability of this and other gas-metallicity diagnostics as a function of redshift, both for ground-based instruments and for JWST, is summarized in Figure 23.5.

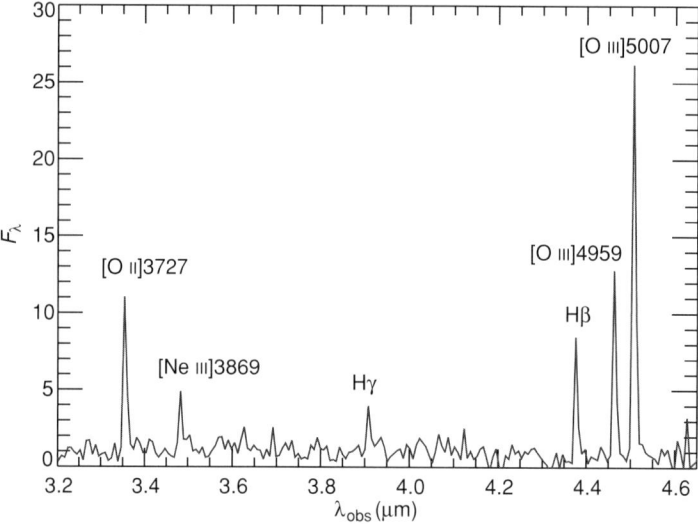

Figure 23.4. A simulated NIRSpec spectrum of an $m_{AB} = 27$ galaxy at $z = 8$, with a metallicity of $Z \sim 0.1 Z_\odot$, observed with the MSA configuration at $R = 1,000$ and with an integration time of 10^5 s. The main, detected emission lines are labeled.

In principle JWST could identify spectroscopically the first generation of stars (Population III), free of metals, through the detection of the He II emission line at $\lambda_{rest} = 1,640$ Å, which is expected to have an equivalent width of a few tens of ngström units in extremely metal-poor systems ($Z < 10^{-4} Z_{Sun}$) (Schaerer 2003). However, as discussed in Panagia (2004), the expected faintness of the primordial galaxies will probably limit the capability to detect the He II line, except for the bright end of the population (or for lensed objects). The latter may be the case for a few strong Lyα emitters at $z > 6$ detected in wide-area surveys, and characterized by a large EW(Lyα), which is suggestive of very young and extremely metal-poor galaxies (Nagao *et al.* 2006b). Followup observations of these objects have so far failed to detect the He II line, but with loose constraints due to the limited sensitivity of ground-based observations (Nagao *et al.* 2005). The JWST–NIRSpec combination will provide much deeper spectra of these objects and, it is hoped, detect the He II line or, alternatively, detect metal lines that would disprove their Population-III/metal-free nature.

Note that, insofar as normal/star-forming galaxies at high redshift are concerned, JWST will be able to measure global metallicities, while relative elemental abundances will likely remain unconstrained. This is because the observable, strong emission lines at $z > 4$ do not provide additional information beside the global metallicity (especially at $z > 6.6$, where even the faint [N II] is outside the NIRSpec range). For a few exceptionally bright (e.g. lensed) sources it will be possible

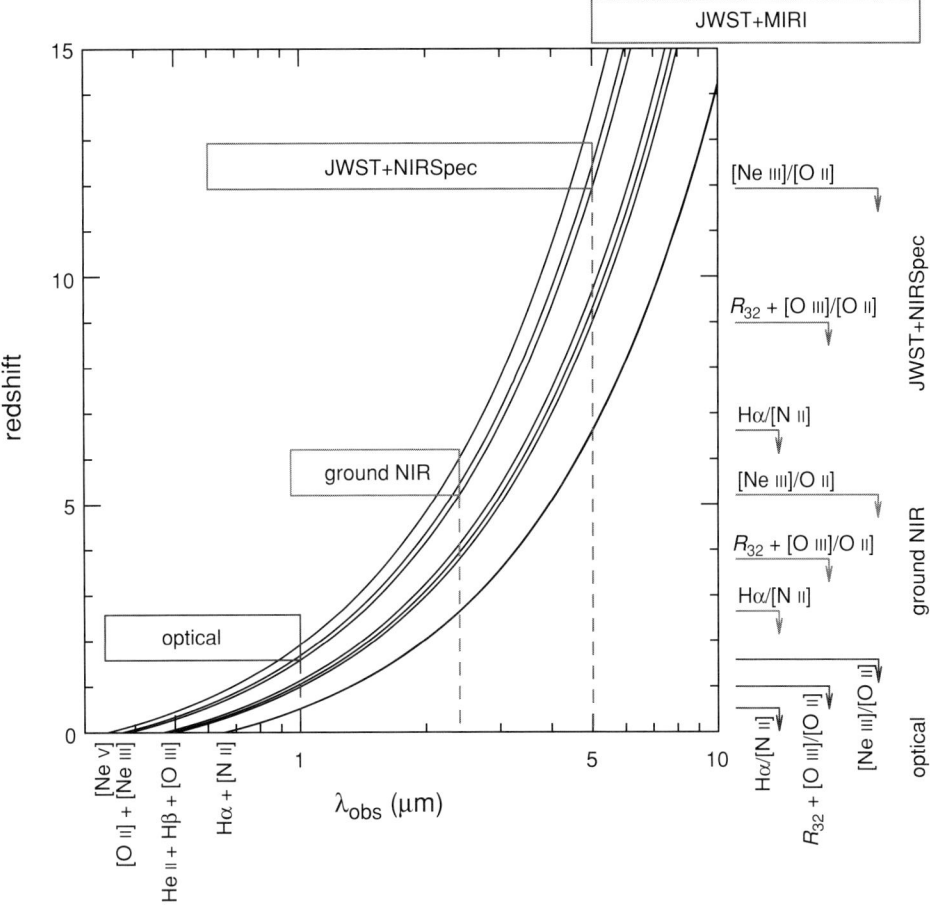

Figure 23.5. Availability of some of the gas-metallicity diagnostics as a function of redshift, both for ground-based facilities and for JWST.

to determine abundance patterns from the detection of stellar and ISM absorption features (Pettini *et al.* 2002; Rix *et al.* 2004; Mehlert *et al.* 2006), but these will be relatively rare cases.

Quasi-stellar objects (QSOs) are an alternative class of objects that can provide important information on the relative abundance of elements at high redshift. Even if high-z QSOs are rare, and probing overdense regions, their near-IR spectra (UV–rest frame) are rich in emission and absorption features whose relative strength provides crucial information on the star-formation history in the early Universe. This is especially true for the prominent iron emission features, relative to emission lines of α-elements, which provide a measure of the Fe/α ratio. The latter gives important constraints on the epoch of initial star formation, since α-elements are mostly produced by Type-II SNe, whereas iron is mostly produced by SN Ia, which

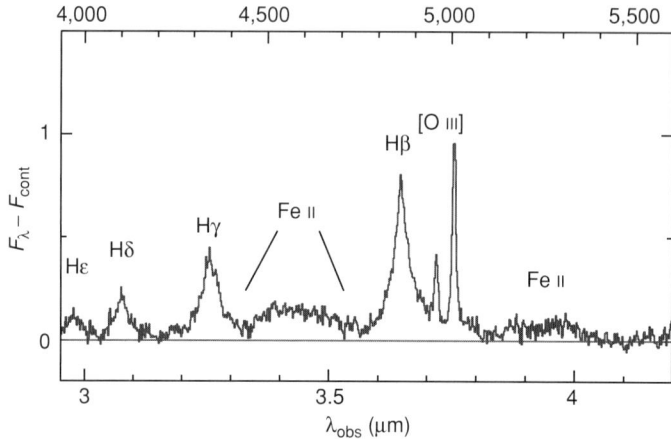

Figure 23.6. A simulated JWST–NIRSpec $R = 1,000$ (continuum-subtracted) spectrum of a QSO at $z = 6.5$, with a brightness comparable to that of the faintest QSOs found in the SDSS at $z > 6$ ($m_{AB} = 21$), for an integration time of 20 min.

provide a delayed enrichment relative to the former; hence the Fe/α ratio is regarded as a sensitive "clock" of star formation. Various authors have tried to investigate the Fe/α ratio in high-z QSOs (e.g. Dietrich *et al.* 2003; Maiolino *et al.* 2003; Iwamuro *et al.* 2005). However, at $z > 4$ ground-based observations can access only the UV iron emission blend at 2,700–3,050 Å, which is not a good tracer of iron abundance, yielding inconclusive results. The optical iron emission at $4,200 \text{ Å} < \lambda_{rest} < 5,400 \text{ Å}$ (the "optical Fe hump") is a better indicator of iron abundance (Verner *et al.* 2004; Verner & Peterson 2004), but it is shifted outside of the K band at $z > 4$, and therefore unobservable at the interesting redshifts approaching re-ionization. JWST will easily detect and measure the intensity of the optical Fe emission in the spectra of QSOs at $z > 6$, as illustrated in Figure 23.6, thus providing crucial information on the epoch of star formation in these objects.

References

Boisson, C., Joly, M., Pelat, D., & Ward, M. J. (2004), *A&A* **428**, 373
Brandl, B. R., Devost, D., Higdon, S. J. U. *et al.* (2004), *ApJS* **154**, 188
Bressan, A., Panuzzo, P., Buson, L. *et al.* (2006), *ApJ* **369**, L55
Dasyra, K. M., Tacconi, L. J., Davies, R. I. *et al.* (2006), *ApJ* **651**, 835
Dietrich, M., Hamann, F., Appenzeller, I., & Vestergaard, M. (2003), *ApJ* **596**, 817
Egami, E., Dole, H., Huang, J.-S. *et al.* (2004), *ApJS* **154**, 130
Erb, D. K., Shapley, A. E., Pettini, M. *et al.* (2006), *ApJ* **644**, 813
Frayer, E., Chapman, S. C., Yan, L. *et al.* (2004), *ApJS* **154**, 137
Genzel, R., & Cesarsky, C. J. (2000), *ARA&A* **38**, 761
Gardner, J. P., Mather, J. C., Clampin, M. *et al.* (2006), *Space Sci. Rev.* **123**, 485
Helmi, A., Irwin, M. J., Tolstoy, E. *et al.* (2006), *ApJ* **651**, L121
Horner, S. D., & Rieke, M. J. (2004), *Proc. SPIE* **5487**, 628–634

Imanishi, M., Dudley, C. C., & Maloney, P. R. (2006), *ApJ* **637**, 114
Iwamuro, F., Kimura, M., Eto, S. *et al.* (2004), *ApJ* **614**, 69
Kulkarni, V. P., Fall, S. M., Lauroesch, J. T. *et al.* (2005), *ApJ* **618**, 68
Larsen, S. S., Origlia, L., Brodie, J. P., & Gallagher, J. S. (2006), *ApJ* **634**, 1293
Liang, S., Hammer, F., & Flores, H. (2006), *A&A* **447**, 113
Maiolino, R., Juarez, Y., Mujica, R., Nagar, N. M., & Oliva, E. (2003), *ApJ* **596**, L55
Mehlert, D., Tapken, C., Appenzeller, I., Noll, S., de Mello, D., & Heckman, T. M. (2006), *A&A* **455**, 835
Nagao, T., Motohara, K., Maiolino, R. *et al.* (2005), *ApJ* **631**, L5
Nagao, T., Maiolino, R., & Marconi, A. (2006), *A&A* **459**, 85
Nagao, T., Murayama, T., Maiolino, R. *et al.* (2007), *A&A* **468**, 877
Origlia, L., Ferraro, F. R., Fusi Pecci, F., & Oliva, E. (1997), *A&A* **321**, 859
Origlia, L., Goldader, J. D., Leitherer, C., Schaerer, D., & Oliva, E. (1999), *ApJ* **514**, 96
Origlia, L., Valenti, E., Rich, R. M., & Ferraro, F. R. (2006), *ApJ* **646**, 499
Panagia, N. (2004), in *IMF50: The Stellar Initial Mass Function Fifty Years Later*, eds. E. Corbelli, F. Palla, & H. Zinnecker (Dordrecht, Springer), pp. 457–479
Pentericci, L., Fan, X., Rix, H.-W. *et al.* (2002), *ApJ* **123**, 2151
Pettini, M., Shapley, A. E., Steidel, C. C. *et al.* (2001), *ApJ* **554**, 981
Pettini, M., Rix, S. A., Steidel, C. C. *et al.* (2002), *ApJ* **569**, 742
Prochaska, J. X., Gawiser, E., Wolfe, A. M., Castro, S., & Djorgovski, S. G. (2003), *ApJ* **595**, L9
Rich, R. M., & Origlia, L. (2005), *ApJ* **634**, 1293
Risaliti, G., Maiolino, R., Marconi, A. *et al.* (2003), *ApJ* **595**, L17
Rix, S. A., Pettini, M., Leitherer, C. *et al.* (2004), *ApJ* **615**, 98
Rowlands, N., Evans, C., Greenberg, E., *et al.* (2004), *Proc. SPIE* **5487**, 676–687
Savaglio, S., Glazebrook, K., Le Borgne, D. *et al.* (2005), *ApJ* **625**, 260
Schaerer, D. (2003), *A&A* **397**, 527
Shi, Y., Rieke, G. H., Hines, D. C. *et al.* (2006), *ApJ* at press (astro-ph/0608645)
Smith, J. D. T., Dale, D. A., Armus, L. *et al.* (2004), *ApJS* **154**, 199
Spoon, H., Armus, L., Cami, J. *et al.* (2004), *ApJS* **154**, 184
te Plate, M., Holota, W., Posselt, W. *et al.* (2005), *Proc. SPIE* **5904**, 185
Valenti, E., Ferraro, F. R., & Origlia, L. (2004), *MNRAS* **351**, 1204
Valenti, E., Origlia, L., & Ferraro, F. R. (2005), *MNRAS* **361**, 272
Verner, E., Bruhweiler, F., Verner, D., Johansson, S., Kallman, T., & Gull, T. (2004), *ApJ* **611**, 780
Verner, E., & Peterson, B. A. (2004), *ApJ* **608**, L85
Willner, S. P., & Nelson, K. (2002), *ApJ* **568**, 679
Wright, G. S., Rieke, G. H., Colina, L. *et al.* (2004), *Proc. SPIE* **5487**, 653

Part IV

Stellar populations and mass functions

24

The stellar initial mass function of metal-rich populations

Pavel Kroupa

Argelander-Institut für Astronomie, Universität Bonn, Auf dem Hügel 71, D-53121 Bonn, Germany

Does the initial mass function (IMF) vary? Is it significantly different in metal-rich environments versus metal-poor ones? Theoretical work predicts this to be the case, but in order to provide robust empirical evidence for this, the researcher *must* understand *all* possible biases affecting the derivation of the stellar mass function. Apart from the very difficult observational challenges, this turns out to be highly non-trivial, relying on an exact understanding of how *stars evolve*, how stellar populations in galaxies are assembled dynamically and how individual star clusters and associations evolve. *N-body modelling* is therefore an unavoidable tool in this game: the case can be made that without complete dynamical modelling of star clusters and associations any statements about the variation of the IMF with physical conditions are most probably wrong. The calculations that do exist demonstrate time and again that the IMF is invariant: there exists no statistically meaningful evidence for a variation of the IMF on going from metal-poor to metal-rich populations. This means that currently existing star-formation theory fails to describe the stellar outcome. Indirect evidence, based on chemical evolution calculations, however, indicates that the extreme starbursts that assembled bulges and elliptical galaxies may have had a top-heavy IMF.

1 Introduction

The stellar initial mass function (IMF), $\xi(m)dm$, where m is the stellar mass, is the parent distribution function of the masses of stars formed in *one* event. Here, the number of stars in the mass interval $m, m + dm$ is $dN = \xi(m)dm$. Salpeter (1955) inferred the IMF from Solar-neighbourhood star-counts by applying corrections for stellar evolution and Galactic-disc structure, finding $\xi(m) \approx km^{-\alpha}$, $\alpha \approx 2.35$,

The Metal-rich Universe, eds. G. Israelian and G. Meynet. Published by Cambridge University Press.
© Cambridge University Press 2008.

for $0.4 \lesssim m/M_\odot \lesssim 10$. Miller & Scalo (1979) and Scalo (1986) derived the IMF for $(0.1-60)M_\odot$ using better data and a more sophisticated analysis establishing that the IMF flattens or turns over at small masses. Results from modern studies of Solar-neighbourhood star-count data in which the authors also applied detailed corrections for unknown multiple systems in the star-counts confirm that $\alpha_2 = 2.2 \pm 0.3$ for $0.5 \lesssim m/M_\odot \lesssim 1$ (Kroupa *et al.* 1993; Kroupa 1995b; Reid *et al.* 2002). Using spectroscopic star-by-star observations, Massey (2003) found the same *slope or index* $\alpha_2 = 2.3 \pm 0.1$ for $m \gtrsim 10 M_\odot$ in many OB associations and star clusters in the Milky Way (MW) and in the Large and Small Magellanic Clouds (LMC and SMC, respectively). It is therefore suggested that one should refer to $\alpha_2 = 2.3$ as the *Salpeter/Massey slope* or *index*. It is valid for $m \gtrsim 0.5 M_\odot$.

The IMF is, strictly speaking, a hypothetical construct because any *observed* system of N stars merely constitutes a particular *representation* of a distribution function. The probable existence of a unique $\xi(m)$ can be inferred from observations of many ensembles of such N systems (e.g. Massey 2003). If, after corrections for

(a) stellar evolution,
(b) unknown multiple stellar systems and
(c) stellar-dynamical biases,

the individual distributions of stellar masses are similar *within the statistical uncertainties*, then we (the astrophysical community) deduce that the hypothesis that the stellar mass distributions are not the same can be excluded. That is, we make the case for a *universal*, *standard* or *canonical stellar IMF* under the physical conditions probed by the relevant physical parameters (metallicity, density, mass) of the populations at hand.

This canonical IMF is a two-part power law, the only structure with confidence found so far being the change of index from the Salpeter/Massey value to a smaller one near $0.5 M_\odot$:

$$\begin{aligned} \alpha_1 &= 1.3 \pm 0.5, & 0.08 \lesssim m/M_\odot \lesssim 0.5, \\ \alpha_2 &= 2.3 \pm 0.5, & 0.5 \lesssim m/M_\odot \lesssim 150. \end{aligned} \quad (24.1)$$

It has been fully corrected for unknown multiple stellar systems in the low-mass ($m < M_\odot$) regime, while multiplicity corrections in the high-mass regime have still to be done. The evidence for a universal upper mass cut-off near $150 M_\odot$ (Weidner & Kroupa 2004; Figer 2005; Oey & Clarke 2005; Koen 2006) seems to be rather well established for populations with metallicities ranging from that of the LMC ($Z \approx 0.008$) to the supersolar Galactic Centre ($Z \gtrsim 0.02$) such that the stellar mass function (MF) simply stops at that mass. This mass needs to be understood theoretically; see the discussion in Kroupa & Weidner (2005).

Chabrier (2003) offers a log-normal form[1] that fits the canonical form quite well (e.g. Romano *et al.* 2005).

There is substantial evidence that, below the hydrogen-burning limit, the IMF flattens further to $\alpha \approx 0.3 \pm 0.7$ (Kroupa 2001a; 2002; Chabrier 2003). Therefore, the canonical IMF most probably has a peak at $0.08 M_\odot$. Brown dwarfs, however, comprise only a few per cent of the mass of a population and are therefore dynamically irrelevant. Note that the logarithmic form of the canonical IMF, $\xi_L(m) = \ln(10) m \xi(m)$, which gives the number of stars in $\log_{10} m$ intervals, also has a peak near $0.08 M_\odot$. However, the *system* IMF (of stellar companions per binary combined to give system masses) has a maximum in the mass range $(0.4-0.6) M_\odot$ (Kroupa *et al.* 2003).

The above form has been derived from detailed considerations of local starcounts, thereby representing an *average* IMF: for low-mass stars it is a mixture of stellar populations spanning a large range of ages $(10-0\,\mathrm{Gyr})$ and metallicities $([\mathrm{Fe/H}] \lesssim 0)$, while for the massive stars it constitutes a mixture of different metallicities $([\mathrm{Fe/H}] \gtrsim -1.5)$ and star-forming conditions (from OB associations to very-dense starburst clusters: R136 in the LMC). Therefore it can be taken to be a canonical form, and the aim is to test whether even more extreme star-forming conditions such as are found in super-metal-rich environments or super-dense regions may deviate from it. Any *systematic* deviations of the IMF with physical conditions of the environment would constrain our understanding of star formation, would give us a prescription of how to set up stellar-dynamical systems and, last but not least, would allow more precise galaxy-formation and evolution calculations.

2 The expectation: the IMF must depend on the star-formation environment and in particular on the metallicity

There are two basic arguments suggesting that the IMF ought to be dependent on the physical conditions of star formation.

2.1 The Jeans-mass argument

(A) A region of a molecular cloud undergoing gravitational collapse will have overdense sub-regions within it which are also Jeans-unstable, collapsing independently to form smaller structures that may themselves undergo further fragmentation (e.g.

---

[1] Note that the log-normal form is physically better motivated in the sense that a physical process will not abruptly change a slope as in the canonical IMF, but it also needs to be extended by a power-law above $M_\odot$ to meet the needs of the observational data, thereby losing its advantage. A reason why the author prefers to use the canonical form is mathematical simplicity and the ease with which parts of it can be changed without affecting other parts.

Zinnecker 1984). Ultimately stars result. The essence of this concept is that a region spanning a Jeans length that contains at least a Jeans mass undergoes gravitational collapse. The Jeans mass depends on the temperature and density of the cloud, $M_{\rm Jeans} \propto T^{3/2}\rho^{-1/2}$ (e.g. Larson 1998; Bonnell *et al.* 2006). Now, in metal-rich environments there is more dust and therefore the collapsing gas can cool more effectively, reducing T and increasing ρ. Thus,

$$[{\rm Fe/H}] \uparrow \implies \text{fragment masses } \downarrow. \tag{24.2}$$

(B) The fact that the IMF is not a featureless power law but has structure in the mass range $0.08 \lesssim m/M_\odot \lesssim 0.5$ suggests there to be a characteristic mass of a few times $0.1 M_\odot$.

Bonnell *et al.* (2006) suggest that this characteristic mass of fragmentation may be a result of the coupling of gas to dust such that there is a change from a cooling equation of state, where $T \propto \rho^{-0.25}$ while the density increases, to one with slight heating at high densities, $T \propto \rho^{0.1}$. Again, this implies a dependence of the characteristic mass on the metallicity through the cooling rate:

$$[{\rm Fe/H}] \uparrow \implies \text{fragment masses } \downarrow. \tag{24.3}$$

There seems to be observational evidence supporting the notion that stellar masses are derived from Jeans-unstable mass fragments: pre-stellar cloud cores are found to be distributed like the canonical IMF in low-mass star-forming regions (Motte *et al.* 1998; Testi & Sargent 1998; Motte *et al.* 2001), but see Nutter & Ward-Thompson (2006).

While the concept of the Jeans mass is very natural and allows one to visualise the physical process of fragmentation nicely, it has the problem that the densest regions of a pre-star-cluster cloud core ought to have the smaller fragment masses, but instead the most massive stars are seen to form in the densest regions.

2.2 The self-limitation argument

A rather convincing physical model of the IMF that avoids the problem with the Jeans-mass argument has been suggested by Adams & Fatuzzo (1996) and Adams & Laughlin (1996). The argument here is that the Jeans mass has virtually nothing to do with the final mass of a star because structure in a molecular cloud exists on all scales. Therefore, no characteristic density can be identified, and 'no single Jeans mass exists'. When a cloud region becomes unstable a hydrostatic core forms after the initial free collapse. This core then accretes at a rate dictated by the physical conditions in the cloud. The metallicity influences the accretion rate through the sound velocity (higher sound velocity, larger accretion rate) by steering the cooling rate (more metals $\Rightarrow$ more dust $\Rightarrow$ lower T $\Rightarrow$ smaller sound velocity). The many physical variables describing the formation of a single star have distributions, and folding these together yields finally an IMF in broad agreement with a log-normal shape, as was also shown by Zinnecker (1984). This fails at large masses, though, for which the IMF is a power law and additional physical processes probably play a role (coagulation, competitive accretion). The important point, however, is that this theory *also* predicts a variation of the resulting characteristic mass with metallicity as above:

$$[\text{Fe/H}] \uparrow \implies \text{characteristic mass} \downarrow. \tag{24.4}$$

2.3 Robust implication?

Thus, both theoretical lines of argument seem to suggest the same qualitative behaviour, namely that the IMF ought to shift to smaller characteristic masses with increasing metallicity. This would therefore seem to suggest a *very robust* if not *fundamental* expectation of star-formation theory.

Is it born out by empirical evidence? The best way to test this theoretical expectation is to measure the IMF in metal-poor environments and to compare this with the shape seen in metal-rich environments.

A measurement of the IMF in a population with supersolar abundance would be especially important. Unfortunately this is extremely difficult, because populations with supersolar abundances are exceedingly rare. Only one star cluster with $[\text{Fe/H}] \approx +0.04$ is known, NGC 6791, but its ancient age of $\approx 10\,\text{Gyr}$ implies it to be dynamically very old. Evaporation of low-mass stars will thus have significantly affected the shape of the present-day mass function, which has not been measured yet (King *et al.* 2005). Other metal-rich environments constitute the central region of the MW (Section 5.2) and galactic spheroids (Section 5.3), the latter allowing only indirect evidence on the IMF through their chemical properties.

In general, measurements of the IMF are hard, because stellar masses can be inferred only indirectly through their luminosity, l, which also depends on age, metallicity and the star's spin-angular-momentum vector, collectively producing a distribution of l for a given m. The empirical knowledge that can be gained about the IMF is discussed next.

3 The shape of the IMF

3.1 The local stellar sample

The best knowledge about the IMF we glean from the local volume-limited stellar sample. The basic technique is to construct the stellar luminosity function (LF),

$\Psi(M_P)$, where M_P is the stellar absolute magnitude in some photometric pass band (we are stuck with magnitudes rather than working with the physically more intuitive luminosities due to our historical inheritance). The number of stars within the complete volume in the magnitude interval M_P, $M_P + dM_P$ is then $dN = \Psi(M_P)\,dM_P$. These are the same stars as enter $dN = \xi(m)\,dm$ above, and thus results our master equation

$$\Psi(M_P) = -\frac{dm}{dM_P}\xi(m). \tag{24.5}$$

The observable is Ψ, which we get from the sky. Our target is ξ and the hurdle is the *derivative of the stellar mass–luminosity relation*, dm/dM_P. This is quite problematical, because we can get at dm/dM_P only by either constructing observational mass–luminosity relations using extremely well-observed binary stars with known Kepler solutions that nevertheless have uncertainties that are magnified when considering the derivative, or resorting to theoretical stellar models, which give us well-defined derivatives but depend on theoretically difficult processes within stellar interiors (opacities, convection, rotation, magnetic fields, the equation of state, nuclear-energy-generation processes).

There are two basic local LFs.

(I) We can count all stars within a trigonometrically defined distance limit such that the stellar sample is complete, i.e. we can see *all* stars of magnitudes in the range M_P, $M_P + dM_P$ within a distance r_t. The volume-limited sample for Solar-type stars having excellent parallax measurements extends to $r_t \approx 20-30$ pc, while for the faintest M dwarfs $r_t \approx 5-8$ pc (Reid et al. 2002). Tests of completeness are made by comparing the number of stars with M_P within r_t with the number of such stars in a volume element further out (Henry et al. 1997), finding that faint stars remain to be discovered even within a distance of 5 pc. For this *nearby LF*, Ψ_{nearby}, the stars are well scrutinised on an individual-object basis, and geometric distances are known to within about 10%. At the faint end Ψ_{nearby} is badly constrained, resting on only a few stars.

(II) Prompted by the 'discovery' of large amounts of dark matter in the MW disc (Bahcall 1984),[2] novel deep surveys were pioneered by Reid & Gilmore (1982). This second type of sampling can be obtained by performing deep, pencil-beam-photographic or CCD-imaging surveys through the Galactic disc. From the 10^5 images the typically 100 or so main-sequence stars need to be gleaned using automatic image-, colour- and brightness-recognition systems. The distances of the stars are determined using the method of *photometric parallax*, which relies on estimating the absolute luminosity of a star from its colour and then calculating its

---

[2] The evidence for dark matter within the Solar vicinity disappeared on closer scrutiny (Kuijken & Gilmore 1991).

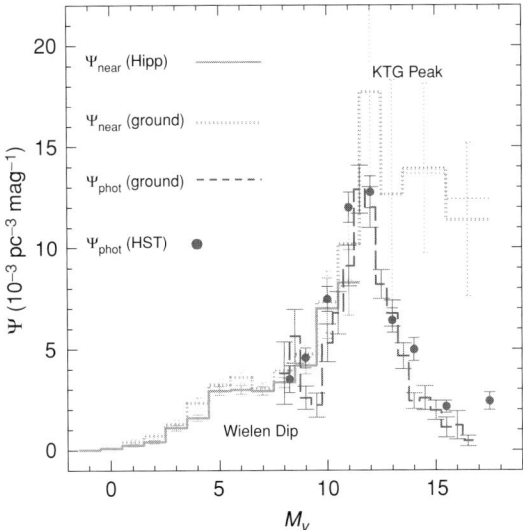

Figure 24.1. Stellar luminosity functions (LFs) for Solar-neighbourhood stars. The photometric LF corrected for Malmquist bias and at the mid-plane of the Galactic disc (Ψ_{phot}) is compared with the nearby LF (Ψ_{near}). The average, ground-based $\overline{\Psi}_{phot}$ (dashed histogram, data pre-dating 1995, Kroupa (1995a)) is confirmed by Hubble Space Telescope star-count data, which pass through the entire Galactic disc and are thus less prone to Malmquist bias (solid dots, Zheng et al. (2001)). The ground-based volume-limited trigonometric-parallax sample (dotted histogram) systematically overestimates Ψ_{near} due to the Lutz–Kelker bias, thus lying above the improved estimate provided by the Hipparcos-satellite data (solid histogram, Jahreiß & Wielen (1997) and Kroupa (2001b)). The depression/plateau near $M_V = 7$ is the *Wielen dip*. The maximum near $M_V \approx 12$, $M_I \approx 9$ is the *KTG peak*. The thin dotted histogram at the faint end indicates the level of refinement provided by recent stellar additions (Kroupa 2001b), demonstrating that even our star-count for the immediate neighbourhood within 5.2 pc of the Sun probably remains incomplete at the faintest stellar luminosities.

distance from the distance modulus. The resulting flux-limited sample of stars has photometric-distance limits within which the counts are complete. The distance limits are smaller for fainter stars.

Clearly, while only *one nearby LF* exists, *many photometric LFs* can be constructed for different fields of view. Each observation yields a few dozen to hundreds of stars, so the overall sample size becomes very significant. The various surveys have shown Ψ_{phot} to be invariant with direction. This should, of course, be the case, since the Galactic-field stars with an average age of about 5 Gyr have a velocity dispersion of about 25–50 pc Myr^{-1} such that within 200 Myr a volume with a dimension of the survey volumes (a few hundred parsecs) is completely mixed.

It therefore came as a surprise that Ψ_{near} and Ψ_{phot} are significantly different at faint luminosities (Figure 24.1).

3.2 The mass–luminosity relation

When not understanding something the best strategy to continue is sometimes simply to 'forget' the problem and continue along the path of least resistance. Thus, while the problem $\Psi_{near} \neq \Psi_{phot}$, $M_V > 12$ could not be explained immediately, it turned out to be constructive to ascertain first which LF shape must be the correct one by using entirely different arguments.

In Equation (24.5) the slope of the stellar mass–luminosity relation of stars enters, giving a clue. Figure 2 in Kroupa (2002) shows the mass–luminosity data of binary stars with Kepler orbits, and demonstrates that there exists a non-linearity near 0.33 $M_\odot$ such that a pronounced peak in $-dm/dM_V$ appears at $M_V \approx 11.5$, with an amplitude and width essentially identical to the maximum seen in the photometric LF at this luminosity (Figure 24.1). This agreement of

(a) the location of the maximum *and*
(b) the amplitude *and*
(c) the width of the extremum

convincingly suggests stellar astrophysics to be the origin of the peak in the LF, rather than the MF. The Wielen dip (Figure 24.1) similarly results from subtle structure in the mass–luminosity relation. *Thus, simply by counting stars on the sky we are able to direct our gaze within their interiors.* It is the internal constitution of stars which changes with changing main-sequence mass and this is what drives the structure in the mass–luminosity relation.

Having thus established that the peak in the LF *must, in fact* be there where it is also found as a result of fundamental astrophysical processes, we can test this result using star clusters that constitute single-age, single-metallicity and equal-distance stellar samples. Figure 24.2 does exactly this, and a very pronounced peak is indeed evident at exactly the right location and with the right width and height (Kroupa 2002, Figure 24.1). We can therefore trust the peak in Ψ_{phot}. In Figure 24.1 it can be seen that Ψ_{near} also shows evidence for this peak, by noting that the peak is smeared apart in the local stellar sample because the stars have a wide spread in metallicities. The metallicity-dependence of the peak has been shown to be in agreement with the LFs of star clusters over a large range of metallicity (von Hippel *et al.* 1996; Kroupa & Tout 1997).

Kroupa *et al.* (1993) then performed a trick to get at the correct mass–luminosity relation without having to resort to theoretical or purely empirical relations, which are very uncertain in their derivatives (fig. 2 in Kroupa 2002): The Malmquist-corrected Ψ_{phot} is used to define the amplitude, width and location of the extremum in the single-age, single-metallicity average dm/dM_V, and integrat-

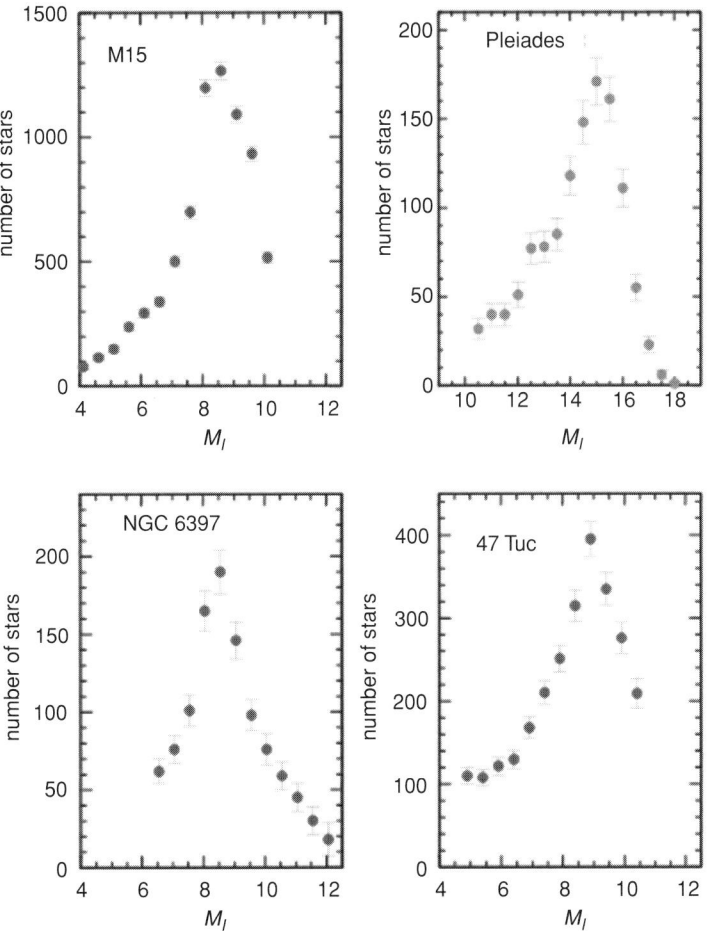

Figure 24.2. I-band LFs of stellar *systems* in four star clusters: globular cluster (GC) M15 (de Marchi & Paresce (1995a), distance modulus $\Delta m = m - M = 15.25$ mag); GC NGC 6397 (Paresce *et al.* (1995), $\Delta m = 12.2$); the young Galactic cluster Pleiades (Hambly *et al.* (1991), $\Delta m = 5.48$); and GC 47 Tuc (de Marchi *et al.* (1995b), $\Delta m = 13.35$).

ing the resulting constraint leads to a semi-empirical $m(M_V)$ relation. It is a semi-empirical relation because we used theoretical stellar models to place, so to speak, zeroth-order constraints on this relation, i.e. to prove the existence of the extremum and to estimate its location, width and amplitude. With this theoretical knowledge in hand we then used the LF to constrain the detailed run of $m(M_V)$. The result is in amazing agreement even with the most-recent high-quality binary-star constraints published by Delfosse *et al.* (2000). With the so-obtained $m(M_V)$ relation, which has the correct derivative, it is now possible to take another step towards constraining the MF.

But first the problem $\Psi_{\text{near}} \neq \Psi_{\text{phot}}$, $M_V > 12$, needs to be addressed.

3.3 Unresolved binary systems

One bias clearly affecting Ψ_{phot} arises from unresolved multiple systems; in constructing Ψ_{near} all stellar companions have been counted individually, while in reality Ψ_{phot} consists of 'stars' too far away to be resolved if they are multiple. Also, a faint companion star might not be seen because it lies below the flux limit while the primary enters the formal photometric distance limit. Furthermore, an unresolved binary system appears redder unless the two stars are of equal mass and this affects the calculated stellar space density because the photometric distance is misjudged. All the effects can and must be corrected for, this having been done thoroughly for the first time by Kroupa *et al.* (1993).

The bias due to unresolved binaries affecting Ψ_{phot} can be nicely demonstrated using realistic numbers for the multiplicity in the Galactic field (Goodwin *et al.* 2006). Suppose that the observer sees 100 systems on the sky. Of these 40 are binaries, 15 triples and 5 quadruples. The multiplicity fraction is $f = (40 + 15 + 5)/100 = 0.6$. The observer will construct the LF using these 100 systems, but in fact 85 stars are missed. All of these are much fainter than their primaries, implying a non-linear depression at the faint end of the LF.

3.4 The standard Galactic-field IMF

Together with a thorough modelling of the star-formation history of the Solar neighbourhood (low-mass stars take a long time to descend to the main sequence), local Galactic-disc structure and metallicity spreads, as well as of fractions of multiple stars and photometric and trigonometric distance uncertainties, Kroupa *et al.* (1993) performed a multi-dimensional optimisation for the parameters of a two-part power-law MF. The result is one MF that incorporates both Ψ_{near} and Ψ_{phot}. It is given by Equation (1) for $m \lesssim M_\odot$ (Kroupa *et al.* 1993; Reid *et al.* 2002). Upon adopting this IMF as being correct, Kroupa (1995b) also showed that the observed LFs are obtained if stars are born as binaries in clusters that evolve and disperse their stellar content into the Galactic field. Chabrier (2003) further points out that non-linearities in the colour–magnitude relation used in photometric parallax add to the underestimate of the stellar densities at the faint end of Ψ_{phot}.

Scalo (1986) had performed a very detailed analysis of the stellar mass function, which was superseded for $m \lesssim M_\odot$ through new star-count data and the new modelling described above, but for $m \gtrsim M_\odot$ his analysis remains valid. Using star-counts out to kiloparsec distances for early-type stars, he modelled their spatial distribution and took account of their production through various star-formation rates (SFRs), arriving at an estimate of the Galactic-field IMF adopted in Kroupa *et al.* (1993):

$$\alpha_3 = 2.7 \pm 0.7, \qquad 1 \lesssim m/M_\odot. \tag{24.6}$$

Together with Equation (1) this is the *standard Galactic-field IMF*. Elmegreen & Scalo (2006) then showed that there can be artificial features in the massive-star part of the Galactic-field IMF when it is deduced from the present-day field MF under conditions of varying SFRs, frustrating attempts to attribute any possible structure there to star-formation physics.

3.5 The canonical IMF

As already stated in Section 1, the IMF for early-type stars can also be inferred for individual OB associations and star clusters. Massey (2003 and previous papers) has shown the IMF for massive stars to be independent of density ($\rho_{\text{central}} \lesssim 10^{5.9}$ stars per pc^3, the densest cluster being R136 in the LMC) and metallicity ($Z \gtrsim 0.002$, the population with the lowest metal abundance being in the SMC). It always has $\alpha = 2.3 \pm 0.1$, thus leading to the standard or canonical stellar IMF, Equation (24.1).

The question now emerges as to why $\alpha_{\text{Scalo}} = 2.7 > \alpha_{\text{Massey}} = 2.3$ for $m \gtrsim M_\odot$. This is referred to as the 'Scalo versus Massey discrepancy'.

Considering the uncertainties, it would be valid to discard the difference. However, using the standard Galactic-field IMF in Galactic-evolution models would yield about ten times fewer O stars than using the canonical IMF. So it is important to know α_3. We will return to this later on, but, as before, when not immediately seeing a possible solution it often proves useful to ignore the problem until new insights open new avenues for exploration.

3.6 Massive stars seldom come alone

Spectroscopic and speckle-interferometric observations of nearby massive-star-forming regions have shown that OB stars prefer more than one partner (Zinnecker 2003; Goodwin *et al.* 2006). In particular, Preibisch *et al.* (1999) present the most detailed multiplicity study of the Orion Nebula cluster (ONC), demonstrating that the rule is triples rather than binaries as amongst T Tauri stars.

This allows us to set up the following *hypothesis*:

$$\frac{\alpha_{\text{Massey}}}{\alpha_{\text{Scalo}}} < 1 \quad \text{because} \quad f_{\text{Scalo}} < 1 \quad \text{while} \quad f_{\text{Massey}} \gtrsim 1. \tag{24.7}$$

That is, Massey was observing dynamically less-evolved and therefore more-binary-rich populations whereas Scalo saw the field population, which is, to a significant extent, 'contaminated' by run-away OB stars, which are mostly single (Stone 1991). This would be a neat solution to our problem: fainter OB stars are hidden among their brighter primaries, flattening the Massey IMF, while the mostly single OB-field stars in the Scalo sample would be a more correct representation of the true IMF.

Thus we may be led to conclude that in actuality the true (binary-corrected) IMF is Equation (24.6) ($\alpha_3 \approx 2.7$), while Equation (24.1) ($\alpha_3 \approx 2.3$) is wrong. This would, of course, have important implications for Galactic astrophysics as well as cosmology; most galaxy-formation and evolution calculations are being done with a Salpeter IMF.

4 Lessons from star clusters

Star clusters pose star-formation events that are extremely well correlated in space, time and chemistry, so studying the MF within clusters avoids most of the problems we need to deal with in the Galactic field. However, there are some serious issues that essentially annihilate any advantage one might have gained by having an equal-age, -distance and -metallicity population.

4.1 Young clusters

To avoid dynamical evolution complicating the star-counts in a cluster, young clusters are chosen. However, uncertainties in pre-main-sequence tracks make the calculation of stellar masses from stellar luminosities and colours or spectra very uncertain for ages less than a few Myr. Errors of about 50% would not be untypical. These errors are not randomly distributed though, but cause systematic offsets from the true masses of stars. Thus unphysical structures appear in the MF. This is discussed in more depth in Kroupa (2002). Very young clusters also have a high binary fraction, which again would require corresponding corrections. Young clusters evolve very rapidly during the first 1 Myr as a result of expulsion of residual gas. For example, the ONC with an age of about 1 Myr is already between 5 and 15 initial crossing times old, implying a rapid dynamical evolution of the binary population (Figure 24.1 in Kroupa (2000); Figure 9 in Kroupa *et al.* (2001)) *and* an expansion of the cluster such that a substantial fraction of its initial population has probably been lost (Kroupa *et al.* 2001). This translates into a selective stellar-mass loss *if* the cluster was initially significantly mass-segregated (Moraux *et al.* 2004). It is therefore never clear exactly which corrections have to be applied, and an inferred stellar MF is very likely not to be a good representation of the IMF.

4.2 Old clusters

To avoid the issues above (uncertain pre-main-sequence theory, rapidly evolving stellar and binary populations), older clusters within which stars are on or close to the main sequence would appear to be useful. They are definitely useful, but such clusters are heavily evolved dynamically, so the MF will not represent the IMF even for clusters with an age near 100 Myr. Unresolved multiple systems remain a problem; for example, for the Pleiades, Stauffer (1984) showed that the fraction of photometric binaries is 26% (these are stars that are more luminous than a single star of the same colour), whereas Kähler (1999) noted that a true binary fraction between 60% and 70% may be possible. Even globular clusters appear to have a sizable binary population (e.g. Rubenstein & Bailyn 1997).

A star cluster that has managed to survive its initial gas-expulsion phase re-virialises on a timescale of tens of Myr (Kroupa *et al.* 2001). Its binary population is depleted by then such that binaries with orbital Kepler velocities smaller than the velocity dispersion in the pre-expansion cluster will have been disrupted. The remaining binaries are mostly inert; sometimes energetic binary–single-star or binary–binary encounters near the cluster core eject 1–3 stars (e.g. a single star and the binary) from the cluster. The cluster's evolution is, however, well described by classical dynamical-evolution tracks, i.e. the surviving binaries are dynamically not important, as shown explicitly by Kroupa (1995c). During this long-lived phase the cluster evolves towards energy equipartition, which implies that the low-mass stars gain energy and thus move outwards, becoming lost to the Galactic tide, while the more massive stars lose energy and segregate towards the centre of the potential. This process never stops, i.e. a cluster never actually reaches energy equipartition and dynamical equilibrium. The consequence is that, as the cluster ages, its MF becomes increasingly depleted in low-mass stars. This is nicely evident in low-mass clusters in terms of a flattening of the LF, while unresolved binaries remain a significant bias despite their dynamical unimportance (Figure 10 in Kroupa (1995c)). Baumgardt & Makino (2003) studied the evolution of the MF in much detail for massive-star clusters without binaries. In their Figure 7 they nicely show how the MF changes its slope as a function of the cluster's dynamical age expressed in terms of the fraction of the time until cluster disruption: clusters that have passed 50% of their disruption time have essentially a flat ($\alpha \approx 0$) MF for $0.1 \lesssim m/M_\odot \lesssim 0.8$, whereas for older clusters α becomes negative.

4.3 Massive stars in clusters

Two processes compete for all stellar masses, but are particularly pronounced for massive stars because of the energetics involved.

4.3.1 Mass segregation

If massive stars form throughout a forming star cluster, energy equipartition will force them to segregate to the centre within a timescale $t_{\rm msgr} \approx 2(m_{\rm av}/m_{\rm m})t_{\rm relax}$, where $m_{\rm av}$ and $m_{\rm m}$ are the masses of the average star and the massive stars and $t_{\rm relax}$ is the median two-body relaxation time, as detailed in Kroupa (2004). For example, for the ONC, $t_{\rm relax} \approx 0.6$ Myr and $m_{\rm av}/m_{\rm m} \approx 10^{-2}$ such that $t_{\rm msgr} \approx 0.1$ Myr, which is much less than the age of the ONC. No wonder that the ONC sports a beautiful Trapezium, although it could also have formed at the centre (Bonnell & Davies 1998).

4.3.2 Core decay

The massive stars at the cluster centre, which may have been born there or segregated there, form a short-lived small-$N_{\rm m}$ core. It decays on a timescale $t_{\rm decay} \approx N_{\rm m} t_{\rm cross,core}$, where $N_{\rm m}$ is the number of massive stars in the core and $t_{\rm cross,core}$ is its crossing time. Again, for the ONC, we find $t_{\rm decay} \approx 10^4$–$10^5$ yr (Kroupa 2004), which is much less than its age. So why does the Trapezium still exist?

4.4 A highly abnormal 'IMF' in the ONC

Detailed N-body computations confirm this timescale problem (Pflamm-Altenburg & Kroupa 2006). The ONC also has a highly abnormal 'IMF' – only 10 stars more massive than $5M_\odot$ are found in it, while 40 would be needed to allow the very final remnant of the central core to be still visible today *and* to account for the mass of the most-massive star, given the cluster-mass–maximal-mass relation observed to exist among young clusters (Weidner & Kroupa 2006).

This appears to be the rule rather than the exception. Pflamm-Altenburg & Kroupa (2006) discuss the older Upper Scorpius OB association which also exhibits a deficit of massive stars relative to the canonical IMF. *Naively this may be taken as good evidence for a bottom-heavy present-day IMF in a relatively metal-rich environment*, which would even be consistent qualitatively with the theoretical arguments of Section 2 (higher-metallicity environments making lower-mass stars on average).

Further work, however, shows this to be an illusion: Pflamm-Altenburg & Kroupa (2006) apply a specially designed chain-regularisation code (CATENA) to study the dynamical stability of ONC-type cores, finding that, if the ONC and Upper Scorpius OB association, which is understood to be an evolved version of the ONC, were born with 40 OB stars, then the observations are well accounted for. Indeed, Stone (1991), among others, has shown that 46% of all O stars are probably run-aways, of which about 18% have line-of-sight velocities $\gtrsim 30\,\mathrm{pc}\,\mathrm{Myr}^{-1}$. Energetic dynamical encounters in cluster cores therefore have a very significant effect on the shape of the 'IMF' in clusters and OB associations. Furthermore, massive stars are rapidly dispersed from their birth sites, with corresponding implications for the spreading of heavy elements and feedback-energy deposition throughout a galaxy.

The implication of these studies is therefore that the massive-star content of star clusters and OB associations is not a measure of the true initial content, unless the full dynamical history of the cluster or OB association is taken into account.

4.5 Summary

It would thus appear that clusters and OB associations of any age are a *horrible* place to study the IMF. This does not mean, however, that these systems are useless. Quite the contrary: they are indeed our only samples of stars formed together and are therefore the only existing samplings from the IMF before dispersion into the field. The point of the above pessimistic note is to instill the fact that any measurement of the 'IMF' in a cluster or association *must* be accompanied by a dynamical investigation *before* conclusions about the possible variation of the IMF can be drawn. An excellent example follows in Section 5.2.

5 Evidence for variation of the IMF

The derivation of MFs in many clusters and associations leads to many values of α measured over various stellar-mass ranges. The alpha-plot, $\alpha(m)$, can therefore give us information about the true shape of the putative underlying IMF and also of the scatter about a mean IMF (Scalo 1998). This scatter can then be studied theoretically using the methods mentioned above, i.e. N-body calculations of star clusters with various properties. Thus, Section 4.5 needs to be invoked in the context of some recent interesting results from observational studies suggesting that the IMF varies at low masses (e.g. Oasa *et al.* 2006) and at high masses (Gouliermis *et al.* 2005).

One important aspect here is the purely statistical scatter resulting from finite-N sampling from a putative universal IMF (Section 1). Elmegreen (1997, 1999) showed that the observed scatter is consistent with the variations of $\alpha(m)$ as a result of sampling from the IMF. Stellar-dynamical and unresolved binary-star biases lead to systematic deviations of $\alpha(m)$ from the true value, and Kroupa (2001b, 2002) has shown the observed alpha-values to be consistent with the canonical IMF (Equation (1)) for a large variety of stellar populations using theoretical models of initially binary-rich star clusters.

As already deduced with some force by Massey (2003 and previous papers), there is no evidence for systematic shifts of $\alpha(m)$ either with metallicity or with density. At one extreme end of the scale of physical conditions, globular clusters have MFs very similar to those of young Galactic clusters and the Galactic field, while constraints on the massive-star content of globular clusters are such that the IMF could not have been too top heavy, since otherwise the mass loss from evolving stars would have unbound the clusters (Kroupa 2001b). At the other extreme, the supersolar Arches cluster near the Galactic Centre has a massive-star IMF again very similar to the Salpeter/Massey one (Section 5.2 below).

Kroupa (2002) studied the distribution, $f_\alpha(\alpha)$, of measured α values for stars more massive than a few $M_\odot$, finding it to be describable by two Gaussians, one somewhat broad, but with symmetric wings, and the other exceedingly narrow, both centred on the Salpeter/Massey value $\alpha = 2.35$. This is very surprising because the various $\alpha(m)$ values were obtained by different research groups observing different objects. An equivalent theoretical data set contains a somewhat broader unsymmetric distribution offset from the input Salpeter/Massey value. This needs further investigation, because one would expect the theoretical data to be better 'behaved' than the observational ones which include all the complications of inferring masses from observed luminosities. Concerning this question, Maíz Apellániz & Ubeda (2005) point out that binning the star-counts into mass bins of equal mass leads to biases that distort the shape of the IMF when the number of stars is small, which is often the case at high masses.

Returning to the empirically determined existence of an upper mass limit to stars near 150 $M_\odot$ (Section 1), it remains to be understood why it does not seem to depend on metallicity. Theoretical concepts would suggest that feedback ought to play a role in limiting the masses of stars, but this implicitly contains a metallicity dependence since the coupling of radiation to the accreting gas is more efficient in metal-rich environments. Thus, we are left yet again with two questions to answer.

5.1 Back to the 'Scalo versus Massey discrepancy'

One important implication from this analysis of $\alpha(m)$ values for massive stars is that the above hypothesis (Equation (7)) must be discarded: if unresolved multiple systems were the origin of the Scalo versus Massey discrepancy, then $f_\alpha(\alpha)$ would have an asymmetric distribution stretching from the single-star value ($\alpha = 2.7$) to the unresolved binary-star value ($\alpha = 2.3$). This is not the case. The Scalo versus Massey discrepancy therefore is either nonexistent (perhaps because Scalo's analysis requires an update), or it has another, hitherto unknown, origin. The proposition by Elmegreen & Scalo (2006) that the IMF calculated from the present-day field

MF may be steeper than the canonical IMF if the field was created during a lull in the star-formation rate deserves some attention in this context.

5.2 An example: the Galactic central region

There have been reports that in extreme environments there is evidence for a top-heavy IMF. These often come from indirect arguments based on luminosity and available mass for star formation in starbursts, for example: scaling the canonical IMF to the observed luminosity would lead to a mass in stars far superior to dynamical mass constraints from kinematics, requiring either a top-heavy (smaller α_3) IMF, or an IMF truncated typically near or above $M_\odot$. Paumard et al. (2006) report evidence for a top-heavy IMF in both circum-nuclear discs in the MW. Although the observations are quite convincing, Section 4.5 needs to be invoked before taking such results as affirmative evidence for a variable stellar IMF.

A case in point is the Arches cluster at the Galactic Centre, where the conditions for star formation are very different from those prevalent in the Solar vicinity and in the disc of the LMC, being in general warmer (≈ 100 K instead of 10 K) and having a supersolar metallicity ($-0.15 \lesssim$ [Fe/H] $\lesssim +0.34$).

Klessen et al. (2006) performed excellent state-of-the art calculations of star formation under such conditions. They state in their abstract that 'In the solar neighbourhood, the mass distribution of stars follows a seemingly universal pattern. In the centre of the Milky Way, however, there are indications for strong deviations and the same may be true for the nuclei of distant star-burst galaxies. Here we present the first numerical hydro-dynamical calculations of stars formed in a molecular region with chemical and thermodynamic properties similar to those of warm and dusty circum-nuclear star-burst regions. The resulting initial mass function is top-heavy with a peak at $\approx 15 M_\odot$, a sharp turn-down below $\approx 7 M_\odot$ and a power-law decline at high masses. We find a natural explanation for our results in terms of the temperature dependence of the Jeans mass, with collapse occurring at a temperature of ≈ 100 K and an H_2 density of a few 10^5 cm^{-3}, ...' Theirs would be in agreement with the previously reported apparent decline of the stellar MF in the Arches cluster below about 6 $M_\odot$.

Shortly after the above theoretical study appeared, Kim et al. (2006) published their observations of the Arches cluster and performed the necessary state-of-the-art N-body calculations of the dynamical evolution of this young cluster, revising our knowledge significantly. Quoting from their abstract: 'We find that the previously reported turnover at 6 $M_\odot$ is simply due to a local bump in the mass function (MF), and that the MF continues to increase down to our 50% completeness limit (1.3$M_\odot$) with a power-law exponent of $\Gamma = -0.91$ for the mass range of $1.3 < m/M_\odot < 50$. Our numerical calculations for the evolution of the Arches cluster show that the Γ

values for our annulus increase by 0.1–0.2 during the lifetime of the cluster, and thus suggest that the Arches cluster initially had a Gamma of $-1.0 - -1.1$, which is only slightly shallower than the Salpeter value.' ($\Gamma = 1 - \alpha$).

This serves to illustrate the general observation that, when detailed high-resolution observations *and* accurate N-body calculations are combined, claims of abnormal IMFs tend to vanish, irrespective of what theory prefers to have. Another example, the ONC, has already been noted above.

5.3 Spheroids: old, very extreme starbursts?

An interesting, albeit indirect, result by Francesca Matteucci and her collaborators based on multi-zone photo-chemical-evolution studies of the MW, its bulge and elliptical galaxies shows the MW to be reproducible by the standard Galactic-field IMF (Equation (6), $\alpha_3 \approx 2.7$) but not by the canonical IMF (Equation (1)). This is supported by the independent work by Portinari *et al.* (2004). In contrast, the MW bulge may require a rapid phase of formation lasting 0.01–0.1 Gyr and a top-heavy IMF with $\alpha \approx 1.95$ (Romano *et al.* 2005; Ballero *et al.* 2006). For elliptical galaxies a similar scenario seems to hold, whereby the IMF needed to account for the colours and chemical properties is the canonical one ($\alpha_3 \approx 2.3$), but flattening (decreasing α_3) slightly with increasing galaxy mass (Pipino & Matteucci 2004). An experiment in which the yields for massive stars are changed as much as possible without violating available constraints confirms this result – it seems not to be possible to understand the metallicity distribution of MW bulge stars with a standard Galactic-field IMF. Instead, $\alpha_3 \approx 1.9$ seems to be required (Ballero *et al.* 2007). Nevertheless, uncertainties remain. Thus, Samland *et al.* (1997) found the bulge to have formed over a time-span of about 4 Gyr with a Salpeter IMF using a two-dimensional chemo-dynamical code, whereas Immeli *et al.* (2004) obtained formation timescales near 1 Gyr using a three-dimensional chemo-dynamical code. Finally, Zoccali *et al.* (2006) find the 'MW bulge to be similar to early-type galaxies, in being α-element enhanced, dominated by old stellar populations, and having formed on a timescale shorter than $\approx$1 Gyr'. 'Therefore, like early-type galaxies the MW bulge is likely to have formed through a short series of starbursts triggered by the coalescence of gas-rich mergers, when the universe was only a few Gyr old'. For low-mass stars, Zoccali *et al.* (2000) find the bulge to have a similar MF to those of the disc and globular clusters.

5.4 Composite populations

The 'Scalo versus Massey discrepancy' found in Section 3.5 and its proposed solution in Section 3.6 in terms of unresolved multiple systems was found, in Section 5.1, not to be possible. Again, as in Section 3.5, we might argue that trying to solve the discrepancy is not worth the effort, given the uncertainties. Indeed, the probable solution drops out through an entirely different line of thought.

It needs to be remembered that star clusters are the fundamental building blocks of galaxies. Therefore, to be exact, all the stellar IMFs in all the clusters forming during one 'star-formation epoch' (Weidner *et al.* 2004) of a galaxy must be added to calculate the global, galaxy-wide 'integrated galactic IMF' (IGIMF). This was pointed out by Vanbeveren (1982, 1984), discussed again by Scalo (1986) and formulated for an entire galaxy by Kroupa & Weidner (2003, 2005), noting that the star clusters are distributed as a power-law embedded-cluster mass function with an exponent $\beta \approx 2 \pm 0.4$, where the Salpeter value would be 2.35. Also, there exists an empirical relation limiting the mass of the most-massive star in a cluster with stellar mass M_{ecl}: $m \leq m_{max}(M_{ecl})$ (Weidner & Kroupa 2006), enhancing the *IGIMF $\neq$ IMF effect*.

Putting this together, an integral over all clusters formed together in one epoch results, and the composite IMF turns out to be steeper with $\alpha_{3,IGIMF} > \alpha_3$ (the canonical IMF value) for $m \gtrsim 1.6~M_\odot$ with a steep downturn at a mass $m_{max,gal}$. Both $\alpha_{3,IGIMF}$ and the galaxy-wide maximum star mass, $m_{max,gal}$, depend on the star-formation rate of the galaxy, so the situation becomes complex, with important implications for galactic spectrophotometric and chemical-evolution studies given that the number of supernovae of Type II is depressed significantly per unit stellar mass formed when compared with a universal Salpeter IMF, which is often assumed in cosmological applications (Goodwin & Pagel 2005; Weidner & Kroupa 2005). Koeppen *et al.* (2007) show that this IGIMF-Ansatz naturally explains the observed form of the mass–metallicity relation of galaxies without the need to invoke selective outflows (the possibility of which is not excluded, though).

Applied to the MW, Scalo's $\alpha_3 \approx 2.7$ comes out naturally for an input canonical IMF. Is this, then, the solution to the 'Scalo versus Massey discrepancy'?

Elmegreen (2006), however, opposes the notion that the IGIMF is different from the stellar IMFs found in individual clusters, mostly because the star-cluster mass function is not steep enough ($\beta < 2$), in which case IGIMF $\approx$ IMF as already shown by Kroupa & Weidner (2003). *Much therefore rests on the detailed shape of the star-cluster initial mass function*! Also, the existence of a physical $m_{max}(M_{ecl})$ relation is of central importance to the IGIMF $\neq$ IMF argument, but opposing views are being voiced (e.g. Elmegreen 2006). An argument against IGIMF $\neq$ IMF is that results from some studies suggest these to be equal, e.g. in elliptical galaxies (Section 5.4). However, an appreciable fraction of the research which Elmegreen fields in support of his conjectured equality actually supports the conjecture of inequality. So this issue is still very much subject to debate and its resolution may need observational improvement by (i) better constraining the true initial star-cluster mass function and (ii) improving the empirical $m_{max}(M_{ecl})$ relation. Observational work to test the IGIMF $\neq$ IMF notion is under way, and, for example, Selman & Melnick

(2005) find the composite IMF to be flatter than the stellar IMF in the 30 Dor star-forming region, as predicted, but only with low statistical significance.

6 Concluding remarks

It would have seemed that, when two mutually more or less exclusive theories of the origin of stellar masses make the same basic prediction concerning the variation of the average stellar mass with physical conditions, this expected variation would be very robust and borne out in observational data. Alas, the observational data on the IMF are resilient – they do not yield what we desire to see. The stellar IMF is invariant and can be best described by a two-part power-law form (Equation (1)). This holds true for metallicities ranging from those of globular clusters to supersolar values near the Galactic Centre and for densities less than about 10^6 stars per pc^3. Even the maximal stellar mass of about 150 $M_\odot$ seems to be independent of metallicity for $Z \gtrsim 0.008$.

Given the quite total failure to account for the resilience of the stellar IMF towards changes, it is clear that this IMF conservatism poses some rather severe challenges to liberal star-formation theory.

Only indirect arguments based on chemical-evolution work seem to suggest the IMF to have been somewhat flatter for massive stars during the extreme physical conditions existing during the formation of the MW bulge and elliptical galaxies. Variations of the galaxy-wide IMF (the IGIMF) among galaxies and also with time can be understood today despite the existence of a universal canonical IMF, but this issue is still very much a matter of debate (between Elmegreen (2006) and Weidner & Kroupa (2006)). The evidence noted in Section 5.3 appears to suggest that perhaps $\alpha_{3,\mathrm{IGIMF}} \approx 2.7$ for the MW disc, $\alpha_{3,\mathrm{IGIMF}} \approx 2.3$ for elliptical galaxies and $\alpha_{3,\mathrm{IGIMF}} \approx 1.9$ for massive elliptical galaxies and the MW bulge. Is this true?

Acknowledgments

I would like to thank the organisers sincerely for a most enjoyable meeting and my collaborators for important contributions. This research was supported by an Isaac Newton and a Senior Rouse Ball Studentship in Cambridge in the UK, and by the Heisenberg programme of the DFG and DFG research grants in Germany. This article was mostly written in Vienna and I thank Christian Theis and Gerhardt Hensler for their hospitality.

References

Adams F. C., & Fatuzzo M. (1996), *ApJ* **464**, 256
Adams F. C., & Laughlin G. (1996), *ApJ* **468**, 586
Maíz Apellániz, J., & Ubeda, L. (2005), *ApJ* **629**, 873
Bahcall, J. N. (1984), *ApJ* **287**, 926

Ballero, S., Matteucci, F., & Origlia, L. (2006), astro-ph/0611650
Ballero, S., Kroupa, P., & Matteucci, F. (2007), *A&A* **467**, 117
Baumgardt H., & Makino J. (2003), *MNRAS* **340**, 227
Bonnell, I. A., & Davies, M. B. (1998), *MNRAS* **295**, 691
Bonnell, I. A., , Larson, R. B., & Zinnecker, H. (2007), in *Proto Stars and Planets V*, eds. B. Reipurth, D. Jewitt & K. Keil (Tucson, AZ, University of Arizona Press), pp. 149–164
Chabrier G. (2003), *PASP* **115**, 763
Delfosse, X., Forveille, T., Ségransan, D. *et al.* (2000), *A&A* **364**, 217
de Marchi G., & Paresce F. (1995a), *A&A* **304**, 202
    (1995b), *A&A* **304**, 211
Elmegreen B. G. (1997), *ApJ* **486**, 944
    (1999), *ApJ* **515**, 323
    (2006), *ApJ* **648**, 572
Elmegreen, B. G., & Scalo, J. (2006), *ApJ* **636**, 149
Figer D. F. (2005), *Nature* **434**, 192
Goodwin, S. P., & Pagel, B. E. J. (2005), *MNRAS* **359**, 707
Goodwin, S. P., Kroupa, P., Goodman, A., & Burkert, A. (2006), in *Proto Stars and Planets V*, eds. B. Reipurth, D. Jewitt & K. Keil (Tucson, AZ, University of Arizona Press), pp. 133–147
Gouliermis, D., Brandner, W., & Henning, T. (2005), *ApJ* **623**, 846
Hambly N. C., Jameson R. F., & Hawkins M. R. S. (1991), *MNRAS* **253**, 1
Henry, T. J., Ianna, P. A., Kirkpatrick, J. D., & Jahreiß, H. (1997), *AJ* **114**, 388
Immeli, A., Samland, M., Gerhard, O., & Westera, P. (2004), *A&A* **413**, 547
Jahreiß H., & Wielen R. (1997), in *ESA SP-402: Hipparcos – Venice '97. The Impact of HIPPARCOS on the Catalogue of Nearby Stars. The Stellar Luminosity Function and Local Kinematics*, eds. B. Battrick, M.A.C. Perryman & P. L. Bernacca, pp. 675–680
Kähler, H. (1999), *A&A* **346**, 67
Kim, S. S., Figer, D. F., Kudritzki, R. P., & Najarro, F. (2006), *ApJL* astro-ph/0611377
King, I. R., Bedin, L. R., Piotto, G., Cassisi, S., & Anderson, J. (2005), *AJ* **130**, 626
Klessen, R. S., Spaans, M., & Jappsen, A.-K. (2007), *MNRAS* **374**, L29
Koen, C. (2006), *MNRAS* **365**, 590
Koeppen, J., Weidner, & C., Kroupa, P. (2007), *MNRAS* **375**, 673
Kroupa P. (1995a), *ApJ* **453**, 350
    (1995b), *ApJ* **453**, 358
    (1995c), *MNRAS* **277**, 1522
    (2000), *New Astronomy* **4**, 615
    (2001a), *MNRAS* **322**, 231
    (2001b), in *Dynamics of Star Clusters and the Milky Way: The Local Stellar Initial Mass Function*, eds. S. Deiters, B. Fuchs, R. Spurzem, A. Just & R. Wielen (San Francisco, CA, Astronomical Society of the Pacific), p. 187
    (2002), *Science* **295**, 82
    (2004), *New Astronomy Review* **48**, 47
Kroupa P., & Weidner C. (2003), *ApJ* **598**, 1076
    (2005), *IAU Symposium* **227**, 423, astro-ph/0507582
Kroupa P., & Tout C. A. (1997), *MNRAS* **287**, 402
Kroupa P., Aarseth S., & Hurley J. (2001), *MNRAS* **321**, 699
Kroupa P., Bouvier J., Duchêne G., & Moraux E. (2003), *MNRAS* **346**, 354
Kroupa P., Tout C. A., & Gilmore G. (1993), *MNRAS* **262**, 545
Kuijken, K., & Gilmore, G. (1991), *ApJL* **367**, L9

Larson R. B. (1998), *MNRAS*, **301**, 569
Massey P. (2003), *AR&A* **41**, 15
Miller G. E., & Scalo J. M. (1979), *ApJS* **41**, 513
Moraux E., Kroupa P., & Bouvier J. (2004), *A&A* **426**, 75
Motte, F., André, P., & Neri, R. (1998), *A&A* **336**, 150
Motte, F., André, P., Ward-Thompson, D., & Bontemps, S. (2001), *A&A* **372**, L41
Nutter, D., & Ward-Thompson, D. (2006), astro-ph/0611164
Oasa, Y., Tamura, M., Nakajima, Y. *et al.* (2006), *AJ* **131**, 1608
Oey, M. S., & Clarke, C. J. (2005), *ApJL* **620**, L43
Paresce F., de Marchi G., & Romaniello M. (1995), *ApJ* **440**, 216
Paumard, T., Genzel, R., Martins, F. *et al.* (2006), *ApJ* **643**, 1011
Pflamm-Altenburg, J., & Kroupa, P. (2006), *MNRAS* **373**, 295
Pipino, A., & Matteucci, F. (2004), *MNRAS* **347**, 968
Portinari, L., Sommer-Larsen, J., & Tantalo, R. (2004), *MNRAS* **347**, 691
Preibisch T., Balega Y., Hofmann K., Weigelt G., & Zinnecker H. (1999), *New Astronomy* **4**, 531
Reid, N., & Gilmore, G. (1982), *MNRAS* **201**, 73
Reid, I. N., Gizis, J. E., & Hawley, S. L. (2002), *AJ* **124**, 2721
Romano, D., Chiappini, C., Matteucci, F., & Tosi, M. (2005), *A&A* **430**, 491
Rubenstein, E. P., & Bailyn, C. D. (1997), *ApJ* **474**, 701
Samland, M., Hensler, G., & Theis, C. (1997), *ApJ* **476**, 544
Salpeter, E. E. (1955), *ApJ* **121**, 161
Scalo, J. M. (1986), *Fundamentals of Cosmic Physics* **11**, 1
Scalo J. (1998), in *The Stellar Initial Mass Function (38th Herstmonceux Conference). The IMF Revisited: A Case for Variations*, eds. G. Gilmore & D. Howell (San Francisco, CA, Astronomical Society of the Pacific), p. 201
Selman, F. J., & Melnick, J. (2005), *A&A* **443**, 851
Stauffer, J. R. (1984), *ApJ* **280**, 189
Stone, R. C. (1991), *AJ* **102**, 333
Testi, L., & Sargent, A. I. (1998), *ApJL* **508**, L91
Vanbeveren, D. (1982), *A&A* **115**, 65
 (1984), *A&A* **139**, 545
von Hippel, T., Gilmore, G., Tanvir, N., Robinson, D., & Jones, D. H. P. (1996), *AJ* **112**, 192
Weidner, C., & Kroupa, P. (2004), *MNRAS* **348**, 187
 (2005), *ApJ* **625**, 754
 (2006), *MNRAS* **365**, 1333
Weidner, C., Kroupa, P., & Larsen, S. S. (2004), *MNRAS* **350**, 1503
Zheng, Z., Flynn, C., Gould, A., Bahcall, J. N., & Salim, S. (2001), *ApJ* **555**, 393
Zinnecker, H. (1984), *MNRAS* **210**, 43
 (2003), *IAU Symposium* **212**, 80
Zoccali, M., Cassisi, S., Frogel, J. A. (2000), *ApJ* **530**, 418
Zoccali, M., Lecureur, A., Barbuy, B. *et al.* (2006), *A&A* **457**, L1

25

Initial-mass-function effects on the metallicity and colour evolution of disc galaxies

Pieter Westera, Markus Samland, Roland Buser & Karin Ammon

Astronomical Institute. Department of Physics and Astronomy, Universität Basel, Venusstrasse 7, CH-4102 Binningen, Switzerland

In this contribution, we use chemo-dynamical models to investigate the influence of the initial mass function (IMF) on the evolution of disc galaxies, in particular of their metallicities and colours. We find that 'bottom-light' IMFs (IMFs with a high high-to-low-mass-stars ratio) lead to higher metallicities both in the stellar content and in the interstellar gas than do 'top-light' IMFs, and also to a higher star-formation rate (SFR) beginning ~ 5 Gyr after the galaxy's birth.

Unfortunately, in terms of integrated colours and magnitudes, these two effects work in the opposite sense, the higher SFR turning the galaxy brighter and bluer, but the higher gas metallicity increasing the extinction and turning it fainter and redder, which complicates making statements about the IMF from these observables. The most likely wavelength region in which to detect IMF effects is the infrared (i.e. JHK), where the absorption overcompensates for SFR effects.

1 Introduction

The initial mass function (IMF) tells us in which relation stars of various birth masses are formed at the birth of a stellar population. Ever since Salpeter's first IMF (Salpeter 1955), astronomers have been trying to find out its exact shape and whether it is universal (independent of parameters such as metallicity and environment). So far, there is no convincing evidence for a variable IMF, and the (therefore-universal) IMF seems to have something like a Kroupa (2001) shape. However, many stellar populations cannot be resolved into stars, in particular the ones (i.e. at high redshifts) that could be different from the populations that were used to determine the IMF(s). Therefore, we cannot be sure that the IMF is universal

The Metal-rich Universe, eds. G. Israelian and G. Meynet. Published by Cambridge University Press.
© Cambridge University Press 2008.

as long as there is no method by which to determine the IMFs of unresolved stellar systems.

In this work, we use two fully self-consistent chemo-dynamical models by M. Samland (private communication; a description of the models can be found in Westera et al. (2007)), which have identical border conditions (cosmology, gas-infall history, etc.), but in which star formation takes place according to two different IMFs, the Salpeter (1955) and the Kroupa 'universal' IMF (2001), respectively. The goal is to find out whether we can infer the IMF of a population from integrated properties (colours, magnitudes, spectra) and, if so, which ones are the most suitable for this purpose.

Since the IMF influences not only the light from the stars but also the entire evolution of a galaxy, we hope to find some influences of the IMF on these observables. Using an evolutionary code (Bruzual & Charlot, 2003) combined with the Padova 1994 stellar evolutionary tracks (Fagotto et al. 1994; Girardi et al. 1996) and the BaSeL 3.1 stellar Spectral Energy Distribution (SED) library (Westera 2001; Westera et al. 2002a, 2002b), integrated spectra, colours and magnitudes were calculated (in the SLOAN Digital Sky Survey (SDSS) *ugriz* system, the Hubble Space Telescope (HST) system and others), as were colour images. These observable quantities were also calculated for two artificial models (the same models, but omitting the gas absorption), in order to quantify absorption effects.

2 The Salpeter and Kroupa models

The two IMFs differ in the sense that, when two populations have the same total mass, the one with the Kroupa IMF has more high-mass stars than does the one with the Salpeter IMF. Therefore, it is brighter and bluer, produces more stellar wind and also has more metals.

As a consequence, the model galaxy with the Kroupa IMF implemented, henceforth called the Kroupa model, has a lower star-formation rate (SFR) in the beginning, when stellar winds keep the gas from collapsing. After ~5 Gyr, however, due to there being more gas available, the SFR overtakes that of the model with the Salpeter IMF implemented, logically called the Salpeter model.

As could also be expected, the Kroupa-model galaxy contains more gas, a higher gas metallicity and thus more 'dust' right from the beginning.

3 Results

Unfortunately, the calculated colour images of the two model galaxies do not look very different. The Salpeter galaxy forms a bulge a bit earlier, at an age of 4 Gyr

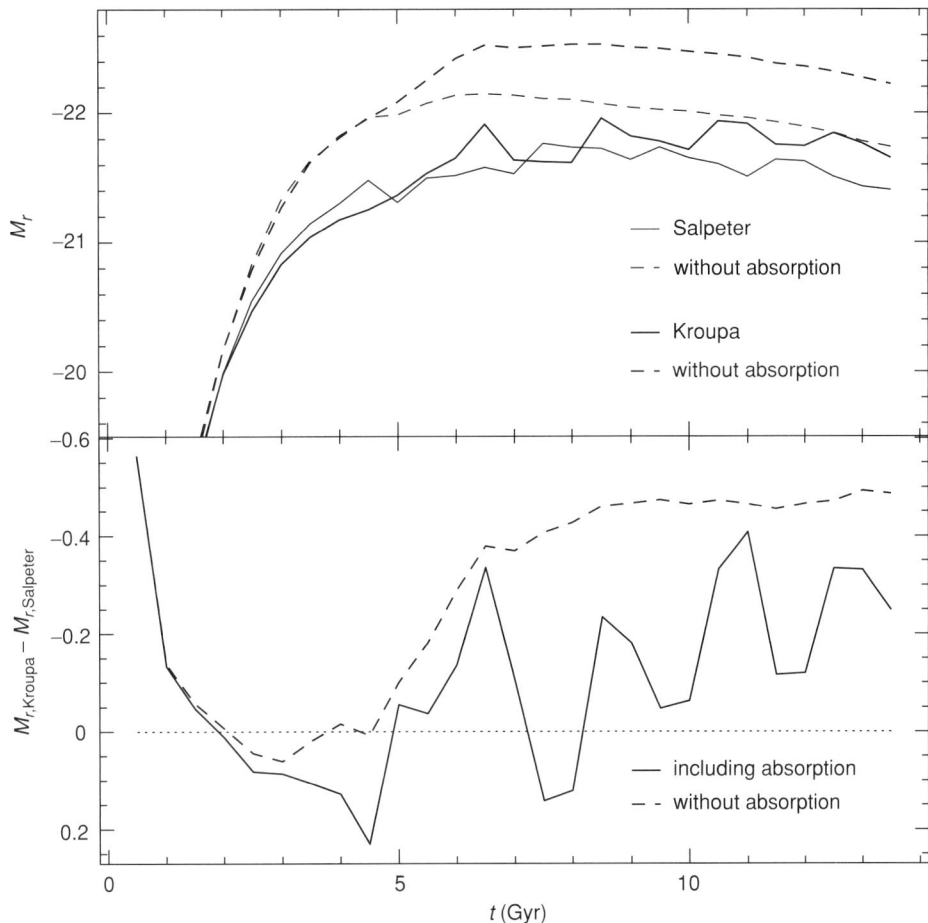

Figure 25.1. Top panel: absolute r magnitude evolution of the Salpeter (thin) and Kroupa (thick) models, both including (solid) and omitting (dashed) gas absorption. Bottom panel: differences in absolute r magnitude between the Salpeter and Kroupa models, both including (solid) and omitting (dashed) gas absorption. The thin dotted line shows the zero level.

($z = 1.59$) versus 6 Gyr ($z = 0.94$) for the Kroupa galaxy, but, apart from that, the two galaxies look disappointingly similar throughout their evolution.

This is confirmed on comparing the integrated magnitudes and colours of the two models. Owing to the higher SFR, the stellar light (which is calculated in the absorptionless models) is brighter (by up to ~ 0.5 mag, depending on the filter band used) and bluer (by up to ~ 0.2 mag, depending on the colour) in the Kroupa model from 5 Gyr onwards. Unfortunately, when one includes the gas absorption in the colour analysis, this effect is counter-balanced by the higher gas absorption due to the higher gas and 'dust' content, thereby reducing the differences in magnitude and

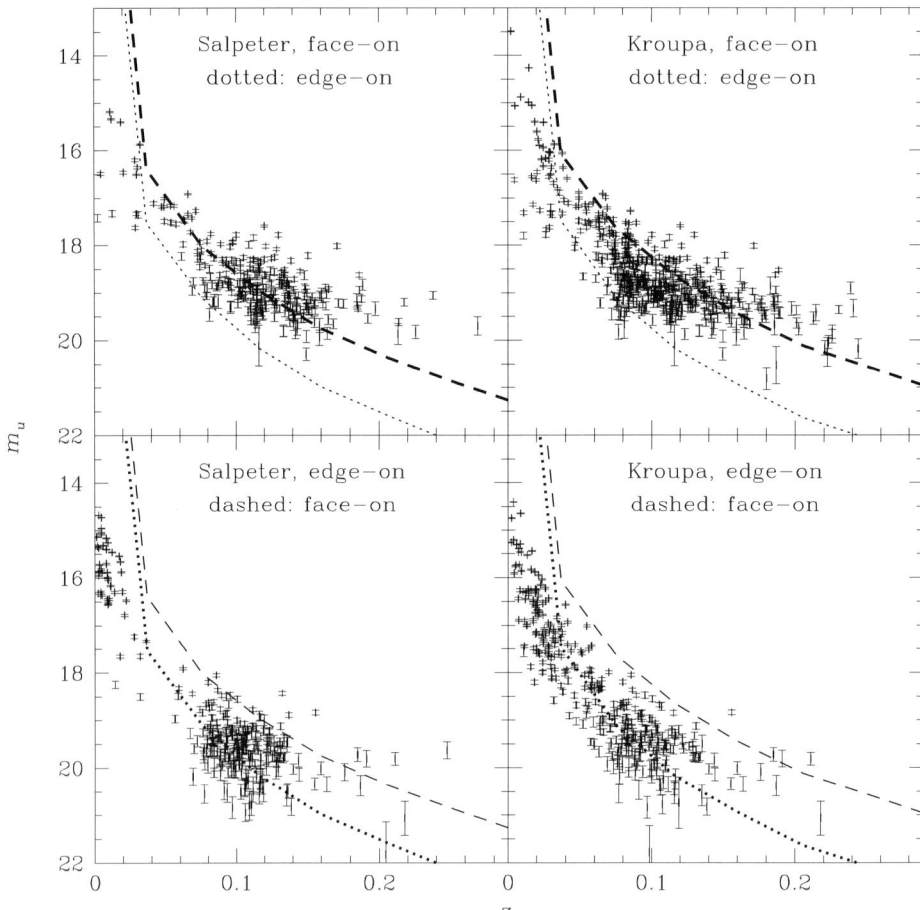

Figure 25.2. Apparent SLOAN u magnitudes of the Salpeter and Kroupa models as seen face-on (dashed) and edge-on (dotted) as a function of redshift z (not to be confused with the SLOAN z passband). Overlaid are the SDSS DR4 data with their observational errors of galaxies that have similar sizes, concentration parameters and orientation to the models.

colour between the two models almost to zero. We call this unfortunate conspiracy the 'IMF degeneracy'. It is illustrated in Figure 25.1 for the r-magnitude.

The magnitudes and colours which offer the best chance of breaking this 'degeneracy' lie in the infrared (JHK, etc.), but even in this wavelength range, the differences between the two models are too small to be measured.

4 Comparison with observations

In Figure 25.2, we compare the model apparent u magnitudes with those of SDSS Data Release 4 (DR4) galaxies of similar orientation (that is, face-on or edge-on), angular size and concentration parameters as a function of redshift z. There seem to be more galaxies matching the Kroupa models than the Salpeter models in terms of these parameters, which is an indication, but not a proof, that the former might be more realistic. However, the model magnitudes agree well with the SDSS magnitudes and the differences between face-on and edge-on are well reproduced, thereby solidifying our model results. This holds true in all colour bands.

The colours of the selected galaxy sample, too, are well reproduced by the models, especially at longer wavelengths (*riz*). For colours too, the differences between the edge-on and face-on samples (the edge-on galaxies are redder) are well reproduced by the models.

5 Summary and conclusions

The main points and conclusions of this work are the following.

- We tried to find out whether the IMF of a stellar system (i.e. a galaxy) can be inferred from integrated observables, such as spectra, magnitudes and colours, using chemo-dynamical models.
- For this purpose, we produced two disc-galaxy models, with identical boundary conditions but different IMFs (Salpeter versus Kroupa).
- Since the Kroupa model produces more high-mass stars, it has brighter and bluer starlight, but also a higher gas yield. Therefore, the Kroupa galaxy contains more, metal-richer gas and has a higher SFR from 5 Gyr onwards.
- Unfortunately, the brighter and bluer starlight due to the higher SFR and the absorption from this gas largely cancel each other out, so we could not find any significant differences in integrated colours, magnitudes and other derived quantities (such as structure parameters, scale heights and lengths). We call this unfortunate coincidence the 'IMF degeneracy'.
- So, for the time being, we still need to resolve populations into individual stars in order to infer their IMFs.
- The question of whether the IMF is universal cannot be answered using our models.
- On the positive side, our model magnitudes and colours agree well with those of galaxies of the same types from the SDSS DR4, especially for the Kroupa model, which is an indication that the models are realistic and could be used for other purposes.

References

Bruzual A. G., & Charlot, S. (2003), *MNRAS* **344**, 1000
Fagotto F., Bressan A., Bertelli G., & Chiosi C. (1994), *A&AS* **105**, 39
Girardi L., Bressan A., Chiosi C., Bertelli G., & Nasi E. (1996), *A&AS* **117**, 113
Kroupa, P. (2001), *MNRAS* **322**, 231
Salpeter, E. E. (1955), *ApJ* **121**, 161

Westera, P. (2001), The BaSeL 3.1 models: metallicity calibration of a theoretical stellar spectral library and its application to chemo-dynamical galaxy models, unpublished Ph.D. thesis, Universität Basel
Westera, P., Lejeune, T., Buser, R., Cuisinier, F., & Bruzual A. G. (2002a), *A&A* **381**, 524
Westera, P., Samland, M., Buser, R., & Gerhard, O. E. (2002b), *A&A* **389**, 761
Westera, P., Samland, M., Kautsch, S. J., Buser, R., & Ammon, K. (2007), *A&A* **465**, 417

26

The metallicity of circumnuclear star-forming regions

Ángeles I. Díaz,[1] Elena Terlevich,[2] Marcelo Castellanos[1]
& Guillermo Hägele[1]

[1]*Universidad Autónoma de Madrid, Madrid, Spain*
[2]*Instituto Nacional de Astrofísica, Óptica y Electrónica, Puebla, Mexico*

We present a spectrophotometric study of circumnuclear star-forming regions (CNSFRs) in the early-type spiral galaxies NGC 2903, NGC 3351 and NGC 3504, all of them of over Solar metallicity according to standard empirical calibrations. A detailed determination of their abundances is performed after careful subtraction of the very prominent underlying stellar absorption. It is found that most regions exhibit the highest abundances in H II-region-like objects. The relative N/O and S/O abundances are discussed. It is also shown that CNSFRs, as a class, segregate from the disk H II region family, clustering around smaller "softness parameter" – η' – values, and therefore higher ionizing temperatures.

1 Introduction

The inner parts of some spiral galaxies have higher star-formation rates than usual and this star formation is frequently arranged in a ring or pseudo-ring pattern around their nuclei. In general, circumnuclear star-forming regions (CNSFRs), also referred to as "hotspots," are alike luminous and large disk H II regions, but look more compact and have higher peak surface brightness (Kennicut *et al.* 1989). In many cases they contribute substantially to the UV emission of the entire nuclear region (e.g. Colina *et al.* 2002). Their Hα luminosities overlap with those of H II galaxies, being typically higher than 10^{39} erg s^{-1}, which indicates that they are relatively massive ionizing star clusters. These regions are expected to have a high metallicity corresponding to their position near the galactic bulge. They have considerable weight in the determination of abundance gradients, which in turn are widely used to constrain chemical-evolution models and constitute excellent laboratories to study how star formation proceeds in high-metallicity environments.

The Metal-rich Universe, eds. G. Israelian and G. Meynet. Published by Cambridge University Press.
© Cambridge University Press 2008.

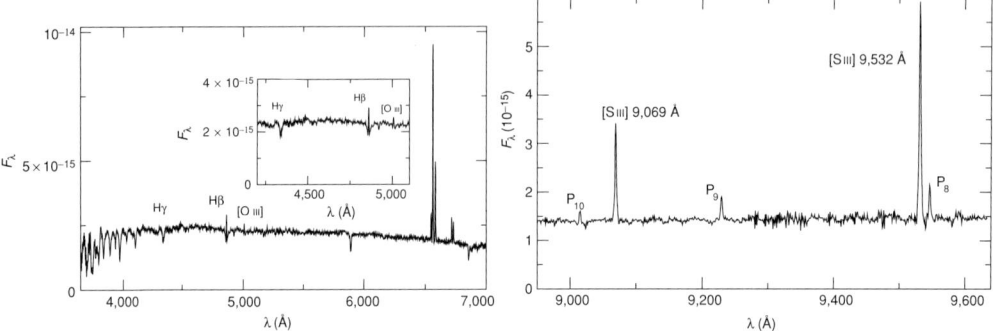

Figure 26.1. Left panel: the blue spectrum of region 4 in NGC 2903. Right panel: the red spectrum of region 3 in NGC 3351.

2 Observations and reductions

Twelve CNSFRs were observed with the 4.2-m WHT at the Roque de los Muchachos Observatory using the ISIS double spectrograph, with the EEV12 and TEK4 detectors in the blue and red arms respectively, with the dichroic filter at $\lambda 7500$ Å. Gratings R300B in the blue arm and R600R in the red arm were used, covering the spectral ranges $\lambda 3650 - \lambda 7000$ Å in the blue and $\lambda 8850 - \lambda 9650$ Å in the near infrared, yielding spectral dispersions of 1.73 Å per pixel in the blue arm and 0.79 Å per pixel in the red arm. A slit width of 1 arcsec was used. The nominal spatial sampling was 0.4 arcsec per pixel in each frame and the average seeing for this night was ~ 1.2 arcsec.

The data were reduced using the Image Reduction and Analysis Facility (IRAF) package following standard procedures. The high spectral dispersion used in the near infrared allowed the almost-complete elimination of the night-sky OH emission lines and, in fact, the observed $\lambda 9532/\lambda 9069$ ratio is close to the theoretical value of 2.48 in all cases. Telluric absorptions were removed from the spectra of the regions by dividing them by a relatively featureless continuum of a sub-dwarf star observed during the same night.

3 Results and discussion

Examples of blue and red spectra are shown in Figure 26.1. Emission-line fluxes were measured using the IRAF SPLOT software package. The presence of a conspicuous underlying stellar population, which is more evident in the blue spectra, in most observed regions complicates the measurements. An example of underlying absorption can be seen in the inset to the left panel of Figure 26.1. A two-component (emission and absorption) Gaussian fit was performed in order to correct

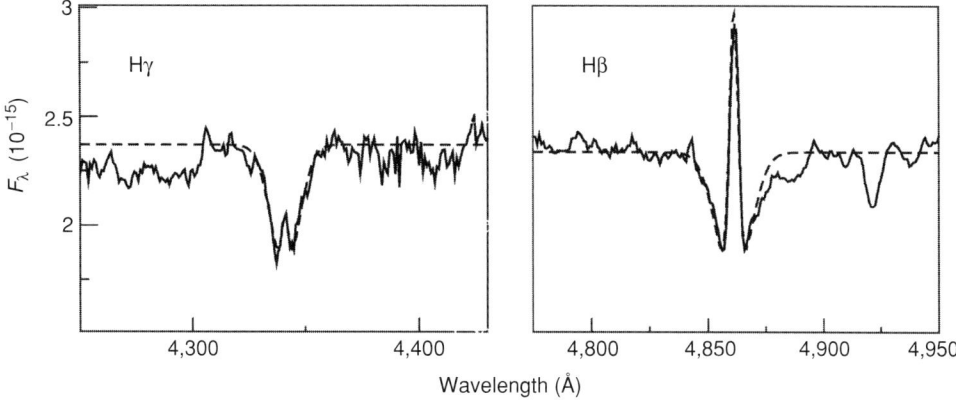

Figure 26.2. An example of the fitting procedure used in order to correct the Balmer emission-line intensities for the presence of the underlying stellar population.

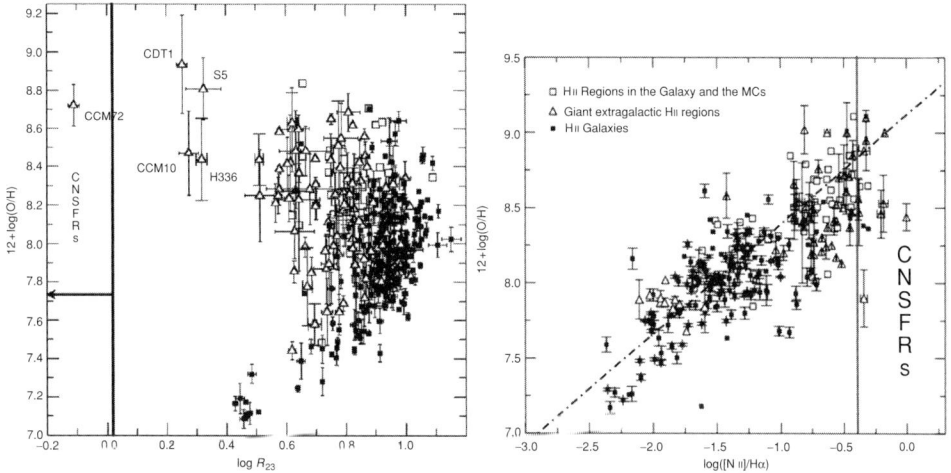

Figure 26.3. Left panel: the calibration of the R_{23} abundance parameter. Right panel: the calibration of the N2 abundance parameter. The location of CNSFRs is indicated.

the Balmer lines for underlying absorption. An example of this procedure is shown in Figure 26.2.

The low excitation of the regions, as indicated by the weakness of the [H III] $\lambda 5007$ Å line (see the left panel of Figure 26.1), precludes the detection and measurement of the auroral [H III] $\lambda 3463$ Å necessary for the derivation of the electron temperature. It is therefore impossible to obtain a direct determination of the oxygen abundances. Empirical calibrations have to be used instead. In the left panel of Figure 26.3 we show the calibration of oxygen abundance by use of the commonly

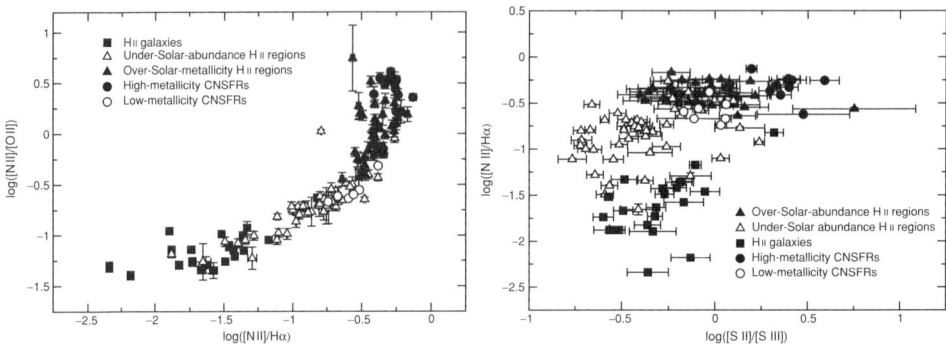

Figure 26.4. Left panel: an empirical version of the N/O versus O/H relation. Right panel: the N2 abundance parameter as a function of excitation, as measured by the [S II]/[S III] ratio.

used R_{23} parameter defined as ([O II] λλ3727, 3729 + [O III] λλ4959, 5007)/Hβ (Pagel et al. 1979). Data on H II galaxies, disk H II regions, and CNSFRs are shown. The H II-region sample (Pérez-Montero & Díaz 2005) has been divided into under-Solar (open triangles) and over-Solar (filled triangles) according to the empirical criterion given by Díaz & Pérez-Montero (2000), i.e. $R_{23} \leq 0.47$ and $-0.5 \leq S_{23} \leq 0.28$.[1] The H II-galaxy data (filled squares) come from Pérez-Montero & Díaz (2003). The CNSFRs are represented by circles, solid ones for our observed objects and open ones for regions in NGC 3310 and NGC 7714 known to have under-Solar abundances (Pastoriza et al. 1993; González-Delgado et al. 1995). As can be seen, the calibration is two-folded and has considerable scatter, and its high-abundance end is not well sampled. The positions of the observed CNSFRs are indicated. Their observed R_{23} values, lower than those of the lowest-abundance galaxy known (IZw18), indicate that CNSFRs belong to the high-abundance branch of the calibration, having possibly the highest metallicities shown by H II-region-like objects. In the right panel of Figure 26.3 the positions of the regions are indicated in the N2 ([N II]/Hα) abundance-calibration diagram (Denicoló et al. 2002), which reveals a linear behaviour. Again CNSFRs appear to have the highest oxygen abundances.

The left panel of Figure 26.4 shows the [N II]/[O II] ratio versus the N2 abundance parameter. Since a good correlation has been found to exist between the [N II]/[O II] ratio and the N^+/O^+ ionic abundance ratio, which in turn can be assumed to measure the N/O ratio, this graph is the observational equivalent of the N/O versus O/H diagram. We can see that a very tight correlation exists for all of the objects

---

[1] The sulfur-abundance parameter S_{23} is defined as ([S II] + [S III])/Hβ.

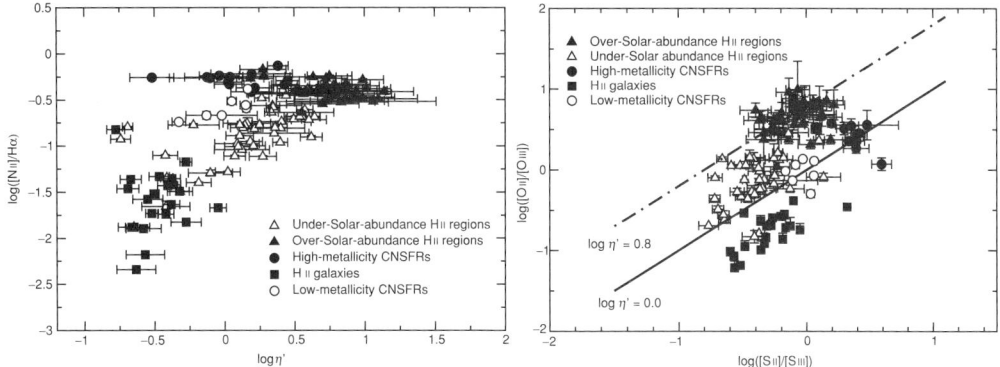

Figure 26.5. Left panel: the N2 versus η' relation. Right panel: the logarithmic relation between the [O II]/[O III] and [S II]/[S III] line ratios.

represented: H II galaxies, low- and high-abundance H II regions, and CNSFRs. Again our observed regions have the highest N/O ratios of the sample.

The right panel of Figure 26.4 shows the run of the degree of excitation with metallicity for the observed regions through the [S II]/[S III] ratio, which has been shown to be a good ionization-parameter indicator (Díaz et al. 1991), and the N2 parameter. It can be seen that the observed CNSFRs have the lowest excitation of the sample.

A hint on the ionizing temperature of the regions can be obtained through the use of the η' parameter, which is a measure of the softness of the ionizing radiation (Vílchez & Pagel 1988) and increases with decreasing ionizing temperature. The left panel of Figure 26.5 shows the run of η' with metallicity as parametrized by N2. Unexpectedly, CNSFRs have values of η' higher than those of over-Solar disk H II regions. This is better appreciated in the right panel of the figure, where CNSFRs are seen to segregate from disk H II regions in the [O II]/[H III] versus [S II]/[S III] diagram. The former cluster around the value of log $\eta' = 0.0$ ($T_{ion} \sim 40,000$ K) while the latter cluster around log $\eta' = 0.8$ ($T_{ion} \sim 35,000$ K).

Acknowledgment

This work was supported by Spanish projects DGICYT- AYA-2004–08262-C03–03 and CM-ASTROCAM.

References

Colina, L., González-Delgado, R., Mas-Hesse, J. M., & Leitherer, C. (2002), *ApJ* **579**, 545–553
Denicoló, G., Terlevich, R., & Terlevich, E. (2002), *MNRAS* **330**, 69–74
Díaz, A. I., & Pérez-Montero, E. (2000), *MNRAS* **135**, 130–138
Díaz, A. I., Terlevich, E., Vílchez, J. M., Pagel, B. E. J., & Edmunds, M. G. (1991), *MNRAS* **253**, 245

González-Delgado, R., Pérez, E., Díaz, A. I., García-Vargas, M. L., Terlevich, E., & Vílchez, J. M. (1995), *ApJ* **439**, 604–622
Kennicut, R. C., Keel, W. C., & Blaha, C. A. (1989), *AJ* **97**, 1022–1035
Pagel, B. E. J., Edmunds, M. G., Edmunds, M. G., Blackwell, D. E., Chun, M. S., & Smith, G. (1979), *MNRAS* **185**, 95–113
Pastoriza, M. G., Dottori, H. A., Terlevich, E., Terlevich, R., & Díaz, I. (1993), *MNRAS* **260**, 177–190
Pérez-Montero, E., & Díaz, A. I. (2003), *MNRAS* **346**, 105–118
 (2005), *MNRAS* **361**, 1063–1076
Vílchez, J. M., & Pagel, B. E. J. (1988), *MNRAS* **231**, 257–267

27

The stellar population of bulges

P. Jablonka

*Ecole Polytechnique Fédérale de Lausanne & Université de Genève,
Observatoire, Chemin des Maillettes 51, CH-1290 Sauverny, Switzerland*

This review summarizes the properties of the stellar population in bulges observed in nearby and distant spiral galaxies. Particular emphasis is placed on comparison with elliptical galaxies, when possible. The sample-selection criteria and choices in data analysis are addressed when they may be involved in discrepancies among published results.

1 Introduction

Studying bulges of spirals is not restricted to a particular class of galaxies, or even to their central regions. There is a growing body of evidence that knowledge of these bulges is crucial to understanding galaxy formation in general. Indeed, the light distribution of most large galaxies is dominated by two components, a bulge and a disk. Even very-early-type galaxies harbor a variety of luminosity profiles, which are interpreted as due to a varying contribution of a disk component (Saglia *et al.* 1997; de Jong *et al.* 2004). Along the Hubble sequence, the variation in bulge magnitude is twice that of the disk (Simien & de Vaucouleurs 1986; de Jong 1996), i.e. the properties of bulges are keys to inferring the nature of the Hubble sequence.

Despite the prospect of their yielding crucial information on galaxy formation and galactic assembly history, bulges have received significantly less attention than elliptical galaxies. This is a direct consequence of the considerable challenge of avoiding contamination from disk light. Figure 27.1 illustrates this point. The data for galaxies, nearby and face-on, are from Jablonka *et al.* (1996). It appears clearly that, within a fixed aperture, classically of the order $R \sim 1\text{--}2$ arcsec for integrated spectroscopy, it is nearly impossible to get rid of all of the disk light. Even more importantly, one can get very different bulge-to-disk light ratios from one galaxy to another, prejudicing our understanding of trends with physical quantities.

The Metal-rich Universe, eds. G. Israelian and G. Meynet. Published by Cambridge University Press.
© Cambridge University Press 2008.

262   *The stellar population of bulges*

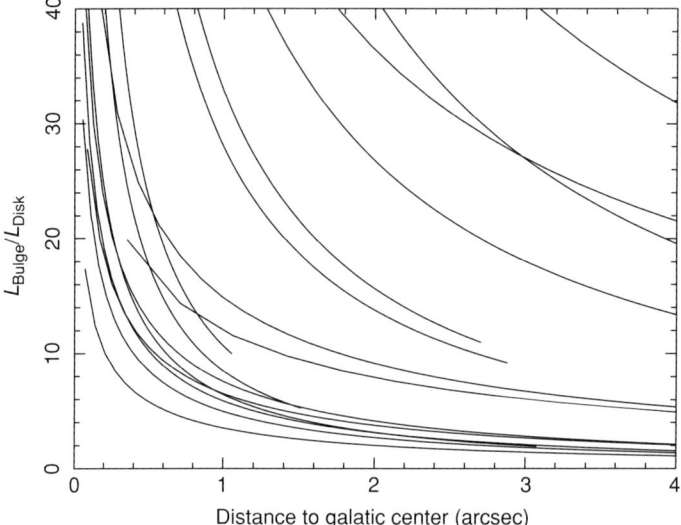

Figure 27.1. The ratio between the bulge and disk luminosities as a function of the distance from the galaxy center for a sample of nearby face-on galaxies.

Nevertheless, observational efforts are intensifying, improving our vision of bulges' properties. Although no definitive certainties have emerged yet, new lines of research are now under way.

2 Metallicity distributions

It will be a long time before it is possible to get spectra of individual stars in our closest spiral neighbor, M31. Jablonka *et al.* (2000), using the MCS deconvolution technique, counted ∼40 bright RGB stars per arcsec^2 1–1.5 kpc from the galaxy's center. Therefore, attempts to derive a metallicity distribution function (MDF) in the bulge of M31 have to be based on high-spatial-resolution images and analyzed with isochrones. Thus far, there have been two studies addressing this issue, one in the optical (Sarajedini & Jablonka 2005) and the other in the infrared (Olsen *et al.* 2006), both using HST data. Sarajedini & Jablonka analyzed a field located about 1.5 kpc from the nucleus of M31 in V and I bands. The MDF that they derived is presented in Figure 27.2 and compared with the metallicity distribution of the bulge of the Milky Way from Zoccali *et al.* (2003). Within 0.1–0.2 dex, the range of metallicity covered by the two galaxies' bulges is the same, and so are the peaks of the distributions. Had bulges straightforwardly reflected the differences between the halos of the two galaxies, one would have expected a differential shift of at least 1 dex between their MDFs (Ryan & Norris 1991; Durrell Harris & Pritchet 2001). On the contrary, it seems that the bulge of M31 does not know about the

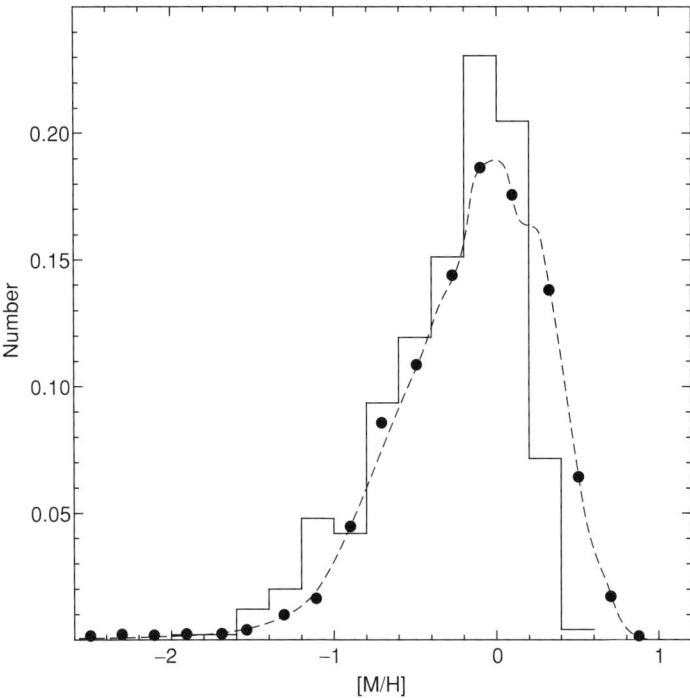

Figure 27.2. A comparison between the metallicity distribution of the bulge of M31 (filled points and dashed line) and that of the Milky Way (plain line), in regions at comparable distances from the galaxy centers.

metal-richness of its halo. Besides, the M31 bulge MDF exhibits a total absence of metal-poor stars, just like the Milky Way bulge does. This is a secure result, since the lowest metallicities identified ($\sim -1.5\,\text{dex}$) are well away from the gray zone of the very low metallicities where isochrones are degenerated in colors.

In contrast to Sarajedini and Jablonka, Olsen *et al.* (2006) left the age of the stellar population as a free parameter in their analysis. Also, instead of trying to reproduce the complete color–magnitude diagrams, they fit model stellar populations to the K luminosity functions of their fields, using a maximum-likelihood method. They found the stellar population mix of their 12 fields to be dominated by old (defined as having ages $\geq 6\,\text{Gyr}$), nearly Solar-metallicity stars. This old population seems to dominate the star-formation history at all radii, irrespective of the relative contributions of bulge and disk stars. In their Figure 19, Olsen *et al.* show the population integrated over all their fields. Neglecting the possibly spurious intermediate-age metal-poor component, which possibly arose due to crowding, they measured an MDF that is a bit more sharply peaked than that of Sarajedini and Jablonka, but still in excellent qualitative agreement.

Assembling these observational facts suggests that the first stars in bulges formed from an already pre-enriched gas. It remains unknown whether this resulted from the first stellar generations in the halo or is due to the location of the observed fields. Indeed, as we will see later, bulges do exhibit radial gradients in metallicity and one might not yet have probed the outermost regions of the M31 and Milky Way bulges. In any case, the bulk of the bulge formation must have taken place before the mergers whose traces are witnessed today (Ferguson *et al.* 2002; Ibata *et al.* 1994; Yanny *et al.* 2003) could influence the bulges' evolution. Otherwise, the large difference between the M31 and Milky Way halos should have been reflected in the properties of their bulges.

3 Scaling relations – central indices

With the exception of the Milky Way and M31, in which we can resolve individual stars, studies of bulges have to deal with integrated properties.

Spectroscopic studies of the central parts of bulges were pioneered by Bica (1988). He gave the first evidence for a relative independence of the bulges' spectral properties with respect to the morphological type of the parent galaxies. He also showed that changes in age/metallicity were linked to the luminosity of the galaxy. During the following years, a central metallicity–luminosity (Z–L) relation for bulges was more firmly established and authors stressed its similarity to the relation derived for ellipticals (Jablonka *et al.* 1996; Idiart *et al.* 1996). This similarity appears both in the slope of the Z–L relation and in similar [α/Fe] ratios. Interestingly, the authors of both of the above studies observed face-on spirals and varied their integration apertures, either by adapting their spectroscopic apertures at the time of the observations, fixing a low and constant bulge-to-disk light ratio for all galaxies, or by inspecting the light profiles along the slit width when extracting the spectra.

Authors of subsequent works sampled inclined galaxies and advocated distinctions between late-type and early-type spiral bulges. Prugniel *et al.* (2001) found bulges located below the Mg_2–σ relation obtained for ellipticals. Falcón-Barroso *et al.* (2002) found a 20% steeper slope than for ellipticals and S0 galaxies. Proctor & Sansom (2002) reported that small bulges (low σ) depart from the relation between spectral indices and σ obeyed by large bulges: while large bulges populate the same region as elliptical galaxies, the smaller ones have relatively lower spectral indices. However, Thomas & Davies (2006), by re-analyzing Proctor and Sansom's sample, showed that this apparent discrepancy vanishes when the same range of central velocity dispersion is considered for both types of system, i.e. when low-σ bulges are compared with low-σ ellipticals.

Figure 4 in Falcón-Barroso *et al.* (2002) could serve as a warning: the dispersion in data from the various studies is rather large, likely due to the various observational strategies. In particular, it is of the order of the difference claimed to exist between different types of bulges and with elliptical galaxies. Nevertheless, there are true points of convergence among the studies quoted here, which can be summarized as follows. There is a range of properties of the bulge stellar populations as sampled by their inner regions. They are related to the bulge mass or maybe even more to the total gravitational potential of the parent galaxy. Indeed, Prugniel *et al.* (2001) and more recently Moorthy & Holtzman (2006) found a tighter relation between Mg_2 and the galaxy-rotation velocity than with the central bulge-velocity dispersion, for example. The bulge central-luminosity-weighted metallicities range from ~ -0.5 to $\sim +0.5$ dex and the luminosity-weighted [α/Fe] ranges from the Solar value to ~ 0.4 dex. Ages are more subject to debate, but a broad consensus would certainly be reached for a range between the age of a very old stellar population and a few gigayears younger.

4 Spatial distribution

We would dramatically limit our understanding of bulges if we confined the analyses to their central regions. Substantial progress has been brought about by spatially resolved spectroscopy, which enables radial gradients of stellar populations to be measured. Investigations of such radial gradients using large surveys have until recently been addressed only to early-type galaxies, i.e. elliptical and lenticular galaxies, with only very modest and rare excursions into the case of later-type galaxies (e.g. Sansom *et al.* 1994; Proctor *et al.* 2000; Ganda *et al.* 2006).

Moorthy & Holtzman (2006) recently published a large study of 38 bulges, composed of about equal numbers of nearly face-on spirals and highly inclined ones. For most of their bulges they found steady decreases in metallicity-sensitive indices with radius and a positive increase in [α/Fe], with the exception that their small bulges have generally weak gradients or no gradient, sometimes positive ones. While age gradients are generally absent for their sample galaxies, some exhibit positive ones, the majority of those being barred spirals. Very interestingly, they found a correlation between line-strength gradients in the bulge and in the disk.

Most of these qualitative results have been confirmed by Jablonka *et al.* (2007), who chose a different strategy. In order to get rid of the disk population totally, they selected 32 genuinely (or nearly) edge-on spiral galaxies with Hubble types from S0 to Sc. They obtained spectra along the bulge minor axes, out to the bulge effective radius and often much beyond. Most of their bulges do present radial stellar-population gradients. The outer parts of bulges do have weaker metallic absorption lines than the inner regions. The distribution of the gradient amplitudes is generally well peaked, but they also display a real intrinsic dispersion, implying a variety of star-formation histories.

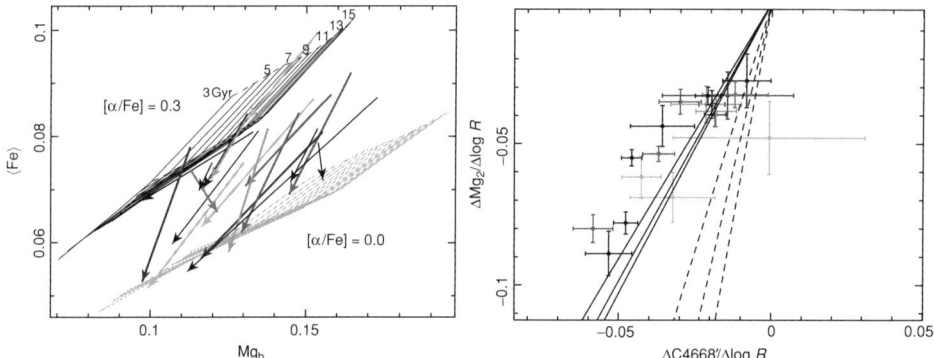

Figure 27.3. Characteristics of bulges, showing clear radial changes in their spectral indices. Colors code Hubble types from S0 to Sbc. Left panel: the arrows join the values of the indices ⟨Fe⟩ and Mg_b (expressed in terms of magnitude) at the bulge effective radius and in a central ($r = 2$ arcsec) aperture. Thomas et al. (2003) grids of single-stellar-population models are shown at [α/Fe] = 0.0 and 0.3, for ages between 3 Gyr and 15 Gyr and metallicities from -0.33 to $+0.35$. Right panel: model lines of variations in pure metallicity (plain lines) and pure age (dotted lines) are derived. They directly predict the relation between pairs of index gradients as observed in the bulge spectra.

The left panel of Figure 27.3 illustrates the decrease in [α/Fe], from the bulge effective radii to their central regions, for galaxies exhibiting clear gradients. Just like in the case of the central spectral properties, the morphology of the parent galaxy is not a driving parameter of the gradient amplitudes. The right panel of Figure 27.3 shows the models of pure metallicity (plain lines) and pure age variation (dotted lines) from Thomas et al. (2003). Bulges populate the region close to the pure-metallicity lines, leaving some room, but of a much smaller magnitude, for [α/Fe] and age variations. A quantitative analysis indicates that radial gradients in luminosity-weighted mean metallicity are twice (on a logarithmic scale) the gradients in age. While [Fe/H] at the bulge effective radii is on average 0.4 dex lower than in the bulge central regions, the age difference is of the order of 1.5 Gyr, the inner regions being younger. The variations in [α/Fe] are small (of the order of 0.1 dex) and rather constant among bulges. These various points indicate that the outer regions of bulges reveal their earliest stages of star formation. Interestingly, the sensitivity of the gradients to the central velocity dispersion is very different from what has been reported for the bulge central indices. Literally, there is no correlation between the gradient amplitude and the bulge central-velocity dispersion. Instead, one sees that bulges with large velocity dispersions can exhibit either strong or negligible gradients. The probability of a strong gradient diminishes with decreasing velocity until it reaches zero. The same had been observed for elliptical galaxies.

This gradual build-up of the index versus σ relation can be clearly observed only for indices with large dynamical ranges, such as Mg_2 and Mg_1.

5 High redshift

There have been very few studies focusing on bulges at high redshift, and only one addressing the comparison between bulges in field and cluster environments.

Ellis *et al.* (2001) analyzed a sample of early-type and spiral galaxies from the northern and southern Hubble Deep Fields. They compared the central (inner 5%) colors of spirals with clearly visible bulges with the integrated colors of ellipticals in their sample up to a redshift of $\sim$1. They found that both ellipticals and bulges exhibit a dispersion in their colors at a given redshift, but that their distributions are different: a smaller fraction of ellipticals is blue. It seems that there is an almost total absence of bulges as red as those predicted by the model of a passive evolution, while this scenario provides on the contrary a good description for the majority of the early-type population. The authors concluded that the optical-luminosity-weighted ages of bulges, to a redshift of at least 0.6, are less than those of the reddest ellipticals. At even higher redshifts some bulges are found to be as red as ellipticals though, suggesting that some kind of rejuvenation is at play at intermediate redshift.

Koo *et al.* (2005a) presented a sample of ellipticals and bulges from the DEEP Groth Strip Survey with redshift between $\sim$0.7 and $\sim$1. This time, the images are decomposed into bulge and disk components by fitting a de Vaucouleurs light profile for the former and an exponential one for the latter. They find that red bulges (85% of them) are nearly as red as, or redder than, the integrated color of either local early-type or distant cluster galaxies. The color–magnitude relations have similarly shallow slope and small scatter. Blue bulges are among the least-luminous ones, and are of similarly low surface brightness to local bulges of similar size. The authors consider that they cannot be genuine proto-bulges and are instead mostly residing in morphologically peculiar galaxies. Interestingly, in most red objects, they detect emission lines indicative of continuous star formation, albeit at a low level.

As stated by Koo *et al.* themselves, the fact that Ellis *et al.*'s sample encompasses faint bulges, whereas their sample is restricted to luminous ones, together with the very different ways of deriving the bulge colors, might be at the origin of the contradictory results of the two works.

Koo *et al.* (2005b) presented an analysis of luminous bulges (MB $<$ 19.5) in a cluster and in the field at redshift $\sim$0.8. They demonstrated that the rest-frame colors, slope, and dispersion of the color–magnitude relation of cluster and field bulges are nearly the same. This also means that they are no larger than in samples at lower redshift. This is in sharp contrast with some theoretical expectations of an increasing fraction of recent star formation with redshift and/or a longer timescale of formation in the field than in clusters. However, here again the consequence of the selection of luminous bulges must be investigated.

6 Conclusion

This review definitely concentrated on the observational properties of bulges. These have been determined with increasing accuracy by an ever-growing number of bulge studies. Here we reviewed three major areas of bulge characterization: resolved stellar populations, line indices from integrated spectra, and direct lookback studies of bulge colors at redshifts out to 1. Despite these diverse approaches, however, there is still a large number of viable models for bulge formation, e.g. gradual accretion of disk material through the action of the bars, gravitational collapse (or, nearly equivalently, early and fast mergers), and the accretion of dwarf systems in the center of disks. These models are not mutually exclusive and in reality bulge growth may combine some of these modes. The lack of agreement between the conclusions drawn from some analyses mentioned here serves to guide us towards questions that remain to be answered. Among them are the following. At which point in their history do the disks influence their central bulges? If they do, then in what proportion? Could this influence become dominant in small systems? Do we have/use the appropriate comparison samples of elliptical galaxies? Which gravitation potential governs most closely the formation of bulges, their own or the total one for the galaxy?

References

Bica, E. (1988), *A&A* **195**, 76
Durrell, P. R., Harris, W. E., & Pritchet, C. J. (2001), *AJ* **121**, 2557
Ellis, R. S., Abraham, R. G., & Dickinson, M. (2001), *ApJ* **551**, 111
Ferguson, A. M. N., Irwin, M. J., Ibata, R. A., Lewis, G. F., & Tanvir, N. R. (2002), *AJ* **124**, 1452
Falcón-Barroso, J., Peletier, R. F., & Balcells, M. (2002), *MNRAS* **335**, 741
Ganda, K., Falcón-Barroso, J., Peletier, R. F. *et al.* (2006), *MNRAS* **367**, 46
Ibata, R. A., Gilmore, G., & Irwin, M. J. (1994), *Nature* **370**, 194
Idiart, T. P., de Freitas Pacheco, J. A., & Costa, R. D. D. (1996), *AJ* **112**, 2541
Jablonka, P., Martin, P., & Arimoto, N. (1996), *AJ* **112**, 1415
Jablonka, P., Courbin, F., Meylan, G., Sarajedini, A., Bridges, T. J., & Magain, P. (2000), *A&A* **359**, 131
Jablonka, P., Gorgas, J., & Goudfrooij, P. (2007), *A&A* astro-ph/0707.0561
de Jong, R. S. (1996), *A&A* **313**, 45
de Jong, R. S., Simard, L., Davies, R. L. *et al.* (2004), *MNRAS* **355**, 1155
Koo, D. C., Simard, L., Willmer, C. N. A. *et al.* (2005a), *ApJS* **157**, 175
Koo, D. C., Datta, S., Willmer, C. N. A., Simard, L., Tran, K.-V., & Im, M. (2005b), *ApJL* **634**, L5
Moorthy, B. K., & Holtzman, J. A. (2006), *MNRAS* **371**, 583
Olsen, K. A. G., Blum, R. D., Stephens, A. W. *et al.* (2006), *AJ* **132**, 271

Proctor, R. N., Sansom, A. E., & Reid, I. N. (2000), *MNRAS* **311**, 37
Proctor, R. N., & Sansom, A. E. (2002), *MNRAS* **333**, 517
Prugniel, P., Maubon, G., & Simien, F. (2001), *A&A* **366**, 68
Ryan, S. G., & Norris, J. E. (1991), *AJ* **101**, 1865
Saglia, R. P., Bertschinger, E., Baggley, G. *et al.* (1997), *ApJS* **109**, 79
Sansom, A. E., Peace, G., & Dodd, M. (1994), *MNRAS* **271**, 39
Sarajedini, A., & Jablonka, P. (2005), *AJ* **130**, 1627
Simien, F., & de Vaucouleurs, G. (1986), *ApJ* **302**, 564
Thomas, D., Maraston, C., & Bender, R. (2003), *MNRAS* **343**, 279
Thomas, D., & Davies, R. L. (2006), *MNRAS* **366**, 510
Yanny, B., Newberg, H. J., Grebel, E. K. *et al.* (2003), *ApJ* **588**, 824
Zoccali, M., Renzini, A., Ortolani, S. *et al.* (2003), *A&A* **399**, 931

28

The metallicity distribution of the stars in elliptical galaxies

Antonio Pipino & Francesca Matteucci

Dipartimento di Astronomia, Università di Trieste, Trieste, Italy

Elliptical galaxies probably host the most metal-rich stellar populations in the Universe. The processes leading to both the formation and the evolution of such stars are discussed in terms of a new multi-zone photo-chemical-evolution model, taking into account detailed nucleosynthetic yields, feedback from supernovae, Population-III stars and an initial infall episode. Moreover, the radial variations in the metallicity distributions of these stars are investigated using G-dwarf-like diagrams.

By comparing model predictions with observations, we derive a picture of galaxy formation in which the higher the mass of the galaxy, the shorter are the infall and the star-formation timescales. Therefore, the stellar component of the most massive and luminous galaxies might attain a metallicity $Z \geq Z_\odot$ in only 0.5 Gyr.

Each galaxy is created outside-in, i.e. the outermost regions accrete gas, form stars and develop a galactic wind very quickly, in contrast to the central core in which star formation can last up to $\sim$1.3 Gyr. This finding will be discussed in the light of recent observations of the galaxy NGC 4697 which clearly exhibits a strong radial gradient in the mean stellar [⟨Mg/Fe⟩] ratio.

1 Introduction

Metallicity gradients are characteristic of the stellar populations inside elliptical galaxies. Evidence for this comes from the increase of line-strength indices (e.g. Carollo *et al.* 1993; Davies *et al.* 1993; Trager *et al.* 2000) and the reddening of the colours (e.g. Peletier *et al.* 1990) towards the centres of the galaxies. The study of such gradients provides insights into the mechanism of galaxy formation, particularly regarding the duration of the chemical-enrichment process at each radius.

The Metal-rich Universe, eds. G. Israelian and G. Meynet. Published by Cambridge University Press.
© Cambridge University Press 2008.

Metallicity indices, in fact, contain information on the chemical composition and age of the single-stellar populations (SSPs) inhabiting a given galactic zone. Pipino & Matteucci (2004, henceforth PM04) showed that a galaxy-formation process in which the most-massive objects form faster and more efficiently than the less-massive ones can explain the photo-chemical properties of ellipticals, in particular the increase of [Mg/Fe] ratio in stars with galactic mass (see Chapter 44 and references therein). The authors of PM04 suggested that a single galaxy should form outside-in, namely the outermost regions form earlier and faster than the central parts. A natural consequence of this model and of the time-delay between the production of Fe and that of Mg is that the mean [Mg/Fe] abundance ratio in the stars should increase with radius. Pipino *et al.* (2006, henceforth PMC06) compared the best model results in PM04 with the very recent observations for the galaxy NGC 4697 (Méndez *et al.* 2005), and found them to be in excellent agreement.

2 The model

The chemical code adopted here is described in full detail in PM04 and PMC06, where the reader can find more details. This model is characterized by the Salpeter (1955) initial mass function, Thielemann *et al.* (1996) yields for massive stars, Nomoto *et al.* (1997) yields for Type-Ia SNe and van den Hoek & Groenewegen (1997) yields for low- and intermediate-mass stars (the case with η_{AGB} varying with metallicity). Here we present our analysis of a $\sim 10^{11} M_\odot$ galaxy (PM04 model IIb), considered representative of a typical elliptical, unless stated otherwise.

The model assumes that the galaxy assembles by merging of gaseous lumps (infall) on a short timescale and suffers a strong starburst that injects into the interstellar medium a large amount of energy able to trigger a galactic wind, this occurring at different times at different radii. After the development of the wind, the star formation is assumed to stop and the galaxy evolves passively with continuous mass loss.

3 Results and discussion

From the comparison between our model predictions (Figure 28.1) and the observed G-dwarf-like diagrams derived for various radii by Harris & Harris (2002, Figure 18) for the elliptical galaxy NGC 5128, we can derive some general considerations. The qualitative agreement is remarkable: we can explain the slow rise in the [Z/H] distribution as the effect of the infall, whereas the sharp truncation at high metallicities is the first direct evidence of a sudden and strong wind that stopped the star formation. The suggested outside-in formation process is reflected in a more-asymmetric

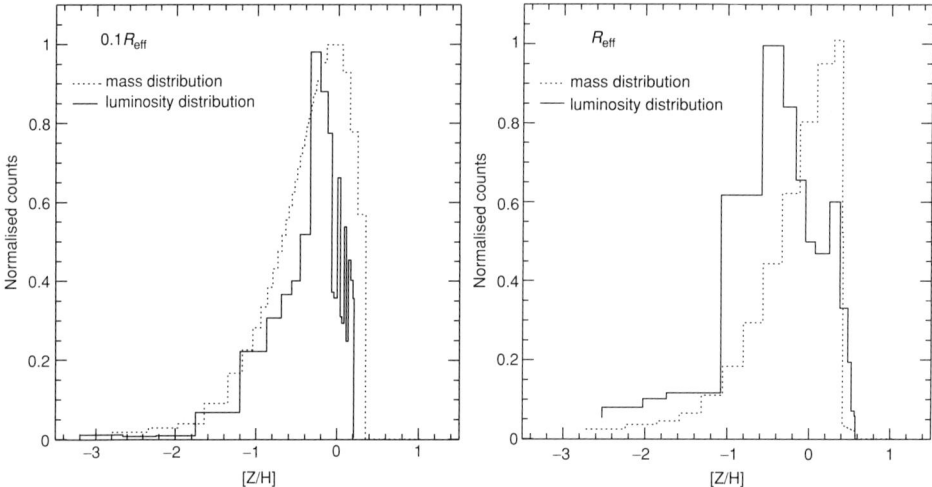

Figure 28.1. 'G-dwarf' distributions for [Z/H] in luminosity (solid line) and mass (dotted line). Left panel: values at $0.1 R_{\rm eff}$. Right panel: values at $1 R_{\rm eff}$. The plots are presented to the same scale in order to allow the reader better to appreciate the differences among the distributions.

shape of the (G-dwarf diagram at larger radii, where the galactic wind occurs earlier (i.e. closer to the peak of the star-formation rate), with respect to the galactic centre.

From a quantitative point of view, properties such as the stellar metallicity distribution of the combined-stellar population (CSP) inhabiting the galactic core, allow us to study the creation of a mass–metallicity relation (see Chapter 44), which is typically inferred from the spectra taken at ~ 0.1 times the effective radius. In Figure 28.2 we have plotted the time evolution of the mass–metallicity relation of the stars (which reflects the average chemical enrichment of the galactic core as seen at the present day; dashed line) and that in the gas (which, instead, is closer to the composition of the youngest SSPs, thus being more indicative of a high-redshift object; solid line). The mean Fe abundance in the stellar component can reach the Solar value in only 0.5 Gyr, making ellipticals among the most metal-rich objects of the Universe.

On the other hand, at variance with the G-dwarf-like diagrams as a function of [Z/H] (and [Fe/H]), abundance ratios such as [α/Fe] have narrow and almost-symmetric distributions. This means that, also from a mathematical point of view, the [$\langle\alpha/\text{Fe}\rangle$] ratios are representative of the whole CSP (PMC06). The robustness of the [α/Fe] ratios as constraints for the galactic formation history is testified to by the fact that [$\langle\alpha/\text{Fe}\rangle$] $\simeq$ [$\langle\alpha/\text{Fe}\rangle_V$], with very similar distributions. In particular, we find that the skewness parameter is much larger for the [Z/H] and [Fe/H] distributions than for the case of the [α/Fe] one, by more than an order of magnitude. Moreover, the asymmetry increases on going to large radii (see Figure 28.1, right panel), by up to a factor of ~ 7 with respect to the inner regions. Therefore, it is not

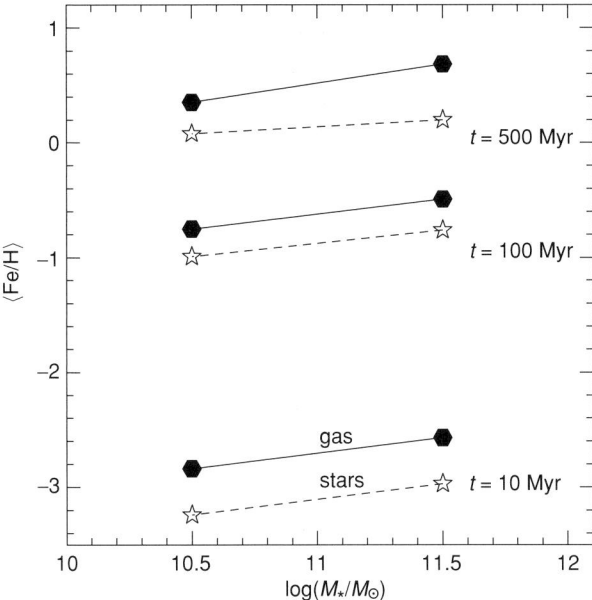

Figure 28.2. The temporal evolution of the mass–metallicity relation for the two galactic components studied (stars and gas).

surprising that the [⟨Z/H⟩] value does not represent the galaxy at large radii. Hence, we stress that care should be taken when one wants to infer the real abundances of the stellar components for a galaxy by comparing the observed indices (related to a CSP) with the theoretical ones (predicted for a SSP). Only the comparison based on the [⟨α/Fe⟩] ratios seems to be robust.

Another possible source of discrepancies is the fact that luminosity-weighted averages (which are more closely related to the observed indices) and mass-weighted averages (which represent the real distributions of the chemical elements in the stellar populations) might differ more in the most external zones of the galaxy (compare the panels in Figure 28.1). All these considerations result in the fact that the chemical-abundance pattern used by modellers to build their SSPs might not necessarily reflect the real trends. Therefore, the interpretation of line-strength indices in term of abundances can be seriously flawed (see PMC06 for further details).

The analysis of the radial variation in the CSPs inhabiting elliptical galaxies seems to be promising as a powerful tool to study ellipticals. Pipino, Puzia & Matteucci (in preparation) make use of the G-dwarf-like distributions predicted by PMC06 to explain the multimodality in the globular-cluster (GC) metallicity distribution as well as their high α enhancement (Puzia *et al.* 2006). In particular, preliminary results show that the GC distribution as a function of [Fe/H] for the whole galaxy can be constructed simply by combining distributions such as those of Figure 28.1 (typical of various radii), once they have been rescaled by means of a suitable function (of time and metallicity) that links the global star-formation rate to that of GC creation. Neither an enhanced rate of GC formation during mergers nor a strong role of the accretion of external objects seems to be required in order to explain the various features of the GC metallicity distributions.

Since GCs are the closest approximation to a SSP, we expect that this technique will be a very helpful one to probe the properties of the stellar populations in spheroids, thus avoiding the uncertainties typical of analysis based on their integrated spectra.

4 Concluding remarks

A detailed study of the chemical properties of the CSPs inhabiting elliptical galaxies, as well as the change of their properties as a function of both time and radius, allowed us to gather a wealth of information. Our main conclusions are as follows.

- Both observed and predicted G-dwarf-like distributions for ellipticals exhibit a sharp truncation at high metallicities that, in the light of our models, might be interpreted as the first direct evidence for the occurrence of the galactic wind in spheroids.
- The stellar component of the most massive and luminous galaxies might attain a metallicity $Z \geq Z_\odot$ in only 0.5 Gyr.
- The best model prediction in PM04 of increasing [$\langle\alpha/\text{Fe}\rangle$] ratio with radius is in very good agreement with the observed gradient in [α/Fe] of NGC 4697. This strongly suggests an outside-in galaxy-formation scenario for elliptical galaxies with strong gradients.
- By comparing the radial trend of [$\langle Z/H \rangle$] with the *observed* one, we noticed a discrepancy, which is due to the fact that a CSP behaves differently from an SSP. In particular, the predicted gradient of [$\langle Z/H \rangle$] is flatter than the observed one at large radii. Therefore, this should be taken into account when estimates for the metallicity of a galaxy are derived from the simple comparison between the observed line-strength index and the predictions for a SSP, a method currently adopted in the literature.
- Abundance ratios such as [Mg/Fe] are less affected by the discrepancy between the SSPs and a CSP, since their distribution functions are narrower and more symmetric. Therefore, we stress the importance of such a ratio as the most robust tool with which to estimate the duration of the galaxy-formation process.
- Our conclusions are strengthened by the comparison between our G-dwarf diagrams and metallicity distributions of the globular clusters residing in ellipticals.

Acknowledgments

The work was supported by MIUR under COFIN03 protocol 2003028039. A.P. thanks the organizers for having provided financial support facilitating his attendance of the conference.

References

Carollo, C. M., Danziger, I. J., & Buson, L. (1993), *MNRAS* **265**, 553
Davies, R. L., Sadler, E. M., & Peletier, R. F. (1993), *MNRAS* **262**, 650
Harris, W. E., & Harris, G. L. H. (2002), *AJ* **123**, 3108
Méndez, R. H., Thomas, D., Saglia, R. P., Maraston, C., Kudritzki, R. P., & Bender, R. (2005), *ApJ* **627**, 767
Nomoto, K., Hashimoto, M., Tsujimoto, T. *et al.* (1997), *Nuclear Physics A* **621**, 467
Peletier, R. F., Davies, R. L., Illingworth, G. D., Davis, L. E., & Cawson, M. (1990), *AJ* **100**, 1091
Pipino, A., & Matteucci, F. (2004), *MNRAS* **347**, 968 (PM04)
Pipino, A., Matteucci, & F. Chiappini, C. (2006), *ApJ* **638**, 739 (PMC06)
Puzia, T. H., Kissler-Patig, M., & Goudfrooij, P. (2006), *ApJ* **648**, 383
Salpeter, E. E. (1955), *ApJ* **121**, 161
Thielemann, F. K., Nomoto, K., & Hashimoto, M. (1996), *ApJ* **460**, 408
Trager, S. C., Faber, S. M., Worthey, G., & González, J. J. (2000), *AJ* **119**, 1654
van den Hoek, L. B., & Groenewegen, M. A. T. (1997), *A&AS* **123**, 305

29

Wolf–Rayet populations at high metallicity

Paul A. Crowther

*Department of Physics & Astronomy, University of Sheffield, Hicks Building,
Hounsfield Road, Sheffield S3 7RH, UK*

Observed properties of Wolf–Rayet (WR) stars at high metallicity are reviewed. Wolf-Rayet stars are more common at higher metallicity, as a result of stronger mass-loss during earlier evolutionary phases with late-WC-subtypes signatures of Solar metallicity or higher. Similar numbers of early (WC4–7) and late (WC8–9) stars are observed in the Solar neighbourhood, whilst late subtypes dominate at higher metallicities, such as Westerlund 1 in the inner Milky Way and in M83. The observed trend to later WC subtype within metal-rich environments is intimately linked to a metallicity dependence of WR stars, in the sense that strong winds preferentially favour late subtypes. This has relevance to (a) the upper mass limit in metal-rich galaxies such as NGC 3049, due to softer ionizing fluxes from WR stars at high metallicity; and (b) the fact that evolutionary models including a WR metallicity dependence provide a better match to the observed $N(WC)/N(WN)$ ratio. The latter conclusion partially rests upon the assumption of constant line luminosities for WR stars, yet observations and theoretical atmospheric models reveal higher line fluxes at high metallicity.

1 Introduction

Wolf–Rayet (WR) stars are the final phase in the evolution of very massive stars prior to core-collapse, in which the H-rich envelope has been stripped away via either stellar winds or close binary evolution, revealing products of H-burning (WN sequence) or He-burning (WC sequence) at their surfaces, i.e. He, N or C, O (Crowther 2007).

Stellar winds of WR stars are significantly denser than those of O stars, as illustrated in Figure 29.1, so their visual spectra are dominated by broad emission lines,

---

The Metal-rich Universe, eds. G. Israelian and G. Meynet. Published by Cambridge University Press.
© Cambridge University Press 2008.

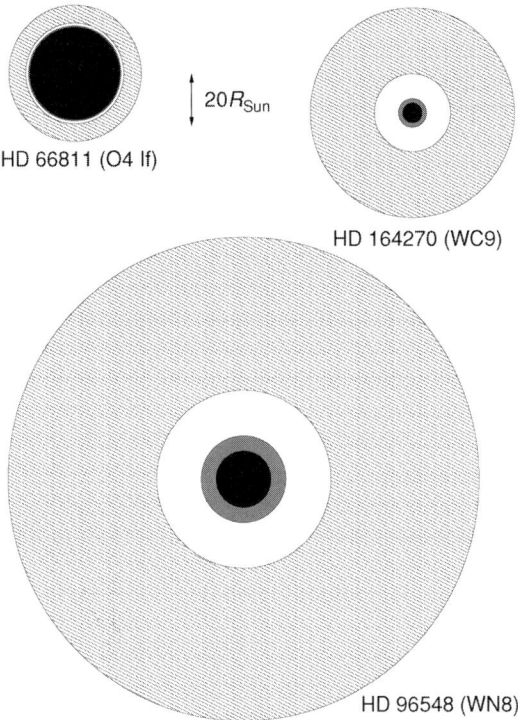

Figure 29.1. Comparisons between stellar radii at Rosseland optical depths of 20 (R_*, black) and 2/3 ($R_{2/3}$, grey) for HD 66811 (O4 If), HD 96548 (WN8) and HD 164270 (WC9), shown to scale, together with the wind region corresponding to the primary optical wind line-forming region, 10^{11} cm$^{-3} \leq n_e \leq 10^{12}$ cm^{-3} (hatched) in each case, illustrating the highly extended winds of WR stars with respect to O stars (Crowther 2007).

notably He II λ4686 (WN stars) and C III λ4647–51, C III λ5696, and C IV λ5801–12 (WC stars). The spectroscopic signature of WR stars may be seen individually in Local Group galaxies (e.g. Massey & Johnson 1998), within knots in local star-forming galaxies (e.g. Hadfield & Crowther 2006) and in the average rest-frame UV spectrum of Lyman-break galaxies (Shapley *et al.* 2003).

In the case of a single massive star, the strength of stellar winds during the main-sequence and blue supergiant phases scales with the metallicity (Vink *et al.* 2001). Consequently, one expects a higher threshold for the formation of WR stars at lower metallicity, and indeed the SMC has a smaller ratio of WR to O stars than is found in the Solar neighbourhood. Alternatively, the H-rich envelope may be removed during the Roche-lobe overflow phase of close binary evolution, a process that is not expected to depend upon metallicity.

Wolf–Rayet stars are the prime candidates for Type-Ib/c core-collapse supernovae and long, soft gamma-ray bursts (GRBs). This is due to their immediate

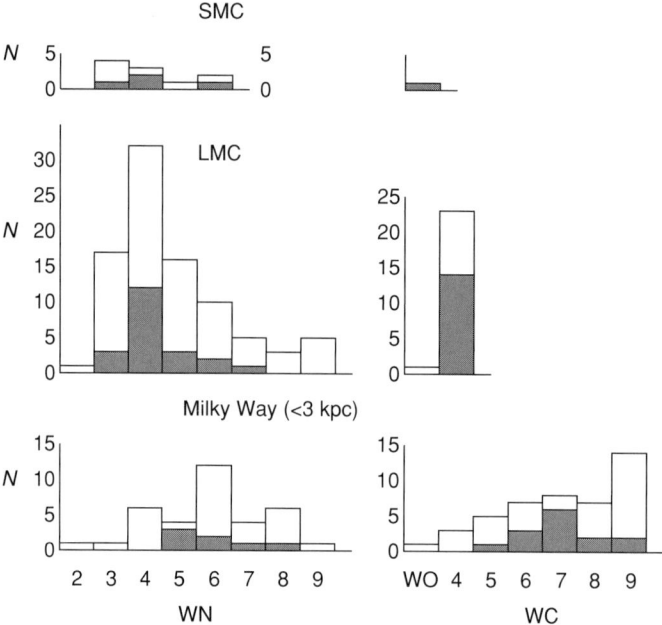

Figure 29.2. Subtype distribution of Milky Way (<3 kpc), LMC and SMC WR stars, in which known binaries are shaded (Crowther 2007).

progenitors being associated with young massive-stellar populations, compact in nature and deficient either in hydrogen (Type Ib) or in both hydrogen and helium (Type Ic). For the case of GRBs, some of which have been associated with Type-Ic hypernovae (Galama et al. 1998; Hjorth et al. 2003), a rapidly rotating core is a requirement for the collapsar scenario in which the newly formed black hole accretes via an accretion disc (MacFadyen & Woosley 1999). Indeed, WR populations have been observed within local GRB host galaxies (Hammer et al. 2006).

In this review article, the observed properties of WR stars at high metallicity are presented and discussed.

2 The distribution of Wolf–Rayet subtypes in the Milky Way and Magellanic Clouds

Historically, the wind properties of WR stars have been assumed to be independent of metallicity (Langer 1989), yet there is a well-known observational trend towards later, lower-ionization, WN and WC subtypes at high metallicity, as illustrated in Figure 29.2.

Within the Milky Way, it is well known that late-type WC stars are restricted to within the Solar circle (Conti & Vacca 1990). Indeed, Hopewell et al. (2005)

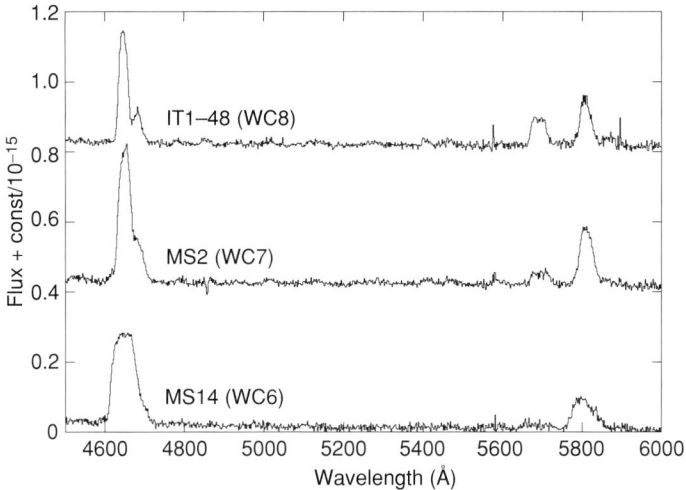

Figure 29.3. Optical WHT/ISIS spectroscopy of representative WC stars in M31.

discovered five new WR stars in the inner Milky Way using the AAO/UKST Hα survey – all were found to be WC9 stars.

In addition, Westerlund 1 (Clark *et al.* 2005) – located at the edge of the Galactic bar, for which a metallicity ~60% higher than that of Orion is expected – possesses eight WC stars with a bias towards late (WC8–9) subtypes according to recent near-IR spectroscopy (Crowther *et al.* 2006). Most of these possess hot dust, indicative of massive binaries, in common with the Quintuplet members of the Galactic Centre Quintuplet cluster (Figer, private communication). Early-type WN stars are also absent from Westerlund 1, with equal numbers of mid (WN5–6) and late (WN7–10) subtypes, of which most also appear to be massive binaries, as a result of hard X-ray fluxes.

3 Wolf–Rayet populations in M31, M83 and beyond

Within the Local Group, M31 (Andromeda) is the only other candidate metal-rich galaxy, although studies are hindered by its orientation on the sky. Wolf–Rayet populations in M31 were studied by Moffat & Shara (1983, 1987) and Massey *et al.* (1986). As in the Milky Way, WC7–8 stars were located at smaller galactocentric distances (7 ± 3 kpc) than WC5–6 stars (11 ± 3 kpc). Representative examples of M31 WC stars obtained with WHT/ISIS are presented in Figure 29.3. These results suggest a weak metallicity gradient for M31, albeit a rather less metal-rich situation than in the Milky Way, according to its observed WC population; see also Trundle *et al.* (2002).

Further afield, the WR population of M83 (NGC 5236) has been studied by Hadfield *et al.* (2005). The galaxy M83 is well suited to optical imaging surveys for

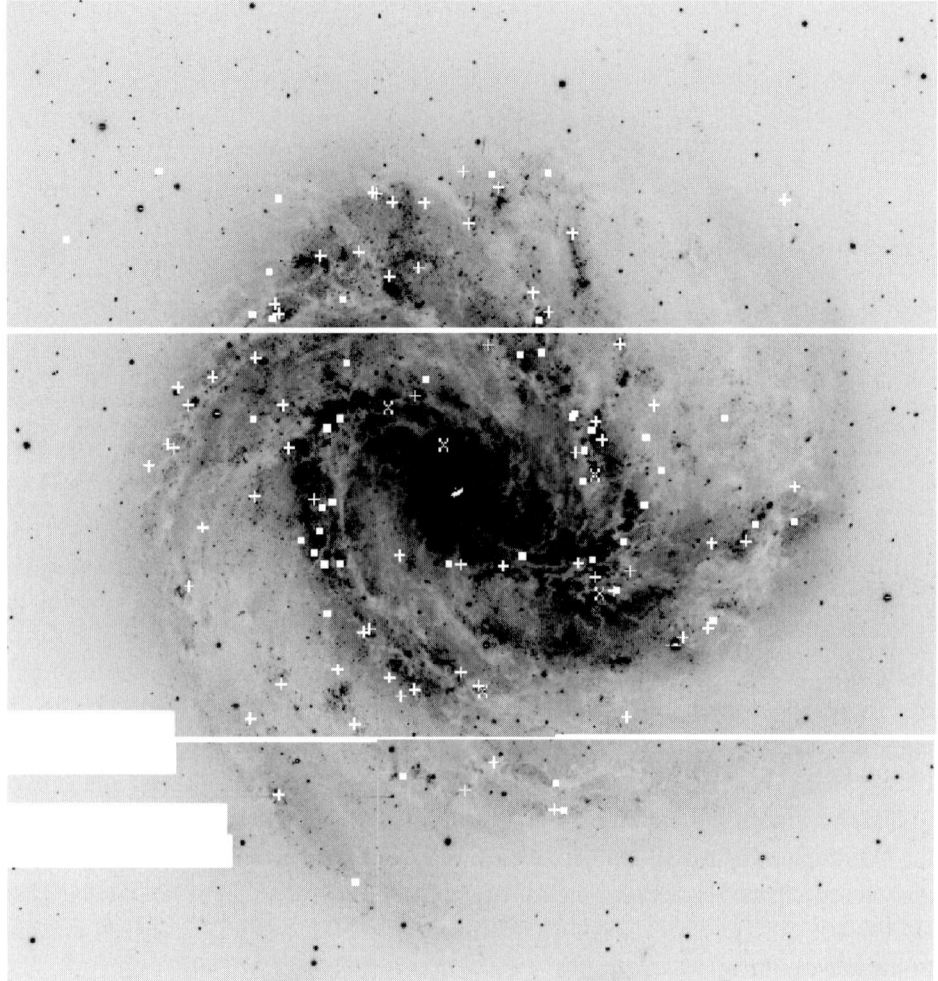

Figure 29.4. A composite 12′ × 12′ VLT FORS2 image (λ4685) of M83 (NGC 5236), indicating the locations of regions containing WN (squares), WC (plus symbols) and both WN and WC (crosses) stars. North is up and east is to the left. Regions to the southeast are masked to avoid saturation by bright foreground stars (Hadfield et al. 2005).

WR stars at high metallicity (though see Chapter 17 in these proceedings), since it is face-on, nearby (4.5 Mpc distant) and possesses a high star-formation rate.

VLT/FORS2 revealed a total of 280 (non-nuclear) regions for which the presence of WR stars was inferred from an excess of λ4685 (He II 4686) narrow-band imaging versus λ4781 (continuum) – see Figure 29.4. Spectroscopic follow-up of 198 regions confirmed a total of 132 sources, hosting in excess of 1000 WR stars according to the standard calibration of Schaerer & Vacca (1998). For M83

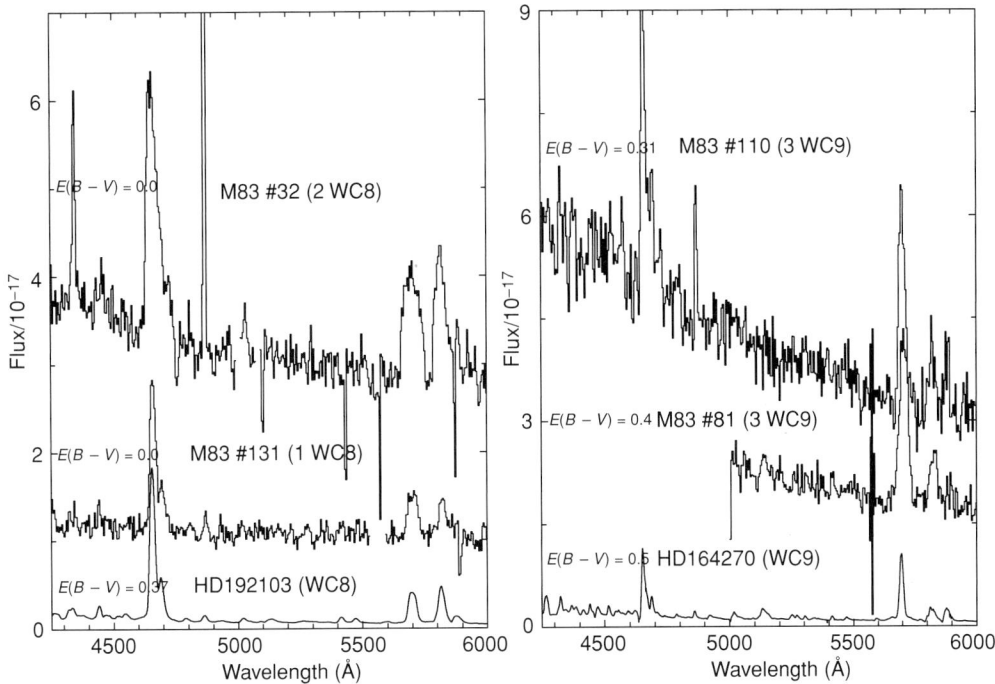

Figure 29.5. Optical VLT/FORS2 spectrosocpy of typical regions of M83 hosting late WC8–9 stars, together with individual Milky Way WC stars (WR135 and WR103) scaled to a distance of 4.5 Mpc (Hadfield et al. 2005).

N(WC)/N(WN) ∼ 1.2, confirming the trend towards a higher ratio at high metallicity as indicated in Figure 29.5. Notably, the WC population of M83 is totally dominated by late subtypes, i.e. N(WC8–9)/N(WC4–7) ∼ 9, in contrast with ∼1 for the Solar neighbourhood, for which representative examples are presented in Figure 29.5.

Wolf–Rayet populations are unresolved in more-remote metal-rich galaxies, but data also confirm the presence of late-type WC stars, as originally discovered by Phillips & Conti (1992) for NGC 1365 within Fornax and confirmed by Pindao et al. (2002) for NGC 4254 (M88) within Virgo.

4 Why late WC stars at high metallicity?

The ubiquitous detection of late WC stars within metal-rich environments (and their absence at low metallicity) requires explanation. The observed trend for WC subtypes in the LMC versus the Milky Way was initially believed to originate from a difference in carbon abundances (Smith & Maeder 1991), yet quantitative analysis reveals similar carbon abundances (Koesterke & Hamann 1995; Crowther et al.

2002). Alternatively, late WC stars might evolve preferentially from relatively low-mass OB stars that enter the WR phase owing to stronger stellar winds at earlier evolutionary phases. This scenario is not supported by cluster studies (e.g. Massey *et al.* 2000; Crowther *et al.* 2006).

The most compelling evidence suggests that late WC stars are favoured in the case of high wind densities (Crowther *et al.* 2002). Consequently, the presence of late WC stars within metal-rich galaxies favours metallicity-dependent winds for Wolf–Rayet stars. The impact of a metallicity dependence for WR winds upon spectral types is as follows. At high metallicity, recombination from high to low ions (early to late subtypes) is very effective in very dense winds, whilst the opposite is true for low-metallicity, low-density winds. Stellar temperatures further complicate this picture, such that the spectral type of a WR star results from a subtle combination of ionization and wind density, in contrast with the case for normal stars.

Theoretically, Nugis & Lamers (2002) argued that the iron opacity peak was the origin of the wind driving in WR stars, which Gräfener & Hamann (2005) supported via a hydrodynamic model for an early-type WC star in which lines of Fe IX–XVII deep in the atmosphere provided the necessary radiative driving. Vink & de Koter (2005) applied a Monte Carlo approach to investigate the metallicity dependence for cool WN and WC stars, revealing $\dot{M} \propto Z^\alpha$, where $\alpha = 0.86$ for WN stars and $\alpha = 0.66$ for WC stars for $0.1 \leq Z \leq Z_\odot$. The weaker WC dependence originates from an increasing Fe content and constant C and O content at high metallicity. Empirical results for the Solar neighbourhood, LMC and SMC are broadly consistent with theoretical predictions, although detailed studies of individual WR stars within galaxies with a broader range in metallicity would provide stronger constraints.

A metallicity dependence of WR winds affects evolutionary-model calculations as follows. Recent evolutionary models of Meynet & Maeder (2005) allow for rotational mixing, but not a metallicity dependence of WR winds. Improved agreement with respect to earlier models is achieved, but the ratio of WC versus WN stars for continuous star formation does does not reproduce that observed at high metallicity, as illustrated in Figure 29.6. In contrast, recent (non-rotating) evolutionary models by Eldridge & Vink (2006) in which the Vink & de Koter (2005) WR metallicity dependence has been implemented provide a much better match to observations.

With regard to the inferred WR populations at high metallicity, a note of caution is necessary. At high metallicity, WR optical recombination lines will (i) increase in equivalent width, since their strength scales with the square of the density; and (ii) increase in line flux, since the lower wind strength will reduce the line blanketing, resulting in an increased extreme-UV continuum strength at the expense of the UV and optical (Crowther & Hadfield 2006). Indeed, the equivalent widths of optical

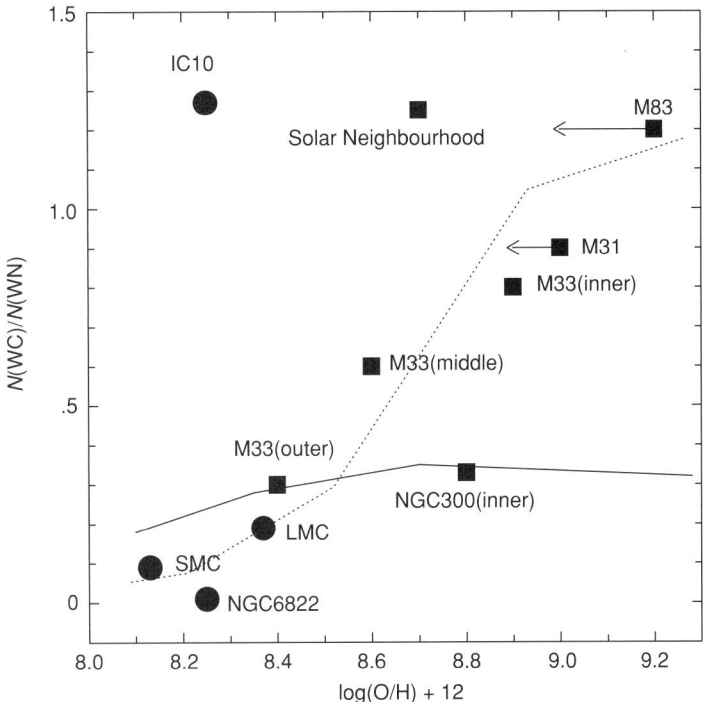

Figure 29.6. Ratios of subtype distribution of WC to WN stars for nearby galaxies, updated from Massey & Johnson (1998) to include M83 (Hadfield et al. 2005). Evolutionary predictions from Meynet & Maeder (2005) (solid line) and Eldridge & Vink (2006) (dotted line) are included.

emission lines of WN stars in the Milky Way and LMC are well known to be higher than those of their SMC counterparts (Conti et al. 1989).

To date, the standard approach for the determination of unresolved WR populations in external galaxies has been to assume metallicity-independent WR line fluxes – obtained for Milky Way and LMC stars (Schaerer & Vacca 1998) – regardless of whether the host galaxy is metal-rich (Mrk 309) (Schaerer et al. 2000) or metal-poor (I Zw 18) (Izotov et al. 1997). Ideally, one would wish to use WR template stars appropriate to the metallicity of the galaxy under consideration. Unfortunately, this is feasible only for the LMC, SMC and Solar neighbourhood, since it is challenging to isolate individual WR stars from ground-based observations of more distant galaxies, which span a larger spread in metallicity.

Enhanced WR line fluxes are also predicted for WR atmospheric models at high metallicity if one follows the metallicity dependence from Vink & de Koter (2005), such that WR populations inferred from Schaerer & Vacca (1998) at high metallicity may overestimate actual populations.

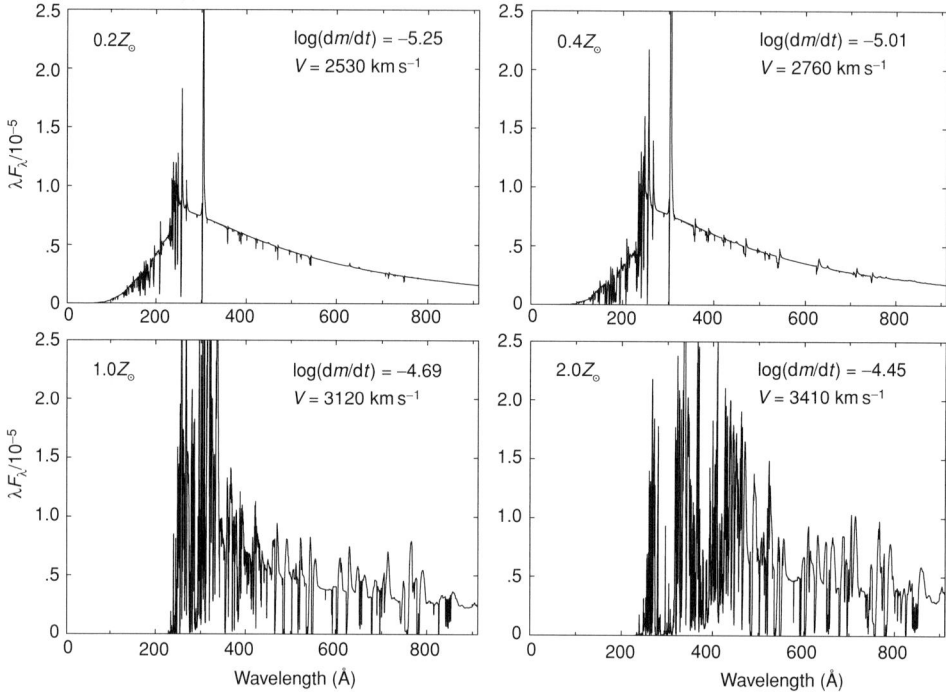

Figure 29.7. Predicted Lyman continuum ionizing fluxes for model reference WN #11 ($T_* = 100$ kK, $L = 10^{5.48} L_\odot$) from Smith *et al.* (2002), illustrating harder ionizing fluxes at lower metallicity, notably below $\lambda = 228$ Å due to weaker stellar winds.

5 Ionizing fluxes

Results from indirect H II-region studies have suggested a low upper mass limit, M_{upper}, for H II regions at high metallicity (e.g. Thornley *et al.* 2000), yet Schaerer *et al.* (2000) claim $M_{\text{upper}} > 40 M_\odot$ due to large WR populations in high-metallicity galaxies such as Mrk 309. However, Schmutz *et al.* (1992) demonstrated that the ionizing fluxes from WR stars soften as the wind density increases. Consequently, a metallicity dependence for WR wind strengths implies that WR ionizing-flux distributions soften with increasing metallicity, as demonstrated by Smith *et al.* (2002), and naturally resolves this apparent discrepancy. For example, high-temperature WN models predict a strong ionizing flux shortwards of the He II Lyman edge at 228 Å at low metallicities, but negligible hard ionizing fluxes at Solar metallicities or above, as illustrated in Figure 29.7.

To illustrate this, we use the example of the metal-rich WR galaxy NGC 3049 (Schaerer *et al.* 1999). González Delgado *et al.* (2002) identified a compact nuclear starburst cluster as the principal origin of WR emission within NGC 3049. A nebular analysis based upon the hard (metallicity-independent) WR ionizing fluxes of Schmutz *et al.* (1992) confirmed previous results, i.e. $M_{\rm upper} < 40 M_\odot$. In contrast, use of the revised (metallicity-dependent), softer WR ionizing fluxes from Smith *et al.* (2002) led to a normal $M_{\rm upper} \geq 100 M_\odot$, in agreement with UV-spectral-synthesis studies.

6 Summary

Wolf–Rayet stars are more common at higher metallicity, as a result of stronger mass-loss during earlier evolutionary phases. Late WC subtypes appear to be signatures of Solar metallicity or higher, as witnessed within the Westerlund 1 and Quintuplet clusters in the inner Milky Way, and in M31, M83, NGC 3049 and Mrk 309. The observed trend towards later WC subtype is intimately linked to a metallicity dependence of WR stars, in the sense that strong winds preferentially favour late subtypes (Crowther *et al.* 2002). This has relevance to (a) the upper mass limit in metal-rich galaxies, due to softer ionizing fluxes from WR stars at high metallicity (González Delgado *et al.* 2002); and (b) the fact that evolutionary models including a WR metallicity dependence provide a better match to the observed N(WC)/N(WN) ratio (Eldridge & Vink 2006). The latter item relies in part upon the assumption of constant line luminosities for WR stars (e.g. Schaerer & Vacca 1998), yet observations and theoretical atmospheric models reveal higher line fluxes at high metallicity (Crowther & Hadfield 2006).

Acknowledgment

Many thanks are due to Lucy Hadfield, with whom the majority of the results on M83 and Westerlund 1 presented here were obtained, and to Chris Evans, who carried out the M31 WHT/ISIS observing run. P.A.C. acknowledges financial support from the Royal Society.

References

Clark, J. S., Negueruela, I., Crowther, P. A., & Goodwin, S. P. (2005), On the massive stellar population of the super star cluster Westerlund 1, *A&A* **434**, 949
Conti, P. S., & Vacca, W. D. (1990), The distribution of massive stars in the Galaxy. I – The Wolf–Rayet stars, *AJ* **100**, 431
Conti, P. S., Garmany, C. D., & Massey P. (1989), Spectroscopic studies of Wolf–Rayet stars. V – Optical spectrophotometry of the emission lines in Small Magellanic Cloud stars, *ApJ* **341**, 113
Crowther, P. A. (2007), Physical properties of Wolf–Rayet stars, *ARA&A* **45**, 177
Crowther, P. A., & Hadfield, L. J. (2006), Reduced Wolf–Rayet line luminosities at low metallicity, *A&A* **449**, 711
Crowther, P. A., Dessart, L., Hillier, D. J., Abbott, J. B., & Fullerton, A. W. (2002), Stellar and wind properties of LMC WC4 stars. A metallicity dependence for Wolf–Rayet mass-loss rates, *A&A* **392**, 653

Crowther, P. A., Hadfield, L. J., Clark, J. S., Negueruela, I., & Vacca, W. D. (2006), A census of the Wolf–Rayet content in Westerlund 1 from near-infrared imaging and spectroscopy, *MNRAS* **372**, 1407

Eldridge, J. J., & Vink, J. S. (2006), Implications of the metallicity dependence of Wolf–Rayet winds, *A&A* **452**, 295

Galama, T. J., Vreeswijk, P. M., van Paradijs, J. *et al.* (1998), An unusual supernova in the error box of the gamma-ray burst of 25 April 1998, *Nature* **395**, 670

González Delgado, R. M., Leitherer, C., Stasińska, G., & Heckman, T. M. (2002), The massive stellar content in the starburst NGC 3049: a test for hot-star models, *ApJ* **580**, 824

Gräfener, G., & Hamann, W.-R. (2005), Hydrodynamic model atmospheres for WR stars. Self-consistent modeling of a WC star wind, *A&A* **432**, 633

Hadfield, L. J., & Crowther, P. A. (2006), How extreme are the Wolf–Rayet clusters in NGC 3125? *MNRAS* **368**, 1822

Hadfield, L. J., Crowther, P. A., Schild, H., & Schmutz, W. (2005), A spectroscopic search for the non-nuclear Wolf–Rayet population of the metal-rich spiral galaxy M83, *A&A* **439**, 265

Hammer, F., Flores, H., Schaerer, D., Dessauges-Zavadsky, M., Le Floc'h, E., & Puech, M. (2006), Detection of Wolf–Rayet stars in host galaxies of gamma-ray bursts (GRBs): are GRBs produced by runaway massive stars ejected from high stellar density regions? *A&A* **454**, 103

Hjorth, J., Sollerman, J., Moller, P. *et al.* (2003), A very energetic supernova associated with the γ-ray burst of 29 March 2003, *Nature* **423**, 847

Hopewell, E. C., Barlow, M. J., Drew, J. *et al.* (2005), Five WC9 stars discovered in the AAO/UKST Hα survey, *MNRAS* **363**, 857

Izotov, Y. I., Foltz, C. B., Green, R. F., Guseva, N. G., & Thuan, T. X. (1997), I Zw 18: a new Wolf–Rayet galaxy, *ApJ* **487**, L37

Koesterke, L., & Hamann, W.-R. (1995), Spectral analyses of 25 Galactic Wolf–Rayet stars of the carbon sequence, *A&A* **299**, 503

Langer, N. (1989), Mass-dependent mass loss rates of Wolf–Rayet stars, *A&A* **220**, 135

MacFadyen, A. I., & Woosley, S. E. (1999), Collapsars: gamma-ray bursts and explosions in "failed supernovae," *ApJ* **524**, 262

Massey, P., & Johnson, O. (1998), Evolved massive stars in the Local Group. II. A new survey for Wolf–Rayet stars in M33 and its implications for massive star evolution: evidence of the "Conti scenario" in action, *ApJ* **505**, 793

Massey, P., Armandroff, T. E., & Conti, P. S. (1986), Massive stars in M31, *AJ* **92**, 1303

Massey, P., Waterhouse, E., & DeGioia-Eastwood, K. (2000), The progenitor masses of Wolf–Rayet stars and luminous blue variables determined from cluster turnoffs. I. Results from 19 OB associations in the Magellanic Clouds, *AJ* **119**, 2214

Meynet, G., & Maeder, A. (2005), Stellar evolution with rotation. XI. Wolf–Rayet star populations at different metallicities, *A&A* **429**, 581

Moffat, A. F. J., & Shara, M. M. (1983) Wolf–Rayet stars in the Local Group galaxies M31 and NGC 6822, *ApJ* **273**, 544

(1987) Wolf–Rayet stars in the Andromeda Galaxy, *ApJ* **320**, 266

Nugis, T., & Lamers, H. J. G. L. M. (2002), The mass-loss rates of Wolf–Rayet stars explained by optically thick radiation driven wind models, *A&A* **389**, 162

Phillips, A. C., & Conti, P. S. (1992), Detection of WC9 stars in NGC 1365, *ApJ* **395**, L91

Pindao, M., Schaerer, D., González Delgado, R. M., & Stasińska, G. (2002), VLT observations of metal-rich extra galactic H II regions. I. Massive star populations and the upper end of the IMF, *A&A* **394**, 443

Schaerer, D., & Vacca, W. D. (1998), New models for Wolf–Rayet and O star populations in young starbursts, *ApJ* **497**, 618

Schaerer, D., Contini, T., & Kunth, D. (1999), Populations of WC and WN stars in Wolf–Rayet galaxies, *A&A* **341**, 399

Schaerer, D., Guseva, N. G., Izotov, Y. I., & Thuan, T. X. (2000), Massive star populations and the IMF in metal-rich starbursts, *A&A* **362**, 53

Schmutz, W., Leitherer, C., & Gruenwald, R. (1992), Theoretical continuum energy distributions for Wolf–Rayet stars, *PASP* **104**, 1164

Shapley, A. E., Steidel, C. S., Pettini, M., & Adelberger, K. L. (2003), Rest-frame ultraviolet spectra of $z \sim 3$ Lyman break galaxies, *ApJ* **588**, 65

Smith, L. F., & Maeder, A. (1991), Comparison of predicted and observed properties of WC stars – explanation of the subtype gradient in galaxies, *A&A* **241**, 77

Smith, L. J., Norris, R. P. F., & Crowther, P. A. (2002), Realistic ionizing fluxes for young stellar populations from 0.05 to $2Z_\odot$, *MNRAS* **337**, 1309

Thornley, M. D., Schreiber, N. M. F., Lutz, D. *et al.* (2000) Massive star formation and evolution in starburst galaxies: mid-infrared spectroscopy with the ISO Short Wavelength Spectrometer, *ApJ* **539**, 641

Trundle, C., Dufton, P. L., Lennon, D. J., Smartt, S. J., & Urbaneja, M. A. (2002), Chemical composition of B-type supergiants in the OB8, OB10, OB48, OB78 associations of M31, *A&A* **395**, 519

Vink, J. S., & de Koter, A. (2005), On the metallicity dependence of Wolf–Rayet winds, *A&A* **442**, 587

Vink, J. S., de Koter, A., & Lamers, H. J. G. L. M. (2001), Mass-loss predictions for O and B stars as a function of metallicity, *A&A* **369**, 574

30

The stellar populations of metal-rich starburst galaxies: the frequency of Wolf–Rayet stars

João Rodrigo Souza Leão,[1] Claus Leitherer,[1] Fabio Bresolin[2]
& Roberto Cid Fernandes[3]

[1]*Space Telescope Science Institute, Baltimore, MD, USA*
[2]*IfA, University of Hawai'i, Honolulu, HI, USA*
[3]*Universidade Federal de Santa Catarina, Florianópolis, SC, Brazil*

We conducted an optical survey (Keck Telescope, 3,700–7,000 Å) of 24 high-metallicity (Z) starburst galaxies to investigate whether high-Z environments favor the formation of Wolf–Rayet (WR) stars. We searched for the presence of the He II 4686 Å line produced by the massive WR stars. We detected this feature in six galaxies (25% of the sample). We also used a stellar-population-synthesis code to determine their ages. We find that (i) all galaxies hosting considerable numbers of WR stars are very young systems, with ages $\log(t) < 8$, with t in years; (ii) not all young star-forming galaxies host WR stars, or at least that population cannot be detected in their integrated spectra; and (iii) most galaxies hosting WR populations are found in interacting systems. We for the first time detect WR populations in galaxies ESO 485-G003, NGC 6090, and NGC 2798.

1 Introduction

Our goal is to understand the frequency of WR stars in metal-rich starburst galaxies. The WR phase is a natural phase in the evolution of massive stars with masses above $40 M_\odot$. In the optical, the broad emission bump around 4686 Å is a very characteristic signature of these stars (Conti 1991). This feature alone proves the existence of stars in the mass range $(40–60) M_\odot$, since only these stars can evolve into the WR phase. This emission bump is a very unique signature because it does not coincide with any other gas emission features. This allow us to study the presence of WR stars using stellar tracers rather than indirect diagnostics based on nebular emission lines.

Galaxies hosting WR stars are relatively rare, mainly because of the observational difficulties (i.e. spectral resolution, exposure times) in detecting the WR bump. A compilation of WR galaxies by Schaerer *et al.* (1999) lists 139 members.

The Metal-rich Universe, eds. G. Israelian and G. Meynet. Published by Cambridge University Press.
© Cambridge University Press 2008.

Interestingly, most of these galaxies are found in metal-poor environments, reflecting their selection criteria based on emission lines that favor the selection of metal-poor systems. However, from stellar-evolution theory we know that metal-rich environments favor the mass-loss mechanism that may increase the chances of a particular O star reaching the WR phase (Leitherer *et al.* 1992). Can we then expect more WR stars in metal-rich environments? If so, why are most of the known WR galaxies found in metal-poor systems?

2 Sample selection and observations

To address these questions we selected 24 starburst galaxies with luminosities in the range $10 < \log L_{IR} < 11.5$. This luminosity range statisticaly favors the selection of relatively large, metal-rich systems that are capable of keeping the products of stellar evolution gravitationally bound. Indeed, our galaxies are metal-rich systems with $8.6 < 12 + \log(O/H) < 9.2$. They are also luminous infrared galaxies (LIRGs) and the luminosity range considered guarantees that star formation is the primary energy-generating mechanism, with little or no contamination from active galactic nuclei (AGN) (Kim *et al.* 1998).

To properly detect and resolve the WR bump a signal-to-noise ratio of 50 and a spectral resolution of 4 Å are required. This typically translates into observing times of 30 min with the 10-m-class Keck telescope.

In some cases (e.g. NGC 1614, NGC 3690, NGC 2798) we also detect the presence of a companion. This, and the occurrence of recent star-formation (Section 3), is an indication that mergers or interactions play an important role in the formation and evolution of these systems.

3 Stellar-population synthesis

We used the STARLIGHT code (Cid Fernandes *et al.* 2005) to fit the spectrum of our galaxies and to understand the distribution in ages and metallicities. It also allows us to subtract the underlying spectrum of older stellar populations to better quantify the numbers of O stars and WR stars.

The fitting of the observed spectra of our galaxies is best illustrated by Figure 30.1, where we show the fit and the corresponding age distribution found for galaxy NGC 2798. Similar fits were produced for all galaxies in the sample.

We find that most galaxies are dominated by young stellar populations, indicating recent bursts of star formation. We also find that an underlying older stellar population is present, indicating that past episodes of star formation occurred. We also recover a very good correlation between the infrared luminosity and the average age of burst, i.e. younger systems are also more luminous.

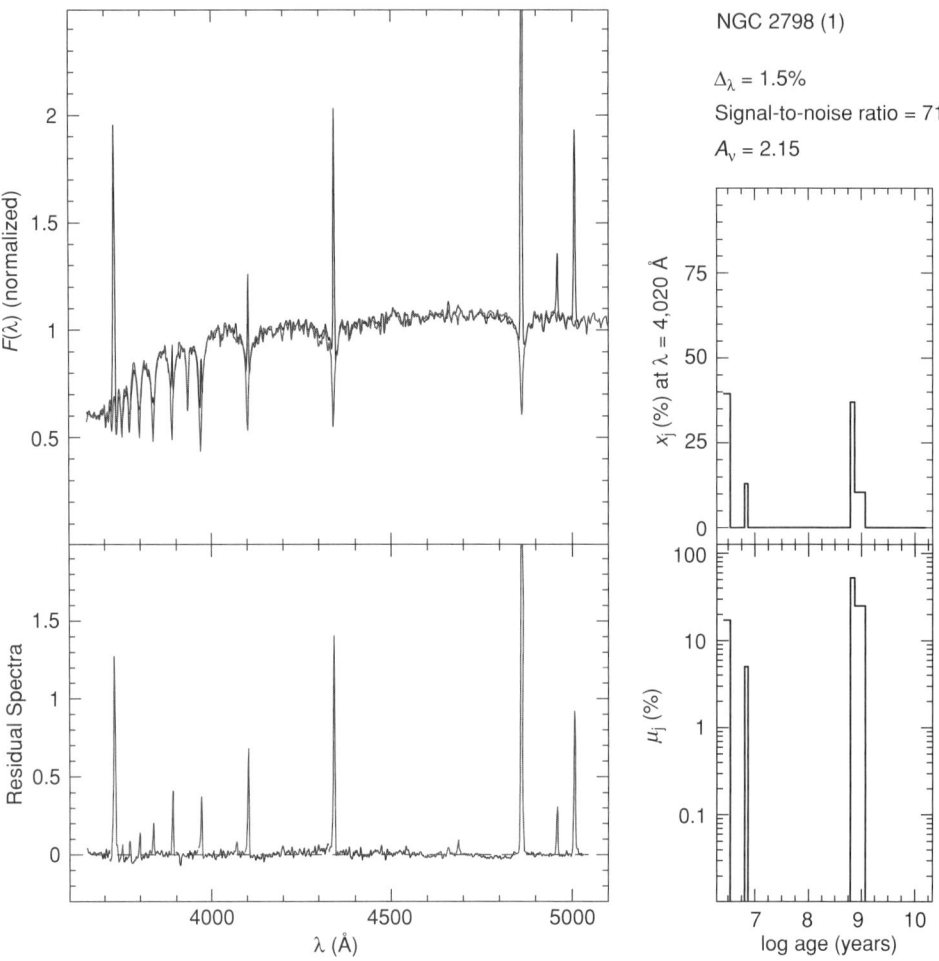

Figure 30.1. The optical spectrum of NGC 2798. Upper left: the optical spectrum of NGC 2798 and the corresponding fit. Lower left: the residual spectrum. Lower right: star-formation history (mass-fraction). Upper right: star-formation history (light).

4 The statistics of massive O stars and Wolf–Rayet stars

We measured the WR bump in the residual spectra described above. To measure the blue WR bump and the red WR bump at 5808 Å, we employed a method similar to that described by Guseva *et al.* (2000). In six galaxies (25% of the sample) we could clearly detect the blue WR bump. However, we measured the WR bumps in all other galaxies, thus establishing upper limits for the numbers of observed WR stars. The numbers of O stars were calculated using the Hβ luminosity and the calibration proposed by Leitherer & Heckman (1995).

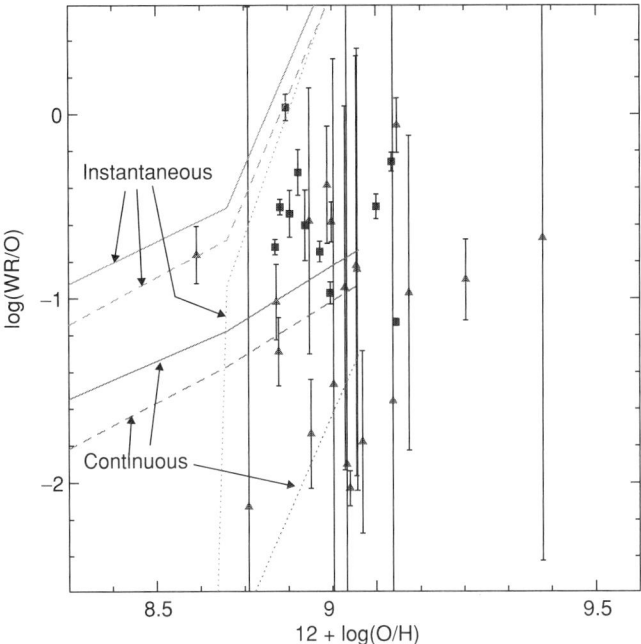

Figure 30.2. The observed WR/O-star ratio versus metallicity. Multiple extractions are shown for each galaxy, when available. Wolf–Rayet galaxies are represented as squares and all others (upper limits) are plotted as triangles. The lines correspond to STARBURST99 models with two different star-formation scenarios: continuous and instantaneous. Three different mass functions are plotted for each scenario: $\alpha = 2.35$, $M_{\text{upp}} = 100 M_\odot$ (solid lines); $\alpha = 3.3$, $M_{\text{upp}} = 100 M_\odot$ (dashed lines); and $\alpha = 2.35$, $M_{\text{upp}} = 30 M_\odot$ (dotted lines). Most of our galaxies present a combination of the two star-forming scenarios.

We for the first time categorize galaxies ESO 485-G003, NGC 6090, and NGC 2798 as WR galaxies, on the basis of the presence of the WR bump and the calculated numbers of WR stars.

In Figure 30.2 we compare the WR/O-star ratio with the metallicity of the host galaxy and with the STARBURST99 models. In this figure, all galaxies for which WR bumps have been detected are plotted as squares; all others are the upper limits, plotted as triangles. A combination of star-formation regimes (continuous and instantaneous) explains the observed WR/O-star ratios.

5 Concluding remarks and future work

Our main conclusions are sumarized below.

(a) All WR galaxies are very young systems, with $\log(t) < 8$, with t in years, undergoing spectacular bursts of star formation. However, not all young star-forming systems host WR stars.

(b) Most WR populations are found in interacting systems that have undergone both recent and older interactions. The age of the dominant stellar population of a galaxy is a key factor governing the detection of WR stars.
(c) Mixed star-formation scenarios are needed in order to explain the observed WR/O-star ratios.
(d) While metal-rich galaxies may indeed favor the formation of WR stars, their complex star-forming histories and the presence of older stellar populations may hide the young WR stars.

We are currently using the Spitzer Space Telescope to probe the energy-generating mechanism of LIRGs and ultra-luminous infrared galaxies. The goal is to use diagnostic techniques to separate AGN from starbursts and to understand the relative frequency of these two possible energy sources as a function of merger stage.

For a more detailed discussion of the results presented here, please refer to Leão *et al.* (2008).

References

Cid Fernandes, R., Mateus, A., Sodré, L., Stasińska, G., & Gomes, J. M. (2005), *MNRAS* **358**, 363–378
Claus, L., & Heckman, T. M. (1995), *ApJS* **96**, 9L
Conti, P. S. (1991), *ApJ* **377**, 115–125
Guseva, N. G., Izotov, Y. I., & Thuan, T. X. (2000), *ApJ* **531**, 776G
Kim, D. C., Veilleux, S., & Sanders, D. B. (1998), *ApJ* **508**, 627–647
Leão, J. R. S., Leitherer, C., Bresolin, F., Chandar, R., & Cid Fernandes, R. (2008), *ApJ*
Leitherer, C. (1992), *ApJ* **401**, 596–617
Schaerer, D., Contini, T., & Pindao, M. (1999), *A&AS* **136**, 35–52

Part V

Physical processes at high metallicity

31

Stellar winds from Solar-metallicity and metal-rich massive stars

Joachim Puls

Universitätssternwarte München, Scheinerstrasse 1, D-81679 München, Germany

We discuss theoretical predictions and observational findings obtained for radiatively driven winds of massive stars, with emphasis on their dependence on metallicity. If these winds are not strongly clumped or the clumping properties are independent of metallicity z, theory and observations agree very well, and mass-loss rates and terminal velocities scale as $\dot M \propto z^{0.62 \pm 015}$ and $v_\infty \propto z^{0.13}$, respectively. This dependence could be validated only for winds with Solar and subsolar abundances, due to the lack of supersolar-metallicity test cases. The actual values for the mass-loss rates are uncertain, due to unknown clumping properties of the wind, and currently accepted numbers might be overestimated by factors in between ~ 2 and 10.

1 Introduction

Massive stars and their winds are crucial for the chemical and dynamical evolution of galaxies through their input of radiation, energy, momentum and nuclear processed material. In the distant Universe, massive stars dominate the integrated UV light of very young galaxies (e.g. Steidel *et al.* 1996; Pettini *et al.* 2000), and even earlier they are the suspected sources of the re-ionisation of the Universe (Bromm *et al.* 2001).

Together with rotation, stellar winds control the specific evolution in the upper part of the HR diagram ($M_{\rm zams} \gtrsim 10 M_\odot$), by affecting timescales, chemical profiles, surface abundances and luminosities. For example, changing the mass-loss rates of massive stars by only a factor of two has a dramatic effect on their evolution (Meynet *et al.* 1994).

The Metal-rich Universe, eds. G. Israelian and G. Meynet. Published by Cambridge University Press.
© Cambridge University Press 2008.

In the following, we will review important theoretical and observational aspects of these winds, with emphasis on their dependence on metallicity. Finally, we will comment on recent evidence indicating that currently accepted mass-loss rates may need to be revised downwards, by as much as a factor of ten.

2 Radiation-driven winds: theoretical predictions

Stellar winds from hot stars are accelerated by radiative *line*-driving, with typical mass-loss rates, $\dot{M}$, of the order of $(0.1–10) \times 10^{-6} M_\odot \, \text{yr}^{-1}$ (resulting in a significant mass fraction being lost during their evolution) and terminal wind velocities, v_∞, of the order of $200 \, \text{km s}^{-1}$ (A-supergiants) to $3000 \, \text{km s}^{-1}$ (hot O-dwarfs). In parallel with the line absorption of stellar photons, radial momentum is transferred to the absorbing ions (mostly metals), which is redistributed to the bulk matter (H/He) via Coulomb collisions. In order to become efficient, this process requires a large number of stellar photons, i.e. a high luminosity, $L \propto R_\star^2 T_{\text{eff}}^4$ (supergiants or hot dwarfs), and a large number of absorbing lines close to the flux maximum with high interaction probabilities. The latter constraint inevitably leads to a *metallicity dependence* of mass loss. Pioneering investigations were performed by Lucy & Solomon (1970) and Castor *et al.* (1975), with improvements regarding a quantitative description/application by Friend & Abbott (1986) and Pauldrach *et al.* (1986). The latest review on this topic was published by Kudritzki & Puls (2000).

The *total radiative line acceleration*, $g_{\text{rad}}^{\text{lines}}$, consists of the individual contributions, $\Sigma_i g_{\text{rad}}^i$, where the corresponding line transitions can be either optically thin or thick, with

$$g_{\text{rad}}^{\text{thin}} \propto L_\nu^i k^i, \qquad g_{\text{rad}}^{\text{thick}} \propto L_\nu^i \frac{dv/dr}{\rho},$$

where L_ν^i is the luminosity at the frequency of line i, $k^i \propto \chi^i/\rho$ the dimensionless line-strength, χ the frequency-integrated line opacity and ρ the density. Because of the large number of lines present *and needed* to accelerate the wind (our database comprises roughly 4 million lines from 150 ions), a statistical approach is well suited, and the above sum can be approximated by an integral over the line-strength distribution, $N(k)$. From early on, it turned out that this distribution closely follows a power law, $dN(k)/dk \approx k^{\alpha-2}$, where α is of the order of 0.6–0.7 under typical conditions (see Figure 31.1, left). Details can be found in Puls *et al.* (2000). With this distribution and integrating over all (optically thin and thick) lines, the line acceleration turns out to depend on the luminosity and the spatial velocity gradient! Namely,

$$g_{\text{rad}}^{\text{lines}} = \sum_i g_{\text{rad}}^i \to \int\int g_{\text{rad}}^i(\nu, k) dN(\nu, k) \propto N_{\text{eff}} L \left(\frac{dv/dr}{\rho}\right)^\alpha,$$

with N_{eff} the so-called effective (flux-weighted) number of lines.

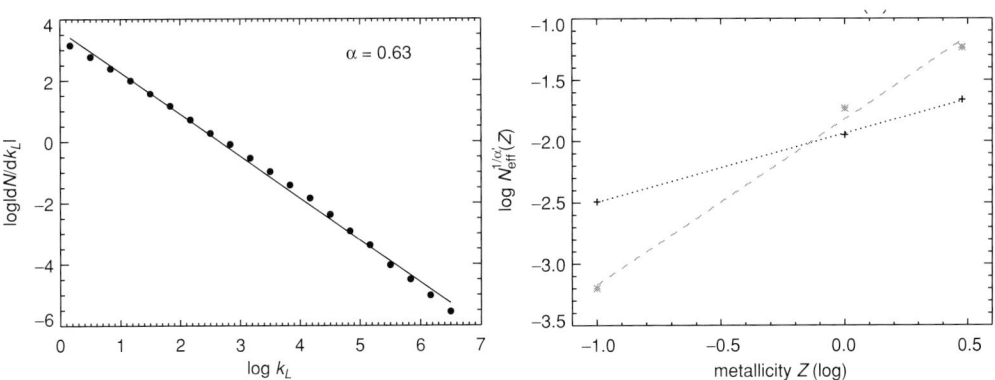

Figure 31.1. Left panel: a logarithmic plot of the line-strength distribution function for an O-type wind of 40 000 K and the corresponding power-law fit. Right panel: predictions from line statistics of the dependence of mass-loss $\dot{M}$ (via $N_{\text{eff}}^{1/\alpha'}$, cf. Equation (31.1)) on metallicity Z, for $T_{\text{eff}} = 40$ kK (crosses) and 10 kK (asterisks). The slopes are 0.56 and 1.35, respectively. Adapted from Puls *et al.* (2000).

Scaling laws

By inserting this expression into the hydrodynamical equation of motion, the latter can be solved, together with the equation of continuity, (almost) analytically (e.g. Kudritzki *et al.* 1989), and the resulting wind parameters obey the following scaling laws (with rotation neglected):

$$\dot{M} \propto N_{\text{eff}}^{1/\alpha'} L^{1/\alpha'} [M(1-\Gamma)]^{1-1/\alpha'}, \quad (2.1)$$

$$v_\infty \approx 2.25 \frac{\alpha}{1-\alpha} v_{\text{esc}}, \quad v_{\text{esc}} = \left(\frac{2GM(1-\Gamma)}{R_\star}\right)^{1/2}, \quad (2.2)$$

$$v(r) = v_\infty \left(1 - \frac{R_\star}{r}\right)^\beta, \quad \beta \approx 0.8 \text{ (O stars) to 2 (BA-supergiants)}, \quad (2.3)$$

where Γ is the Eddington factor (Thompson scattering diminishes the effective gravity), α the power-law index of the line-strength distribution function and $\alpha' = \alpha - \delta$, with $\delta \approx 0.1$ the so-called ionisation parameter.

The wind-momentum–luminosity relation

Exploiting the fact that α' is close to 2/3, the so-called modified wind-momentum rate, D_{mom} (Kudritzki *et al.* 1995; Puls *et al.* 1996), becomes almost independent of mass:

$$D_{\text{mom}} = \dot{M} v_\infty (R_\star/R_\odot)^{1/2} \propto N_{\text{eff}}^{1/\alpha'} L^{1/\alpha'}, \quad (2.4)$$

$$\log D_{\text{mom}} \approx \frac{1}{\alpha'} \log L + \text{constant}(Z, \text{spectral type}). \quad (2.5)$$

This *wind-momentum–luminosity relation* (WLR) constitutes one of the most important predictions of radiation-driven-wind theory, and can be applied at least in two ways. (i) From spectral analyses of large samples of massive stars, one can construct *observed* WLRs, and calibrate them as a function of spectral type and metallicity Z (N_{eff} and α' depend on both parameters). The derived relations can then be used as an independent tool to measure extragalactic distances from the wind properties, effective temperatures and metallicities of distant stellar samples. (ii) Observed WLRs can be compared with *theoretical* predictions in order to test the validity of the theory itself.

Predictions from line statistics

Since a higher metallicity translates into higher opacities, an increase in abundance leads to a larger number of lines which can accelerate the wind. Denoting the *global* metallicity by Z (normalized with respect to the Solar value), it turns out that the effective number of lines scales via $N_{\text{eff}} \propto Z^{1-\alpha}$ and thus, via Equation (2.1),

$$\dot{M}, D_{\text{mom}} \propto Z^{(1-\alpha)/(\alpha')}. \qquad (2.6)$$

For O-type winds then ($\alpha \approx 2/3$), this means a scaling with $Z^{0.6}$, whereas for A-supergiant winds ($\alpha \approx 0.4\ldots0.5$) a dependence of $Z^{1.3\ldots 2}$ is predicted (see Figure 31.1, right). *For* (so-far-hypothetical) *massive-star winds with $z = 2$, this implies mass-loss rates a factor of 1.5 . . . 2.8 higher than for winds of Solar composition.*

Not only the global abundance but also the specific composition affects the wind. Owing to their different line-strength statistics, lines from Fe-group elements and light elements (CNO and similar) have different impacts on the wind properties. In particular, Fe-group elements dominate the acceleration of the lower wind, and thus determine $\dot{M}$, whereas lines from light elements dominate the acceleration of the outer wind, and thus determine v_∞. For details, see Puls *et al.* (2000), but also similar results from hydrodynamical calculations by Pauldrach (1987), Vink *et al.* (1999, 2001) and Krticka (2006).

Predictions from hydrodynamical models

The most frequently quoted predictions for the wind properties of OB-type stars result from the hydrodynamical models provided by Vink *et al.* (2000), summarized by them in terms of a 'mass-loss recipe'. These predictions are in very good agreement with independent models by Kudritzki (2002) ($v_\infty \propto Z^{0.12}$), Puls *et al.* (2003) and Krticka & Kubat (2004). Similar approaches have been used to predict the metallicity dependence. In particular, Vink *et al.* (2001) derived $\dot{M} \propto Z^{0.69}$ for O stars and $\dot{M} \propto Z^{0.64}$ for B-supergiants, and Krticka (2006) found $\dot{M} \propto Z^{0.67}$,

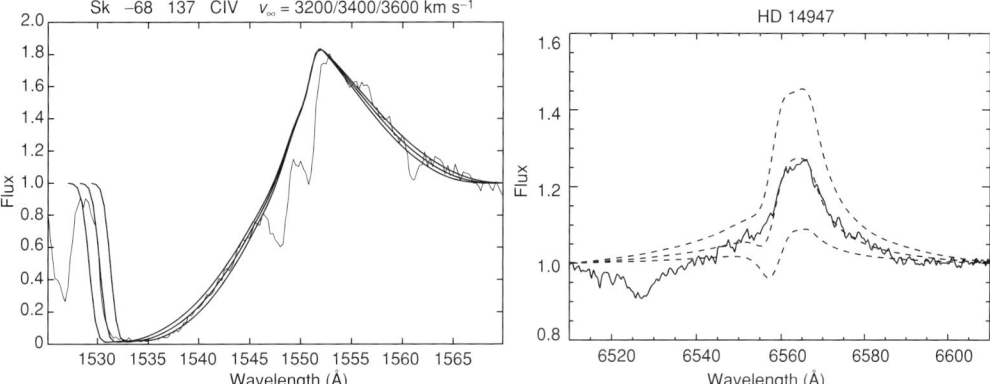

Figure 31.2. Left panel: derivation of v_∞ from UV P-Cygni lines (here, from CIV 1548/50). From Kudritzki (1998). Right panel: $\dot{M}$ from Hα. Synthetic profiles (dashed) from models varied by $\pm 30\%$ in $\dot{M}$. Figure adapted from Puls et al. (1996).

$v_\infty \propto Z^{0.06}$ for O stars. Note that these results are in very good agreement with the results from line statistics alone (see above). With respect to the winds of Wolf–Rayet stars, which depend strongly on the Fe-content, we refer the reader to the work by Gräfener & Hamann (2005) (see also Chapter 33 of this volume), Vink & de Koter (2005), Crowther & Hadifield (2006) and Chapter 29 of this volume.

3 Results derived from observations

The derivation of stellar/wind parameters from observations via *quantitative spectroscopy* requires the assumption of a physical model (incorporated into the atmosphere codes which synthesise the spectra). Most of the results presented in the following are based on a standard, one-dimensional description with a *smooth* wind (but see Section 4). The parameters have to be derived by using suitable diagnostics:

- photospheric parameters, $T_{\rm eff}$, $\log g$ and helium content from optical lines and NLTE atmospheres (including wind);
- wind parameters, v_∞ (Figure 31.2 left), $\dot{M}$ and the velocity law from UV P-Cygni lines and/or optical/IR (emission) lines (Hα – Figure 31.2, right, He II 4686, Brα), $\dot{M}$ also from the radio free–free excess;
- stellar radius, from distance, V-band magnitude and theoretical fluxes (reddening!); and
- metallicity, from spectra (UV and optical) and models.

During the last decade, many such spectroscopic investigations have been conducted. In Table 31.1 we have compiled important contributions regarding OBA stars *and their winds* (excluding Galactic Centre objects; see Chapter 13 of this

Table 31.1. *Quantitative spectroscopy of OBA stars and their winds in the Galaxy and the MCs, by means of spherically extended model atmospheres*

Diagnostic	Reference	Atmospheric model	Sample
Hα	Lamers & Leitherer (1993)	Approximate	Galactic O stars
	Puls et al. (1996)	Approximate	Galactic/LMC/SMC O stars
	Kudritzki et al. (1999)	NLTE/unblanketed	Galactic BA-supergiants
	Markova et al. (2004)	Approximate	Galactic O stars
UV	Bianchi & Garcia (2002)	NLTE/WM-basic	Galactic O stars
	Garcia & Bianchi (2004)	NLTE/WM-basic	Galactic O stars
	Martins et al. (2004)	NLTE/CMFGEN	SMC O-dwarfs
UV + optical	Crowther et al. (2002)	NLTE/CMFGEN	LMC/SMC O-supergiants
	Hillier et al. (2003)	NLTE/CMFGEN	SMC O-supergiants
	Bouret et al. (2003)	NLTE/CMFGEN	SMC O-dwarfs
	Martins et al. (2005)	NLTE/CMFGEN	Galactic O-dwarfs
	Bouret et al. (2005)	NLTE/CMFGEN	Galctic O stars
Optical	Herrero et al. (2002)	NLTE/FASTWIND	Cyg-OB2 OB stars
	Repolust et al. (2004)	NLTE/FASTWIND	Galactic O stars
	Trundle et al. (2004)	NLTE/FASTWIND	SMC B-supergiants
	Trundle & Lennon (2005)	NLTE/FASTWIND	SMC B-supergiants
	Massey et al. (2004/2005)	NLTE/FASTWIND	LMC/SMC O stars
	Mokiem et al. (2005)	NLTE/FASTWIND	Galactic O stars
	Crowther et al. (2006)	NLTE/CMFGEN	Galactic B-supergiants
	Mokiem et al. (2006a/b)	NLTE/FASTWIND	SMC/LMC OB stars

CMFGEN (Hillier & Miller 1998), WM-basic (Pauldrach et al. 2001), FW (Puls et al. 2005).

volume). Most of them were performed using *line-blanketed* NLTE atmosphere codes.

Taken together, the most important results of these investigations can be summarised as follows. Compared with results from previous investigations, the $T_{\rm eff}$ scale has become lower, due to line-blanketing effects. The mass-loss rates in the SMC (and in the LMC, see below) are systematically smaller than in the Galaxy, though the scatter is large. The observed WLRs meet the theoretical predictions, except for (i) O-supergiants with rather dense winds (which might be explained by wind-clumping, see Section 4), (ii) low-luminosity O-dwarfs with observed wind momenta much lower than predicted (the reason for this is unknown) and (iii) a large fraction (but not all) of B2/3 supergiants, for which again the observed wind momenta are 'too low" (this remains unexplained as well).

Wind-momentum rates of O stars from the FLAMES survey of massive stars

Using the FLAMES multi-object spectrograph attached to the VLT, a large collaboration conducted a programme to investigate the stellar contents of clusters of various ages in the Galaxy and the Magellanic Clouds. For the introductory publication, see Evans *et al.* (2005). With respect to massive stars, roughly 60 O/early-B stars from the SMC/LMC have been analysed in a *homogeneous and objective* way. This has been achieved by means of an 'automatic' analysis method combining a genetic algorithm (PIKAIA) (Charbonneau 1995) used to obtain the optimum fit and FASTWIND (see Table 31.1) to calculate the synthetic spectra. The method itself has been presented (and tested) by Mokiem *et al.* (2005), and the analysis of the SMC/LMC data is described in Mokiem *et al.* (2006, 2007a).

By combining these data with data from previous investigations, the 'observed' metallicity dependences of $\dot{M}$ and $D_{\rm mom}$ could be derived with unprecedented precision,

$$\dot{M} \propto Z^{0.62 \pm 0.15}, \qquad v_\infty \propto Z^{0.13},$$

where the scaling-law for v_∞ had been obtained much earlier, by Leitherer *et al.* (1992). Both results are in very good agreement with the theory (cf. Section 2 and Figure 31.3).

Beyond the Local Group

The availability of 8-m-class telescopes allows us to extend our investigations to objects in more distant galaxies, not only in the Local Group but also beyond, when concentrating on the visually brightest stars in the sky, the A-supergiants. Examples of recent results are given in Figure 31.4, which shows that the slope

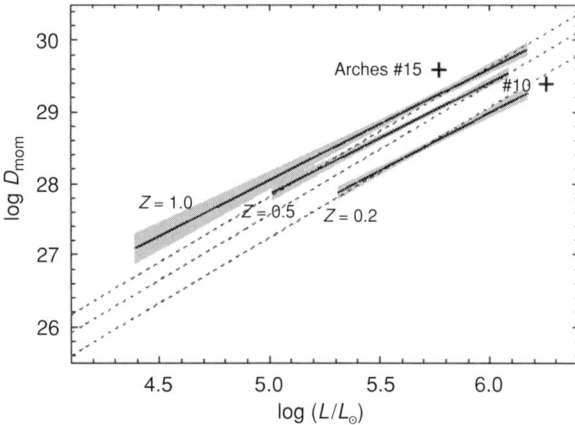

Figure 31.3. Observed WLRs for the Galaxy and the MCs, as derived by Mokiem et al. (2007b); see also de Koter (2007). The grey shaded areas denote the 1σ confidence interval, and the dashed lines represent the theoretical predictions from Vink et al. (2000, 2001). For comparison, we have overplotted the wind-momentum rates from two Of$^+$ stars in the Arches cluster, both of them with $T_{\rm eff} \approx 30$ kK, as analysed by Najarro et al. (2004). Original figure from Mokiem et al. (2007b).

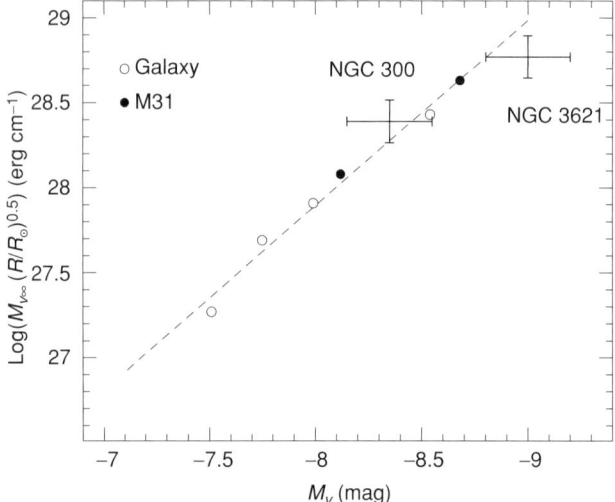

Figure 31.4. The WLR for Galactic and extragalactic A-supergiants. Data from Kudritzki et al. (1999) for the Galaxy, McCarthy et al. (1997) for M31, Bresolin et al. (2001) for M31 and NGC3621 and Bresolin et al. (2002) for NGC 300). Regarding the M31 objects, see also Chapter 35 in this volume. Figure from Bresolin et al. (2002).

of the corresponding WLR is consistent with the theoretical prediction in this parameter range ($\alpha' \approx 0.4$). So far, no object providing *clear* evidence for supersolar metallicity has been found (see Chapter 6 of this volume). Further work has been presented by Bianchi *et al.* (1996) (UV analyses of M31/M33 OB-supergiants), Smartt *et al.* (2001) (UV/optical analyses of M31 B-supergiants), Urbaneja *et al.* (2003) (optical analysis of NGC 300 B-supergiants), Urbaneja *et al.* (2005) (optical analyses of M33 B-supergiants) and Lennon and Trundle (Chapter 6 of this volume, M31 B-supergiants).

4 The impact of wind-clumping

From what has been discussed so far, it seems that the outflows from luminous OBA stars are well understood. There is, however, accumulating evidence that currently accepted mass-loss rates *may* need to be revised downwards *by as much as a factor of ten*, as a consequence of previously neglected wind-'clumping', which affects most mass-loss diagnostics.

Such revisions, of course, would have dramatic consequences, not only for the stellar evolution (Section 1) but also regarding the feedback from massive stars. In the following, we will summarise the status quo, whereas a more detailed discussion can be found in Puls *et al.* (2006, 2007 and references therein).

The present hypothesis states that clumping (if present) is a matter of *small-scale* density inhomogeneities in the wind, which redistribute the matter into clumps of enhanced density embedded in a rarefied, almost-void medium. The amount of clumping is conveniently quantified by the so-called clumping factor, $f_{cl} \geq 1$, which is a measure of the overdensity inside the clumps (relative to a smooth flow of identical average mass-loss rate). Diagnostics that are linearly dependent on the density (e.g. UV resonance lines) are insensitive to clumping, whilst those sensitive to ρ^2 (such as Hα and free–free radio emission) *will tend to overestimate the mass-loss rate of a clumped wind, by a factor* $\sqrt{f_{cl}}$. For further details, see Abbott *et al.* (1981), Lamers & Waters (1984), Schmutz (1995) and Puls *et al.* (2006).

Until now, the most plausible physical process responsible for small-scale structure formation in massive-star winds is the so-called line-driven instability, which was found in the first time-dependent hydrodynamical simulations of such winds (Owocki *et al.* 1988). Nevertheless, it took some while to incorporate clumping into the *atmospheric models* of massive stars, firstly for Wolf–Rayet-star atmospheres, in particular to explain the strength of the observed electron-scattering wings of emission lines (Hillier 1991) and the presence and variability of substructures in these lines (e.g. Moffat & Robert 1994).

The diagnostics of OB-star winds, on the other hand, did not require (significant) clumping until recently, particularly because of the very good agreement between the theoretically predicted and observed WLRs (see above). Purely coincidental agreement seemed rather unlikely.

Though there is still hardly any *direct* observational evidence (but see Eversberg *et al.* (1998)), to date a number of indirect indications favour the presence of wind-clumping in OB-star winds, so we would like to stress here three important aspects.

(i) From detailed investigations of large samples of Galactic O stars, Puls *et al.* (2003), Markova *et al.* (2004) and Repolust *et al.* (2004) found that supergiants with Hα emission lie above the theoretical WLR (see Section 3), whereas the rest fits almost perfectly. Since the WLR should be independent of luminosity class (e.g. Puls *et al.* 1996), this discrepancy was interpreted in terms of clumpy winds, with $f_{cl} \approx 5$, and mass-loss rates reduced by factors between 2 and 3. Indeed, an analogous correction has been applied in the investigation by Mokiem *et al.* (2007b); see Figure 31.3).

(ii) A compelling, independent indication of clumping comes from analyses of the UV P-Cygni P v 1118/28 resonance line doublet (Massa *et al.* 2003; Fullerton *et al.* 2006) observed by FUSE. Because phosphorus has a low cosmic abundance, this doublet never saturates in normal OB stars, providing useful estimates of $\dot{M}$ when P^{4+} is the *dominant ion* – as is implied to be the case at least for mid-O-star winds (Puls *et al.* 2007). These mass-loss rates turned out to lie considerably below those inferred from other (clumping-sensitive) diagnostics such as Hα and radio emission. The most reasonable way to reconcile these two results is to invoke extreme clumping in the wind ($f_{cl} \approx 100$), with actual mass-loss rates being *much* lower than previously thought, by a factor $\gtrsim 10$.

(iii) If clumping were indeed present, there is, of course, the additional question regarding the radial stratification of f_{cl}. To this end, Puls *et al.* (2006) performed a self-consistent analysis of Hα, IR, millimetre-wave and radio fluxes, thus sampling the lower, intermediate and outer wind in parallel, on the basis of a sample of 19 well-known Galactic O-type supergiants and giants. A major result of this investigation is that in *weaker* winds the clumping factor is the same in the inner ($r < 2R_\star$) and outermost regions. However, for *stronger* winds, the clumping factor in the inner wind is larger than that in the outer one, by factors of 3–6. This finding indicates that there is a physical difference between the clumping properties of weaker and stronger winds, and is consistent with the arguments outlined in (i) above and earlier findings by Drew (1990).

Unfortunately, the latter analysis is hampered by one severe restriction. Since *all* employed diagnostics have a ρ^2 dependence, only *relative* clumping factors could be derived, normalised with respect to the values in the outermost, radio-emitting region. In other words, $\dot{M}(REAL) \leq \dot{M}(radio)$, since the clumping in the radio-emitting region still remains unknown. Only if $f_{cl}(radio)$ were unity would we have $\dot M(\mathrm{REAL}) = \dot M(\mathrm{radio})$. *Thus, the issue of absolute values for $\dot M$ remains unresolved.*

Implications

If, on the one hand, the latter hypothesis were true, i.e. the outer winds were unclumped, the results obtained by Puls *et al.* (2006) would be consistent with theoretical WLRs. In this case then, one would meet a severe dilemma with the results from the far UV, which one might hope to explain by invoking additional effects from X-rays emitted due to clump–clump collisions (Feldmeier *et al.* 1997; Pauldrach *et al.* 2001). *If*, on the other hand, the far-UV values were correct, the outer wind would have to be significantly clumped, and the present match of 'observed' and predicted WLRs would indeed be merely coincidental. This scenario would imply a number of severe problems, not only for radiation-driven-wind theory but, most importantly, also concerning the stellar evolution in the upper HRD and related topics. A possible way out has been suggested by Smith & Owocki (2006), namely that the 'missing' mass-loss in the O-star phase might be compensated for by a higher mass-loss in the LBV phase during brief eruptions.

Acknowledgment

Part of this work was supported by the Spanish MEC through project AYA2004–08271-CO2, which is gratefully acknowledged.

References

Abbott, D. C., Bieging, & J. H., Churchwell, E. (1981), *ApJ* **250**, 645
Bianchi, L., & Garcia, M. (2002), *ApJ* **581**, 610
Bianchi, L., Hutchings, J. B., & Massey, P. (1996), *AJ* **111**, 2303
Bouret, J.-C., Lanz, T., Hillier, D. J. *et al.* (2003), *ApJ* **595**, 1182
Bouret, J.-C., Lanz, T., & Hillier, D. J. (2005), *A&A* **438**, 301
Bresolin, F., Kudritzki, R. P., Méndez, R. H., *et al.* (2001), *ApJL* **548**, 159
Bresolin, F., Gieren, W., Kudritzki, R. P. *et al.* (2002), *ApJ* **567**, 277
Bromm, V., Kudritzki, R. P., & Loeb, A. (2001), *ApJ* **552**, 464
Castor, J. I., Abbott, D. C., & Klein, R. I. (1975), *ApJ* **195**, 157
Charbonneau, P. (1995), *ApJS* **101**, 309
Crowther, P. A., & Hadfield, L. J. (2006), *A&A* **449**, 711
Crowther, P. A., Hillier, D. J., Evans, C. J. *et al.* (2002), *ApJ* **579**, 774
Crowther, P. A., Lennon, D. J., & Walborn, N. R. (2006), *A&A* **446**, 279
de Koter, A. (2007), ASP conference series, "Mass Loss from Stars and the Evolution of Stellar Clusters", eds. A. de Koter, L. J. Smith & L. B. F. M. Waters, at press
Drew, J. E. (1990), *ApJ* **357**, 573
Evans, C. J., Smartt, S. J., Lee, J.-K. *et al.* (2005), *A&A* **437**, 467
Eversberg, T., Lepine, S., & Moffat, A. F. J. (1998), *ApJ* **494**, 799
Feldmeier, A., Puls, J., & Pauldrach, A. W. A. (1997), *A&A* **322**, 878
Friend, D. B., & Abbott, D. C. (1986), *ApJ* **311**, 701
Fullerton, A. W., Massa, D. L., & Prinja, R. K. (2006), *ApJ* **637**, 1025

Garcia, M., & Bianchi, L. (2004), *ApJ* **606**, 497
Gräfener, G., & Hamann, W.-R. (2005), *A&A* **432**, 633
Herrero, A., Puls, J., & Najarro, F. (2002), *A&A* **396**, 949
Hillier, D. J. (1991), *A&A* **247**, 455
Hillier, D. J., & Miller, D. L. (1998), *ApJ* **496**, 407
Hillier, D. J., Lanz, T., Heap, S. R. *et al.* (2003), *A&A* **588**, 1039
Krticka, J. (2006), *MNRAS* **367**, 1282
Krticka, J., & Kubat, J. (2004), *A&A* **417**, 1003
Kudritzki, R. P. (1998), in *Proc. 8th Canary Winter School*, eds. A. Aparicio, A. Herrero & F. Sanchez (New York, Cambridge University Press), p. 149
Kudritzki, R. P. (2002), *ApJ* **577**, 389
Kudritzki, R. P., & Puls, J. (2000), *ARA&A* **38**, 613
Kudritzki, R. P., Pauldrach, A., Puls, J., & Abbott, D. C. (1989), *A&A* **219**, 205
Kudritzki, R. P., Lennon, D. J., & Puls, J. (1995), in *ESO Astrophysics Symposia, Science with the VLT*, eds. J. R. Walsh & I. J. Danziger (Berlin, Springer), p. 246
Kudritzki, R. P., Puls, J., Lennon, D. J. *et al.* (1999), *A&A* **350**, 970
Lamers, H. J. G. L. M., & Waters, L. B. F. M. (1984), *A&A* **138**, 25
Lamers, H. J. G. L. M., & Leitherer, C. (1993), *A&A* **412**, 771
Leitherer, C., Robert, C., & Drissen, L. (1992), *ApJ* **401**, 596
Lucy, L. B., & Solomon, P. M. (1970), *ApJ* **159**, 879
Markova, N., Puls, J., Repolust, T. *et al.* (2004), *A&A* **413**, 693
Martins, F., Schaerer, D., & Hillier, D. J. *et al.* (2004), *A&A* **420**, 1087
Martins, F., Schaerer, D., & Hillier, D. J. (2005), *A&A* **436**, 1049
Massey, P., Kudritzki, R. P., Bresolin, F. *et al.* (2004), *ApJ* **608**, 1001
Massey, P., Puls, J., Pauldrach, A. W. A. *et al.* (2005), *ApJ* **627**, 477
Massa, D., Fullerton, A. W., Sonneborn, G. *et al.* (2003), *A&A* **586**, 996
McCarthy J. K., Kudritzki R. P., Lennon D. J. *et al.* (1997), *ApJ* **482**, 757
Meynet, G., Maeder, A., Schaller, G. *et al.* (1994), *A&AS* **103**, 97
Moffat, A. F. J., & Robert, C. (1994), *ApJ* **421**, 310
Mokiem, M. R., de Koter, A., Puls, J. *et al.* (2005), *A&A* **441**, 711
Mokiem, M. R., de Koter, A., Evans, C. J. *et al.* (2006), *A&A* **456**, 1131 (2007a), *A&A* **465**, 1003
Mokiem, M. R., de Koter, A., Vink, J. S., & Puls, J. (2007b), *A&A* astro-ph, arXiv:0708.2042v1
Najarro, F., Figer, D. F., Hillier, D. J. *et al.* (2004), *ApJ* **611**, 105
Owocki, S. P., Castor, J. I., & Rybicki, G. B. (1988), *ApJ* **335**, 914
Pauldrach, A. (1987), *A&A* **183**, 295
Pauldrach, A. W. A., Puls, J., & Kudritzki, R. P. (1986), *A&A* **164**, 86
Pauldrach, A. W. A., Hoffmann, T. L., & Lennon, M. (2001), *A&A* **375**, 161
Pettini, M., Steidel, C. C., Adelberger, K. L. *et al.* (2000), *ApJ* **528**, 96
Puls, J., Kudritzki, R. P., Herrero, A. *et al.* (1996), *A&A* **305**, 171
Puls, J., Springmann, U., & Lennon, M. (2000), *A&AS* **141**, 23
Puls, J., Repolust, T., Hoffmann, T. *et al.* (2003), in *A Massive Star Odyssey: From Main Sequence to Supernova*, eds. K. A. van der Hucht, A. Herrero & C. Esteban (San Francisco, CA, Astronomical Society of the Pacific), p. 61
Puls, J., Urbaneja, M. A., Venero, R. *et al.* (2005), *A&A* **435**, 669
Puls, J., Markova, N., Scuderi, S. *et al.* (2006), *A&A* **454**, 625
Puls, J., Markova, N., & Scuderi, S. (2007), in *Mass Loss from Stars and the Evolution of Stellar Clusters* (San Francisco, CA, Astronomical Society of the Pacific), astro-ph/0607290

Repolust, T., Puls, J., & Herrero, A. (2004), *A&A* **415**, 349
Schmutz, W. (1995), in *Wolf–Rayet Stars: Binaries; Colliding Winds; Evolution*, eds.
   K. A. van der Hucht & P. M. Williams (Dordrecht, Kluwer), p. 127
Smartt, S., Crowther, P. A., Dufton, P. L. *et al.* (2001), *MNRAS* **325**, 907
Smith, N., & Owocki, S. P. (2006), *ApJL* **645**, 45
Steidel, C. C., Giavalisco, M., Pettini, M. *et al.* (1996), *ApJL* **462**, L17
Trundle, C., Lennon, D. J., Puls, J. *et al.* (2004), *A&A* **417**, 217
Trundle, C., & Lennon, D. J. (2005), *A&A* **434**, 677
Urbaneja, M. A., Herrero, A., Bresolin, F. *et al.* (2003), *ApJL* **584**, 73
Urbaneja, M. A., Herrero, A., Kudritzki, R. P. *et al.* (2005), *ApJ* **635**, 311
Vink, J. S., de Koter, A., & Lamers, H. J. G. L. M. (1999), *A&A* **350**, 181
Vink, J. S., de Koter, A., & Lamers, H. J. G. L. M. (2000), *A&A* **362**, 295
   (2001), *A&A* **369**, 574
Vink, J. S., & de Koter, A. (2005), *A&A* **442**, 587

32

On the determination of stellar parameters and abundances of metal-rich stars

Yoichi Takeda

*National Astronomical Observatory of Japan,
2-21-1 Osawa, Mitaka, Tokyo 181-8588, Japan*

Several topics of interest involved with precise determination of surface abundances and stellar parameters in the metal-rich regime are reviewed. The main emphasis is placed upon Solar-type F–G dwarfs, though K giants are also mentioned briefly. In particular, in connection with the problem of the validity of the hypothesis of LTE, recent spectroscopic studies of Hyades-cluster stars are discussed together with our own results. Some further discussion concerns age determination using evolutionary tracks, in connection with the existence of old metal-rich stars and the high metallicity of planet-host stars.

1 Introduction: derivation of stellar physical parameters

It is generally of importance in observational stellar astronomy to precisely establish the parameters of stars (e.g. atmospheric parameters such as $T_{\rm eff}$, $\log g$, $v_{\rm t}$, [Fe/H] for constructing model atmospheres; or fundamental stellar parameters such as M or *age* that are closely related to the current status of a star), the typical derivation processes of which are schematically depicted in Figure 32.1.

An accurate determination of surface chemical abundances is particularly important in the study of metal-rich stars since they may hold the key to the origin of metal-richness (e.g. the "primordial versus acquired" controversy regarding the tendency of planet-host stars to be metal-rich). Above all, unlike metal-poor stars, the precision in the abundance determination should be a vital factor in this case, since the extent of the chemical peculiarity is rather delicate (typically the enrichment is of the order of ∼0.2–0.4 dex at most), which in turn leads to the strong necessity of knowing the parameters of atmospheric models precisely.

The Metal-rich Universe, eds. G. Israelian and G. Meynet. Published by Cambridge University Press.
© Cambridge University Press 2008.

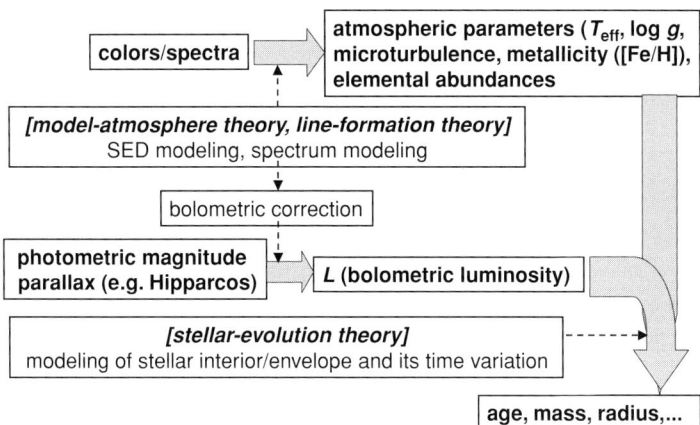

Figure 32.1. A schematic overview for the derivation of stellar physical parameters.

There are various ways of deriving the stellar atmospheric parameters necessary for constructing model atmospheres; for example, the determination of $T_{\rm eff}$ may be implemented by using colors, Balmer-line profiles, the requirement of excitation equilibrium, line-depth ratios, etc.; while $\log g$ may be obtained via the direct method (from M and R), the requirement of ionization equilibrium, or the fitting of strong-line wings.

Among these, the traditionally well-known spectroscopic method using the strengths of Fe I and Fe II lines (Takeda et al. 2002) may be preferable in the context of the abundance determination, because the necessary parameters can be derived only from the "same" spectrum as that from which the abundances are to be derived. In this approach, however, the condition of LTE (i.e. application of the Saha–Boltzmann equation for computing level populations) is usually assumed, which greatly simplifies the problem. Hence, its *practical* validity (though it is surely unrealistic from a *strict* point of view) should be crucial for this spectroscopic method of parameter determination.

2 The ionization equilibrium of Fe from a theoretical viewpoint

In this respect, our concern is that an appreciable departure from LTE (Saha) ionization equilibrium (i.e. so-called overionization) may be more or less predicted because of the imbalance between the photoionization rate R ($\propto J$), determined mainly by the mean radiation field (J), and the photo-recombination rate R^* ($\propto B$), determined by the Planck function (B) corresponding to the local electron temperature. Namely, since the radiation from hot deeper layers contributes mostly to J, the inequality of $J > B$ turns out to hold in the optically thin

310 *Determination of stellar parameters and abundances*

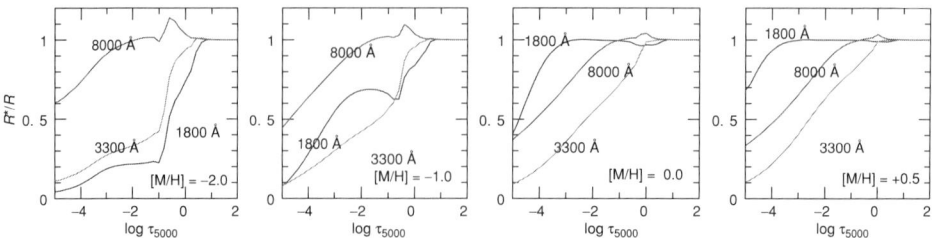

Figure 32.2. The photo-recombination-to-photoionization ratio (R^*/R) with depth at three Fe I ionization edges for models with various metallicities ($T_{\text{eff}} = 6{,}000$ K and $\log g = 4.0$).

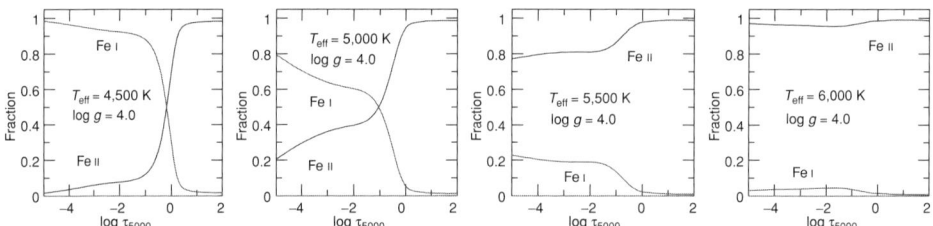

Figure 32.3. The neutral (Fe I) and singly-ionized (Fe II) fractions relative to the total amount of Fe atoms with depth, for the Solar-metallicity $\log g = 4.0$ models of $T_{\text{eff}} = 4{,}500$, $5{,}000$, $5{,}500$, and $6{,}000$ K.

atmospheric layer, especially for the T-sensitive UV wavelength region within which the important Fe I ionization edges are located. In effect, this reduces the Fe I/Fe II ratio compared with that expected from the LTE Saha equation since (Fe I/Fe II)$_{\text{NLTE}}$ < (Fe I/Fe II)$_{\text{LTE}}$.

There are several key points requiring attention in discussing how and when the NLTE effect caused by this overionization becomes quantitatively significant.

(1) First is the metallicity which affects the opacity; i.e. the radiation field tends to be thermalized as the metallicity becomes higher. This situation is depicted in Figure 32.2, which shows that the departure of R^*/R from unity (overionization) becomes progressively insignificant with increasing [M/H], especially at the important UV ionization edge of ground or low-excited Fe I levels ($\lambda \sim 1{,}800$ Å).

(2) Second is the population fraction of each ionization stage relative to the total number of Fe atoms, which determines the significance of the overionization effect on the abundance determination. As a general rule, influences on abundances derived from lines of the dominant stage are negligible, while those for the trace species are sensitively affected. In the case of unevolved dwarfs, most Fe atoms are singly ionized and only a small fraction remains neutral in the atmospheres of G stars and stars in earlier stages, whereas neutral Fe becomes predominant in K stars (Figure 32.3). In the former case,

therefore, Fe I lines appreciably suffer the NLTE overionization effect while Fe II lines do not, and vice versa for the latter case.

(3) Third, one should pay attention to the excitation potential of the levels under question. That is, the populations which suffer appreciable overionization are generally those of lower excitation levels, and highly excited levels often do not conform to it because of a comparatively stronger coupling to the continuum (i.e. the situation is closer to the condition of LTE). Some empirically derived tendency, e.g. that found by Ruland *et al.* (1980) for Fe I and Ti I in their analysis of Pollux, may be interpreted in this connection.

Accordingly, in the case of the F–G stars (most Fe are in the Fe II state) commonly used for galactic chemical-evolution studies, the fact that a lower metallicity favors an appreciable NLTE effect causes the following trend which is theoretically predicted.

- The LTE abundances from Fe I lines tend to be underestimated while those from Fe II lines are hardly affected.
- Quantitatively, however, this effect becomes progressively more appreciable as we go into the very-metal-poor regime of halo stars; while it is almost negligible for disk stars ($-1 <$ [Fe/H]), not to mention for metal-rich stars with [Fe/H] > 0 (Thévenin & Idiart 1999, Figure 9).

3 The validity of the hypothesis of LTE questioned from the observational side

Therefore, from the theoretical point of view mentioned above, there should be no need to worry about the NLTE overionization effect for Fe (and presumably also for other Fe-group elements) in the metal-rich or near-Solar metallicity stars of our present concern. Yet, a few groups have recently reported that appreciable inconsistencies were observed in the abundances obtained under the assumption of LTE even for disk stars of around solar metallicity.

For example, Bodhagee *et al.* (2003) and Gilli *et al.* (2006) found in their analysis of planet-host and comparison stars of early K through late F types that Fe-normalized abundances [X/Fe] derived from lines of several neutral elements (e.g. Ti I, V I, Mn I, Co I) exhibit systematic tendencies ([X/Fe] increasing with decreasing T_{eff}) over a wide T_{eff} range from $\sim$5,000 K to $\sim$6,000 K, which they attributed to some NLTE effect. We feel, however, that such an apparent T_{eff} dependence as they claimed might be due to some spurious effect (e.g. blending with unknown lines whose importance would progressively increase toward a lower T_{eff}), because we could not confirm such a tendency in our recent analysis (Takeda 2007) of 160 F, G, and K stars including 27 planet-host stars similar to theirs, as

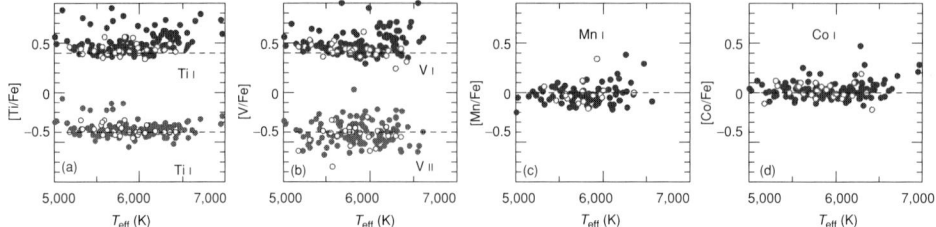

Figure 32.4. Plots of [Ti/Fe], [V/Fe], [Mn/Fe], and [Co/Fe] with $T_{\rm eff}$, constructed from the results of Takeda's (2007) analysis of 160 F, G, and K stars. The zero levels are indicated by dashed lines. Open and filled symbols denote planet-host stars and non-planet-host stars, respectively.

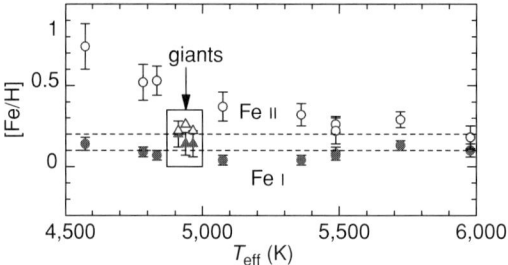

Figure 32.5. The Fe abundances of Hyades G–K dwarfs obtained by Schuler et al. (2006b), re-plotted by ourselves as functions of $T_{\rm eff}$ using the data presented in Table 5 in their paper. Abundances from Fe I and Fe II lines are indicated by filled and open symbols, respectively. Together with the results for dwarfs (circles), those for three giants are also shown by triangles.

shown in Figure 32.4. In any case, further investigation of this matter should be carried out.

Another item of evidence for a definite breakdown of LTE has recently emerged from analyses of G–K dwarfs in Hyades. Although almost the same abundances should in principle be derived for such cluster members irrespective of $T_{\rm eff}$, Yong et al. (2004) reported a general tendency of $A({\rm Fe\,I}) < A({\rm Fe\,II})$ together with a manifest rise (toward a lower $T_{\rm eff}$) of $A({\rm Fe\,II})$ at $T_{\rm eff} < 5{,}000$ K. A similar result was obtained also by Schuler et al. (2006b) as shown in Figure 32.5, which we have redrawn using their data in order to facilitate comparison with Figure 4 of Yong et al. (2004). This result strongly suggests that NLTE overionization prevails for Hyades K dwarfs of $T_{\rm eff} < 5{,}000$ K, which raises the Fe II population (while the concomitant reduction of the Fe I population is inconspicuous because Fe I is the dominant fraction in K dwarfs).

Meanwhile, we should clarify the cause of such an overionization in K dwarfs' atmospheres, since it cannot be explained within the framework of the classical model atmosphere, which predicts that UV radiation fields are essentially thermalized due to the enormously enhanced (atomic as well as molecular) line opacities. We suspect that the key may be the chromospheric activity. Schuler *et al.* (2006a, 2006b) found that the oxygen abundances of Hyades dwarfs derived from the infrared O I 7773 triplet also undergo a conspicuous rise with decreasing $T_{\rm eff}$ at $T_{\rm eff} < 5,000$ K (cf. Schuler *et al.* 2006b, Figure 10), just like the trend of Fe II abundances. That is, neither the O I 7773 triplet nor Fe II lines can yield reliable abundances (i.e. values are significantly overestimated) under the assumption of LTE in Hyades K dwarfs. Recalling here that the strength of the O I 7773 triplet is quite sensitive to the chromospheric temperature rise (e.g. Takeda 1995), we may speculate as a possibility that Hyades K dwarfs become progressively more active toward a lower $T_{\rm eff}$, and that the enhanced chromospheric UV radiation is the cause for the NLTE overionization.

If this is the case, such an appreciable NLTE effect in K stars might not simply apply to any stars in general; i.e. it may be comparatively younger stars that are mainly affected here, since the activity is closely related to the stellar age through a secular breaking of rotation due to loss of angular momentum. Actually, the Fe overionization for the case of M34 (age $\sim 2 \times 10^8$ yr), which is younger than Hyades (age $\sim 8 \times 10^8$ yr), appears to be noticeable already at $T_{\rm eff} \sim 5,500$ K (cf. Schuler *et al.* 2003, Figure 5), which is earlier than in the case of Hyades (at $T_{\rm eff} \sim 5,000$ K). We hence suspect that this kind of NLTE effect in Fe may be less significant for most of the disk stars in general (whose ages are typically from $\sim 10^9$ to $\sim 10^{10}$ yr). In this connection, it would be very interesting to study G–K dwarfs of old open clusters (such as M67, which is as old as the Sun).

4 Our analysis based on Okayama spectra

4.1 A validity check of LTE on Hyades G dwarfs

As mentioned above, it is almost certain that the assumption of LTE cannot be applied to K dwarfs with $T_{\rm eff}$ lower than $\sim 5,000$ K, at least for comparatively young stars. This, however, would not cause a very serious problem from the viewpoint of chemical-abundance studies, since spectra of K dwarfs are anyhow unsuitable for such a purpose because of their numerous mutually blended lines with strong damping wings. On the other hand, we should care about G-type stars more earnestly, since their spectra are commonly used in abundance determinations. In this respect, the results of Yong *et al.* (2004) and Schuler *et al.* (2006b) are not necessarily consistent with each other; i.e. the departure between $A({\rm Fe\,I})$ and $A({\rm Fe\,II})$ over the range of $\sim 5,000$–$6,000$ K is not appreciably $T_{\rm eff}$-dependent in the former whereas it becomes progressively larger in the latter, though the average difference amounts to up to ~ 0.2 dex for both cases.

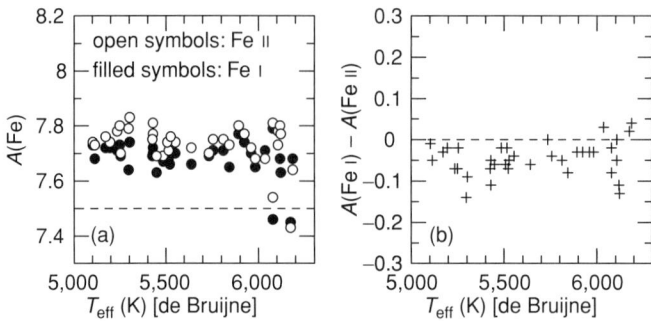

Figure 32.6. (a) Plots of Fe abundances versus $T_{\rm eff}$ for Hyades G dwarfs, which we obtained by using the model atmospheres of [de Bruijne] parameters. Abundances from neutral [A(Fe I)] and ionized [A(Fe II)] lines are indicated by filled and open circles, respectively. (b) Differences A(Fe I) $-$ A(Fe II) (shown in panel (a)) plotted against $T_{\rm eff}$.

In order to check on this situation, our Japanese group carried out a systematic study of atmospheric parameters and Fe abundances for 37 Hyades G-type dwarfs at $5{,}000\,{\rm K} < T_{\rm eff} < 6{,}000\,{\rm K}$, using the equivalent widths of Fe lines measured from the high-dispersion spectra ($R \sim 70{,}000$, signal-to-noise ratio $\sim$100–200, 6,000–7,200 Å) obtained with the 188-cm reflector and the HIDES spectrograph at Okayama Astrophysical Observatory. Two independent approaches were adopted, as described below.

First, the Fe abundances were determined separately from the lines of neutral and singly ionized species by using model atmospheres constructed with "actual" $T_{\rm eff}$ and $\log g$ derived by de Bruijne *et al.* (2001) that are based on Hipparcos parallaxes, colors, and evolutionary tracks. Then the behaviors of A(Fe I) and A(Fe II) were investigated in terms of their $T_{\rm eff}$ dependences or mutual discrepancies. Regarding the model atmospheres, we used Kurucz's (1993) ATLAS9 models with convective overshooting (our test calculations revealed that the differences caused by use of "non-overshooting" models were insignificant). The microturbulence was computed with the help of the empirical formula derived from the linear-regression analysis of Takeda *et al.*'s (2005) v_t results:

$$v_t\,({\rm km\,s^{-1}}) = 9.9 \times 10^{-4} T_{\rm eff}\,({\rm K}) - 0.41 \log g\,({\rm cm\,s^{-2}}) - 2.92 \quad (T_{\rm eff} < 5{,}800\,{\rm K})$$
$$= 5.6 \times 10^{-4} T_{\rm eff}\,({\rm K}) - 0.31 \log g\,({\rm cm\,s^{-2}}) - 0.79 \quad (T_{\rm eff} > 5{,}800\,{\rm K}).$$

We denote these adopted parameters and the derived abundances by using a notation/suffix of "[de Bruijne]." The results are shown in Figure 32.6, from which we can conclude the following.

– Qualitatively, there is surely a tendency of A(Fe I) $<$ A(Fe II) as reported by Yong *et al.* (2004) and Schuler *et al.* (2006b).

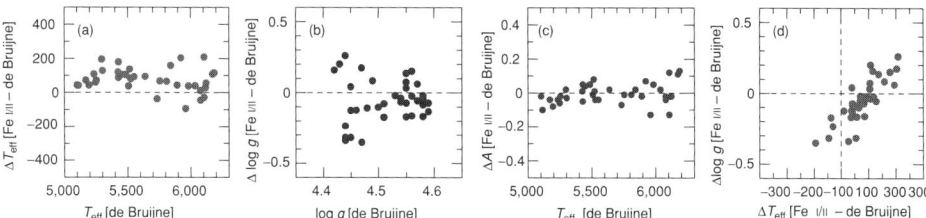

Figure 32.7. In (a)–(c) Δ values, defined as the "[Fe I/II] – [de Bruijne]" differences for $T_{\rm eff}$, log g, and A(Fe), are plotted against the [de Bruijne] parameters. (d) The mutual correlation between $\Delta T_{\rm eff}$ and $\Delta \log g$.

- However, the typical difference is only $<\sim 0.1$ dex and not very significant.
- Also, we cannot observe any clear trend of progressively increasing Fe I–Fe II discrepancy toward lower $T_{\rm eff}$ such as was derived by Schuler et al. (2006b) (Figure 32.5).

Next, all atmospheric parameters ($T_{\rm eff}$, log g, $v_{\rm t}$, [Fe/H]) were determined spectroscopically in the same way as was done by Takeda et al. (2005). We denote such derived parameters by using a notation/suffix of "[Fe I/II]." In this case, A(Fe I) = A(Fe II) naturally holds, since it is one of the requirements to be satisfied. From the resulting solutions, the "[Fe I/II] – [de Bruijne]" differences (Δ) of the parameters or of the Fe abundance (Fe abundances from neutral and ionized lines were averaged in the [de Bruijne] case) are plotted in Figure 32.7. The following trends between the "true" [de Bruijne] and "spectroscopic" [Fe I/II] parameters can be seen from this figure.

- $T_{\rm eff}$(spec) tends to be higher than $T_{\rm eff}$(true) by ~ 100 (± 100) K.
- There are no clear systematic differences between log g(spec) and log g(true), which scatter around zero with a dispersion of ~ 0.2 dex. Note that this result appears to disagree with what Meléndez & Ramírez (2005) recently reported; namely, they concluded that there is a systematic underestimation of log g(spec) by ~ 0.2 dex.
- However, there is a positive correlation between $\Delta T_{\rm eff}$ and $\Delta \log g$.
- The resulting Fe abundances for the two cases are consistent within $<\sim 0.1$ dex.

From what has been described above, we would tentatively conclude as follows in response to the question of whether or not the hypothesis of LTE is safely applicable to the determination of Fe abundance and spectroscopic parameters in G-type dwarfs. Though a sign of slight NLTE overionization for Fe is detectable, its effect on the spectroscopic determination of atmospheric parameters appears marginal; e.g. the difference between A(true) and A(spec) is $<\sim 0.1$ dex. *If one is content with this precision, why not invoke LTE?*

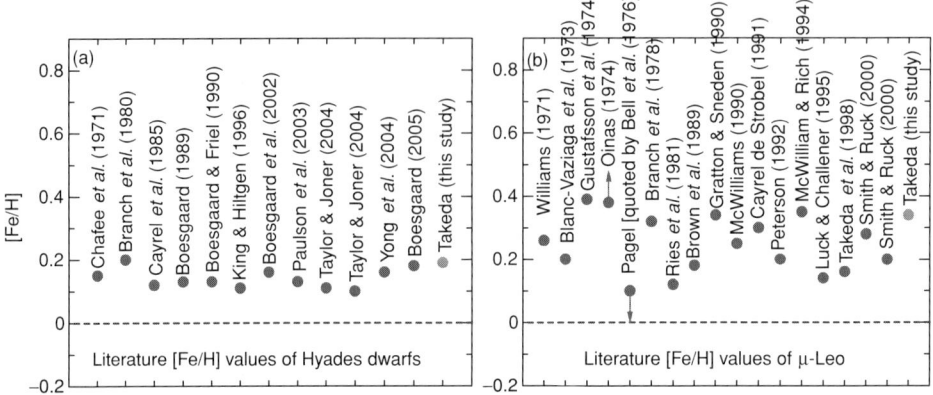

Figure 32.8. Published [Fe/H] values taken from the literature: (a) Hyades (dwarfs) and (b) μ Leo. (The original references are cited here.)

4.2 On the metallicity of the Hyades cluster

As a by-product of our study on Hyades G stars, it would be appropriate to touch briefly upon some topics that are closely related to the subject of metal-rich stars.

The metallicity ([Fe/H]) of the Hyades cluster is of profound importance, since it is involved with the definition of super-metal-rich (SMR) stars. While its values reported so far have a scatter of between +0.1 and +0.2 (Figure 32.8(a)), our result suggests A(Fe) $\sim$ 7.7 (Figure 32.6(a)) or [Fe/H] $\sim$ +0.2 (the solar Fe abundance is 7.5). In order to confirm this, we carried out a complete differential analysis relative to the Sun by applying the technique described in Takeda (2005) to 16 near-solar-type stars, which were selected from our sample of 37 Hyades G-type stars by application of the criterion that the difference of $T_{\rm eff}$ from the solar value (5,780 K) had to be within ±250 K. The results are displayed in Figure 32.9, from which we derived the average [Fe/H] to be +0.19 ($\sigma = 0.05$). Therefore, we would lend support to the "high" scale of [Fe/H] $\simeq$ +0.2 as the metallicity of Hyades cluster, which means that the widely adopted criterion of [Fe/H] > 0.2 is surely reasonable for the definition of SMR stars.

Incidentally, the fact that the resulting standard deviation of 0.05 is larger than the formally estimated errors may reflect the intrinsic metallicity dispersion within the Hyades-cluster stars. From this viewpoint, the apparent abundance dispersion of 0.05–0.06 dex for five heavy elements recently derived by De Silva *et al.* (2006) in their analysis of Hyades-cluster F–K stars (from which they concluded that there is little or no intrinsic scatter within the Hyades, though) might be real. Further follow-up studies should be awaited.

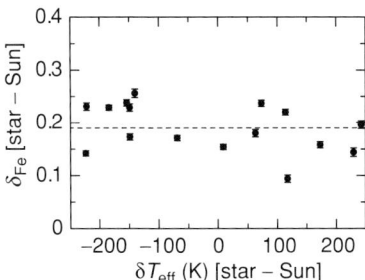

Figure 32.9. The differential metallicity (relative to the Sun) plotted as a function of the differential effective temperature ($\delta T_{\rm eff}$; relative to the Sun), on the basis of results of the complete differential analysis for 16 selected near-Solar-type Hyades G dwarfs.

4.3 Hyades giants and μ Leo

We have discussed only dwarfs so far, but there are four late-G or early-K giants in Hyades (γ Tau, δ^1 Tau, ε Tau, and θ^1 Tau); and Figure 32.5 suggests that the NLTE effect is much less significant for giants than it is for dwarfs. We tried to determine the atmospheric parameters of these stars spectroscopically by using Fe I and Fe II lines as was done for dwarfs, and found that the metallicities of these giants ([Fe/H] = +0.19, +0.15, +0.10, and +0.13, respectively) spread between +0.1 and +0.2. While these results compare more or less well to the published [Fe/H] values for dwarfs (Figure 32.8(a)), the accuracies of abundance/parameter determinations in this case appear to be comparatively low (e.g. the solutions are sensitive to the choice of lines).

Incidentally, in connection with the topic of giants, we also carried out a similar analysis for μ Leo (K2 III, a prototype SMR star concerning whose metallicity has been controversial regarding whether it is really SMR; see Figure 32.8(b), which is based on Table 3 of Taylor (2001)) and obtained [Fe/H] = +0.34. Therefore, we may safely conclude that the metallicity of μ Leo manifestly exceeds that of Hyades, and thus that it deserves the appellation SMR.

5 Evolution of surface-polluted stars and age determination

Finally, it would be worth briefly mentioning another important stellar parameter: the age. Very intriguing as well as puzzling is the origin of "old metal-rich stars" (e.g. Feltzing & González 2001), the existence of which apparently contradicts the galactic chemical evolution. While this might partly be attributed to an inhomogeneity of the gas from which stars formed, it is unlikely that all cases can be explained in this manner. Here, attention should be paid to two points that are actually related to each other, i.e. (i) the reality of the (spectroscopically derived) metallicity and (ii) the reliability of the age (determined mostly with the help of theoretical stellar evolutionary tracks).

Regarding the first point, we must recall that a great many planet-harboring stars are metal-rich and the reason for this fact has been a controversial matter throughout this decade (e.g. González 2003): one explanation is the hypothesis of so-called "surface enrichment" due to infall of H-depleted solid planetesimals; and the other is that of the "primordial origin" (i.e. formation of planets favors metal-rich environments). Although the latter interpretation appears to be more advantageous than the former on the basis of currently accumulated observational facts (e.g. Ecuvillon *et al.* 2006 and references therein), the possibility of surface-polluted enrichment may still be worthy of consideration. In this case, the metal-richness is confined entirely to the surface convection zone, while the metallicity of the stellar interior remains normal.

On the basis of a working hypothesis that the high metallicity of metal-rich stars is of enrichment origin and merely superficial, Y. Katsuta and M. Fujimoto at Hokkaido University investigated the evolution of such surface-polluted stars, in order to assess errors in the age determination expected when the "ordinary" tracks of homogeneous metallicity are inadequately used. The results of some representative cases (Y. Katsuta, private communication) are depicted in Figure 32.10(a); two consequences have tentatively been concluded.

- If the polluted track and the standard homogeneous track with the same surface (enriched) metallicity (e.g. $Z = 0.04$) are compared with each other, the former is situated on the bluer side relative to the latter; accordingly, the true age of a superficially metal-rich star may be older than that obtained from the standard tracks.
- Such a superficially metal-enriched star would be situated below (leftward of) the main sequence (like subdwarfs), so it would be impossible to determine its age by using the ordinary tracks/isochrones. Therefore, such "subdwarf-like metal-rich stars" (should they be found) may be promising candidates for surface-polluted stars.

Unfortunately, the former implication does not solve the problem of old metal-rich stars; i.e. the direction of the age correction is if anything the opposite. Also, in our check on the positions of the HR diagram for the SMR ([Fe/H] > 0.2) stars selected from Valenti & Fisher's (2005) extended sample we could not confidently nominate such clear subdwarf-like stars (Figure 32.10(b)). Accordingly, we can not state that such a surface-polluted model (at least in the present simple version) successfully applies for explaining the problems currently being confronted.[1] This,

---

[1] Yet, further improvement of this model is being carried out by Katsuta and Fujimoto while examining the effects of other parameters, such as the convection mixing length and the He abundance.

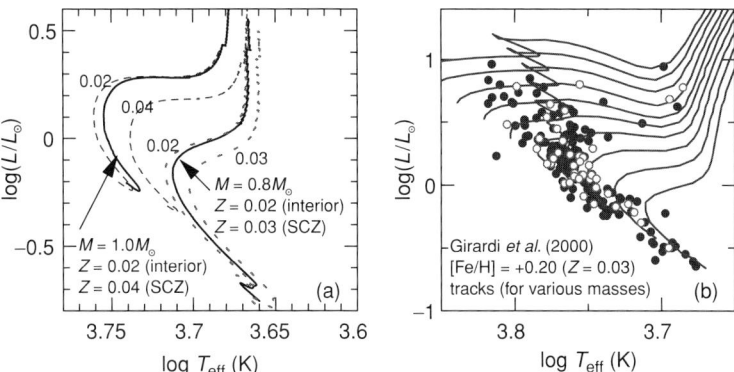

Figure 32.10. (a) Evolutionary tracks for homogeneous-metallicity (normal) stars and surface-polluted stars, calculated by Katsuta & Fujimoto (in preparation) for the representative $M = 0.8 M_\odot$ and $1.0 M_\odot$ cases. The numbers marked on each of the normal tracks, where the dotted lines are for $M = 0.8 M_\odot$ as the dashed lines are for $M = 1.0 M_\odot$, are the corresponding Z values (metallicity). The tracks of the surface-polluted stars are shown by the solid lines. (b) Super-metal-rich stars ([Fe/H] > 0.2) selected from Valenti & Fisher's (2005) sample plotted on the theoretical HR diagram, where Girardi et al.'s (2000) metal-rich ($Z = 0.03$) tracks for various masses from $0.8 M_\odot$ to $1.7 M_\odot$ with a step of $0.1 M_\odot$ are also shown. The open and filled symbols denote planet-host stars and non-planet-host stars, respectively.

in turn, suggests that the enrichment scenario may be less promising for the origin of high metallicity in planet-harboring stars.

6 Conclusion

(a) For F–G stars with T_{eff} higher than ∼5,000 K and K giants, the NLTE effect is not so significant in spite of a marginal sign of overionization and LTE would still remain a practically useful approximation.
(b) Meanwhile, use of LTE had better be avoided for K dwarfs with $T_{eff} < 5,000$ K because classical modeling is likely to break down (is this activity-related?).
(c) Regarding the metallicity of the Hyades cluster, our analysis of G dwarfs suggested a preference for the "high" scale of [Fe/H] ∼ 0.2.
(d) The age of a polluted (superficially) metal-rich star derived from standard tracks would be underestimated, and it seems difficult to explain the mystery of old metal-rich stars by invoking this kind of model (unless other parameters are tuned appropriately).
(e) A pollution-enriched star would be detected as a "subdwarf-like" metal-rich star, though such candidates appear to be rather few, if any exist at all.

References

Bodaghee, A., Santos, N. C., Israelian, G., & Mayor, M. (2003), *A&A* **404**, 715
de Bruijne, J. H. J., Hoogerwerf, R., & de Zeeuw, P. T. (2001), *A&A* **367**, 111

De Silva, G. M., Sneden, C., Paulson, D. B. *et al.* (2006), *AJ* **131**, 455
Ecuvillon, A., Israelian, G., Santos, N. C., Mayor, M., & Gilli, G. (2006), *A&A* **449**, 809
Feltzing, S., & González, G. (2001), *A&A* **367**, 253
Gilli, G., Israelian, G., Ecuvillon, A., Santos, N. C., & Mayor, M. (2006), *A&A* **449**, 723
Girardi, L., Bressan, A., Bertelli, G., & Chiosi, C. (2000), *A&AS* **141**, 371
González, G. (2003), *Rev. Modern Phys.* **75**, 101
Kurucz, R. L. (1993), Kurucz CD-ROM, No. 13 (Harvard-Smithsonian Center for Astrophysics)
Meléndez, J., & Ramírez, I. (2005), in *Cosmic Abundances as Records of Stellar Evolution and Nucleosynthesis in Honor of David L. Lambert*, ed. T. G. Barnes III & F. N. Bash (San Francisco, CA, Astronomical Society of the Pacific) pp. 343ff
Ruland, F., Biehl, D., Holweger, H., Griffin, R., & Griffin, R. (1980), *A&A* **92**, 70
Schuler, S. C., Hatzes, A. P., King, J. R., Kürster, M., & The, L.-S. (2006b), *AJ* **131**, 1057
Schuler, S. C., King, J. R., Fisher, D. A., Soderblom, D. R., & Jones, B. F. (2003), *AJ* **125**, 2085
Schuler, S. C., King, J. R., Terndrup, D. M., Pinsonneault, M. H., Murray, N., & Hobbs, L. M. (2006a), *ApJ* **636**, 432
Takeda, Y. (1995), *PASJ* **47**, 463
  (2005), *PASJ* **57**, 83
  (2007), *PASJ* **59**, 335
Takeda, Y., Ohkubo, M., & Sadakane, K. (2002), *PASJ* **54**, 451
Takeda, Y., Ohkubo, M., Sato, B., Kambe, E., & Sadakane, K. (2005), *PASJ* **57**, 27 [*Erratum in PASJ* **57**, 415]
Taylor, B. J. (2001), *A&A* **379**, 917
Thévenin, F., & Idiart, T. P. (1999), *ApJ* **521**, 753
Valenti, J. A., & Fisher, D. A. (2005), *ApJS* **159**, 141
Yong, D., Lambert, D. L., Allende Prieto, C., & Paulson, D. B. (2004), *ApJ* **603**, 697

33

Are WNL stars tracers of high metallicity?

G. Gräfener & W.-R. Hamann

Institut für Physik, Universität Potsdam, Am Neuen Palais 10, D-14469 Potsdam, Germany

We present new atmosphere models for Wolf–Rayet (WR) stars that include a self-consistent solution of the wind hydrodynamics. We demonstrate that the formation of optically thick WR winds can be explained by radiative driving on Fe-line opacities, implying a strong dependence on metallicity (Z). Our Z-dependent model calculations for late-type WN stars show that these objects are very massive stars close to the Eddington limit, and that their formation is strongly favored for high-metallicity environments.

1 Potsdam Wolf–Rayet hydrodynamic model atmospheres

The Potsdam Wolf–Rayet (PoWR) hydrodynamic model atmospheres combine fully line-blanketed NLTE models with the equations of hydrodynamics (Gräfener & Hamann 2005; Hamann & Gräfener 2003; Koesterke *et al.* 2002; Gräfener *et al.* 2002). The wind structure ($\rho(r)$ and $v(r)$) and the temperature structure $T(r)$ are computed in line with the full set of NLTE populations, and the radiation field in the co-moving frame. In contrast to all previous approaches, the radiative wind acceleration a_rad is obtained by direct integration,

$$a_\text{rad} = \frac{1}{c} \int \chi_v F_v \, dv, \tag{1.1}$$

instead of by making use of the Sobolev approximation. In this way, complex processes such as strong line overlap and the redistribution of radiation are automatically taken into account. Moreover, the models include small-scale wind clumping (throughout this work we assume a clumping factor of $D = 10$, for details see also Hamann & Koesterke (1998)). The models describe the conditions in WR

The Metal-rich Universe, eds. G. Israelian and G. Meynet. Published by Cambridge University Press.
© Cambridge University Press 2008.

atmospheres in a realistic manner, and provide synthetic spectra, i.e. they allow direct comparison with observations.

Utilizing these models, we recently obtained the first fully self-consistent WR wind model, for the case of an early-type WC star with strong lines (Gräfener & Hamann 2005). Moreover, we have examined the mass loss from late-type WN stars and its dependence on metallicity (Gräfener & Hamann 2006, 2007).

2 Spectral analyses of galactic Wolf–Rayet stars

A comprehensive study of Galactic WR stars using spectral analyses with line-blanketed PoWR models (Hamann *et al.* (2006) for WN stars; Barniske *et al.* (2006) for WC stars) revealed a bimodal WR-subtype distribution in the HRD, where the H-rich WNL stars are located to the right of the ZAMS with luminosities above $10^6 L_\odot$, whereas the (mostly H-free) early to intermediate WN subtypes, as well as the WC stars, have lower luminosities and hotter temperatures (see Figure 33.1).

This dichotomy already implies that the H-rich WNL stars are the descendants of very-massive stars, possibly still in the phase of central H-burning, whereas the earlier subtypes (including the WC stars) are more-evolved, less-massive, He-burning objects. Note, however, that distance estimates are available for only a small part of the WNL sample. Some of these objects thus might have lower luminosities and be the direct progenitors of the earlier subtypes.

3 Hydrodynamic atmosphere models for WNL stars

In a recent work we investigated the properties of the luminous, H-rich WNL stars with our hydrodynamic PoWR models (Gräfener & Hamann 2007). The most important conclusion from that work is that WR-type mass loss is primarily triggered by high L/M ratios or, equivalently, Eddington factors $\Gamma_e \equiv \chi_e L_\star/(4\pi c G M_\star)$ approaching unity. Note that high L/M ratios are expected for very-massive stars *and* for He-burning objects, giving a natural explanation for the occurrence of the WR phenomenon.

In Figure 33.2 we show the results from grid computations for WNL stars with a fixed luminosity of $10^{6.3} L_\odot$ and stellar temperatures $T_\star$ in the range 30–60 kK. For the stellar masses, values of $67 M_\odot$ and $55 M_\odot$ are adopted, corresponding to Eddington factors of $\Gamma_e = 0.55$ and 0.67. Notably, the mass loss strongly depends on Γ_e and $T_\star$. The obtained synthetic spectra nicely reflect the observed sequence of *weak-lined* WNL subtypes, starting with WN 6 at 55 kK and extending to WN 9 at 31 kK. From a more-detailed investigation of the WN 7 component in WR 22, an eclipsing WR + O binary system in Car OB1, we infer a stellar mass of $78 M_\odot$ ($\Gamma_e = 0.67$), in agreement with Rauw *et al.* (1996), who obtained $(72 \pm 3) M_\odot$ from

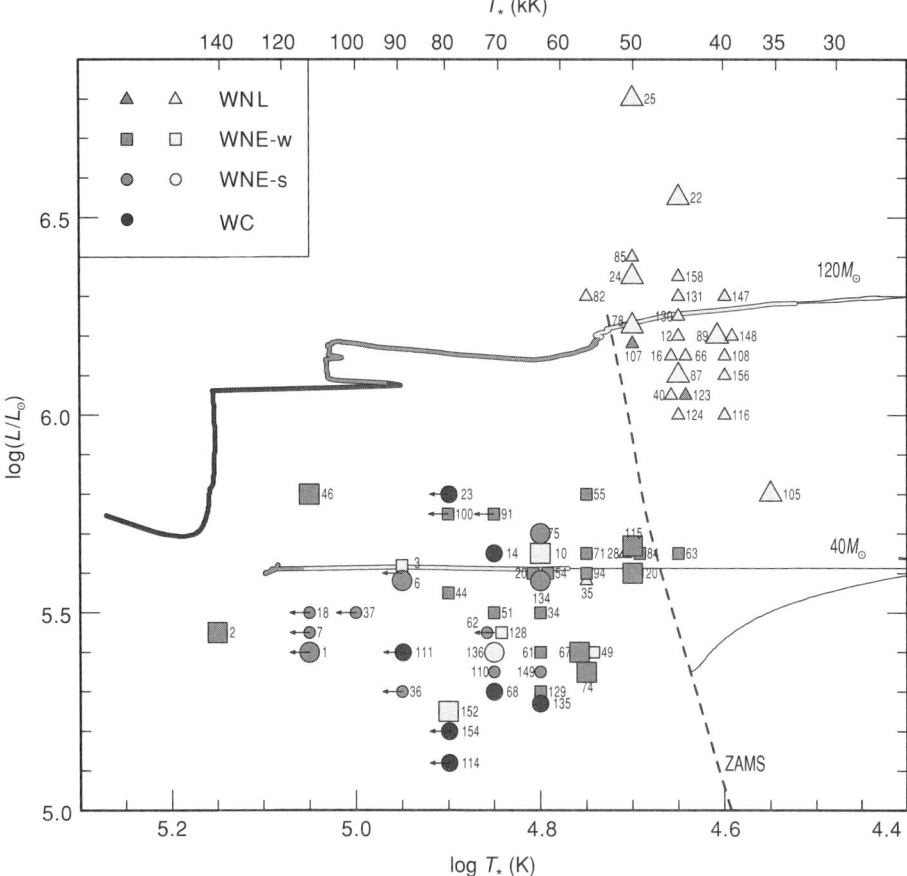

Figure 33.1. Recent spectral analyses of Galactic WR stars with line-blanketed models, according to Hamann et al. (2006) and Barniske et al. (2006): symbols in light gray denote H-rich WR stars, whereas H-free objects are indicated in dark gray, and WC stars are indicated in black; w, weak lines; and s, strong lines. For objects with large symbols distance estimates are available (van der Hucht 2001), whereas objects with small symbols were calibrated by their spectral subtype. Evolutionary tracks for non-rotating massive stars (Meynet & Maeder 2003) are shown for comparison.

the binary orbit. Such high stellar masses imply that the weak-lined WNL stars are still in the phase of central H-burning, suggesting an evolutionary sequence of the form O → WNL → LBV → WN → WC for very massive stars.

4 WNL stars at various metallicities

In Figure 33.3 we present results from a grid of models for luminous WNL stars at various metallicities (Gräfener & Hamann 2007). The models were computed for

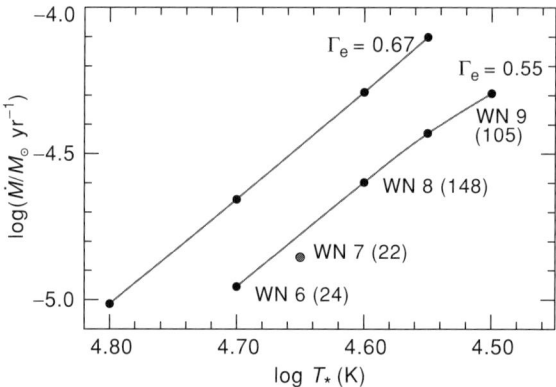

Figure 33.2. Wind models for Galactic WNL stars: mass-loss rates for various stellar temperatures T_* and Eddington factors Γ_e are shown. The corresponding spectral subtypes are indicated, together with WR numbers of specific Galactic objects according to van der Hucht (2001), in brackets, showing the good agreement with the synthetic line spectra. Note that the models were computed for a fixed stellar luminosity of $10^{6.3} L_\odot$. The WN 7 model (WR 22) is slightly offset from the standard grid models because it is calculated with an enhanced hydrogen abundance (see the text).

a fixed luminosity of $10^{6.3} L_\odot$, a stellar temperature of $T_* = 45$ kK, and a hydrogen surface mass fraction of $X_H = 0.4$. In addition to the metal abundances, we varied the Eddington factor Γ_e (or equivalently the stellar mass). Note that we scaled all metals with Z, assuming a CNO-processed (i.e. N-enriched) WN surface composition. The wind-driving in our models is chiefly due to radiation pressure on Fe-group line opacities. In accordance with Vink & Koter (2005) we thus find a strong dependence on Z. However, as in our previous computations, the proximity to the Eddington limit plays an equally important role. We find that optically thick winds with high WR-type mass-loss rates are formed over the whole range of metallicities, from $(1/1,000) Z_\odot$ to $3 Z_\odot$, if the stars are close enough to the Eddington limit. Only the limiting value of Γ_e at which the WR-type winds start to form changes. For solar Z, values of $\Gamma_e = 0.5$–0.6 lead to the formation of weak-lined WNL stars. As we have seen for the case of WR 22, this corresponds to very-massive, slightly over-luminous stars in a late phase of H-burning. Note that it is indeed observed that the most massive stars in very young galactic clusters are in the WNL phase (e.g. Figer *et al.* (2002)) and Najarro *et al.* (2004) for the Arches cluster; Drissen (1999) and Crowther & Dessart (1998) for NGC 3603).

For higher values of Z, the limit for the formation of WR-type winds shifts towards even lower values of Γ_e. For $3 Z_\odot$, we find that stars with $\Gamma_e \approx 0.4$ already have typical WNL mass-loss rates (see Figure 33.3). This corresponds to objects with mass $120 M_\odot$ on the ZAMS. We thus expect that metal-rich stars with very

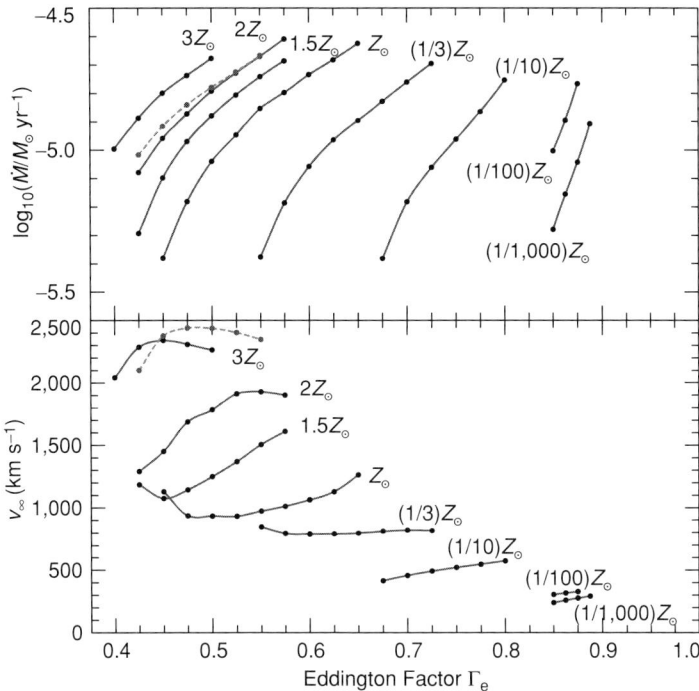

Figure 33.3. Mass loss for WNL stars over a broad range of metallicities (Z): mass-loss rates (top) and terminal wind velocities (bottom) obtained from our hydrodynamic models are plotted against the Eddington factor Γ_e. The solid curves indicate model series for WNL stars, where Γ_e is varied for a given value of Z. The dashed gray lines indicate models with $3Z_\odot$ and $X_H = 0.7$, corresponding to very-massive, metal-rich stars on the ZAMS.

high masses start their lives in the WNL phase, i.e. the occurrence of WNL stars in young massive clusters is strongly favored for high metallicities.

References

Barniske, A., Hamann, W.-R., & Gräfener, G. (2006), in *Stellar Evolution at Low Metallicity: Mass-Loss, Explosions, Cosmology*, ed. H. Lamers, N. Langer, & T. Nugis (San Francisco, CA, Astronomical Society of the Pacific), p. 243
Crowther, P. A. & Dessart, L. (1998), *MNRAS* **296**, 622
Drissen, L. (1999), in *Wolf–Rayet Phenomena in Massive Stars and Starburst Galaxies*, eds. K. A. van der Hucht, G. Koenigsberger, & P. R. J. Eenens (San Francisco, CA, Astronomical Society of the Pacific), p. 403
Figer, D. F., Najarro, F., Gilmore, D. *et al.* (2002), *ApJ* **581**, 258
Gräfener, G., & Hamann, W.-R. (2005), *A&A* **432**, 633
  (2006), in *Stellar Evolution at Low Metallicity: Mass-Loss, Explosions, Cosmology*, ed. H. Lamers, N. Langer, & T. Nugis (San Francisco, CA, Astronomical Society of the Pacific), p. 171
  (2007), *A&A*

Gräfener, G., Koesterke, L., & Hamann, W.-R. (2002), *A&A* **387**, 244
Hamann, W.-R., & Gräfener, G. (2003), *A&A* **410**, 993
Hamann, W.-R., Gräfener, G., & Liermann, A. (2006), *A&A* **457**, 1015
Hamann, W.-R., & Koesterke, L. (1998), *A&A* **335**, 1003
Koesterke, L., Hamann, W.-R., & Gräfener, G. (2002), *A&A* **384**, 562
Meynet, G., & Maeder, A. (2003), *A&A* **404**, 975
Najarro, F., Figer, D. F., Hillier, D. J., & Kudritzki, R. P. (2004), *ApJ* **611**, L105
Rauw, G., Vreux, J.-M., Gosset, E. *et al.* (1996), *A&A* **306**, 771
van der Hucht, K. A. (2001), *New Astronomy Review* **45**, 135
Vink, J. S. & de Koter, A. (2005), *A&A* **442**, 587

34

The observable metal-enrichment of radiation-driven-plus-wind-blown H II regions in the Wolf–Rayet stage

Gerhard Hensler,[1] Danica Kroeger[2] and Tim Freyer[2]

[1]Institut für Astronomie, Universität Wien, A-1180 Wien, Austria
[2]Institut für Theoretische Physik und Astrophysik, Universität Kiel, D-24098 Kiel, Germany

From stellar-evolution models and from observations of Wolf–Rayet stars it is known that massive stars are releasing metal-enriched gas during their Wolf–Rayet phase by means of strong stellar winds. Although H II-region spectra serve as diagnostics to determine the present-day chemical composition of the interstellar medium, it is not yet reliably known to what extent the diagnostic H II gas is already contaminated by chemically processed stellar-wind matter. In a recent paper, we therefore analyzed our models of radiation-driven and wind-blown H II bubbles around an isolated $85 M_\odot$ star of originally Solar metallicity with respect to its chemical abundances. Although the hot stellar-wind bubble (SWB) is enriched with ^{14}N during the WN phase and even more so with ^{12}C and ^{16}O during the WC phase of the star, we found that at the end of the stellar lifetime the mass ratios of the traced elements N and O in the warm ionized gas are insignificantly higher than Solar, whereas an enrichment of 22% above Solar is found for C. The transport of enriched elements from the hot SWB to the cool gas occurs mainly by means of mixing of hot gas with cooler at the back side of the SWB shell.

1 Introduction

H II regions are used as the most reliable targets from which to derive actual abundances in the interstellar medium (ISM). Kunth & Sargent (1986) discussed the problem of determining the heavy-element abundances of very-metal-poor blue compact dwarf galaxies from emission lines of H II regions in the light of local self-enrichment by massive stars but placed more stress on the effect of material ejected by Type-II supernovae (SNe II). However, already during the Wolf–Rayet (WR) phase of massive-stellar evolution the stellar wind peels off the outermost

The Metal-rich Universe, eds. G. Israelian and G. Meynet. Published by Cambridge University Press.
© Cambridge University Press 2008.

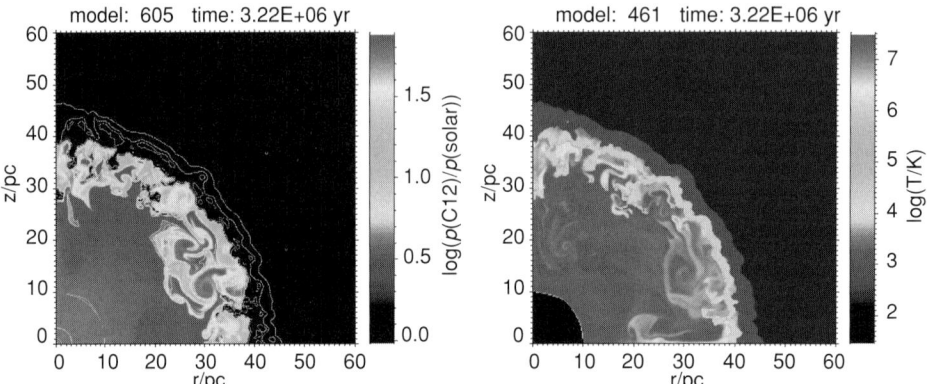

Figure 34.1. Left panel: the ^{12}C distribution within the stellar-wind bubble and the H II region for comparison with the temperature distribution (right panel). Both figures are snapshots at the end of the lifetime of a star of mass $85M_\odot$. All plots cover the whole computational domain of 60 pc × 60 pc.

stellar layers, so that elements from shell-burning regimes are released into the surrounding ISM already at later stages of their normal lifetimes. The stellar-wind energetics let one presume that this gas is deposited into the hot phase only, but it has not yet reliably been investigated in detail how and to what extent the complex structure of the stellar-wind bubble (SWB) could facilitate the cooling of wind material, thereby making observations of it as H II gas attainable.

That WR stars should play an important role in C enrichment of the ISM at Solar metallicity was advocated by Dray *et al.* (2003). Their models predict that the C enrichment by WR stars is at least comparable to that by AGB stars, while the enrichment by N is dominated by AGB stars and the O enrichment is dominated by SNe II. In their investigation, however, they summed over all gas phases and did not evaluate the abundances in specific gas phases with detailed diagnostics, such as in the warm H II gas.

In a series of models of radiation-driven and wind-blown bubbles produced by massive stars we investigated the effects of structuring and energizing the surrounding ISM for stars of masses $15M_\odot$ (Kroeger *et al.* 2007), $35M_\odot$ (Freyer *et al.* 2006), $60M_\odot$ (Freyer *et al.* 2003), and $85M_\odot$ (Kroeger *et al.* 2006b). From these, we could conclude that differently strong but significant structures are formed by the dynamical and radiative processes involving both the SWB and the enveloping H II region (Freyer *et al.* 2003) where hot gas mixes with warm gas and cools further to a "warm" phase. The mixing occurs mainly at the back of the SWB shell with photo-evaporated material and through turbulence in an interface between the SWB and the shell (see e.g. Figure 34.1).

Nevertheless, the WR stage is metal-dependent in the sense that, first, the lower the metallicity the more massive a star has to be in order for it to evolve through the WR stages and, secondly, the lower the metallicity the shorter the WR lifetimes, with the consequence that not all WR stages are reached. The first point means that the number of WR stars decreases with decreasing metallicity. Schaller et al. (1992) found that for a metallicity of $Z = 0.001$ the minimal zero-age main-sequence (ZAMS) mass for a WR star is $>80 M_\odot$, while at Solar $Z = 0.02$ it is $>25 M_\odot$ as discussed by Chiosi & Maeder (1986).

2 The model

The hydrodynamical equations are solved together with the transfer of H-ionizing photons on a two-dimensional cylindrical grid. To provide a refined resolution mainly in the central part around the star, the grid is structured using a nested scheme. The time-dependent ionization and recombination of hydrogen are calculated in each time step and we carefully take stock of all the important energy-exchange processes in the system. A detailed description of the numerical method and further references are given in Freyer et al. (2003).

As initial condition an undisturbed homogeneous background gas with Solar abundances (Anders & Grevesse 1989), hydrogen number density $n_0 = 20$ cm^{-3}, and temperature $T_0 = 200$ K was applied for the reasons described in Freyer et al. (2003). The models were then started with the sudden turn-on of the ZAMS stellar radiation field and stellar wind. Since the gas is assumed to be devoid of molecular material the radiation field commences immediately to ionize the environment outwards of the SWB without an enveloping photo-dissociation region. From the series of models of radiation- and wind-driven H II regions around single massive stars mentioned above, for our purpose the $85 M_\odot$ star (Kroeger et al. 2006b) looks the most appropriate with respect to its self-enrichment. The time-dependent parameters of this star with "standard" mass-loss rate and Solar metallicity ($Z = 0.02$) during its H-burning main-sequence stage and its subsequent evolution are taken from Schaller et al. (1992). The model analysis has already been published (Kroeger et al. 2006a).

The investigation starts not before the onset of the WR stage, but with the onset of the WN stage at an age of $t = 2.83$ Myr. The WR star enriches the combined SWB-H II region with ^{12}C, ^{14}N, and ^{16}O. During its WN phase the star releases $0.143 M_\odot$ of ^{14}N, which is more than half of its total release, but hardly any extra ^{12}C or ^{16}O is supplied. As the condition for observability within the H II region only the "warm" gas $(6.0 \times 10^3 \text{ K} \leq T < 5.0 \times 10^4 \text{ K})$ is accounted for. The mass fractions of ^{12}C, ^{14}N, and ^{16}O with respect to Solar are set according to Anders &

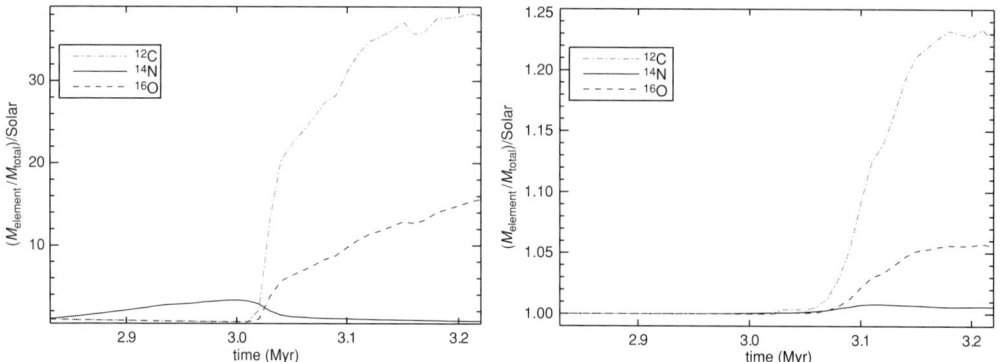

Figure 34.2. Time-dependent abundances of ^{12}C, ^{14}N, and ^{16}O in the hot (left panel) and in the warm gas phase (right panel). The plot does not start before the onset of the WN phase at 2.83 Myr.

Grevesse (1989) to 4.466×10^{-3}, 1.397×10^{-3}, and 1.061×10^{-2}, respectively, normalized with respect to H.

3 Results

At the end of its lifetime at $t = 3.22$ Myr the $85 M_\odot$ star has supplied $0.28 M_\odot$ of ^{14}N, $13.76 M_\odot$ of ^{12}C, and $11.12 M_\odot$ of ^{16}O, which are contained in the combined SWB–H II region. Since N was released initially in the WN stage its abundance increases slightly during this period and is thereafter diluted by the nitrogen-poor gas feed. These facts are discernible in the left-hand panel of Figure 34.2 by noting the rise and subsequent decrease of the N abundance during the hot phase after the transition to the WC stage, when C and O are released. The ^{12}C content increases steeply and reaches an overabundance of 38 times Solar while the enrichment with ^{16}O is weaker (Figure 34.2, left panel).

In Figure 34.1 C is revealed as the element making the largest contribution in comparison with the temperature distribution at the end of the stellar life. Two facts can be easily discerned: (1) the carbon enrichment is, reasonably, largest within the hot SWB; and (2) also regions with the "warm" temperature range are significantly enriched in ^{12}C.

While the mixing and incorporation of ^{14}N from the hot phase into the warm phase becomes only slightly detectable after about 3.1 Mry, but occurs with a time delay after its release of almost 0.2 Myr, the ^{12}C enrichment in the warm phase is clearly perceptible as soon as less than 50,000 yr after the steep rise in ^{12}C abundance of the hot SWB.

At the end of the $85 M_\odot$ star's life the element quantities measurable in emission spectra of the warm H II-region gas amount to 1.22 times Solar for ^{12}C, to less than 1.01 times Solar for ^{14}N, and only 1.05 times Solar for ^{16}O. From this model we conclude that the enrichment of the circumstellar environment with ^{14}N and ^{16}O by WR stars is negligible, if the $85 M_\odot$ star is representative of massive stars passing through the WR stage. Only for ^{12}C is the enrichment of the H II region significant. For a giant H II region containing a full set of massive stars according to a normal initial mass function, with various lifetimes and wind mass-loss rates, the enrichment effect of C should, however, be smaller than that modeled here for a single most-massive star. A comprehensive description and discussion of the model has already been published (Kroeger *et al.* 2006a).

Since the occurrence of a WR phase is strongly metal-dependent, the enrichment with C should also depend on the average metallicity. This would mean that any radial gradient of C abundance of H II regions in galactic disks is steeper than that of O. Indeed, Esteban *et al.* (2005) found $d[\log(C/O)]/dr = -0.058 \pm 0.018$ dex kpc^{-1} for the Galactic disk.

In metal-poor galaxies one would expect less chemical self-enrichment because the stellar mass range of the occurrence of WR stars is diminished and shifted towards higher masses and the WR phases are shorter.

References

Anders, E., & Grevesse, N. (1989), *Geochim. Cosmochim. Acta* **53**, 197
Chiosi, C., & Maeder, A. (1986), *ARA&A* **24**, 329
Dray, L. M., Tout, C. A., Karakas, A. I., & Lattanzio, J. C. (2003), *MNRAS* **338**, 973
Esteban, C., García-Rojas, J., Peimbert, M. *et al.* (2005), *ApJ* **618**, L95
Freyer, T., Hensler, G., & Yorke, H. W. (2003), *ApJ* **594**, 888
   (2006), *ApJ* **638**, 262
Kroeger D., Hensler, G., & Freyer, T. (2006a), *A&A* **450**, L5
Kroeger D., Freyer T., Hensler, G., & Yorke, H. W. (2006b), *A&A* submitted
Kroeger D., Freyer T., Hensler, G., & Yorke, H. W. (2007), *A&A* in preparation
Kunth, D., & Sargent, W. L. W. (1986), *ApJ* **300**, 496
Schaller, G., Schaerer, D., Meynet, G., & Maeder, A. (1992), *A&AS* **96**, 269

35

Metal-rich A-type supergiants in M31

Norbert Przybilla,[1] Keith Butler[2] & Rolf-Peter Kudritzki[3]

[1]*Dr. Karl Remeis-Sternwarte Bamberg, Sternwartstrasse 7, D-96049 Bamberg, Germany*
[2]*Universitäts-Sternwarte München, Scheinerstrasse 1, D-81679 München, Germany*
[3]*Institute for Astronomy, 2680 Woodlawn Drive, Honolulu, HI 96822, USA*

We discuss results of an exploratory NLTE analysis of two metal-rich A-type supergiants in M31. Using comprehensive model atoms we derive accurate atmospheric parameters from multiple indicators and show that NLTE effects on the abundance determination can be substantial (altering results by a factor of 2–3). The NLTE analysis removes systematic trends apparent in the LTE approach and reduces statistical uncertainties. Characteristic abundance patterns of the light elements provide empirical constraints on the evolution of metal-rich massive stars.

1 Introduction

Absorption by interstellar dust in the Galactic plane prohibits the study of the Milky Way in its entirety. A comprehensive global view of a giant spiral galaxy can be obtained only for the nearest neighbour of this class, the Andromeda galaxy (M31). The old stellar population of this, the most luminous galaxy of the Local Group, turns out to be unexpectedly metal-rich (e.g. van den Bergh 1999) and present-day abundances as traced by nebulae (H II regions, supernova remnants) also indicate a metal-rich character and the presence of abundance gradients in the disc (Dennefeld & Kunth 1981; Blair *et al.* 1982).

The current generation of large telescopes allows spectroscopy of luminous stars in M31 to be carried out. Quantitative studies of massive blue supergiants (BSGs) can help to verify results from nebulae. In particular, analyses of bright BA-type supergiants (BA-SGs) facilitate the extension of abundance determinations beyond the light and α-process elements to iron-group and s-process species. This makes BA-SGs highly valuable for constraining the galactochemical evolution of

The Metal-rich Universe, eds. G. Israelian and G. Meynet. Published by Cambridge University Press.
© Cambridge University Press 2008.

M31 empirically by tracing the abundance gradients. However, there is more to gain: observational constraints on the evolution of massive stars in a metal-rich environment, using BSGs as probes for mixing with nuclear-processed matter; and the potential to employ them as distance indicators via application of the flux-weighted gravity–luminosity relationship (FGLR) (Kudritzki et al. 2003).

Only a few studies of individual BSGs in M31 have been published so far (Venn et al. 2000; Smartt et al. 2001; Trundle et al. 2002). These are based either on the assumption of LTE or on the use of unblanketed NLTE model atmospheres. Here, we present results of an exploratory study of two A-SGs in M31. We used a hybrid NLTE analysis technique that considers line blanketing and NLTE line formation and is thus able to provide results of hitherto unachieved accuracy and consistency (Przybilla 2002; Przybilla et al. 2006).

2 Observations and model computations

Spectra of two luminous A-type supergiants in the northeastern arm of M31 were taken in November 2002 with ESI on the Keck II telescope. The Echelle spectra were reduced with the MAKEE package and our own IDL-based routines for order merging and continuum normalisation. Complete wavelength coverage of the spectra between ∼3900 and ∼9300 Å was obtained at high signal-to-noise ratio (>150 in the visual) and moderately high resolution ($R \simeq 8000$). The data set is complemented by Keck I HIRES spectra with a more limited wavelength coverage ($R \simeq$ 35 000, signal-to-noise ratio ∼80) as utilised by Venn et al. (2000).

The model calculations were carried out using a hybrid NLTE approach as discussed in detail by Przybilla et al. (2006). In brief, hydrostatic, plane-parallel and line-blanketed LTE model atmospheres are computed with ATLAS9 (Kurucz 1993) with further modifications (Przybilla et al. 2001). Then, NLTE line formation is performed on the resulting model stratifications. The coupled radiative-transfer and statistical-equilibrium equations are solved and spectrum synthesis with refined line-broadening theories is performed using DETAIL and SURFACE. State-of-the-art NLTE model atoms relying on data from *ab initio* computations, avoiding rough approximations wherever possible, are utilised for the stellar-parameter and abundance determinations.

3 Stellar parameters and elemental abundances

The atmospheric parameters are derived spectroscopically from multiple indicators, following the methodology described by Przybilla et al. (2006). The effective temperature T_{eff} and surface gravity log g are constrained by several NLTE ionization equilibria (C I/II, N I/II, Mg I/II) and from modelling of the Stark-broadened

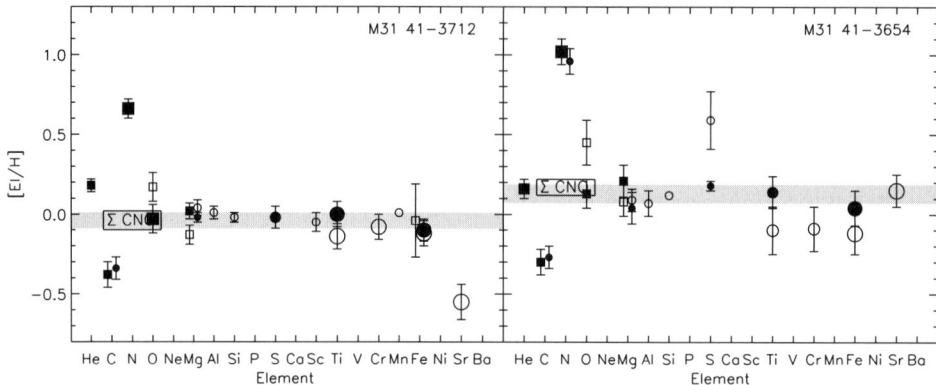

Figure 35.1. Preliminary results from the elemental-abundance analysis for our two sample stars, relative to the Solar composition (on a logarithmic scale) (Grevesse & Sauval 1998). Filled symbols denote NLTE, open symbols LTE results. The symbol size codes the number of spectral lines analysed – small, 1–5; medium, 6–10; large, more than 10. Boxes are for neutral and circles for singly ionized species. The error bars represent 1σ uncertainties from the line-to-line scatter. The grey shaded areas mark the deduced metallicities of the objects within 1σ errors. The NLTE abundance analyses imply a scaled Solar abundance distribution for the M31 objects. An exception is constituted by the light elements which have been affected by mixing with nuclear-processed matter.

profiles of the higher Balmer and Paschen lines. The internal accuracy of the method allows the 1σ uncertainties to be reduced to $\sim 1\%$–2% in $T_{\rm eff}$ and to 0.05–0.10 dex in log g. Several He I lines are used to derive the helium abundance. The stellar metallicity relative to the Solar standard [M/H] (logarithmic values) is determined from the heavier metals (O, Mg, S, Ti, Fe) in NLTE. Microturbulent velocities ξ are obtained in the usual way by requiring abundances to be independent of line equivalent width – consistency is achieved for all NLTE species. Finally, macro-turbulences ζ and rotational velocities $v \sin i$ are determined from line-profile fits. The results are summarised in Table 35.1, where information on the fundamental stellar parameters luminosity L, evolutionary and spectroscopic mass $M_{\rm e}/M_{\rm s}$ and radius R is also given. The photometric data of Massey *et al.* (2006) were adopted.

Elemental abundances were determined for several chemical species, with many of the astrophysically most interesting in NLTE and the remainder in LTE; see Figure 35.1. The two M31 supergiants are more metal-rich than the Galactic BA-SGs studied using the same method (Przybilla *et al.* 2006; Firnstein & Przybilla 2006; Schiller & Przybilla 2006), by up to ~ 0.2 dex. One object is found to have supersolar metallicity. The abundance distribution of the heavier elements follows a scaled Solar pattern, whereas the abundances of light elements have been affected by mixing with CN-cycled matter.

Table 35.1. *Atmospheric and fundamental stellar parameters with uncertainties*

Object	T_{eff} (K)	$\log g$	He	[M/H]	$\xi/\zeta/v \sin i$ (km s^{-1})	$\log(L/L_\odot)$	$M_e/M_\odot$	$M_s/M_\odot$	$R/R_\odot$
41-3654 (A2 Iae)	9200 150	1.00 0.05	0.13 0.02	+0.13 0.06	8/20/36 1/5/5	5.63 0.04	29 4	24 4	257 15
41-3712 (A3 Iae)	8550 150	1.00 0.05	0.13 0.02	−0.04 0.05	8/18/25 1/5/5	5.45 0.04	24 3	22 4	243 14

4 Results and discussion

The NLTE computations reduce random errors and remove systematic trends from the analysis. Inappropriate LTE analyses tend to underestimate iron-group abundances systematically and overestimate the light- and α-process-element abundances by factors of up to 2–3 (this is most notable for M31-41-3654, while M31-41-3712 is less affected). This is because of the different responses of these species to radiative and collisional processes in the microscopic picture, which is explained by fundamental differences of their detailed atomic structure. This is not taken into account in LTE. Contrary to common assumptions, significant NLTE abundance corrections of ~ 0.3 dex can be found even for the weakest lines ($W_\lambda \sim 10$ mÅ). Non-LTE abundance uncertainties amount to typically 0.05–0.10 dex (random) and ~ 0.10 dex (systematic 1σ errors). Note that line-blocking effects increase with metallicity, such that photon mean free paths are reduced in metal-rich environments and the NLTE effects correspondingly. This is the reason why NLTE effects in these objects close to the Eddington limit are similar to those in less-extreme Galactic BA-type supergiants with – on average – lower metallicity.

Fundamental stellar parameters and light-element abundances allow us to discuss the two M31 objects in the context of stellar evolution. The comparison with evolutionary tracks for rotating massive stars is made in Figure 35.2, which also summarises results from a Galactic sample of BA-SGs. The M31 supergiants extend the sample towards higher luminosities/stellar masses than possible in the Galactic study and towards higher metallicity. Both objects appear to cross the Hertzsprung–Russell diagram towards the red supergiant stage for the first time, because of the absence of the extremely high helium abundances and N/C ratios expected for stars entering the Wolf–Rayet phase. The predicted trend of increasing chemical mixing (strong N and moderate He enrichment, C depletion and almost constant O as a result of the action of the CNO-cycle and transport to the stellar surface because of meridional circulation and dynamical instabilities) with increasing stellar mass is qualitatively recovered. Note a group of highly processed stars at $M_0 < 15 M_\odot$, which suggests an extension of blue loops towards higher temperatures than predicted. The observed N/C ratios are generally higher than indicated by theory. Stellar-evolution computations accounting for the interplay of rotation and magnetic fields (e.g. Maeder & Meynet 2005) may resolve this discrepancy since they predict a much higher efficiency for chemical mixing. Also the recent revision of the cross-section for the bottleneck reaction $^{14}N(p,\gamma)^{15}O$ in the CN-branch of the CNO-cycle by almost a factor of 2 (Lemut *et al.* 2006) will be of importance.

Finally, improvements in stellar parameters allow re-evaluation of the two M31 supergiants in the empirical calibration of the FGLR (Kudritzki *et al.* 2003). Two factors play a role in this context: a revision of the previously used photometric

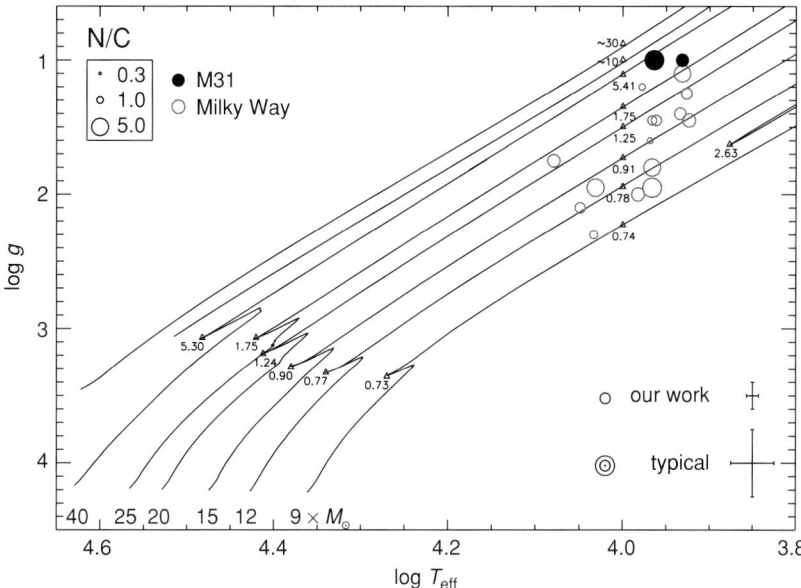

Figure 35.2. Observational constraints on massive-star evolution: N/C ratios as tracers of mixing with nuclear-processed material. Evolution tracks for rotating stars ($v_{\mathrm{ini}} = 300$ km^{-1} s, with marks indicating N/C ratios from the models of Meynet & Maeder (2003)) at Solar metallicity are displayed. An initial N/C $\sim$ 0.3 was adopted in their model computations. Observed N/C ratios (typical values are indicated in the box) for a sample of Galactic BA-SGs (Przybilla *et al.* 2006; Firnstein & Przybilla 2006; Schiller & Przybilla 2006) and for the two (slightly) metal-rich M31 objects of the present work are indicated, all analysed in a homogeneous way. Error bars characteristic for our work and for similar studies from the literature are given.

data of Magnier *et al.* (1992), which have been shown to suffer from systematic uncertainties (Massey *et al.* 2006); and the extended wavelength coverage (and high signal-to-noise ratio) of the ESI spectra, which allows the atmospheric parameters to be constrained more precisely. As a consequence, some of the largest deviations from the empirical relation can be explained.

References

Blair, W. P., Kirshner, R. P., & Chevalier, R. A. (1982), *ApJ* **254**, 50–69
Dennefeld, M., & Kunth, D. (1981), *AJ* **86**, 989–997
Firnstein, M., & Przybilla, N. (2006), *Proceedings of Science*, PoS(NIC-IX)095
Grevesse, N., & Sauval, A. J. (1998), *Space Sci. Rev.* **85**, 161–174
Kudritzki, R. P., Bresolin, F., & Przybilla, N. (2003), *ApJ* **582**, L83–L86
Kurucz, R. L. (1993), Kurucz CD-ROM No. 13 (Cambridge, MA: SAO)
Lemut, A., Bemmerer, D., Confortola, F. *et al.* (2006), *Physics Letters B* **634**, 483–487
Maeder, A., & Meynet, G. (2005), *A&A* **440**, 1041–1049
Meynet, G., & Maeder, A. (2003), *A&A* **404**, 975–990

Magnier, E. A., Lewin, W. H. G., van Paradijs, J. *et al.* (1992), *A&AS* **96**, 379–388
Massey, P., Olsen, K. A. G., Hodge, P. W. *et al.* (2006), *AJ* **131**, 2478–2496
Przybilla, N. (2002), Ph.D. Thesis, Universität München.
Przybilla, N., Butler, K., & Kudritzki, R. P. (2001), *A&A* **379**, 936–954
Przybilla, N., Butler, K., Becker, S. R., & Kudritzki, R. P. (2006), *A&A* **445**, 1099–1126
Schiller, F., & Przybilla, N. (2006), *Proceedings of Science*, PoS(NIC-IX)174
Smartt, S. J., Crowther, P. A., Dufton, P. L. *et al.* (2001), *MNRAS* **325**, 257–272
Trundle, C., Dufton, P. L., Lennon, D. J. *et al.* (2002), *A&A* **395**, 519–533
van den Bergh, S. (1999), *A&AR* **9**, 273–318
Venn, K. A., McCarthy, J. K., Lennon, D. J. *et al.* (2000), *ApJ* **541**, 610–623

Part VI

Formation and evolution of metal-rich stars and stellar yields

36

Massive-star evolution at high metallicity

Georges Meynet,[1] Nami Mowlavi[2] & André Maeder[1]

[1]*Observatoire de Genève, Université de Genève, CH-1290 Sauverny, Switzerland*
[2]*ISDC, Observatoire de Genève, Université de Genève, Chemin d'Ecogia 16, CH-1290 Versoix, Switzerland*

After a review of the many effects of metallicity on the evolution of rotating and non-rotating stars, we discuss the consequences of a high metallicity for massive-star populations and stellar nucleosynthesis. The most striking effect of high metallicity is to enhance the amount of mass lost by stellar winds. Typically, at a metallicity of $Z = 0.001$ only 9% of the total mass returned by non-rotating massive stars is ejected by winds (91% by supernova explosions), whereas at Solar metallicity this fraction may amount to more than 40%. High metallicity favors the formation of Wolf–Rayet stars and Type-Ib supernovae, but militates against the occurrence of Type-Ic supernovae. We estimate empirical yields of carbon on the basis of the observed population of WC stars in the Solar neighborhood, and obtain that WC stars eject 0.2%–0.4% of the mass initially locked into stars in the form of newly synthesized carbon. Models give values well in agreement with these empirical yields. Chemical-evolution models indicate that such carbon yields may have an important impact on the abundance of carbon at high metallicity.

1 General effects of metallicity on the evolution of stars

The metallicity affects the evolution of stars mainly through its impact on the radiative opacities, the equation of state, the nuclear-reaction rates, and the stellar mass-loss rates (Maeder 2002). Metallicity also affects the way in which the various instabilities induced by rotation occur in stars.

Recent calculations ordered by increasing metallicity are briefly presented in Table 36.1. The first column gives the reference; the second to fourth columns indicate respectively the range of initial masses, the values of the initial mass

The Metal-rich Universe, eds. G. Israelian and G. Meynet. Published by Cambridge University Press.
© Cambridge University Press 2008.

Table 36.1. *Grids of high-metallicity massive-star models*

Reference	Initial masses ($M_\odot$)	Y	Z
Claret (1997)	1–40	0.42–0.32–0.22	0.03
Schaerer et al. (1993)	0.8–120	0.340	0.04
Meynet et al. (1994)	12–120	0.320	0.04
Meynet & Maeder (2005)[a]	20–120	0.310	0.04
Eldridge & Vink (2006)[b]	25–120	0.340	0.04
Fagotto et al. (1994a)	0.6–120	0.352	0.05
Fagotto et al. (1994b)	0.6–9	0.475	0.10
Mowlavi et al. (1998a)	0.8–60	0.480	0.10

[a] Models including the effects of rotation.
[b] Models with Z-dependent mass-loss rates during the Wolf–Rayet phase.

fraction of helium, and the value of Z considered in the various grids of stellar models.

In this paper we first review the various effects of metallicity on the evolution of stars, and then discuss the consequences of high metallicity for massive-star populations and their associated nucleosynthesis.

1.1 The $\Delta Y/\Delta Z$ ratio

Let us recall how a change of Z, the mass fraction of heavy elements, affects the initial mass fractions of hydrogen, X, and of helium, Y. In the course of galactic evolution, the interstellar matter is progressively enriched in heavy elements and in helium. The increase of the abundance in helium with respect to Z is conveniently described by a $\Delta Y/\Delta Z$ law such that $Y = Y_0 + (\Delta Y/\Delta Z)Z$ (where Y_0 is the primordial He abundance). At low metallicities ($Z \leq 0.02$), Y remains practically constant, and the stellar properties *as a function of metallicity* are thus expected to be determined mainly by Z. At high metallicities ($Z \geq 0.02$), on the other hand, Y increases (or, alternatively, X decreases) significantly with Z. Under those conditions, both Z and Y (and $X = 1 - Y - Z$) determine the stellar properties as a function of metallicity.

1.2 The κ-effect

In general, the opacities increase with increasing metallicity. This is the case for the bound–free (see Equation (16.107) in Cox & Giuli (1969)) and free–free (see Equation (16.95) in Cox & Giuli (1969)) transitions. Using the well-known mass–luminosity relation $L \propto \sim \mu^4 M^3/\kappa$, where μ is the mean molecular mass, M the mass of the star, and κ the opacity, one immediately deduces that the increase of the opacity produces a decrease of the luminosity. This can be seen in the left-hand part

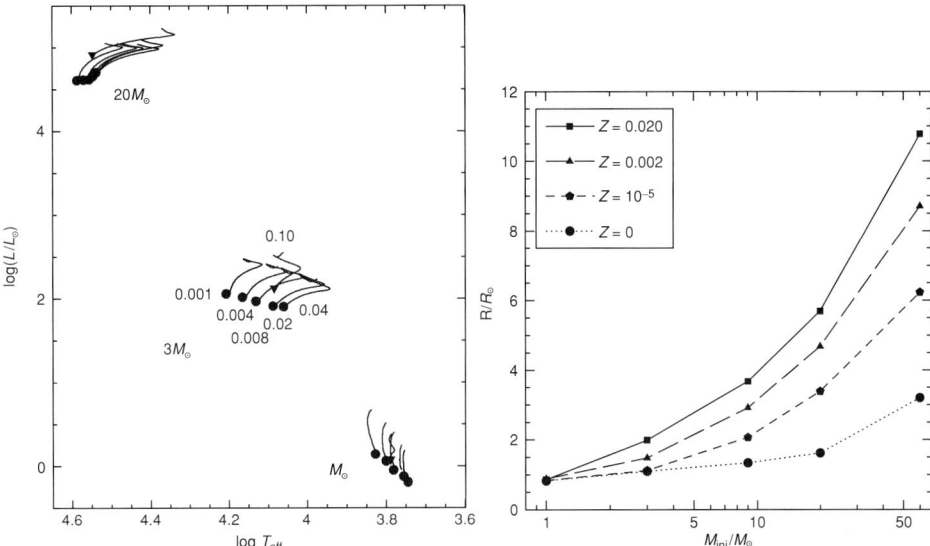

Figure 36.1. Left panel: zero-age main-sequence (ZAMS) locations in the HR diagram of $M_\odot$, $3M_\odot$ and $20M_\odot$ star models for various metallicities. Models for $Z = 0.1$ are identified by triangles. Figure taken from Mowlavi *et al.* (1998b). Right panel: variations of the radius as a function of initial mass, for various metallicities on the ZAMS.

of Figure 36.1 for the $3M_\odot$ and $M_\odot$ stellar models and for metallicities inferior or equal to 0.04. At still higher metallicities the effect of μ becomes more important than the opacity effect and the luminosity increases (see below).

In contrast, for massive stars free electron scattering is the main opacity source. This opacity depends only on X, with $\kappa_e \simeq 0.20(1 + X)$, which is about constant at $Z \leq 0.01$ and decreases with Z at $Z \geq 0.01$. Thus at high Z the luminosity of the $20M_\odot$ model shown in the left-hand part of Figure 36.1 first remains nearly constant and then increases with the metallicity.

In the right-hand part of Figure 36.1 the variation of the radii on the zero-age main sequence (MS) for stars of various initial masses and metallicities is shown. We see that the radii increase with the metallicity. From the relation $L = 4\pi R^2 \sigma T_{\rm eff}^4$ and from the variation of L with Z one can deduce that, when the metallicity increases, the effective temperature decreases. For metallicities superior to about 0.04, the effective temperature and the luminosity both increase due to the μ-effect (see below).

1.3 The μ-effect

When the metallicity increases, μ increases. This tends to make the star more luminous according to the mass–luminosity relation, provided that the opacity does not increase too much. At very high metallicity ($Z > \sim 0.04$), this is what happens for the $M_\odot$ and $3M_\odot$ models. For the $20M_\odot$-stellar-mass model, the μ-effect is also present, as can be seen from Figure 36.1 (left-hand part). The luminosity already begins to increase as a function of Z at $Z \simeq 0.02$ since the κ-effect on L is negligible.

1.4 The ε_{nuc}-effect

The nuclear energy production ε_{nuc} sustains the stellar luminosity. If ε_{nuc} is arbitrarily increased, the central regions of the star expand, leading to decreases in the temperatures and densities, and an increase in the stellar radius. The temperature gradient therefore decreases, leading to decreases both of the luminosity and of the effective temperature.

The way ε_{nuc} depends on the metallicity is closely related to the mode of nuclear burning. When the CNO cycles are the main mode of H-burning, ε_{nuc} increases with Z. In contrast, when the main mode of burning is the pp chain, ε_{nuc} is related to X. It is thus about independent of Z at $Z \leq 0.02$, and decreases with increasing Z at $Z \geq 0.02$. These properties determine the behavior of T_c and ρ_c in MS stars as a function of Z.

Usually, the CNO cycles are the main mode of core H-burning for stars more massive than about $(1.15\text{–}1.30)M_\odot$ (depending on metallicity), whereas the pp chain operates at lower masses. At $Z = 0.1$, however, all stars are found to burn their hydrogen through the CNO cycles, at least down to $M = 0.8M_\odot$ (Mowlavi et al. 1998a).

When the CNO cycle is the main mode of burning, ρ_c and T_c decrease with increasing Z (see above). This is clearly visible in the left-hand part of Fig. 36.2 for the $3M_\odot$ and $20M_\odot$ stars at $Z \leq 0.04$. At $Z = 0.1$, however, the higher surface luminosities of the models relative to those at $Z = 0.04$ result in concomitantly higher central temperatures.

When the pp chain is the main mode of burning, on the other hand, the location in the ($\log \rho_c$, $\log T_c$) plane is not very sensitive to Z. This is well verified for the mass-$M_\odot$ models at $Z < 0.1$.

The combined effects of μ, κ, and ε_{nuc} on the stellar surface properties can be estimated more quantitatively with a semi-analytical approach using homology relations. This is developed in the appendix of Mowlavi et al. (1998b). We just recall below that a variation $\Delta \mu$ of the mean molecular mass of a star of mass M affects its position in the HR diagram according to

$$\Delta \log L = \begin{Bmatrix} 7.3 \\ 7.8 \end{Bmatrix} \Delta \log \mu - \begin{Bmatrix} 1.02 \\ 1.08 \end{Bmatrix} \Delta \log \kappa_0 - \begin{Bmatrix} 0.02 \\ 0.08 \end{Bmatrix} \Delta \log \epsilon_0, \quad (1.1)$$

$$\Delta \log T_{\text{eff}} = \begin{Bmatrix} 1.6 \\ 2.2 \end{Bmatrix} \Delta \log \mu - \begin{Bmatrix} 0.28 \\ 0.35 \end{Bmatrix} \Delta \log \kappa_0 - \begin{Bmatrix} 0.03 \\ 0.10 \end{Bmatrix} \Delta \log \epsilon_0, \quad (1.2)$$

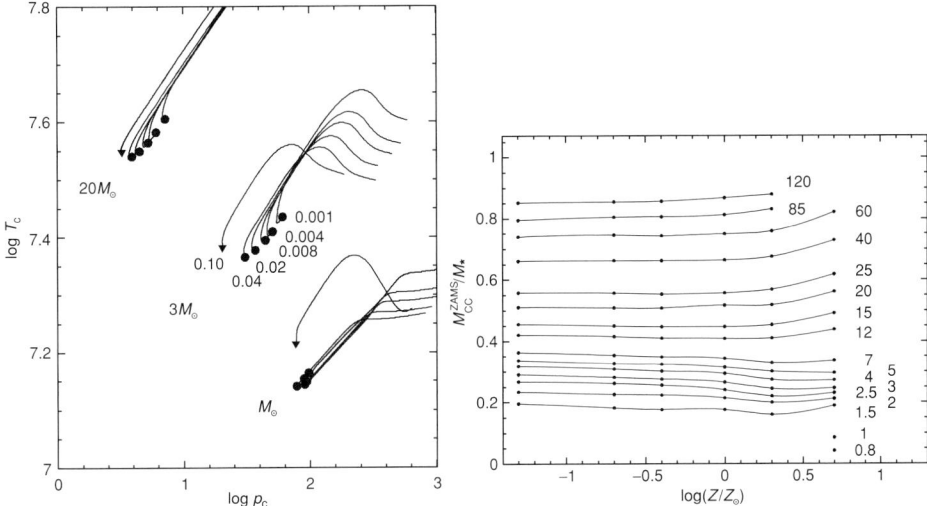

Figure 36.2. Left panel: the same as the left-hand part of Figure 36.1 but for the ($\log \rho_c, \log T_c$) diagram. Right panel: masses of convective cores relative to the stellar masses as labeled (in Solar masses) next to the curves, as a function of metallicity. Figures taken from Mowlavi et al. (1998b).

where ϵ_0 is the temperature- and density-independent coefficient in the relation for the nuclear energy production $\varepsilon_{\text{nuc}} = \epsilon_0 \rho^\lambda T^\nu$ (Cox and Giuli 1969, p. 692), and κ_0 is the opacity coefficient in Kramers' law $\kappa = \kappa_0 \rho T^{-3.5}$. The numbers on the first line in Equations (1.1) and (1.2) apply when the nuclear energy production results from the pp chain, whereas the second line applies when the CNO cycles provide the main source of nuclear energy. Variation of μ is related to variation of Z by

$$\Delta \log \mu = \frac{4.5}{\ln 10} \mu \, \Delta Z.$$

The mass of the convective core is mainly determined by the nuclear energy production and by L. At $Z \leq 0.04$, it slightly decreases with increasing Z in low- and intermediate-mass stars, whereas it remains approximately constant in massive stars (see the right-hand part of Figure 36.2). At $Z \geq 0.04$, on the other hand, it increases with Z at all stellar masses.[1] We recall furthermore that at $Z = 0.1$ all stars with M as low as $0.8 M_\odot$ possess convective cores.

The MS lifetimes for several stellar masses as a function of metallicity are summarized in the left-hand part of Figure 36.3. They can be understood in terms of two factors: the initial H abundance, which determines the quantity of available fuel; and the luminosity of the star, which fixes the rate at which this fuel burns.

---

[1] It is interesting to note that we could have expected smaller convective cores in metal-rich massive stars, for which the main source of opacity is electron scattering (since this opacity is positively correlated with the H content), but the effect of higher luminosity in those stars overcomes that κ-effect, and the core mass increases with increasing Z.

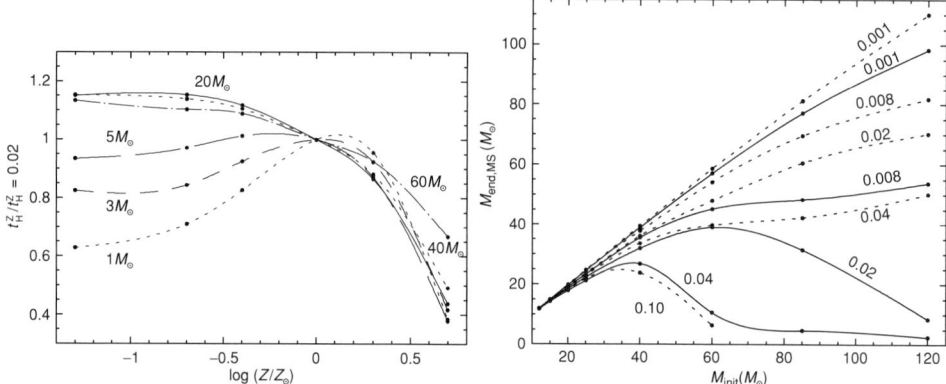

Figure 36.3. Left panel: main-sequence (MS) lifetimes of models of initial masses as labeled on the curves, as a function of their metallicity. The lifetimes are normalized for each stellar mass to their values at $Z = 0.02$. Right panel: stellar masses at the end of the MS phase as a function of initial mass for various metallicities as labeled on the curves. Dotted lines correspond to models computed with moderate mass-loss rates, while thick lines correspond to models computed with twice those mass-loss rates. See Mowlavi *et al.* (1998b) for the references to the models used. Figures taken from Mowlavi *et al.* (1998b).

At $Z \leq 0.02$, t_H is mainly determined by L, being shorter at higher luminosities. The left-hand part of Figure 36.3 confirms, as expected from L (see the left-hand part of Figure 36.1), that t_H increases with Z for low- and intermediate-mass stars, and is about independent of Z for massive stars.

At $Z > 0.02$, on the other hand, the initial H abundance decreases sharply with increasing Z (the H-depletion law is dictated by the adopted $\Delta Y/\Delta Z$ law; for $\Delta Y/\Delta Z = 2.4$ and $Y_0 = 0.24$, X drops from 0.69 at $Z = 0.02$ to 0.42 at $Z = 0.1$). The MS lifetimes are then mainly dictated by the amount of fuel available. This, combined with the higher luminosities at $Z = 0.1$, leads to MS lifetimes that are about 60% shorter at $Z = 0.1$ than at $Z = 0.02$. This result is independent of the stellar mass for $M \leq 40 M_\odot$. Above this mass, however, the action of mass loss (see below) extends the MS lifetime, as can be seen from the $60 M_\odot$ curve in the left-hand part of Figure 36.3.

1.5 The effect of $\dot{M}$

When mass loss is driven by radiation, Kudritzki *et al.* (1987), see also Vink *et al.* (2001,) showed that $\dot{M}$ in O stars is proportional to $Z^{0.5-0.8}$. As a result, the effects of mass loss dominate in more metal-rich stars. The stellar masses remaining at the end of the MS phase for various mass-loss-rate prescriptions are shown in the

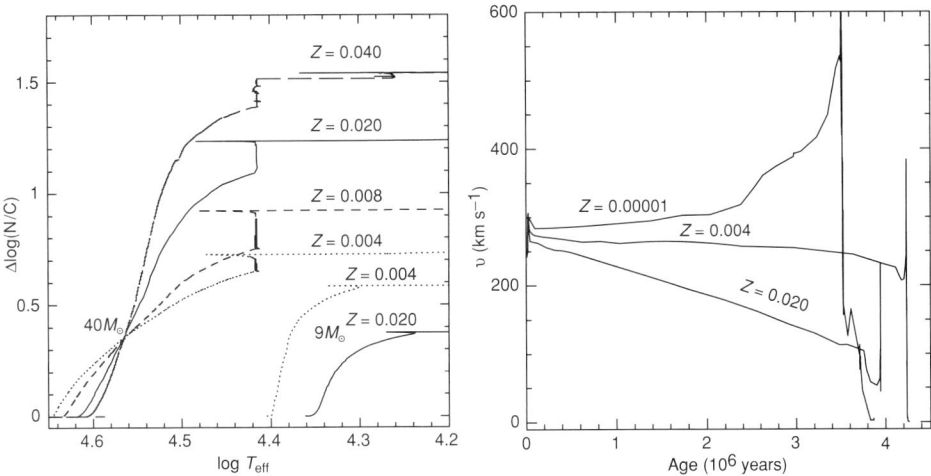

Figure 36.4. Left panel: evolution during the MS phase of the N/C number ratios at the surface of rotating stellar models as a function of the effective temperature. The differences in N/C ratios are given with respect to the initial values. Right panel: evolution of the surface equatorial velocity as a function of time for $60 M_\odot$ stars with $v_{\text{init}} = 300$ km s^{-1} for various initial metallicities.

right-hand part of Figure 36.3. The most striking result at metallicities higher than twice Solar is the rapid evaporation of massive stars with $M \geq 50 M_\odot$, and the consequent formation of Wolf–Rayet (WR) stars during core H-burning.

1.6 The effects of rotation

The effects of shellular rotation on the evolution of massive-star models at high metallicity have been discussed in Meynet & Maeder (2005). As at lower metallicity, rotation induces mixing in the stellar interiors. However, as was discussed in Maeder & Meynet (2001), rotational mixing tends to be less efficient in metal-rich stars. This can be seen by looking at the $9 M_\odot$ stellar model in the left-hand part of Figure 36.4. This comes from the fact that the gradients of Ω are much less steep in the higher-metallicity models, so they trigger less-efficient shear mixing. The Ω gradients are shallower because less angular momentum is transported outwards by meridional currents, whose velocity scales as the inverse of the density in the outer layers; see the Gratton–Öpik term in the expression for the meridional velocity in Maeder & Zahn (1998). On looking at the $40 M_\odot$ stellar model in the left-hand part of Figure 36.4, one sees that the higher-metallicity model presents the highest surface enrichments, in striking contrast with the behavior of the $9 M_\odot$ model. This comes from the fact that the changes occurring at the surface of the $40 M_\odot$ star are due not only to rotation but also to mass loss, which is more efficient at higher Z.

In the high-mass range at high metallicity, the stars have less chance to reach the critical velocity during the MS phase, as can be seen from the right-hand part of Figure 36.4. This is due to the fact that stellar winds remove angular momentum at the surface, thus preventing the outer layers from reaching the critical limit. Let us note that, for smaller initial masses, a high metallicity may favor the approach to the critical limit. Indeed, in that case, the stellar winds are too weak, even at high metallicity, to remove a lot of mass and therefore of angular momentum, while the meridional currents, which bring angular momentum from the inner regions of the star to the surface, are more rapid due to the lower densities achieved in the outer layers of metal-rich stars. Another point that should be kept in mind at this point is that, at very high metallicity, the stars may lose large amounts of mass without losing too much angular momentum. This is due to the fact that, in the high-velocity regime, the stellar winds are polar (Maeder & Meynet 2001). Let us stress, however, that this situation has little chance to be realized at high metallicity. Indeed, the timescales for mass loss are likely shorter than the timescale for the torque due to wind anisotropy to affect the surface velocity.

2 Massive-star populations in high-metallicity regions

At high metallicity, as a result of the metallicity dependence of the mass-loss rates, one expects larger fractions of WR stars (Maeder *et al.* 1980). This can be seen in the left-hand part of Figure 36.5, where the WR lifetimes of rotating models for four metallicities are plotted as a function of the initial mass. The metallicity dependence of the mass-loss rates is responsible for two features. (1) For a given initial mass and velocity the WR lifetimes are greater at higher metallicities. Typically at $Z = 0.040$ and for $M > 60 M_\odot$ the WR lifetime is of the order of 2 Myr, while at the metallicity of the SMC the WR lifetimes in this mass range are in the range 0.4–0.8 Myr. (2) The minimum mass necessary for a single star to evolve into the WR phase is lower at higher metallicity.

Comparisons with observed populations of WR stars are shown in Meynet & Maeder (2005). When the variation with the metallicity of the number ratio of WR to O-type stars is considered, good agreement is obtained provided that models with rotation are used. Models without rotation predict ratios of WR to O-type stars that are much too small, even at high metallicity. This illustrates the fact that, even at high metallicity, for which the effects of the mass-loss rates are dominant, one cannot neglect the effects of rotation.

Models well reproduce the variation with Z of the WC/WN ratio at low metallicity, but underestimate this ratio at high Z (Meynet & Maeder 2005, Figure 11). It might be that the mass-loss rates during the post-MS WNL phase are

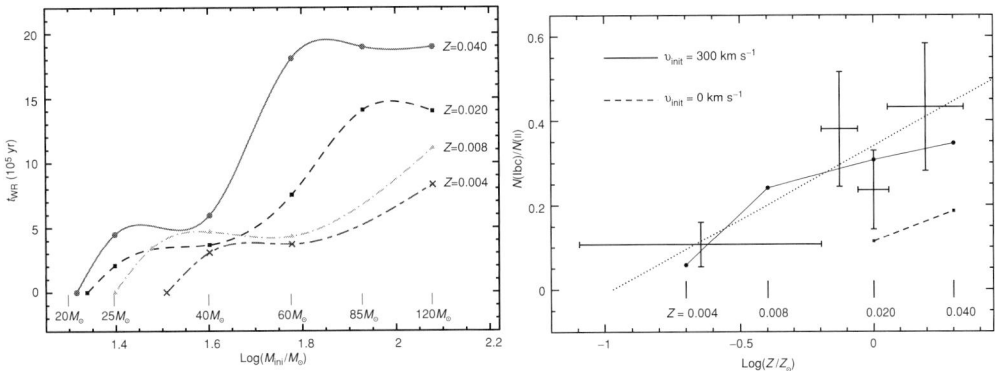

Figure 36.5. Left panel: lifetimes of Wolf–Rayet stars of various initial masses for the four metallicities indicated. All the models begin their evolution with $v_{init} = 300$ km s^{-1} on the ZAMS. Right panel: the variation with metallicity of the number ratio of Type-Ib/Ic supernovae to Type-II supernovae. The crosses with the error bars correspond to the values deduced from observations by Prantzos & Boissier (2003). The dotted line is an analytical fit proposed by those authors. The continuous and dashed line show the predictions of rotating and non-rotating stellar models, respectively. Note that at high metallicity no Type-Ic supernovae are predicted by the models (see the text). Figures taken from Meynet & Maeder (2002).

underestimated. Also, the metallicity dependence of the WR stellar winds may help in resolving this disagreement (Eldridge and Vink 2006).

Current knowledge associates supernovae of Type Ib/Ic with the explosion of WR stars, the H-rich envelope of which has been completely removed by stellar winds and/or by mass transfer through Roche-lobe overflow in a close binary system. If we concentrate on the case of single-star models, theory predicts that the fraction of supernovae progenitors without H-rich envelopes with respect to H-rich supernovae should be higher at higher metallicity. The reason is the same as the one invoked to explain the increasing number ratio of WR to O-type stars with metallicity, namely the growth of the mass-loss rates with Z. Until recently very little observational evidence had been found confirming this predicted behavior. The situation began to change with the work by Prantzos & Boissier (2003). Those authors derived from published data the observed number ratios of Type-Ib/Ic supernovae to Type-II supernovae for various metallicities. The regions considered are regions of constant star-formation rate. Their results are plotted in the right-hand part of Figure 36.5.

On looking at this figure, it becomes clear that rotating models give a much better fit to the observed data than non-rotating models. This comparison can be viewed as a check of the lower initial-mass limit M_{WNE} of the stars evolving into a WR phase without hydrogen. Let us note also at this point that, at high metallicity, no Type-Ic supernovae (supernovae with no trace of H and He) are expected to occur. This is a consequence of the high mass loss which allows the star to enter at an early stage into the WC phase when a lot of helium still has to be transformed into carbon and oxygen. The star thus keeps a high abundance of helium at its surface until the pre-supernova stage and explodes as a Type-Ib supernova. At low metallicities, since the mass-loss rates are weaker, the star may enter (if it does) at a later stage of the core He-burning phase, when most of the helium has already been transformed into carbon and oxygen (Smith & Maeder 1991). In that case the star may explode as a Type-Ic supernova. This might explain why the long soft gamma-ray bursts associated with Type-Ic supernovae occur only in metal-poor regions (Hirschi *et al.* 2005).

3 Massive-star nucleosynthesis in high-metallicity regions

Let us begin this section by recalling a few orders of magnitude. When single stars form with a Salpeter initial mass function, about 14% of the mass locked into stars consists of massive stars, i.e. stars with masses greater than $8M_\odot$, 25% is locked into stars of masses between $M_\odot$ and $8M_\odot$, and 61% is in stars with masses between $0.1M_\odot$ and $M_\odot$. When all the stars with masses above $M_\odot$ have died, about 13% of the mass initially locked into stars is ejected by massive stars (1% remains locked into black holes or neutron stars). The intermediate-mass stars (masses from $M_\odot$ to $8M_\odot$) eject about 18.5% of the mass initially locked into stars (6.5% remains locked in white dwarfs).

Figure 36.6 shows for several models at various metallicities the fraction of the mass in stars which is eventually ejected in the form of new elements. The part ejected by stellar winds is distinguished from that ejected at the time of the supernova explosion. One sees that, on the whole, the fraction of the mass initially locked into stars and transformed into new elements by the massive stars does not depend too much either on the model or on the metallicity. All the results are between 3.5% and 4.5%.

One notes, however, that more variations appear when one looks at the proportions of these new elements ejected by the winds and the supernova explosion. Indeed, at higher metallicity a greater part of the new elements synthesized by stars is ejected by stellar winds, as can be see in Figure 36.6. Typically the non-rotating stellar models of Schaller *et al.* (1992) predict that a stellar population at $Z = 0.001$ ejects in the form of new elements during the supernova explosion a little less than 4% of the total mass used to form the stars. Models at Solar metallicity and higher eject about half of their new elements at the time of the supernova explosion and the rest through stellar winds. This may have a great influence on the final yields, as has been shown by Maeder (1992). Indeed, mass loss removes matter at earlier

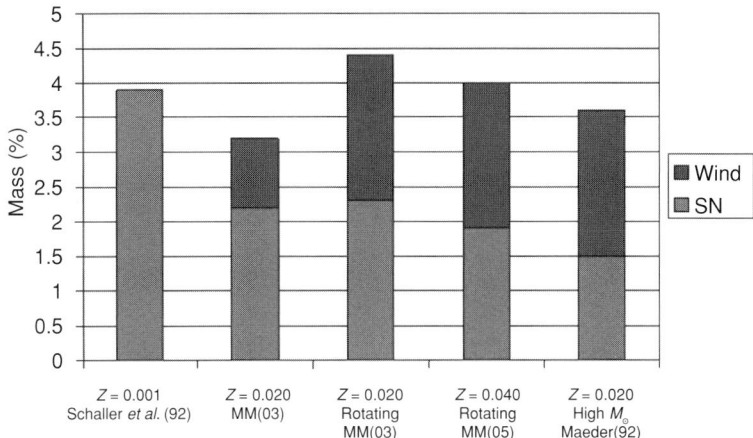

Figure 36.6. Masses of new elements ejected by stars more massive than $8M_\odot$ per unit mass in stars given by five stellar models. A Salpeter initial mass function was used. The labels MM(03) and MM(05) are for Meynet & Maeder (2003) and Meynet & Maeder (2005), respectively.

evolutionary stages. The ejected matter has therefore been partially processed by nuclear burning and has a chemical composition different from the one it would have if the matter had remained locked in the star.

This effect is responsible for many specific enrichments at high metallicity: for instance, because of this effect, massive-star models are expected to be stronger sources of ^{4}He, ^{12}C, ^{22}Ne, and ^{26}Al and to be less important sources of ^{16}O at high metallicity than at low metallicity. The cases of helium, carbon, and oxygen are shown in Figure 36.7. We can see the importance of mass loss for the carbon yields. Indeed, the larger yields are obtained for the models computed with enhanced mass-loss rates. It is interesting here to note that high mass loss is not a sufficient condition for obtaining high carbon yields. High carbon yields are obtained only when the star enters into the WC phase at an early stage of the core He-burning phase. This can be seen by comparing the $60M_\odot$ stellar model at $Z=0.04$ computed by Meynet & Maeder (2005) with the $60M_\odot$ stellar model at $Z=0.02$ with enhanced mass-loss rate computed by Maeder (1992). These two models end their lifetimes with similar final masses, the model of Meynet & Maeder (2005) with $6.7M_\odot$ and the model of Maeder (1992) with $7.8M_\odot$), thus the mass of new carbon ejected at the time of the supernova explosion is quite similar for these two models and is around $0.45M_\odot$. When one compares the predictions for the mass of new carbon ejected by the winds, we obtain $7.3M_\odot$ and $0.16M_\odot$ for the model of Maeder (1992) and that of Meynet & Maeder (2005), respectively. This difference arises from the fact that the model with enhanced mass loss enters the WC phase at a much earlier time of the core He-burning phase, typically when the mass fraction of helium at the

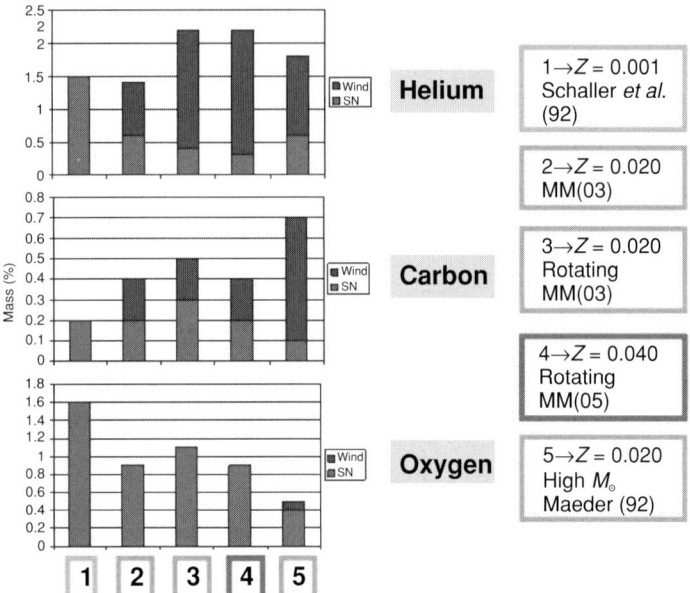

Figure 36.7. Masses of new helium, carbon, and oxygen ejected by stars more massive than $8M_\odot$ per unit mass initially in stars given by five stellar models. A Salpeter initial mass function was used. The labels MM(03) and MM(05) are for Meynet & Maeder (2003) and Meynet & Maeder (2005), respectively.

center, Y_c, is 0.43 and the actual mass of the star is $25.8M_\odot$, whereas the model of Meynet & Maeder (2005) enters the WC stage when Y_c is 0.24 and the actual mass is $8.6M_\odot$.

The example above shows that entry at an early stage into the WC phase is more favored by strong mass loss than by rotation. This comes from the fact that rotation favors an early entry into the WN phase, while the star has still an important H-rich envelope. It takes time for the whole H-rich envelope to be removed and, when this has been done, the star is already well advanced into the core He-burning phase. Of course, this conclusion is quite dependent on the magnitudes of the mass-loss rates. For instance, higher mass-loss rates during (a part of) the WNL phase would favor an early entrance into the WC phase. This would give a better agreement with the number ratio of WC to WN observed and would lead to higher carbon yields.

Let us conclude this paper by estimating an empirical yield in carbon from the WR stars in the Solar neighborhood. From the catalog published by van der Hucht (2001) one obtains that the number of WC stars within 3 kpc of the Sun is 44. The mass-loss rate during the WC phase is estimated to be $10^{-4.8} M_\odot$ per year, and the mass fraction of carbon observed in the WC stellar wind is around 0.35 (Crowther 2006, Table 2). If we consider a star formation rate of $(2-4)M_\odot$ per square parsec and per Gyr, we obtain that the mass of (new) carbon ejected by WC wind per unit mass used to form stars is between 0.25% and 0.5%. These empirical yields are well in line with the range of values given by the models, namely 0.2%–0.6%; see Figure 36.7, the greatest value corresponding to the case of Maeder (1992). Let us note that incorporating the yields of Maeder (1992) into chemical-evolution models has an important impact (e.g. Prantzos *et al.* 1996). If the upper value of the empirical yields is the correct one then this indicates that WC stars are very important sources of carbon in metal-rich regions.

References

Claret, A. (1997), *A&AS* **125**, 439–443
Cox, J. P., & Giuli, R. T. (1969), *Principles of Stellar Structure* (New York, Gordon and Breach)
Crowther, P. (2006), *ARAA* **45**, in press (astro-ph/0610356)
Eldridge, J. J., & Vink, J. S. (2006), *A&A* **452**, 295–301
Fagotto, F., Bressan, A., Bertelli, G., & Chiosi, C. (1994a), *A&AS* **104**, 365–376
  (1994b), *A&AS* **105**, 39–45
Hirschi, R., Meynet, G., & Maeder, A. (2005), *A&A* **443**, 581–591
Kudritzki, R. P., Pauldrach, A. W. A., & Puls, J (1987), *A&A* **173**, 293–298
Maeder, A. (1992), *A&A* **264**, 105–120
Maeder, A. (2002), in *The Evolution of Galaxies. II – Basic Building Blocks*, eds. M. Sauvage, G. Stasińska, & D. Schaerer (Dordrecht, Kluwer Academic Publishers), pp. 223–230.
Maeder, A., & Meynet, G. (2001), *A&A* **373**, 555–571
Maeder, A., & Zahn, J.-P. (1998), *A&A* **334**, 1000–1006
Maeder, A., Lequeux, J., & Azzopardi, M. (1980), *A&AL* **90**, L17–L20
Meynet, G., & Maeder, A. (2002), *A&A* **390**, 561–583
  (2003), *A&A* **404**, 975–990
  (2005), *A&A* **429**, 581–598
Meynet, G., Maeder, A., Schaller, G., Schaerer, D., & Charbonnel, C. (1994), *A&AS* **103**, 97–105
Mowlavi, N., Schaerer, D., Meynet, G., Bernasconi, P. A., Charbonnel, C., & Maeder, A. (1998a), *A&AS* **128**, 471–474
Mowlavi, N., Meynet, G., Maeder, A., Schaerer, D., & Charbonnel, C. (1998b), *A&A* **335**, 573–582
Prantzos, N., & Boissier, S. (2003), *A&A* **406**, 259–264
Prantzos, N., Aubert, O., & Audouze, J. (1996), *A&A* **309**, 760
Schaller, G., Schaerer, D., Meynet, G., & Maeder, A. (1992), *A&AS* **96**, 269–331
Schaerer, D., Charbonnel, C., Meynet, G., Maeder, A., & Schaller, G. (1993), *A&AS* **102**, 339
Smith, L. F., & Maeder, A. (1991), *A&A* **241**, 77–86
van der Hucht, K. A. (2001), *New Astronomy Rev.* **45**, 135–232
Vink, J., de Koter, A., & Lamers, H. J. G. L. M. (2001), *A&A* **369**, 574

37

Supernovae in Galactic evolution: direct and indirect metallicity effects

Carla Fröhlich,[1] Raphael Hirschi,[1] Matthias Liebendörfer,[1]
Friedrich-Karl Thielemann,[1] Gabriel Martínez Pinedo[2] &
Eduardo Bravo[3]

[1]*Department of Physics and Astronomy, Universität Basel, Switzerland*
[2]*Gesellschaft für Schwerionenforschung, Darmstadt, Germany*
[3]*Departament de Física i Enginyeria Nuclear, Universitat Politècnica de Catalunya, Barcelona, Spain*

Galactic chemical evolution witnesses the enrichment of the interstellar medium with elements heavier than H, He, and Li that originate from the Big Bang. These heavier elements can be traced via the surface compositions of low-mass stars of various ages, which have remained unaltered since their formation and therefore measure the composition in the interstellar medium at the time of their birth. Thus, the metallicity [Fe/H] is a measure of the enrichment with nucleosynthesis products and indirectly of the ongoing duration of galactic evolution. For very early times, when the interstellar medium was essentially pristine, this interpretation might be wrong and perhaps we see the ejecta of individual supernovae where the amount of H with which these ejecta mix is dependent on the energy of the explosion and the mass of the stellar progenitor. Certain effects are qualitatively well understood, i.e. the early ratios of alpha elements (O, Ne, Mg, Si, S, Ar, Ca, Ti) to Fe, which represent typical values from Type-II supernova explosions that originate from rapidly evolving massive stars. On the other hand, Type-Ia supernovae, which are responsible for the majority of Fe-group elements and are the products of binary evolution of lower-mass stars, later emit their ejecta and reduce the alpha/Fe ratio. In addition to being a measure of time, the metallicity [Fe/H] also enters stellar nucleosynthesis in two other ways. (i) Some nucleosynthesis processes are of secondary nature, e.g. the s-process, requiring initial Fe in stellar He-burning. (ii) Other processes are of primary nature, e.g. the production of Fe-group elements in both types of supernovae. These explosive nucleosynthesis yields originate in both cases from initial H, which is burned during stellar evolution and in the final explosion, but the question is whether the initial metallicity affects the way of explosive processing (e.g. by changing the neutron-richness of matter measured by Y_e) or influences the stellar evolution and consequently the final nucleosynthesis products. In the present paper we will first outline the general

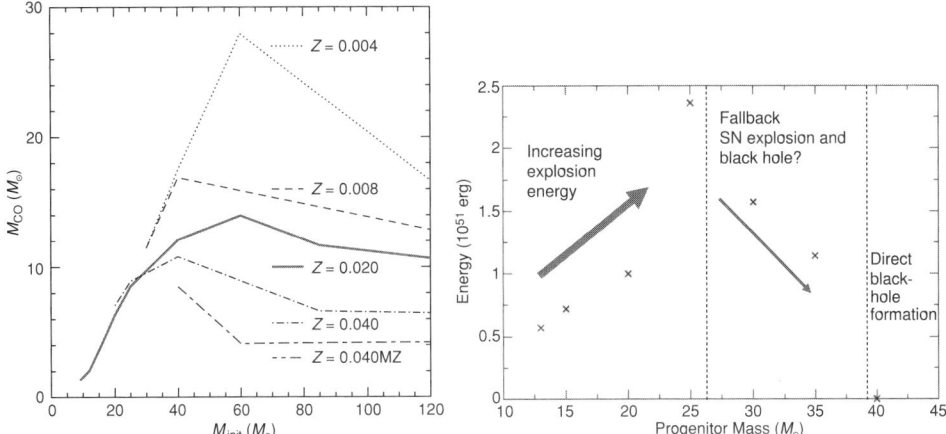

Figure 37.1. Left panel: C–O core masses from stellar-evolution calculations for various metallicities as a function of the initial stellar mass, with $V_{\rm init} = 300\,{\rm km\,s^{-1}}$ (Hirschi *et al.* 2005). Right panel: explosion energy as a function of progenitor mass for core-collapse supernovae (of Solar metallicity) due to neutrino absorption (Liebendörfer *et al.* 2003). The explosion energy peaks around $\sim 25 M_\odot$, increasing with increasing progenitor mass at lower masses and decreasing fast for higher masses. Above $\sim 40 M_\odot$ no explosions are observed.

questions addressed here and then analyze the abundance trend of individual elements.

1 Stellar evolution and explosive end stages

The evolution of single stars can be followed in the Hertzsprung–Russell (HR) diagram (e.g. Iben 1991). Their fate is mainly determined by the initial mass and composition. This implies in principle also the history of mass loss, but in theoretical modeling the mass-loss prescription adds a degree of freedom. Rotation can drastically affect the evolutionary track of a star in the HR diagram due to modification of the surface composition, induced by rotational diffusion, which directly affects the mass loss by stellar winds. The mass-loss rates depend on metallicity via the interaction of radiation transport with the surface composition. The present theoretical and observational knowledge was recently summarized by Meynet and Maeder (2005). In general this leads to low-metallicity stars experiencing less mass loss than high metallicity stars do and therefore possessing larger He cores and H envelopes at the ends of their lives. The same applies to the C–O core size after He-burning, which determines the final fate of the star, due to the much shorter burning timescales of the later burning stages (see Figure 37.1, left panel).

For different initial masses and metallicities, different end stages such as planetary nebulae, supernovae, and hypernovae/gamma-ray bursts can be expected. The lower boundary for stars to form cores massive enough to undergo core collapse is at $\sim 9 M_\odot$. For stars of masses above $10 M_\odot$ core collapse is the only possible fate. In between, stars form O–Ne–Mg cores, which can either collapse and form neutron stars or lose their envelopes and result in white dwarfs. While the size of the C–O core determines the final stellar fate, its relation to the progenitor mass depends on the metallicity. At low metallicities massive stars end their lives as neutron stars (for initial mass of $\sim 10 M_\odot$ to $\sim 25 M_\odot$), as black holes through fallback onto the neutron star (initial mass between $\sim 25 M_\odot$ and $\sim 40 M_\odot$), or directly as black holes (initial mass above $\sim 40 M_\odot$). At higher metallicities mass loss becomes more important, producing smaller He and C–O cores for a given initial mass. For very massive stars the metallicity dependence can be so strong that, with increasing metallicity, the mass loss during stellar evolution is large enough to exclude the possibility of black-hole formation, permitting neutron stars as the only possible type of remnant.

Core collapse with neutron-star formation leads to supernovae. Figure 37.1 (right panel) shows the resulting explosion energy for Solar metallicities from calculations discussed in Section 3 (Liebendörfer *et al.* 2003). If the star still possesses a hydrogen envelope it is visible as a Type-II supernova (containing H lines in its spectra), whereas if the hydrogen envelope was lost this leads to a Type-Ib supernova (no H lines but still He lines). If also the He envelope is lost a Type-Ic supernova (also no He lines) results. For a recent summary see e.g. Heger *et al.* (2003).

Type-Ia SNe exhibit, in addition to the absence of hydrogen in their optical spectra, a specific Si line. They are the only type of supernovae observed in elliptical galaxies and therefore have to originate from an older (revived) stellar population. Their origin is explained in terms of carbon–oxygen white dwarfs in binary stellar systems exploding after having accreted sufficient matter from the companion star for them to undergo thermonuclear runaway. Details are discussed in Section 2.

2 Type-Ia supernovae

There are strong theoretical and observational indications that SNe Ia are thermonuclear explosions of accreting white dwarfs in binary systems (Höflich and Khokhlov 1996; Nugent *et al.* 1997; Nomoto *et al.* 2000; Livio *et al.* 2001; Höflich *et al.* 2003; Thielemann *et al.* 2004; Höflich 2006). The basic idea is simple: a white dwarf in a binary system grows towards the Chandrasekhar mass limit through accretion of material from the companion. The contraction and central carbon ignition cause thermonuclear runaway since the pressure is dominated by a degenerate electron gas and has no temperature dependence. This prevents stable and controlled burning, causing a complete explosive disruption of the white dwarf (Nomoto et al. 1984; Woosley and Weaver 1994). The mass-accretion rate determines the central ignition density in spherically symmetric models. However, the initial white dwarf mass, its C/O ratio, and its metallicity might enter as well (Umeda et al. 1999; Höflich et al. 2000; Domínguez et al. 2001; Höflich et al. 2003; Höflich 2006). In addition, rotation and multi-dimensional effects complicate the situation (e.g. Travaglio et al. 2004b).

The flame front propagates initially at subsonic speed as a deflagration wave due to heat transport across the front (Hillebrandt and Niemeyer 2000; Reinecke et al. 2002; Röpke et al. 2003). The averaged spherical flame speed depends on the development of instabilities of various scales at the flame front. Multi-dimensional hydrodynamic simulations suggest a speed v_{def} as slow as a few percent of the sound speed v_s in the central region of the white dwarf. Electron capture affects the central electron fraction Y_e and depends on (i) the electron-capture rates of nuclei, (ii) v_{def}, influencing the duration of survival of matter at high temperatures (and with it the availability of free protons for electron captures), and (iii) the central density of the white dwarf ρ_{ign} (increasing the electron chemical potential, i.e. the Fermi energy) (Iwamoto et al. 1999; Brachwitz et al. 2000; Langanke and Martínez 2000). After an initial deflagration in the central layers, the deflagration might turn into a detonation (supersonic burning front) at larger radii and lower densities (Khokhlov et al. 1999; Niemeyer 1999). This is debated (Reinecke et al. 2002; Röpke et al. 2003) but leads to a picture that is more consistent with observations (Höflich 2006). The transition from a deflagration to a detonation (the delayed-detonation model) leads to a change in the ratios of Si-burning sub-categories with varying entropies. It also leaves an imprint on the Fe-group composition.

Explosive-nucleosynthesis calculations in spherical symmetry for slow deflagrations followed by a delayed detonation or a fast deflagration are used to investigate the constraints on the parameters ρ_{ign}, v_{def}, and the transition density. Variations in the ignition density ρ_{ign} and the initial deflagration velocity v_{def} affect the Fe-group composition in the central part. The effect of the choice of model on the maximum temperature and density during Si-burning of the central layers, which occurs during the propagation of the burning front, has been discussed extensively (Iwamoto et al. 1999).

The ignition density ρ_{ign} dominates the amount of electron capture and thus Y_e in the central layers. The deflagration speed v_{def} affects the duration of burning in a zone and with it the possible amount of electron captures by free protons and nuclei. It is also responsible for the time delay between the arrival of the information that a burning front is approaching (propagating with the speed of sound and causing an expansion of the outer layers) and the actual arrival of the burning front. Burning

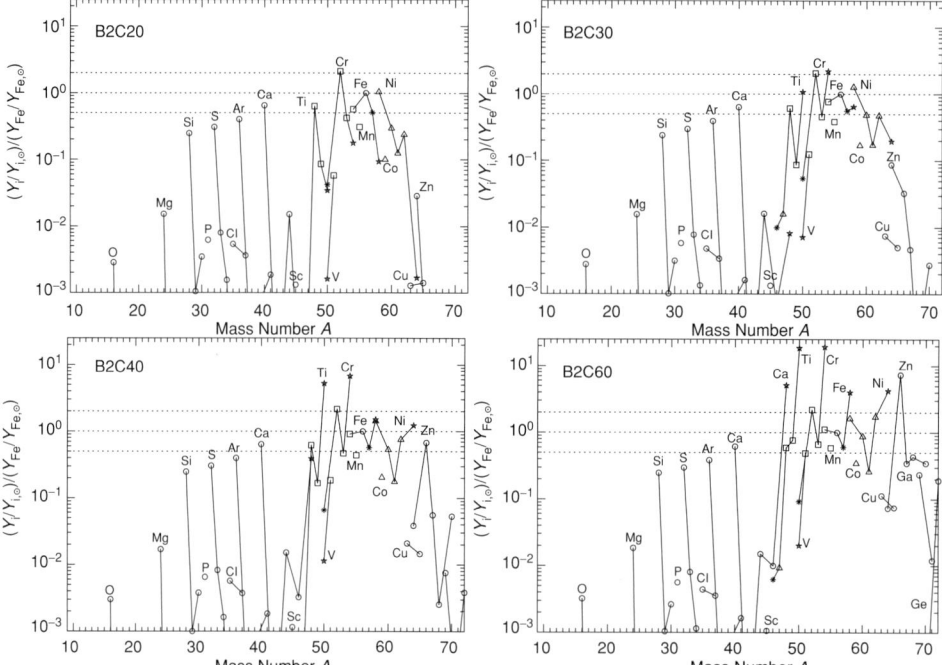

Figure 37.2. Abundance ratios in comparison with Solar for a series of central ignition densities of $(2\text{–}6)\times 10^9$ g cm^{-3} (the models B2C20–60) (Thielemann et al. 2004). While ignition densities in the range $(2\text{–}3)\times 10^9$ g cm^{-3} lead to Solar relative abundance ratios in the Fe-group, higher ignition densities can explain overabundances of ^{48}Ca, ^{50}Ti, ^{54}Cr, ^{58}Fe, ^{64}Ni, and ^{66}Zn. They should happen occasionally to account for the known abundances of these isotopes.

at lower densities causes less electron capture. Thus, v_{def} determines the resulting Y_e gradient as a function of radius (Iwamoto et al. 1999; Brachwitz et al. 2000).

During the burning the central region undergoes electron captures by free protons and Fe-peak nuclei. Similar central densities with higher temperatures lead (via more-energetic Fermi distributions of electrons and hence larger abundances of free protons) to larger amounts of electron captures and therefore smaller central Y_e values.

Most of the central region experiences conditions for complete Si-burning and subsequent normal (or alpha-rich) freeze-out. The main nucleosynthesis products are Fe-group nuclei. The outer part of the central region undergoes incomplete Si-burning (due to lower peak temperatures) and has therefore Ca and other alpha-elements as main products. The total nucleosynthesis yields obtained in slow-deflagration models (Figure 37.2) show that the production of Fe-group nuclei is a factor of 2–3 larger than the production of intermediate nuclei from Si to Ca in comparison with their Solar values (Iwamoto et al. 1999; Brachwitz et al. 2000).

There are some Fe-group contributions from alpha-rich freeze-out and layers with incomplete Si-burning that depend on the deflagration–detonation transition. The mass of the region experiencing incomplete Si-burning (indicated by the production of ^{54}Fe) decreases with decreasing transition density. The region experiencing alpha-rich freeze-out (indicated by the production of ^{58}Ni) increases with decreasing transition density. The isotopes ^{52}Fe (decaying to the dominant Cr isotope, ^{52}Cr) and ^{55}Co (decaying to the only stable Mn isotope, ^{55}Mn) are typical products of incomplete Si-burning. Production of ^{59}Cu (decaying to the only stable Co isotope, ^{59}Co) is a typical feature of an alpha-rich freeze-out.

Generally speaking, to first order we do not expect these main features to change with galactic evolution or metallicity. The main Fe-group composition is determined by the Y_e resulting from electron captures in the explosion. The Y_e of the outer layers, which are not affected by electron captures, depends mildly on the initial CNO (i.e. metallicity). However, secondary effects such as (a) the main-sequence mass distribution of the progenitors (determining the C–O core from core He-burning and the C–O layers from burning during the accretion phase), (b) the accretion history within the progenitor binary system, and (c) the central ignition density (determined by the binary accretion history) can implicitly be affected by the metallicity (Höflich 2006).

3 Core-collapse supernovae

Observations reveal typical kinetic energies of 10^{51} erg in supernova remnants. This permits one to perform light-curve as well as explosive-nucleosynthesis calculations by introducing a shock of appropriate energy into the pre-collapse stellar model (Woosley and Weaver 1995; Thielemann *et al.* 1996; Nomoto *et al.* 1997; Hoffman *et al.* 1999; Nakamura *et al.* 1999; Rauscher *et al.* 2002; Umeda and Nomoto 2005). Such induced calculations lack self-consistency and cannot predict the masses of ^{56}Ni ejected from the innermost explosive Si-burning layers (powering the supernova light curves by the decay chain ^{56}Ni–^{56}Co–^{56}Fe) due to our lack of knowledge about the detailed explosion mechanism and therefore the partition of mass between the neutron star and supernova ejecta. However, the amounts of intermediate-mass elements Si–Ca produced are dependent solely on the explosion energy and the stellar structure of the progenitor star, whereas abundances for elements like O and Mg are essentially determined by the evolution of the stellar progenitor. Thus, on moving in from the outermost to the innermost ejecta of an SN II explosion, we see an increase in the complexity of our understanding, this depending (a) only on stellar evolution, (b) on the stellar evolution and explosion energy, and (c) on stellar evolution and the complete explosion mechanism.

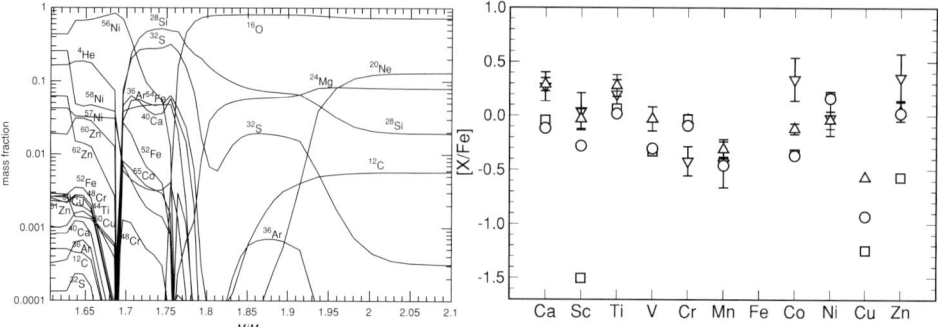

Figure 37.3. Left panel: results for the composition of supernova ejecta from an induced-explosion calculation for a $20M_\odot$ star (Thielemann *et al.* 1996). The change of the Fe-group composition in the innermost ejecta is due to the change in the Y_e of the pre-collapse stellar model. Right panel: Comparison of these abundances (open squares; Thielemann *et al.* 1996) with abundance observations of low-metallicity stars (triangles; Gratton & Sneden 1991; Cayrel *et al.* 2004) that reflect average Type-II supernova ejecta. Calculations that include also neutrino interactions during the explosion (open circles; Fröhlich *et al.* 2006a) lead to proton-rich conditions in the innermost zones and an improved description of the iron-group abundances (see Sc and Zn). Copper is an s-process element and originates from other sources.

The correct prediction of the amount of Fe-group nuclei ejected (which includes also one of the so-called alpha-elements, i.e. Ti) and their relative composition depends directly on the explosion mechanism and the size of the Fe core. Three types of uncertainties are inherent in consideration of the Fe-group ejecta, related to (i) the total amount of Fe-group nuclei ejected and the partition of mass between neutron star and ejecta, mostly measured in terms of ^{56}Ni decaying to ^{56}Fe, (ii) the total explosion energy, which influences the entropy of the ejecta and with it the amount of radioactive ^{44}Ti as well as ^{48}Cr (decaying to ^{48}Ti and being responsible for production of elemental Ti), and (iii) finally the neutron richness or $Y_e = \langle Z/A \rangle$ of the ejecta, which depends on stellar structure, electron captures, and neutrino interactions (Fröhlich *et al.* 2006a). The electron fraction Y_e strongly influences the overall Ni/Fe ratio.

An example for the composition after explosive processing due to an (induced) shock wave is shown in the left panel of Figure 37.3 (Thielemann *et al.* 1996). The outer ejected layers ($M(r) > 2M_\odot$) are unprocessed by the explosion and contain results of prior H-, He-, C-, and Ne-burning during the stellar evolution. The interior parts of SNe II contain products of explosive Si-, O-, and Ne-burning. In the inner ejecta, which experience explosive Si-burning, Y_e changes from 0.4989 to 0.494. The Y_e originates from beta-decays and electron captures in the pre-explosive hydrostatic fuel in these layers. Neutrino reactions during the explosion have not yet been included in these induced-explosion calculations utilizing a thermal-bomb prescription. Huge changes occur in the Fe-group composition for mass zones below $M(r) = 1.63 M_\odot$. There the abundances of ^{58}Ni and ^{56}Ni become comparable. Amounts of all neutron-rich isotopes (^{57}Ni, ^{58}Ni, ^{59}Cu, ^{61}Zn, ^{62}Zn) increase, the effect being strongest for the even-mass isotopes (^{58}Ni, ^{62}Zn). One can also recognize the increase in production of ^{40}Ca, ^{44}Ti, ^{48}Cr, and ^{52}Fe for the inner high-entropy zones, but a decrease in production of the $N = Z$ nuclei in the more neutron-rich layers. More details can be found in extended discussions (Thielemann et al. 1996; Nakamura et al. 1999).

Recent core-collapse supernova simulations with accurate modeling of neutrino transport (Liebendörfer et al. 2001; Buras et al. 2003; Thompson et al. 2005) show the presence of proton-rich neutrino-heated matter, both in the inner ejecta (Liebendörfer et al. 2001; Buras et al. 2003) and in the early neutrino wind from the proto-neutron star (Buras et al. 2003). This matter, part of the initially shock-heated material located between the surface of the proto-neutron star and the shock front expanding through the outer layers, is subjected to a large deposition of neutrino energy, heating the matter. This and the expansion, lifting the electron degeneracy, make it possible for the reactions $\nu_e + n \leftrightarrow p + e^-$ and $p + \overline{\nu}_e \leftrightarrow n + e^+$ (i.e. neutrino and antineutrino captures on free nucleons and their inverse reactions, electron and positron capture) to drive the composition proton-rich (Fröhlich et al. 2005; Pruet et al. 2005; Fröhlich et al. 2006a), i.e. the electron fraction $Y_e > 0.5$. This effect will always occur in a successful explosion with ejected matter irradiated by a strong neutrino flux, irrespective of the details of the explosion. While this matter expands and cools, nuclei can form. This results in a composition dominated by $N = Z$ nuclei, mainly ^{56}Ni and ^{4}He, and protons. Without the further inclusion of neutrino and antineutrino reactions the composition of this matter will finally consist of protons, alpha-particles, and heavy (Fe-group) nuclei, i.e. a proton- and alpha-rich freeze-out that results in enhanced abundances of ^{45}Sc, ^{49}Ti, and ^{64}Zn (Fröhlich et al. 2005; Pruet et al. 2005; Fröhlich et al. 2006a).

Traditional explosive (supernova)-nucleosynthesis calculations did not include interactions with neutrinos and antineutrinos. The heaviest nuclei synthesized in these calculations have a mass number $A = 64$. The matter flow stops at the nucleus ^{64}Ge, which has a small proton-capture probability and a beta-decay half-life (64 s) that is much longer than the expansion timescale (10 s) (Pruet et al. 2005). When reactions with neutrinos and antineutrinos are considered for both free and bound nucleons the situation becomes dramatically different (Fröhlich et al. 2006b; Pruet et al. 2006; Wanajo 2006).

The $N \sim Z$ nuclei are practically inert to neutrino capture (i.e. converting a neutron into a proton) because such reactions are endoergic for neutron-deficient nuclei located away from the valley of stability. The situation is different for antineutrinos, which are captured in a typical time of a few seconds, both by protons and by nuclei, at the distances at which nuclei form ($\sim$1,000 km). Since protons are more

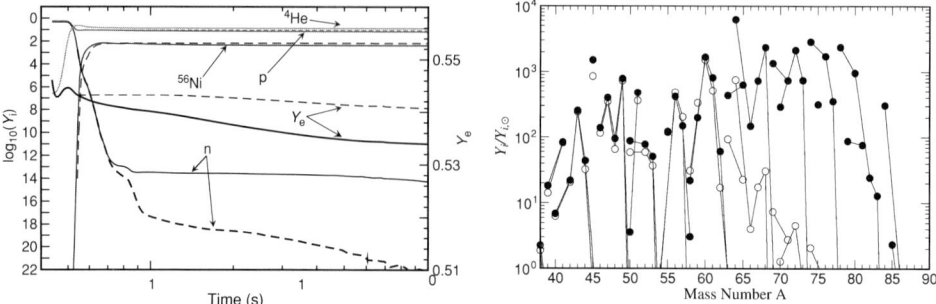

Figure 37.4. Left panel: the evolution of the abundances of neutrons, protons, alpha-particles, and ^{56}Ni in a nucleosynthesis trajectory resulting from model B07 (Fröhlich et al. 2006a). Right panel: isotopic abundances for model B07 relative to Solar abundances (filled circles; Lodders 2003) compared with earlier predictions (open circles; Thielemann et al. 1996). The filled circles represent results from calculations in which (anti)neutrino-absorption reactions are included in the nucleosynthesis, whereas for the open circles neutrino interactions are neglected. The effect of neutrino interactions is clearly seen for nuclei above $A > 64$, for which enhanced abundances are obtained.

abundant than heavy nuclei, antineutrino capture occurs predominantly on protons, causing a residual density of free neutrons of $10^{14} - 10^{15}$ cm^{-3} for several seconds when the temperatures are in the range $(1-3) \times 10^9$ K. This effect is clearly seen in Figure 37.4 (left panel), where the time evolution of the abundances of protons, neutrons, alpha-particles, and ^{56}Ni is shown for a trajectory of the model B07 (Fröhlich et al. 2006a). The solid (dashed) lines display the nucleosynthesis results which include (omit) neutrino- and antineutrino-absorption interactions after nuclei are formed. The abundance of ^{56}Ni serves to illustrate when nuclei are formed. The difference in proton abundances between these two calculations is due to antineutrino captures by protons, producing neutrons that drive the νp process. Without the inclusion of antineutrino captures the neutron abundance soon becomes too small to allow any capture by heavy nuclei.

The neutrons produced via antineutrino absorption by protons can easily be captured by neutron-deficient $N \sim Z$ nuclei (for example ^{64}Ge) that have large neutron-capture cross sections. While proton capture, (p, γ), on ^{64}Ge takes too long or is impossible, the (n, p) reaction dominates, permitting the matter flow to continue to nuclei heavier than ^{64}Ge via subsequent proton captures with freeze-out at close to 1×10^9 K.

Figure 37.4 (right panel) shows the results for the composition of supernova ejecta from one hydrodynamical model (Fröhlich et al. 2006a) that includes neutrino-absorption reactions in the nucleosynthesis calculations (filled circles) that lead initially to proton-rich conditions in the innermost zones, which experience afterwards the νp process. These abundances are compared with those from an older set of nucleosynthesis calculations (open circles) by Thielemann *et al.* (1996) that did not include neutrino interactions and therefore did not produce the proton-rich matter resulting in models with accurate modeling of neutrino transport (Liebendörfer *et al.* 2001; Buras *et al.* 2003; Thompson *et al.* 2005). In later phases of the cooling proto-neutron star, neutrino interactions will cause the emission of neutron-rich ejecta. Whether this permits a weak or strong r-process is still being debated (Thompson 2003).

High-mass stars (as discussed in Section 1) will undergo direct black-hole formation or black-hole formation via fallback onto the initially formed neutron star. Accretion onto stellar-mass black holes (if the surroundings are of sufficiently low density) can cause a fireball behavior and be related to gamma-ray bursts or the collapsar/hypernova phenomenon (Woosley 1993; Paczynski 1998; Nomoto *et al.* 2006). They indicate the occurence of higher explosion energies beyond 10^{52} erg and that large masses of ^{56}Ni are ejected (e.g. Nakamura *et al.* 2001; Umeda and Nomoto 2005). While the general explosive-nucleosynthesis behavior is similar to that of supernovae (see Figure 37.3), the higher explosion energy of hypernovae shifts both the complete Si-burning region ($T_{\text{peak}} > 5 \times 10^9$ K with Co, Zn, V, and some Cr as products) and the incomplete Si-burning region (4×10^9 K $> T_{\text{peak}} > 5 \times 10^9$ K with Cr and Mn as products after decay) outwards in mass. With this outward shift of the boundary the ratio of complete to incomplete Si-burning becomes larger and therefore higher [(Zn, Co, V)/Fe] ratios and lower [(Mn, Cr)/Fe] ratios are obtained (Nomoto *et al.* 2006). Which fraction of high-mass stars leads to such events is still unknown and depends on rotation and magnetic-field effects.

4 Galactic evolution

Explosive-nucleosynthesis yields leave fingerprints in spectra, light curves, X-rays, and radioactivities/decay gamma-rays of individual events for the explosive outbursts as well as their remnants. Galactic chemical evolution is a global test for all contributing stellar yields, especially ejecta of SNe II and SNe Ia (e.g. François *et al.* 2004). In the following we want to confront expectations resulting from the yield predictions presented in Sections 2 and 3 with observational data (see Figure 37.5). Before doing so, we give a short discussion of the expected rates of these events and the possible direct/indirect metallicity dependences of their yields.

4.1 Supernova rates

The chemical evolution of galaxies is dominated by contributions from SNe II and SNe Ia. Early in galactic evolution massive stars will dominate due to their short evolution timescales. The death of these massive stars results correspondingly in a

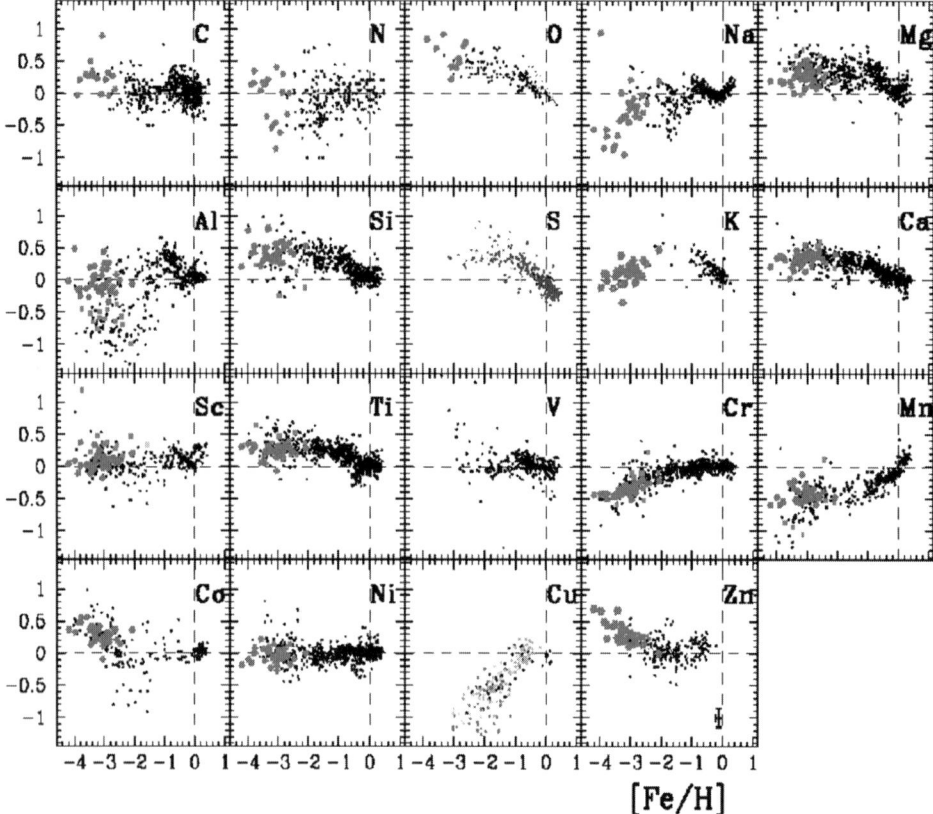

Figure 37.5. Abundance ratios [X/Fe] for various elements from C to Zn as a function of metallicity [Fe/H] in stars of the Milky Way. Small data points represent various observational sources collected by Prantzos (2005 & private communication), large data points at low metallicity are from a recent VLT survey (Cayrel et al. 2004). A typical uncertainty is indicated by the error bar in the last frame.

large Type-II + Type-Ib/c supernova rate. The longer lifetime of intermediate-mass stars (leading to white dwarfs) and the required mass transfer in binary systems delay the onset of Type-Ia supernovae (Timmes et al. 1995). Such chemical-evolution calculations indicate the strong influence of SNe Ia starting at metallicities of about [Fe/H] = −1. Estimates of the present Type-II + Type-Ib/c supernova rate in the Galaxy are 2.4–2.7 per century, while the Type-Ia rate is 0.3–0.6 per century (van den Bergh and McClure 1994). Thus, the total rate amounts to about three per century, while the ratio of core collapse (Type II + Type Ib) to thermonuclear (Type Ia) is about six or inversely that of Type Ia/(II + Ib) = 0.167. This will play a role in determining the relative Fe-group contributions from the two types of supernova.

4.2 The general Fe-group contribution

For a slightly low SN Ia frequency of Type Ia/(II + Ib) = 0.15 and typical W7-type abundances for SNe Ia (Nomoto *et al.* 1984; Iwamoto *et al.* 1999; Brachwitz *et al.* 2000) ^{56}Fe from SNe Ia amounts to about 55% of the total ^{56}Fe. Larger values would increase this contribution. Other estimates for the contributions to Fe and Fe-group elements from SNe II and SNe Ia are about a third and about two thirds, respectively (Prantzos 2005). Other Fe-group elements are co-produced with Fe. Their ratios in SNe Ia and SNe II could differ and could be dependent on metallicities as well. This would reflect the evolution of [X/Fe]. The fact that [(Sc, V, Cr, Co, Ni, Zn)/Fe] remains constant and equal to its Solar value for metallicities down to about -2 indicates that both SNe Ia and SNe II/Ibc reproduce these elements on average in Solar proportions as discussed in Sections 2 and 3. In particular, the abundances of Sc and Zn can now be explained by invoking the effect of neutrinos in the innermost ejected zones of SNe II. Effects of lower-metallicity remain to be discussed.

4.3 Alpha-elements

Abundances of alpha-elements (O, Ne, Mg, Si, S, Ar, Ca) are enhanced in comparison with that of Fe in SNe II (as has been seen in low-metallicity stars and as expected from Section 3). To obtain Solar abundance ratios for combined nucleosynthesis products near Solar metallicity, this overabundance of alpha-elements in SNe II has to be compensated for by a higher Fe to alpha-element ratio in SNe Ia (as seen in Section 2). The ratio of alpha-elements to Fe, $[\alpha/Fe]$, is found to be constant and above its Solar value at low metallicities and to decline slowly to the Solar value at about $[Fe/H] = -1$.

4.4 Indirect metallicity effects

The evolution of a single massive star, i.e. its explosion mechanism and therefore its nucleosynthesis, is determined by the core size, which in turn depends on its initial mass, rotation, and metallicity. For a given initial mass the mass of the C–O core decreases with increasing metallicity due to stronger mass loss; see Figure 37.1 and Hirschi *et al.* (2005). This has the effect that with decreasing metallicity in galactic evolution the explosive-nucleosynthesis results are shifted towards those of larger C–O cores (or, if we speak in simple terms, the explosive-nucleosynthesis products of more-massive supernovae start to dominate).

While the typical explosion energy of a core-collapse supernovae is of the order of 10^{51} erg, very probably individual events will exhibit variations. Results from studies of explosions due to neutrino absorption exhibit a twofold behavior (Figure

37.1): the explosion energy increases steadily with progenitor mass, peaking around progenitor masses of $\sim 25 M_\odot$, with a fast decline thereafter. For progenitors of about $\sim 40 M_\odot$ no explosions are observed, corresponding to direct black-hole formation. If one combines this apparent increase of explosion energy with progenitor mass with the metallicity effect discussed above, one expects on average an increase in explosion energy for the core-collapse supernovae contributing to very-low-metallicity galactic evolution. This would alter nucleosynthesis products on average similarly to the discussion of hypernovae in Section 3, i.e. there would be higher [(Zn, Co, V)/Fe] ratios and lower [(Mn, Cr)/Fe] ratios.

4.5 V, Cr, Mn, Co, and Zn

Manganese seems on average to be somewhat underproduced below [Fe/H] $= -1$ (i.e. by SNe II) and overproduced by SNe Ia in order to attain Solar values at [Fe/H] $= 0$. All the other elements typically exhibit Solar behavior down to [Fe/H] $= -3$. At very low metallicities the deviations discussed in the previous paragraph occur. That is, we seem to see explosive-nucleosynthesis results of explosions with increasing energy. There are three explanations that might act in combination: (a) the implicit metallicity effect discussed above, (b) hypernova nucleosynthesis as the first explosive-nucleosynthesis events, and (c) correlated inhomogeneities. Metallicities of [Fe/H] $= -3$ can be attained if a single supernova pollutes pristine material of the order of $10^4 M_\odot$. At such early times the interstellar medium is not well mixed. We could just see the effects of single supernovae/hypernovae for [Fe/H]<-3. Since, with increasing explosion energy, the amount of interstellar medium (mostly H) with which the ejecta are mixed increases, higher explosion energies cause smaller [Fe/H]. Thus, one would also expect increasing deviations for V, Cr, Mn, Co, and Zn, just as is observed.

4.6 Elements Cu, N, Na, and Al

For these elements secondary processes play an important role. Production of Cu is dominated by the s-process; for N, Na, and Al there are contributions from the CNO, NeNaMg, and MgAlSi cycles, respectively. These are processes acting in H- and He-burning shells and in parts of stellar-wind ejecta that have not been discussed here.

4.7 Elements Sr, Y, and Zr

The production of elements beyond iron is traditionally attributed to two neutron-capture processes: the s-process (slow process) and the r-process (rapid process).

The s-process occurs during He-burning as a secondary process, building about half the nuclides from Fe to Bi, in particular the elements Sr, Y, Zr, Ba to Nd, and Pb (corresponding to the three main s-process peaks). Asymptotic giant branch star nucleosynthesis for a variety of metallicities and ^{13}C pocket efficiencies exhibit a strong decrease in production efficiency with decreasing metallicity (Travaglio *et al.* 2004a). On summing all the contributions to the production of elements Sr, Y, and Zr there are large fractions of the observed Sr, Y, and Zr missing. They are assumed to be of primary origin, i.e. to be formed independently of the r-process, and result from all massive stars (Travaglio *et al.* 2004a). The νp process discussed in Section 3 could explain their origin.

References

Brachwitz, F., Dean, D. J., Hix, W. R. *et al.* (2000), *ApJ* **536**, 934–947
Buras, R., Rampp, M., Janka, H.-T., & Kifonidis, K. (2003), *Phys. Rev. Lett.* **90**(24), 241101
Cayrel, R., Depagne, E., Spite, M. *et al.* (2004), *A&A* **416**, 1117–1138
Domínguez, I., Höflich, P., & Straniero, O. (2001), *ApJ* **557**, 279–291
François, P., Matteucci, F., Cayrel, R., Spite, M., Spite, F., & Chiappini, C. (2004), *A&A* **421**, 613–621
Fröhlich, C., Hauser, P., & Liebendörfer, M. *et al.* (2005), *Nucl. Phys. A* **758**, 27–30
Fröhlich, C., Hauser, P., Liebendörfer, M. *et al.* (2006a), *ApJ* **637**, 415–426
Fröhlich, C., Martínez-Pinedo, G., Liebendörfer, M. *et al.* (2006b), *Phys. Rev. Lett.* **96**(14), 142502
Gratton, R. G., & Sneden, C. (1991), *A&A* **241**, 501–525
Heger, A., Fryer, C. L., Woosley, S. E., Langer, N., & Hartmann, D. H. (2003), *ApJ* **591**, 288–300
Hillebrandt, W., & Niemeyer, J. C. (2000), *ARA & A* **38**, 191–230
Hirschi, R., Meynet, G., & Maeder, A. (2005), *A&A* **443**, 581–591
Hoffman, R. D., Woosley, S. E., Weaver, T. A., Rauscher, T., & Thielemann, F.-K. (1999), *ApJ* **521**, 735–752
Höflich, P. (2006), *Nucl. Phys. A* **777**, 579–600
Höflich, P. & Khokhlov, A. (1996), *ApJ* **457**, 500
Höflich, P., Gerardy, C., Linder, E. *et al.* (2003), *LNP Vol. 635: Stellar Candles for the Extragalactic Distance Scale*, 635, 203–227.
Höflich, P., Nomoto, K., Umeda, H., & Wheeler, J. C. (2000), *ApJ* **528**, 590–596
Iben, I. J. (1991), *ApJ. Suppl.* **76**, 55–114
Iwamoto, K., Brachwitz, F., Nomoto, K. *et al.* (1999), *ApJ Suppl.* **125**, 439–462
Khokhlov, A. M., Höflich, P. A., Oran, E. S., Wheeler, J. C., Wang, L., & Chtchelkanova, A. Y. (1999), *ApJ Lett.* **524**, L107–L110
Langanke, K. & Martínez-Pinedo, G. (2000), *Nucl. Phys. A* **673**, 481–508
Liebendörfer, M., Mezzacappa, A., Thielemann, F.-K., Messer, O. E., Hix, W. R., & Bruenn, S. W. (2001), *Phys. Rev. D* **63**(10), 103004
Liebendörfer, M., Mezzacappa, A., Messer, O. E. B., Martínez-Pinedo, G., Hix, W. R., & Thielemann, F.-K. (2003), *Nucl. Phys. A* **719**, 144
Livio, M., Panagia, N., & Sahu, K. (2001), Supernovae and gamma-ray bursts: the greatest explosions since the Big Bang.
Lodders, K. (2003), *ApJ* **591**, 1220–1247

Meynet, G., & Maeder, A. (2005), *A&A* **429**, 581–598
Nakamura, T., Umeda, H., Iwamoto, K. *et al.* (2001), *ApJ* **555**, 880–899
Nakamura, T., Umeda, H., Nomoto, K., Thielemann, F.-K., & Burrows, A. (1999), *ApJ* **517**, 193–208
Niemeyer, J. C. (1999), *ApJ Lett.* **523**, L57–L60
Nomoto, K. *et al.* (2000), in *Type Ia Supernovae, Theory and Cosmology*, eds. J. C. Niemeyer & J. W. Truran (Cambridge, Cambridge University Press), p. 63
Nomoto, K., Hashimoto, M., Tsujimoto, T. *et al.* (1997), *Nucl. Phys. A* **616**, 79–90
Nomoto, K., Thielemann, F.-K., & Yokoi, K. (1984), *ApJ* **286**, 644–658
Nomoto, K., Tominaga, N., Umeda, H., Kobayashi, C., & Maeda, K. (2006), *Nucl. Phys. A* **777**, 424
Nugent, P., Baron, E., Branch, D., Fisher, A., & Hauschildt, P. H. (1997), *ApJ* **485**, 812
Paczynski, B. (1998), *ApJ* **494**, L45
Prantzos, N. (2005), *Nucl. Phys. A* **758**, 249–258
Pruet, J., Hoffman, R. D., Woosley, S. E., Janka, H.-T., & Buras, R. (2006), *ApJ* **644**, 1028–1039
Pruet, J., Woosley, S. E., Buras, R., Janka, H.-T., & Hoffman, R. D. (2005), *ApJ* **623**, 325–336
Rauscher, T., Heger, A., Hoffman, R. D., & Woosley, S. E. (2002), *ApJ* **576**, 323–348
Reinecke, M., Hillebrandt, W., & Niemeyer, J. C. (2002), *A&A*, **391**, 1167–1172
Röpke, F. K., Niemeyer, J. C., & Hillebrandt, W. (2003), *ApJ* **588**, 952–961
Thielemann, F.-K., Nomoto, K., & Hashimoto, M. (1996), *ApJ* **460**, 408–436
Thielemann, F.-K., Brachwitz, F., Höflich, P., Martínez-Pinedo, G., & Nomoto, K. (2004), *New Astronomy Rev.* **48**, 605–610
Thompson, T. A. (2003), *ApJ* **585**, L33–L36
Thompson, T. A., Quataert, E., & Burrows, A. (2005), *ApJ* 620, 861–877
Timmes, F. X., Woosley, S. E., & Weaver, T. A. (1995), *ApJ Suppl.* **98**, 617–658
Travaglio, C., Gallino, R., Arnone, E., Cowan, J., Jordan, F., & Sneden, C. (2004a), *ApJ* **601**, 864–884
Travaglio, C., Hillebrandt, W., Reinecke, M., & Thielemann, F.-K. (2004b), *A&A* **425**, 1029–1040
Umeda, H., & Nomoto, K. (2005), *ApJ* **619**, 427–445
Umeda, H., Nomoto, K., Yamaoka, H., & Wanajo, S. (1999), *ApJ* **513**, 861–868
van den Bergh, S., & McClure, R. D. (1994), *ApJ* **425**, 205–209
Wanajo, S. (2006), *ApJ* **647**, 1323–1340
Woosley, S. E. (1993), *ApJ* **405**, 273–277
Woosley, S. E., & Weaver, T. A. (1994), in *Supernovae*, eds. S. A. Bludman, R. Mochkovitch & J. Zinn-Justin p. 63.
  (1995), *ApJ Suppl.* **101**, 181

38

Progenitors of Type-Ia supernovae: evolution and implications for yields

Sung-Chul Yoon

Astronomical Institute "Anton Pannekoek," Universiteit Amsterdam, Kruislaan 403, 1098 SJ, Amsterdam, the Netherlands

Recent progress in the study of the various proposed SN Ia-progenitor scenarios is reviewed. We discuss the effects of rotation on the evolution of SN Ia progenitors, in particular the stabilization of helium-shell burning and the increase of the Chandrasekhar limit. The latter may have been confirmed by a recent analysis of an overluminous SN 2003fg. For the evolution of CO white dwarf mergers, we discuss new arguments in favor of obtaining Type-Ia SNe from those systems, in contrast to the previous consensus. We address the issue of SN Ia delay times, and the dependence of average SN Ia properties on the type of host galaxy, in the light of recent observations and progenitor models, and derive implications for SN Ia yields.

1 Introduction

Type-Ia supernovae (SNe Ia) are considered to be the major source of iron in the Universe, which critically determines stellar/galactic properties and evolution. The fairly homogeneous nature of SNe Ia light curves has made it possible to use them as excellent distance indicators, which led to the revelation of the existence of the so-called dark energy that drives the accelerated expansion of the Universe. The SNe Ia are thus of key importance for the study of galactic evolution and cosmology. The SNe Ia originate from thermonuclear explosions of carbon–oxygen white dwarfs near or above the Chandrasekhar limit (e.g. Hillebrandt & Niemeyer 2000). Although a detonation of the degenerate helium envelope in sub-Chandrasekhar-mass ($\sim 0.8 M_\odot$) CO white dwarfs could also lead to explosions like SNe Ia (e.g. Woosley & Weaver 1994; Livne & Arnett 1995), such models fail to explain the observed light curves and spectra of SNe Ia (e.g. Höflich & Khokhlov 1996; Nugent

The Metal-rich Universe, eds. G. Israelian and G. Meynet. Published by Cambridge University Press.
© Cambridge University Press 2008.

Table 38.1. *Most-studied types of SNe Ia-progenitor binary system*

System	References	Rate (y^{-1} in our Galaxy)	Delay time (yr)
MS + WD	1, 2, 3	A few 10^{-3}	$10^8 - (2 \times 10^9)$
Sub-Giant + WD	4, 5	$10^{-5} - 10^{-3}$	$(5 \times 10^8) - 10^{10}$
He-Giant + WD	6, 7	$10^{-5} - 10^{-3}$	$(2-5) \times 10^7$
WD + WD	8, 9	A few 10^{-3}	$(5 \times 10^7) - 10^{10}$

1, Li & van den Heuvel (1997); 2, Langer *et al.* (2000); 3, Han & Podsiadlowski (2004); 4, Hachisu *et al.* (1996); 5, Hachisu *et al.* (1999); 6, Iben & Tutukov (1994); 7, Yoon & Langer (2003); 8, Iben & Tutukov (1984), 9, Webbink (1984).

et al. 1997). In addition, such detonation might not occur at all when rotation is considered, since the degeneracy of the helium envelope, where differential rotation is induced by accretion of mass and angular momentum, is likely to be lifted due to viscous heating (Yoon & Langer 2004b).

Although the evolution of SNe Ia progenitors is not yet well understood, only a limited number of possibilities may exist for a CO white dwarf to grow to the Chandrasekhar limit: steady mass accretion of hydrogen and/or helium onto a CO white dwarf from a non-degenerate companion (the singly-degenerate scenario), or the coalescence of two CO white dwarfs (the doubly-degenerate scenario). Since these two scenarios predict different progenitor ages and populations, the identification of the major channel to SNe Ia is crucial for understanding the role of SNe Ia in the early chemical evolution of galaxies. Furthermore, a detailed knowledge of the SNe Ia-progenitor evolution is essential for understanding the observed diversity of SNe Ia, which has important implications for the use of SNe Ia for probing the expansion history of the Universe. Here we discuss recent theoretical progress in the study of each scenario, and investigate which scenario can explain the observational constraints better. Special attention is devoted to the effects of rotation on the white dwarf structure, and their implications for the diversity of SNe Ia. Chemical yields of SNe Ia are also briefly discussed.

2 Evolution of SNe Ia progenitors

2.1 The singly-degenerate scenario

Three different types of binary system have been most studied within the singly-degenerate (SD) scenario, as summarized in Table 38.1 (see also references in the table): main-sequence (MS) star + white dwarf (WD), sub-giant (SG) + WD, and helium giant (HeG) + WD. In these systems, thermally unstable mass transfer from the non-degenerate companion leads to accretion of hydrogen- or helium-rich matter onto the CO white dwarf at rates suitable for steady hydrogen- or helium-shell burning in the accreting white dwarf. These systems seem to have observational counterparts, such as luminous super-soft X-ray sources (van den Heuvel *et al.* 1992) and recurrent novae, e.g. U Scorpii (Hachisu *et al.* 2000; Thoroughgood *et al.* 2001). Each type of binary system considered predicts a different progenitor age, as shown in Table 38.1. Since observations imply a wide range of SNe Ia-progenitor ages (from a few times 10^7 to 10^{10} yr; see below), a single channel alone cannot explain all observed SNe Ia, but different channels should be considered together.

In MS + WD and SG + WD systems, hydrogen accretion is expected at rates of a few times $10^{-7} M_\odot$ yr^{-1} to allow steady burning in the hydrogen-shell source (Nomoto *et al.* 2007). However, the resulting helium accretion at this rate usually causes strong instabilities in the helium shell, which might induce significant loss of mass from the system (Cassisi *et al.* 1998; Kato & Hachisu 2004). In addition, accretion rates that allow steady hydrogen-shell burning in white dwarfs are limited to a very narrow range. Hence it has been debated whether the SD channel can be the major path to SNe Ia, since it is difficult for a white dwarf to grow to the Chandrasekhar limit by hydrogen accretion. The problem involved with the double-shell sources can be avoided for the HeG + WD system, for which only helium accretion is concerned (Iben & Tutukov 1994). Indeed, Yoon & Langer (2003) demonstrated that the accreting white dwarf can reach the Chandrasekhar limit in a HeG + WD binary system, without suffering strong helium-shell flashes. However, since progenitor ages are limited to $\lesssim 10^8$ yr, HeG + WD systems should be relevant only for so-called prompt SNe Ia (see the discussion below), and the rate of production of SNe Ia from these systems might be very low (10^{-5} yr^{-1}) (L. Yungelson 2004, private communication).

Yoon *et al.* (2004a) suggested that helium-shell burning can be significantly stabilized by rotation (Figure 38.1), in favor of the SD scenario. Nuclear shell burning becomes more stable with a geometrically thicker, less degenerate, and hotter shell source, as shown in Figure 38.2. Accretion of mass and angular momentum spins up the accreting white dwarf, which lifts the degeneracy of the shell source. The shell source is geometrically widened due to rotationally induced chemical mixing between the accreted helium-rich matter and CO-rich layers. These effects militate in favor of stable helium-shell burning. It is a subject of future study to determine how significantly this stabilizing effect of rotation can enhance the mass-accumulation efficiency in H/He-accreting white dwarfs (Kato & Hachisu 2004).

2.2 The doubly-degenerate scenario

A double CO white dwarf system is supposed to lose orbital angular momentum due to gravitational-wave radiation. Once the less-massive white dwarf has filled the Roche lobe, dynamically unstable mass transfer induces complete merging of the

372 *Progenitors of Type-Ia supernovae*

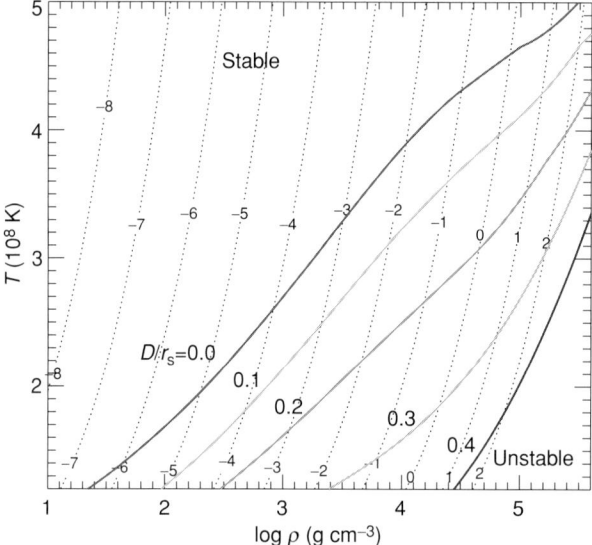

Figure 38.1. Evolution of the nuclear luminosity due to helium burning in helium-accreting white dwarf models with a constant mass-accretion rate of $\dot{M} = 5 \times 10^{-7} M_\odot$ yr^{-1}, as a function of the increasing white dwarf mass, for models with and without rotation as indicated. The initial white dwarf mass is $0.998 M_\odot$.

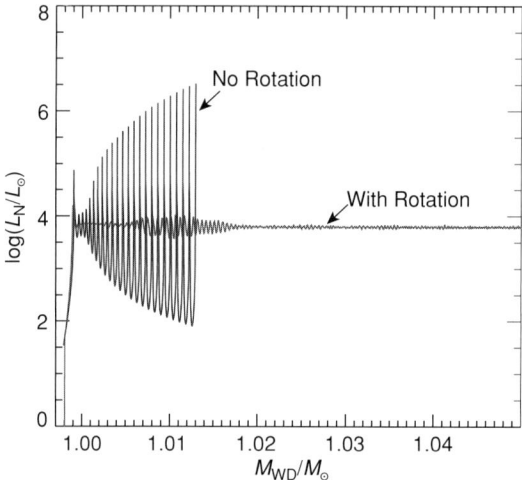

Figure 38.2. Stability conditions for a helium-shell source in the density–temperature plane. The solid lines separate the thermally unstable region from the stable region, for five relative shell-source thicknesses (i.e. $D/r_s = 0.0, 0.1, 0.2, 0.3,$ and 0.4). The dotted contour lines denote the degeneracy parameter $(:= \psi/(kT))$; $X_{\mathrm{He}} = 0.662$ and $X_{\mathrm{C}} = 0.286$ have been assumed. From Yoon et al. (2004b).

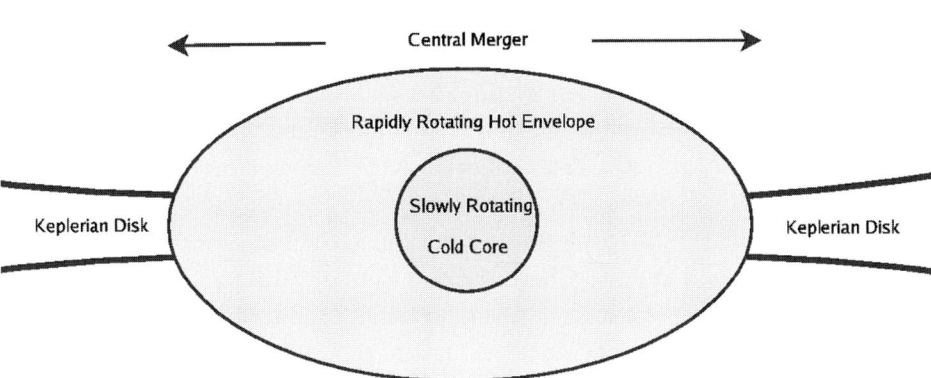

Figure 38.3. A schematic illustration of the merger of double CO white dwarfs at quasi-static equilibrium.

binary components within a few minutes (e.g. Benz *et al.* 1990). Authors of previous theoretical studies assumed that a thick disk is produced around the primary white dwarf, and that mass accretion from the disk onto the central object occurs at a rate near the Eddington accretion rate ($\sim 10^{-5} M_\odot$ yr^{-1}). Formation of a neutron star is the most likely outcome in this case, because off-center carbon ignition due to rapid mass accretion converts the white dwarf into an ONeMg white dwarf, which is susceptible to electron-capture-induced collapse at the Chandrasekhar limit (e.g. Saio & Nomoto 1998).

However, smoothed particle-hydrodynamics calculations indicate that the merger consists of three distinctive regions when it reaches quasi-static equilibrium: a slowly rotating cold core, a rapidly rotating hot envelope, and a centrifugally supported disk (Benz *et al.* 1990; Segretain *et al.* 1997; Yoon *et al.* 2007). Therefore, the merger at quasi-static equilibrium may be better described by a differentially rotating CO giant star surrounded by a Keplerian disk, as illustrated in Figure 38.3, rather than by a CO white dwarf + thick-disk system. Yoon *et al.* (2007) constructed one-dimensional differentially rotating CO giant stellar models, which mimic the central region of the merger. Their evolutionary calculations show that accretion of matter onto the cold core from the hot envelope is controlled by neutrino cooling at the interface where the temperature is highest, and that off-center carbon ignition can be avoided if the temperature peak at the interface during quasi-static equilibrium does not exceed the critical limit for carbon ignition (i.e. $T_{\text{peak}} < \sim 6 \times 10^8$ K), and if the mass accretion from the Keplerian disk onto the envelope occurs slowly enough ($\dot{M} < \sim 10^{-5} M_\odot$ yr^{-1}). This makes it possible for the DD system to be a viable channel to SNe Ia explosion, in contrast to the previous belief that double CO white dwarfs generally undergo accretion-induced collapse.

2.3 On super-Chandrasekhar-mass progenitors

Whatever the evolutionary channel for SNe Ia progenitors may be, the accreting white dwarf is supposed to gain large amounts of angular momentum until it explodes (Langer *et al.* 2000; Piersanti *et al.* 2003; Saio & Nomoto 2004; Yoon & Langer 2004a). Owing to the effect of the centrifugal force on the white dwarf structure, the critical mass limit for thermonuclear explosion or collapse can significantly increase beyond the canonical Chandrasekhar limit of $\sim 1.4 M_\odot$. In particular, differentially rotating white dwarfs can be dynamically stable up to $4.0 M_\odot$, as shown by Ostriker & Bodenheimer (1968). The mass of rigidly rotating white dwarfs cannot exceed $1.48 M_\odot$ since the amount of angular momentum that a rigidly rotating white dwarf can retain is severely limited (e.g. James 1964).

Interestingly, Yoon & Langer (2004a) showed that non-magnetic white dwarfs should rotate differentially, for the following reason. In a rotating flow, the linear condition for the dynamical shear instability (DSI) is given by

$$\frac{N^2}{\sigma^2} < R_{i,c} \approx 0.25, \tag{38.1}$$

where N^2 denotes the Brunt–Väisälä frequency, σ the shear factor $(:= \partial \omega / \partial \ln r)$, and $R_{i,c}$ the critical Richardson number. In other words, shear with $\sigma > \sigma_c := \sqrt{N^2/R_{i,c}}$ is susceptible to the DSI (see Figure 38.4). Although the DSI can reduce the degree of differential rotation on a dynamical time until the rotation profile becomes stable against the DSI ($\sigma \longrightarrow \sigma_c$), further redistribution of angular momentum by the secular shear instability or by Eddington–Sweet circulations can occur only on a thermal timescale ($\sim 10^8$ yr in the core of white dwarfs), which is much longer than the accretion time (10^5–10^6 yr). Consequently, the white dwarf rotates differentially with a shear rate close to the critical value for the onset of the DSI throughout the mass-accretion phase, as shown in Yoon & Langer (2004a).

A thermonuclear explosion may occur when the central density of accreting white dwarfs exceeds about 2×10^9 g cm^{-3} (e.g. Yoon & Langer 2003). The mass needed for a differentially rotating white dwarf to reach this density depends on the total amount of angular momentum retained in the white dwarf, as shown in Figure 38.5. Therefore, white dwarfs exploding as Type-Ia SNe may have masses in the range from $1.38 M_\odot$ to $2.0 M_\odot$ for the SD scenario, or up to $2.4 M_\odot$ in the DD scenario. The upper limit for the critical mass is determined by the available amount of mass in each scenario (Yoon & Langer 2005; Langer *et al.* 2000). This theoretical prediction may have been confirmed by the recent analysis of SNLS-30D3bb (SN 2003fg), which is the most luminous SN Ia ever observed (Howell *et al.* 2006). The nickel mass produced in this supernova appears to be about double ($M_{Ni} \approx 1.3 M_\odot$) that in usual SNe Ia ($M_{Ni} \approx 0.6 M_\odot$), while the velocity of the

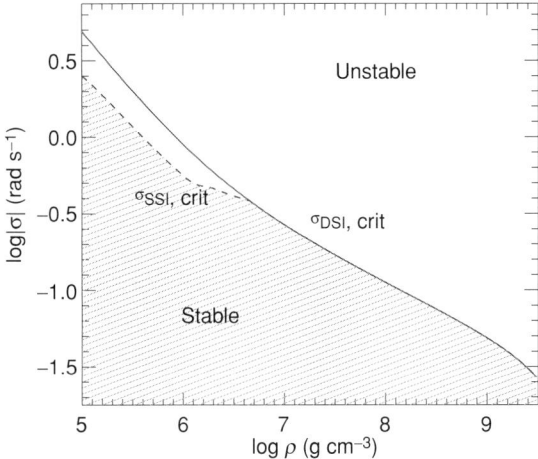

Figure 38.4. The threshold shear factor $\sigma := \partial\omega/\partial\ln r$ for the dynamical (solid line) and the secular (dashed line) shear instability, as a function of the local density. Constant gravity and temperature, i.e. $g = 10^9$ cm s^{-2}, $T = 5 \times 10^7$ are assumed. The chemical composition is also assumed to be constant, with $X_C = 0.43$ and $X_O = 0.54$. For the critical Richardson number, $R_{i,c} = 0.25$ is used; $R_{e,c} = 2,500$ is used for the critical Reynolds number. See Yoon & Langer (2004a) for a detailed discussion.

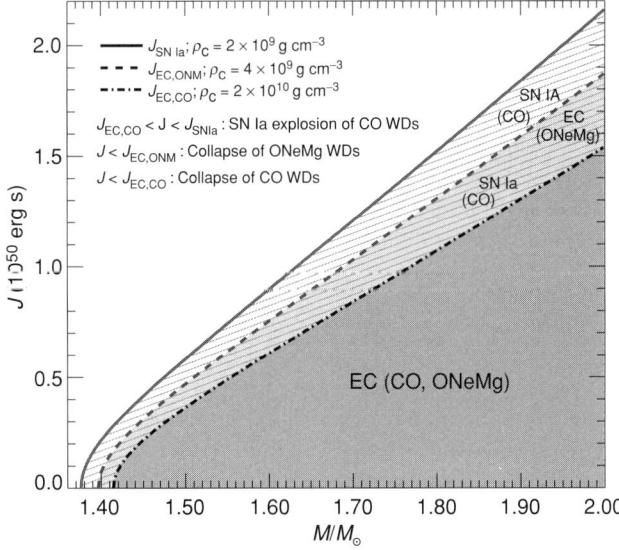

Figure 38.5. The critical angular momentum for thermonuclear explosion of CO white dwarfs ($J_{SN\,Ia}$; solid line), and for electron-capture (EC)-induced collapse of CO white dwarfs ($J_{EC,\,CO}$; dash–dotted line) and ONeMg white dwarfs ($J_{EC,\,ONM}$; dashed line), as a function of the white dwarf mass. A SN Ia explosion is expected when $J_{EC,CO} \lesssim J \lesssim J_{SN\,Ia}$ (shown as the hatched region). Electron-capture-induced collapse is supposed to occur when $J \lesssim J_{EC,CO}$ for CO white dwarfs, and when $J \lesssim J_{EC,ONeMg}$ for ONeMg white dwarfs.

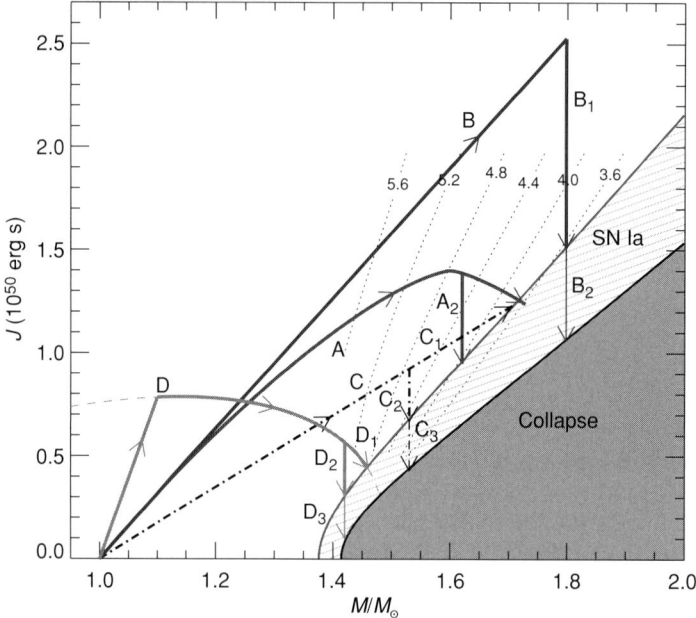

Figure 38.6. A schematic representation of supernova-progenitor evolution for various scenarios in the M–J plane. Thick lines starting at a white dwarf mass of $\sim M_\odot$ with $J = 0$ indicate various possible evolutionary paths of accreting white dwarfs: with accretion and simultaneous loss of angular momentum (A), without angular-momentum loss during the accretion phase (B), with inefficient accretion of angular momentum (C), and with maximum efficiency of internal angular-momentum transport (D) (see the text for details). The region where SN Ia explosion is expected is hatched, and the region where electron-capture-induced collapse is likely is shaded. For $\dot{M} \simeq (10^{-7}$–$10^{-6}) M$ yr^{-1}, a supernova explosion is expected at the upper borderline of the hatched area. The thin dashed line gives the M–J relation of rigidly rotating white dwarfs at critical rotation. Dotted lines denote lines of constant growth time of the r-mode instability (τ_r). See Yoon & Langer (2005) for more details.

ejecta is unusually low, which could be explained only by invoking a super-Chandrasekhar-mass progenitor of mass about $2.0 M_\odot$ (Howell et al. 2006; Jeffrey et al. 2007).

Such an unusually massive white dwarf can exist only due to differential rotation. It should be noted that the merger of a double CO white dwarf per se cannot produce a super-Chandrasekhar white dwarf with a mass above $1.5 M_\odot$ – as is often misunderstood in the literature – unless differential rotation is involved (see Figure 38.6): if the central object is kept rotating rigidly, e.g. by magnetic torques, coalescence of double CO white dwarfs would produce a SN Ia progenitor whose mass is limited to about $1.48 M_\odot$, surrounded by a carbon–oxygen Keplerian disk consisting of a fraction of the disrupted matter of the secondary.

On the other hand, the light curve of SN 2003fg deviates significantly from the standard relation between the maximum brightness and width. Since differentially rotating progenitors of SNe Ia may follow complicated evolutionary paths as illustrated in Figure 38.6 (Yoon & Langer 2005), it should thus be carefully investigated why most SNe Ia in nearby galaxies have a fairly homogeneous nature while some do not, a question that may have important implications for the use of SNe Ia for cosmological probes.

3 Implications for yields

Recent observations indicate that rates of production of SNe Ia are higher in active star-forming galaxies than in weak or non-star-forming ones (Della Valle *et al.* 2005; Mannucci *et al.* 2005; Sullivan *et al.* 2006). This implies that a significant fraction of the SNe Ia is produced *promptly* after the star formation, with delay times less than 10^8 yr. Although the fraction of prompt SNe Ia is not well known, their presence would have accelerated the decrease of the [O/Fe] ratio during the early evolution of galaxies (Matteucci *et al.* 2006). In the SD scenario, the HeG + WD system might explain prompt SNe Ia (see Table 38.1), but the DD scenario could better explain both prompt SNe Ia and the production of SNe Ia from elliptical galaxies whose progenitors must be as old as 10^{10} yr, in terms of the observed SNe Ia rate. In reality, it is most likely that both SD and DD channels contribute to production of SNe Ia. Thus, it should be kept in mind that different channels have different characteristic progenitor ages, and that they may make different contributions in different types of galaxies. It is also likely that the dominant channel for SNe Ia may change during the cosmic evolution (cf. Förster *et al.* 2006).

The possible variation of the SNe Ia-progenitor masses due to rotation may add to the complication. In particular, this may relate to the observation that SNe Ia in star-forming galaxies are systematically brighter than those occuring in elliptical galaxies. The mass budget in binaries belonging to old stellar systems in elliptical galaxies may be too limited to produce a white dwarf that is significantly more massive than the canonical Chandrasekhar limit, and super-Chandrasekhar-mass progenitors, which might be related to over-luminous SNe Ia, should be more easily produced in star-forming galaxies.

In short, the classical simple assumptions that SNe Ia begin to produce iron only $\sim 10^9$ yr after star formation and that all SNe Ia produce the same amount of iron-group elements may lead to a biased result, and should be considered as too simplified. If we want to investigate the effect of SN Ia yields on the chemical evolution of galaxies, all the above-mentioned complications should be considered carefully, which will be an important subject of future study.

Acknowledgments

The author is grateful to Norbert Langer, Philipp Podsiadlowski, and Stephan Rosswog for fruitful collaboration. He would also like to thank Norbert Langer for the improvement of the text. This work was in part supported by the VENI grant (639.041.406) of the Netherlands Organization for Scientific Research (NWO).

References

Benz, W., Cameron, A. G. W., Press, W. H., & Bowers, R. L. (1990), *ApJ* **348**, 647
Cassisi, S., Iben, I. Jr., & Tormanbè, A. (1998), *ApJ* **496**, 376
Della Valle, M., Panagia, N., Padovani, P., Cappellaro, E., Mannucci, F., & Turatto, M. (2005), *ApJ* **629**, 750
Förster, F., Wolf, C., Podsiadlowski, Ph., & Han, Z. (2006), *MNRAS* **368**, 1893
Hachisu, I., Kato, M., & Nomoto, K. (1996), *ApJ* **470**, L97
  (1999), *ApJ* **522**, 487
Hachisu, I., Kato, M., Kato, T., & Matsumoto, K. (2000), *ApJ* **528**, L97
Han, Z., & Podsiadlowski, Ph. (2004), *MNRAS* **350**, 1301–1309
Hillebrandt, W., & Niemeyer, J. C. (2000), *ARA&A* **38**, 191–230
Höflich, P., & Khokhlov, A. (1996), *ApJ* **457**, 500
Howell, D. A., Sullivan, M., Nugent, P. E. *et al.* (2006), *Nature* **443**, 308
Iben, I. Jr., & Tutukov, A. V. (1984), *ApJS* **54**, 335
  (1994), *ApJ* **431**, 264
James, R. A. (1964), *ApJ* **140**, 552
Jeffrey, D. J., Branch, D., & Baron, E. (2007), *ApJ* astro-ph/0609804
Kato, M., & Hachisu, I. (2004), *ApJ* **613**, L129–L132
Langer, N., Deutschmann, A., Wellstein, S., & Hoeflich, P. (2000), *A&A* **362**, 1046
Li, X.-D., & van den Heuvel, E. P. J. (1997), *A&A* **322**, L9
Livne, E., & Arnett, D. (1995), *ApJ* **452**, 62
Mannucci, F., Della Valle, M., Panagia, N. *et al.* (2005), *A&A* **433**, 807
Matteucci, F., Panagia, N., Pipino, A., Mannucci, F., Recchi, S., & Della Valle, M. (2006), *MNRAS* **372**, 265
Nomoto, K., Saio, H., Kato, M., & Hachisu, I. (2007), *ApJ* **663**, 1269
Nugent, P., Baron, E., Branch, D., Fisher, A., & Hauschildt, P. H. (1997), *ApJ* **485**, 812
Ostriker, J. P., & Bodenheimer, P. (1968), *ApJ* **151**, 1089
Piersanti, L., Gagliardi, S., Iben, I. Jr., & Tornambé, A. (2003), *ApJ* **583**, 885
Saio, H., & Nomoto, K. (1998), *ApJ* **500**, 388
  (2004), *ApJ* **615**, 444
Segretain, L., Chabrier, G., & Mochkovitch, R. (1997), *ApJ* **481**, 355
Sullivan, M., Le Borgne, D., Pritcher, C. J. *et al.* (2006) *ApJ* **648**, 868
Thoroughgood, T. D., Dhillon, V. S., Littlefair, S. P., March, T. R., & Smith, D. A. (2001), *MNRAS* **327**, 1323
van den Heuvel, E. P. J., Bhattacharya, D., Nomoto, K., & Rappaport, S. A. (1992), *A&A* **262**, 97
Webbink, R. F. (1984), *ApJ* **277**, 355
Woosley, S. E., & Weaver, T. A. (1994), *ApJ* **423**, 371

Yoon, S.-C., & Langer, N. (2003), *A&A* **412**, L53–L56
 (2004a), *A&A* **419**, 623–644
 (2004b), *A&A* **419**, 645–652
 (2005), *A&A* **435**, 967–985
Yoon, S.-C., Langer, N., & Scheithauer, S. (2004a), *A&A* **425**, 217–228
Yoon, S.-C., Langer, N., & van der Sluys, M. (2004b), *A&A* **425**, 207–216
Yoon, S.-C., Podsiadlowski, Ph., & Rosswog, S. (2007), *MNRAS* **380**, 933

39

Star formation in the metal-rich universe

Ian A. Bonnell

School of Physics and Astronomy, University of St Andrews, St Andrews KY16 9SS, UK

I review current models of star formation and discuss potential effects of high metallicity. Our current paradigm for star formation is that it is a dynamical process in which molecular clouds and regions of star formation form on their local dynamical times. Molecular clouds are characterised by turbulent motions, which, together with gravity, lead to their fragmentation and the formation of individual stars. The resulting distribution of stellar masses can be most easily understood as a combination of fragmentation, continued accretion to form higher-mass stars and dynamical interactions. Regions of high metallicity are likely to differ in terms of their star formation in three main areas: the formation of molecular gas on grains; the cooling processes which determine the characteristic stellar mass; and the higher opacity of dust grains, which increases the effects of radiation pressure in limiting the growth of massive stars by accretion. Characterising star formation in regions of high metallicity will allow accurate determinations of these effects.

1 Introduction

Understanding star formation is central to most branches of modern astronomy since it is star formation that produces the majority of light in the Universe, drives galactic evolution and produces the chemical and kinematic feedback that dominates the interstellar medium (ISM). Our current paradigm of star formation has evolved from models based on single, isolated star formation (Shu *et al.* 1987) to one in which we know that stars form predominantly in groups and clusters (Lada & Lada 2003) and that dynamics and dynamical interactions can play an important role in the formation process (e.g. Larson 2003; Mac Low & Klessen 2004; Bonnell *et al.* 2006).

The Metal-rich Universe, eds. G. Israelian and G. Meynet. Published by Cambridge University Press.
© Cambridge University Press 2008.

Dynamical models for star formation have gained favour at the expense of slower quasi-static models with the realisation that generally star formation occurs on timescales comparable to the dynamical timescales (Elmegreen 2000). The fact that systems of size less than a parsec up to systems over 100 pc in size all obey such a relationship is a powerful indicator that star formation occurs as fast as the physical conditions required are produced. It is also essential that the dynamical models are able to explain the processes involved and reproduce the physical properties of star-forming regions and young stars. Chief amongst these is the stellar initial mass function (IMF).

The major goals for a theory of star formation are that it should be able to answer the following questions. Why and where is star formation initiated? What determines the stellar properties formed and how do these depend on input parameters such as metallicity? Why is star formation such an inefficient process that only a few per cent of the total gas mass is turned into stars every dynamical time? How does feedback affect star formation and the ISM? We are not yet in a position to answer most, if any, of the above questions, but we have achieved a certain insight into the star-formation process, which should lead to a fully developed theory of star formation.

In general, little effort has yet been expended on studying the effects of metallicity on star formation. The little work done so far has concentrated on the effects of very low metallicities and the top-heavy IMFs which could result due to the complete lack of cooling below 100 K (Bromm *et al.* 1999, 2002; Abel *et al.* 2000). This does highlight straightaway one potential effect of high metallicities in that the presence of more metals will increase the cooling and potentially decrease the characteristic mass of star formation. In the following sections I will review our current understanding of star formation and highlight how metal-rich conditions could affect the star-formation process and its products.

2 Molecular-cloud formation

The first stage of the star-formation process is the formation of molecular clouds as a necessary physical condition for star formation. We know that molecular clouds are where stars are currently forming in our Galaxy but their formation mechanism and their lifetimes are still being debated (Blitz & Williams 1999). Molecular-cloud lifetimes have been debated for decades, but what is clear is that, since there are few clouds in the nearby Galaxy that are not undergoing star formation, the timescale for an individual cloud to commence forming stars is short, comparable to its dynamical timescale. Mechanisms have been suggested to form molecular clouds from gravitational instabilities, thermal instabilities and the shock-driven molecular-cloud formation as detailed below. What is clear is that dense physical

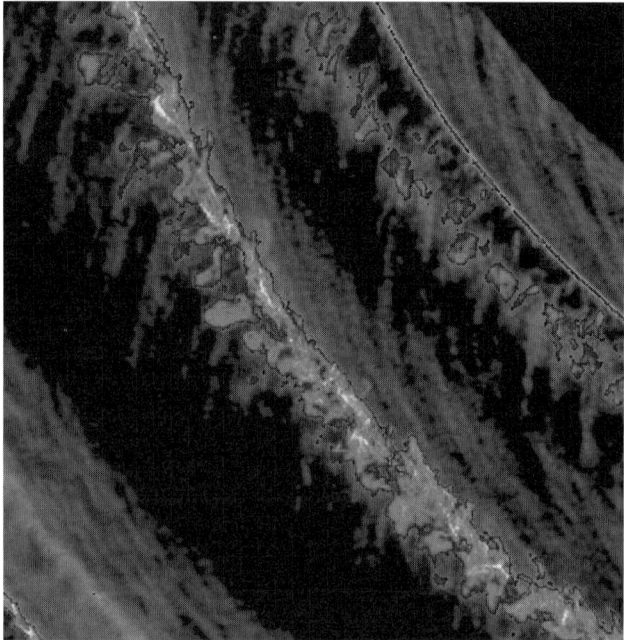

Figure 39.1. The formation of molecular clouds as the gas passes through a spiral shock (Dobbs et al. 2006). Note the spurs and feathering that appear as the dense clumps are sheared away upon leaving the spiral arm.

conditions are necessary in order to get a sufficient collision rate of H atoms onto dust grains.

One way of getting the necessary physical conditions for H_2 formation is through the dense shocks which occur either in turbulent clouds (Glover & Mac Low 2006) or when gas passes through a spiral arm and shocks (Dobbs et al. 2006); see Figure 39.1. In either case, high-Mach-number shocks induce the formation of high-density regions, which can then form molecular gas rapidly. The gas need be cold, possibly due to cooling in the shock itself, in order to attain the high densities required.

Spiral shocks can also explain the large-scale distribution of molecular clouds in spiral arms. Structures in the spiral arms arise due to the shocks that tend to gather material together on converging orbits. Thus, structures grow in time through multiple spiral-arm passages. These structures present in the spiral arms are also found to form the spurs and feathering in the inter-arm region as they are sheared by the divergent orbits when leaving the spiral arms (Dobbs & Bonnell 2006a). In this model, molecular clouds are limited to spiral arms since it is only there that the gas is sufficiently dense to form molecules. These clouds need not be self-gravitating because their formation is independent of self-gravity.

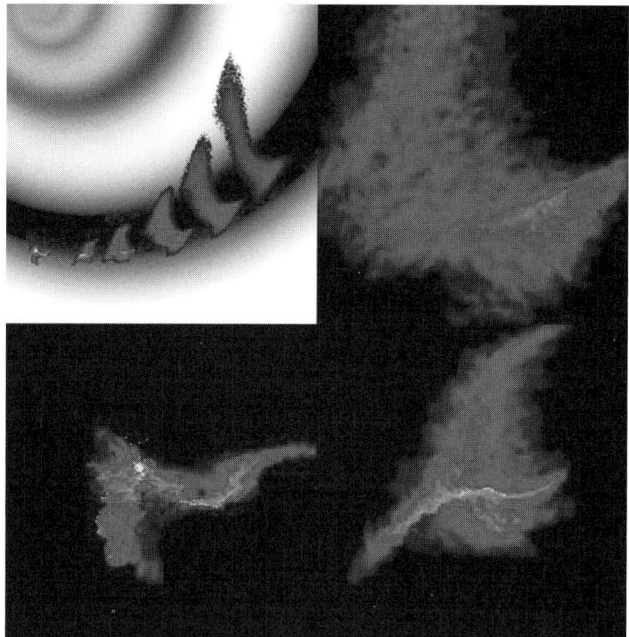

Figure 39.2. The evolution of cold interstellar gas through a spiral arm is shown relative to the spiral potential of the galaxy (upper left). The minimum of the spiral potential is shown as black and the overall galactic potential is not shown for clarity. The three other panels, arranged clockwise, show close-ups of the gas as it is compressed in the shock and sub-regions become self-gravitating. Gravitational collapse leading to star formation occurs within 2×10^6 yr of the gas reaching molecular-cloud densities. The cloud produces stars inefficiently because the gas is not globally bound.

Spiral shocks can also trigger the occurrence of star formation since the shock can produce locally bound regions that undergo collapse to form stars (Roberts 1969; Pringle *et al.* 2001; Bonnell *et al.* 2006). The evolution, over 34 million years, of $10^6 M_\odot$ of gas passing through the spiral potential is shown in Figure 39.2, from Bonnell *et al.* (2006). The initially clumpy, low-density gas ($\rho \approx 0.01 M_\odot$ pc^{-3}) is compressed by the spiral shock as it leaves the minimum of the potential. The shock forms some very-dense ($>10^3 M_\odot$ pc^{-3}) regions, which become gravitatinally bound and thus collapse to form regions of star formation, with masses of typical stellar clusters, namely $(10^2$–$10^4) M_\odot$. Star formation occurs within 2×10^6 yr of the attainment of molecular-cloud densities. The total spiral-arm passage lasts for $\approx 2 \times 10^7$ yr. The gas remains globally unbound throughout the simulation and re-expands in the post-shock region. Star formation in unbound clouds produces low star-formation efficiencies of the order of 10% or less even in the absence of any form of stellar feedback (Clark & Bonnell 2004; Clark *et al.* 2005). In addition, the passage of clumpy gas through a spiral (or other) shock induces a supersonic velocity dispersion that follows the observed $v_{\text{disp}} \propto R^{-0.5}$ relation found in molecular clouds (Bonnell *et al.* 2006; Dobbs & Bonnell 2006b). Thus, a dynamical onset of star formation is able to produce the observable structure and kinematics in molecular clouds.

2.1 Effects of high metallicity

The primary effect of metallicity on molecular-cloud formation and the triggering of star formation is on the rate of H_2 formation. Since this occurs on dust grains, a higher metal content then corresponds to more dust grains and a faster H_2-production rate. Thus in high-metallicity regions molecular clouds should form more rapidly and more efficiently such that the molecular content, and potentially the star-formation rate, should increase. One numerical simulation of H_2 formation in turbulent molecular clouds (S. Glover, personal communication) has shown that a higher metallicity directly corresponds to a higher H_2 fraction in the cloud.

3 Stellar masses: the IMF

One of the main goals for a theory of star formation (Bonnell *et al.* 2006c) is to understand the origin of the stellar IMF. In order to construct a theory for the IMF, we need to explain first the origin of the characteristic stellar mass near $0.3 M_\odot$ and then the decreasing frequency (a power-law distribution) of stars of higher and lower masses. The best explanation to date for the characteristic mass is that it derives from the thermal Jeans mass at the point of fragmentation. It needs to be the thermal Jeans mass because this is the only support mechanism which cannot be fully removed and can therefore be expected to provide a near-universal characteristic mass.

The Jeans mass at the point of fragmentation is most probably set by the thermal physics of the line and dust cooling which counteracts the compressional heating. Larson (2005) has recently re-emphasised that, at low densities, atomic and molecular line cooling results in an effective polytropic $\gamma < 1$ ($\gamma \approx 0.75$), whereas at higher densities, where dust cooling dominates, the effective polytopic γ is $\gamma \approx 1.1$. This transition from a slight cooling to a slow heating regime with compression is likely to set the Jeans mass for fragmentation at $\approx 0.5 M_\odot$, corresponding to the characteristic stellar mass (Jappsen *et al.* 2005). In such a case, the characteristic stellar mass is relatively insensitive to the initial conditions for star formation and thus likely to produce a robust-and-near universal IMF. For example, such an equation of state naturally produces a realistic IMF when an isothermal equation

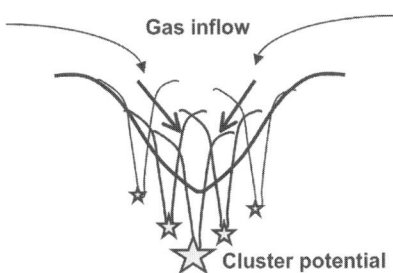

Figure 39.3. A schematic diagram of the physics of accretion in a stellar cluster. The gravitational potentials of the individual stars form a larger-scale potential that funnels gas down to the cluster core. The stars located there are therefore able to accrete more gas and become higher-mass stars. The gas reservoir can be replenished by infall into the large-scale cluster potential.

of state produces an IMF that depends on the exact Jeans mass under the initial conditions (Bonnell *et al.* 2006a). The broad peak can be understood as being due to the dispersion in gas densities and temperature at the point where fragmentation occurs. Lower-mass stars are most probably formed through the gravitational fragmentation of a collapsing region such that the increased gas density allows the production of lower-mass fragments. For them to maintain their low-mass status it is required that they be ejected from their natal environment, or at least accelerated by stellar interactions such that their accretion rates drop to close to zero (Bate *et al.* 2002a).

Higher-mass stars are probably due to continued accretion in a group or cluster environment. This competitive accretion (see Figure 39.3) naturally produces higher-mass stars in the centre of clusters due to the action of the larger-scale gravitational potential funnelling mass down to the centre (Bonnell *et al.* 2001a, 2004). This produces a Salpeter-like IMF (Bonnell *et al.* 2001b) for higher-mass stars due primarily to the increasing gravitational attraction of more-massive stars coupled with the increased gas density due to the cluster potential; see the right-hand panel of Figure 39.4 and Bonnell & Bate (2006). Continued accretion and dynamical interactions can also explain the existence of closer binary stars, and the dependence of properties of binaries on stellar masses (Bate *et al.* 2002b; Bonnell & Bate 2005).

Turbulence has also been invoked to produce a stellar IMF directly from the distribution of clump masses (Padoan & Nordlund 2002). The main difficulties with this possibility are that it relies on a very problematic one-to-one mapping of clump to stellar masses and that it should produce an inverse mass segregation whereby the massive stars are the least likely to form in dense stellar regions (Bonnell *et al.* 2006c).

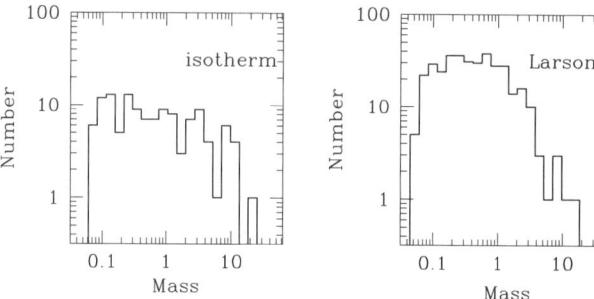

Figure 39.4. The mass functions for the two simulations starting with a Jeans mass of $M_{\rm Jeans} = 5M_\odot$ are shown at $t = 1.55t_{\rm eff}$. The isothermal simulation (left panel) produces an unrealistically flat IMF extending to high masses whereas the Larson (2005)-type barotropic equation of state (right panel) produces a realistic IMF with a characteristic knee at $\approx M_\odot$. This barotropic equation of state effectively reproduces the low-Jeans-mass results from more-general initial conditions. From Bonnell et al. 2006a).

3.1 Effects of high metallicity

The primary effect on the IMF in a metal-rich environment is due to the cooling from atomic and molecular lines and to dust. Increasing the metal content will increase the cooling rate such that the collapsing gas should be cooler at a given density, given the same background radiation field (Spaans & Silk 2000). The increased dust abundance results in a lowering of the Jeans mass according to $M_{\rm Jeans} \propto Z^{-3/8}$ such that a region which has twice Solar metallicity should have a characteristic mass 0.75 times that in the Solar neighbourhood. This decrease in the thermal Jeans mass would result in more fragmentation and a general shift towards lower-mass stars (e.g. Bate & Bonnell 2005). A large increase in the metallicity would result in a significantly different IMF shifted towards lower masses (see Figure 39.5).

A second, although smaller, effect is that the opacity limit should also decrease with increasing metallicity according to $M_{\rm opacity} \propto Z^{-1/7}$ (Low & Lynden-Bell 1974). This small decrease would allow objects of slightly smaller mass to form but is likely to be lost among the uncertainties over the opacity limit in non-spherical geometries (Boyd & Whitworth 2005).

4 Accretion and the formation of massive stars

The model for the IMF discussed above provides a framework in which to understand the formation of massive stars. In this scenario, the massive stars form due to competitive accretion onto the core of the cluster in which the massive star is forming. Numerical simulations have indicated that the vast majority of the mass which comprises the massive stars comes from large distances and is accreted onto

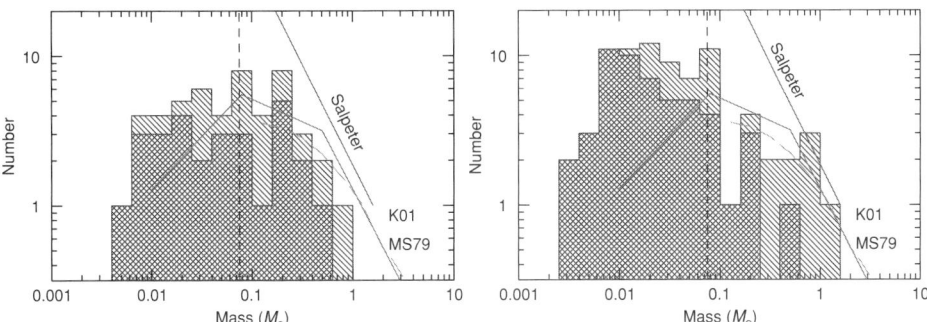

Figure 39.5. A comparison between two calculations with identical initial conditions except for the Jeans mass, which is $M_\odot$ in the left panel and $0.3 M_\odot$ in the right panel. The lower Jeans mass could correspond to an increase in the metallicity by a factor of 25. From Bate & Bonnell (2005).

the star after a stellar cluster has formed (Figure 39.6) (Bonnell *et al.* 2004). The initial fragment mass which forms the star is of low mass, as is typical for the mean stellar mass. The infalling gas then has to pass through the cluster in order for it to be accreted by the central massive star. It should be noted that these simulations neglect the effect of radiation pressure from the massive stars, or equivalently assume that accretion through a disc (Yorke & Sonnhalter 2002) occurs.

An alternative model (McKee & Tan 2003) envisages that massive-star formation is just a scaled-up version of low-mass-star formation whereby an individual massive clump collapses to form a single massive star. The potential problem with this scenario is that massive stars are formed in dense stellar clusters where stars are closely packed together. There is at present no known physical mechanism to produce the necessary initial conditions for a clump that will not be susceptible to fragmentation (Dobbs *et al.* 2005) long before reaching the state of collapsing to form a single star.

In competitive accretion, the infalling gas is accompanied by newly formed stars such that the formation of a massive star is a necessary byproduct of the formation of a stellar cluster. This produces a strong correlation between the number of stars, or the total mass in a cluster, and the mass of the most-massive star therein (Figure 39.7). This correlation follows $M_{\max} \propto M_{\text{stars}}^{2/3}$, where M_{stars} is the total mass in the cluster stars. This can be understood in the following way. Given an effective initial efficiency of fragmentation, for every star that falls into the cluster a certain amount of gas also enters the cluster. This gas joins the common reservoir from which the most-massive star takes the largest share in this competitive environment. Thus, the mass of the most-massive star increases as the cluster grows in numbers of stars. We therefore have a prediction from this model that there should exist a strong

Figure 39.6. The mass that forms the most-massive ($30M_\odot$) star is plotted at three different times during the formation of a stellar cluster. Note that the gas is very distributed when star formation is initiated (left) and when the cluster is growing through the infall of newly formed stars and gas (middle and right). The vast majority of the final mass is due to competitive accretion in a clustered environment (Bonnell *et al.* 2004).

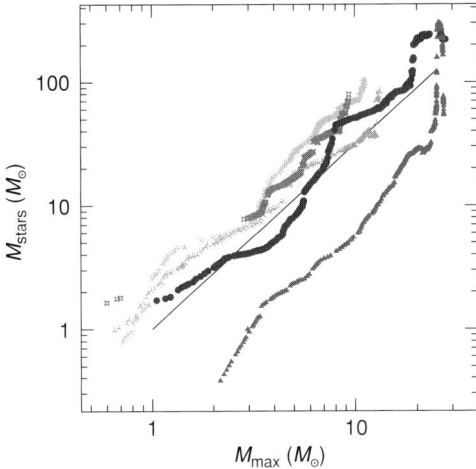

Figure 39.7. The total mass in stars is plotted against the mass of the most-massive star in the system. The hypothesis of competitive accretion predicts that, as the total mass grows, the mass of the most-massive star also increases as $M_{\max} \propto M_{\text{stars}}^{2/3}$ (Bonnell *et al.* 2004).

correlation between the mass of the most-massive star and the number of stars (or the total mass) in the cluster (Figure 39.7) (Bonnell *et al.* 2004).

5 Massive stars

The formation of massive stars, with masses in excess of $10 M_\odot$, is problematic due to the high radiation pressure on dust grains, which may actually halt the infall of gas and thus appears to limit stellar masses. Simulations to date suggest that this sets an upper mass limit to accretion somewhere in the $(10$–$40) M_\odot$ range (Wolfire & Casinelli 1986; Yorke and Sonnhalter 2002; Edgar and Clarke, 2004). Clearly, there needs to be a mechanism for circumventing this problem because stars as massive as $(80$–$150) M_\odot$ exist (Figer 2005). Possible solutions include disc accretion and radiation beaming (Yorke & Sonnhalter 2002), Rayleigh Taylor instabilities in the infalling gas (Krumholz *et al.* 2005a), stellar collisions between single stars (Bonnell *et al.* 1998; Bonnell and Bate 2002; Bally & Zinnecker 2005) and the collisional merger of close binary systems (Bonnell & Bate 2005) (Figure 39.8).

If radiation pressure does prove to be a significant impediment to the formation of massive stars, then a metal-rich environment should prove more difficult for the formation of these objects. The expectation is then that there should be fewer massive stars formed in such environments and especially that a signature in terms of the IMFs should be apparent as a function of metallicity.

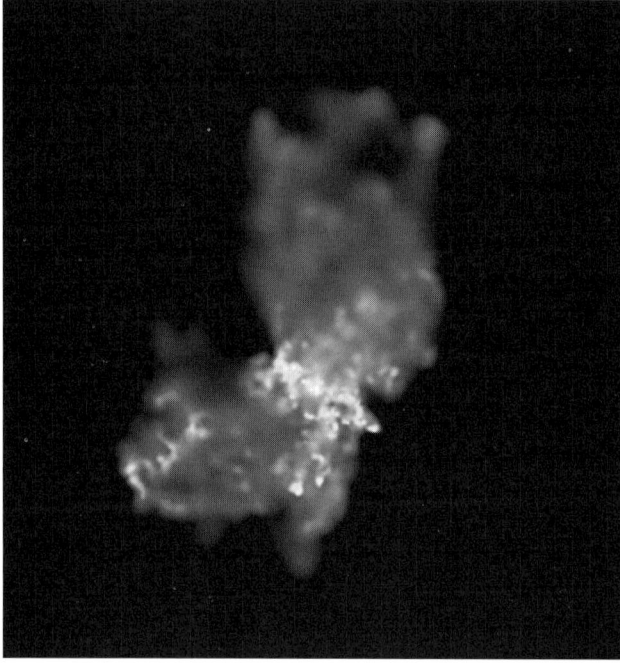

Figure 39.8. The stellar wind from a massive star, collimated by a nearly edge-on circumstellar disc, from a numerical simulation of wind feedback in star formation. Note that the intrinsically spherical wind is collimated by the external distribution of gas in the circum-binary disc.

6 Feedback

We know that feedback from young stars is likely to play an important role in the star-formation process, potentially either stopping or even triggering subsequent star-forming events. To date, there have been few studies of feedback in numerical simulations of star formation. Li & Nakamura (2006) attempted to model the effects of jets, while Dale et al. (2005) included ionisation from O stars. One of the conclusions from the latter study is that the feedback tends to escape along any weak points in the environment, allowing accretion to occur relatively unimpeded unless the feedback is sufficiently strong to evacuate the region completely; see also Krumholz et al. (2005b). Stellar winds, especially from massive stars, are also likely to play an important role in the star-formation process. Since the wind parameters from massive stars depend on the metallicity, the increased kinematical feedback in high-metallicity regions should increase the role of stellar winds.

7 Conclusions

Current work on star formation has been developing a paradigm of star formation on dynamical timescales. In this scenario, molecular clouds form rapidly and start forming stars almost immediately. Fragmentation, combined with thermal physics, determines the characteristic mass of star formation while accretion and dynamical interactions in groups and clusters can explain both the high- and the low-mass tail to the IMF. There are still some significant unknowns, including how star formation is initiated, the exact role of turbulence, the formation of massive stars and the role of feedback. With this in mind, only general considerations can be applied to metal-rich conditions. The largest effect of higher metallicity will be in terms of molecular-cloud formation. High metallicity should also decrease the mean stellar mass and perhaps the low-mass limit, but neither should be affected significantly in the absence of very-metal-rich gas. The formation of high-mass stars is potentially very susceptible to metal-rich conditions if radiation pressure does significantly impede continued accretion. Lastly, feedback in the form of stellar winds will also be more important in the metal-rich Universe, although at present these effects are still very uncertain even at Solar metallicity.

References

Abel, T., Bryan, G. L., & Norman, M. L. (2000), *ApJ* **540**, 39
Bally, J., & Zinnecker, H. (2005), *AJ* **129**, 2281
Bate, M. R., & Bonnell, I. A. (2005), *MNRAS* **356**, 1201
Bate, M. R., & Bonnell, I. A., & Bromm, V. (2002a), *MNRAS* **332**, L65
   (2002b), *MNRAS* **336**, 705
Blitz, L., & Williams, J. (1999), *The Origin of Stars and Planetary Systems*, eds C. J. Lada
   & N. D. Kylafis (Dordrecht, Kluwer), p. 3
Bonnell, I. A., & Bate, M. R. (2002), *MNRAS* **336**, 659
   (2005), *MNRAS* **362**, 915
   (2006), *MNRAS* **370**, 488
Bonnell, I. A., Bate, M. R., Clarke, C. J., & Pringle, J. E. (2001a), *MNRAS* **323**, 785
Bonnell, I. A., Bate, M. R., & Vine, S. G. (2003), *MNRAS* **343**, 413
Bonnell, I. A., Bate, M. R., & Zinnecker, H. (1998), *MNRAS* **298**, 93
Bonnell, I. A., Clarke, C. J., Bate, M. R., & Pringle, J. E. (2001b), *MNRAS* **324**, 573.
Bonnell, I. A., Clarke, C. J., & Bate, M. R. (2006a), *MNRAS* **368**, 1296
Bonnell, I. A., Dobbs, C. L., Robitaille, T. P., & Pringle, J. E. (2006b), *MNRAS* **365**, 37
Bonnell, I. A., Larson, R. B., & Zinnecker, H. (2006c), in *Protostars and Planets V*, eds
   B. Reipurth, *et al.*, in press
Bonnell, I. A., Vine, S. G., & Bate, M. R. (2004), *MNRAS* **349**, 735
Bromm, V., Coppi, P. S., & Larson, R. B. (1999), *ApJ* **527**, L5
   (2002), *ApJ* **564**, 23
Clark, P. C., & Bonnell, I. A. (2004), *MNRAS* **347**, L36
   (2006), *MNRAS* **368**, 1787
Clark, P. C., Bonnell, I. A., Zinnecker, H., & Bate, M. R. (2004), *MNRAS* **359**, 809
Dale, J. E., Bonnell, I. A., Clarke, C. J., & Bate, M. R. (2005), *MNRAS* **358**, 291
Dobbs, C. L., & Bonnell, I. A. (2006a), *MNRAS* **367**, 873
   (2006b), *MNRAS*, (in press) (arXiv:astro-ph/0610720)
Dobbs, C. L., Bonnell, I. A., & Clark, P. C. (2005), *MNRAS* **360**, 2
Dobbs, C. L., Bonnell, I. A., & Pringle, J. E. (2006), *MNRAS* **371**, 1663
Edgar, R., & Clarke, C. (2004), *MNRAS* **349**, 678

Elmegreen, B. (2000), *ApJ* **530**, 277
Figer, D. F. (2005), *Nature* **434**, 192
Glover, S. C. O., & Mac Low, M.-M. (2006), astro, arXiv:astro-ph/0605121
Jappsen, A.-K., Klessen, R. S., Larson, R. B., Li, Y., & Mac Low, M.-M. (2005), *A&A* **435**, 611
Klessen, R., Spaans, M., & Jappsen, A. K. (2006), *MNRAS* in press (arXiv:astro-ph/0610557)
Krumholz, M., Klein, R. I., & McKee, C. F. (2005b), *ApJ* **618**, L33
Krumholz, M., Klein, R. I., & McKee, C. F. (2005a), in *Massive Star Birth: A Crossroads of Astrophysics*, eds R. Cesaroni, *et al.* pp. 231ff
Lada, C. J., & Lada, E. (2003), *ARA&A* **41**, 57
Larson, R. B. (2003), *Rep. Prog. Phys.* **66**, 1651
  (2005), *MNRAS* **359**, 211
Li, Z.-Y., & Nakamura, F. (2006), *ApJ* **640**, L187
McKee, C. F., & Tan, J. C. (2003), *ApJ* **585**, 850
Mac Low, M. M., & Klessen, R. S. (2004), *RvMP* **74**, 125
Padoan, P., & Nordlund, Å. (2002), *ApJ* **576**, 870
Pringle, J. E., Allen, R., & Lubow, S. H. (2001), *MNRAS* **327**, 663
Shu, F. H., Adams, F. C., & Lizano, S. (1987), *ARA&A* **25**, 23
Spaans, M., & Silk, J. (2000), *ApJ* **538**, 115
Wolfire, M. G., & Cassinelli, J. P. (1987), *ApJ* **319**, 850
Yorke, H., & Sonnhalter, C. (2002), *ApJ* **569**, 846

40

Metallicity of Solar-type main-sequence stars: seismic tests

Sylvie Vauclair

Laboratoire d'Astrophysique de Toulouse Tarbes; Observatoire Midi-Pyrénées; Université Paul Sabatier, Toulouse, France

Among the Solar-type stars observed in the Galaxy, many appear to be metal-rich relative to the Sun. The case of exoplanet-host stars is particularly interesting in that respect since they present, on average, an overmetallicity of 0.2 dex. This metallicity is probably original, from the protostellar nebula, but it could also have been increased by accretion of hydrogen-poor material during the early stage of planetary formation. Asteroseismic studies provide an excellent way to determine the internal structure and chemical composition of these stars. Such studies may also establish constraints on the external parameters (gravity, effective temperature, metallicity) that are more precise than the constraints obtained from spectroscopy. After a general discussion on this subject, I present the special cases of three stars: μ-Arae, which was observed with the HARPS spectrograph in June 2004; ι-Horologii, which has been modeled in detail and will be observed with HARPS in November 2006; and finally HD 52265, one of the main targets of the COROT mission, an exoplanet-host star that will be observed with the COROT satellite for five consecutive months.

1 Introduction

Since the first discovery of a planet orbiting around Peg 51 (Mayor & Queloz 1995), nearly 200 exoplanets have been detected. Owing to the bias of the detection techniques (radial velocity or transit methods), the planetary systems observed are different from the Solar System: only planets orbiting close to the central star can give observable effects. Most of them are "hot Jupiters," i.e. Jupiter-like planets at distances of the order of or less than one astronomical unit.

The Metal-rich Universe, eds. G. Israelian and G. Meynet. Published by Cambridge University Press.
© Cambridge University Press 2008.

The central stars of these planetary systems appear to be overmetallic compared with the Sun, at least in their atmospheres. Their average metallicity is ~0.2 dex larger than Solar, whereas the average metallicity of stars for which no planets have been detected is about Solar (Santos *et al.* 2003, 2005; González 2003; Fischer & Valenti 2005). Two scenarios have been proposed to explain these high metallicities (Bazot and Vauclair 2004). In the first scenario, they are the result of a high initial metal content in the proto-stellar gas, whereas in the second scenario the overmetallicity is due to the accretion of hydrogen-poor matter during planetary formation.

The initial-overmetallicity scenario seems more probable than the accretion scenario. The first argument which was given against accretion was that the observed overmetallicity does not vary with the stellar mass while the masses of the outer convective zones may differ strongly. This argument does not hold, however, due to the extra convection induced by the inverse μ-gradient (Vauclair 2004). On the other hand, the huge mass which should be accreted while the star is already on the main sequence (about 100 Earths) does not seem realistic.

An important argument generally given in favor of a primordial overmetallicity is related to planet formation, since the metallic overabundance should help planet condensation. Here we must stress again that the observed planetary systems have "hot Jupiters," which means that giant planets have migrated from their formation site, far from the star, to their present situation. Stars without detected planets may very well host giant planets at large orbits, as in the Solar System: viewed from elsewhere with the present detectors, the Sun would appear to be a star without planets! In this framework, the importance of overmetallicity is not directly related to the fact that planets may be formed, but rather to the timescale of this formation.

Asteroseismic studies constitute an excellent tool to help in determining the structural differences between stars with and without detected planets. Bazot & Vauclair (2004) studied the internal differences of stellar models computed with the overmetallicity and accretion assumptions, but iterated so as to present exactly the same observable parameters ($T_{\rm eff}$, $L/L_\odot$, Z, $\log g$): these models can account for the same observed star, but their modeled interiors are different. Moreover, they may have different masses while their observed parameters are the same. It is also possible that, of two models representing the same observed star, one has a convective core while the other does not.

In the following sections, I first give some generalities about the modeling procedures, and then present the cases of three particular stars: HD 160691 (alias μ-Arae), which was observed for seismology with the HARPS spectrograph in June 2004; HD 17051 (alias ι-Horologii), which we have studied in detail and which will be observed with HARPS in November 2006; and finally HD 52265, one of the main

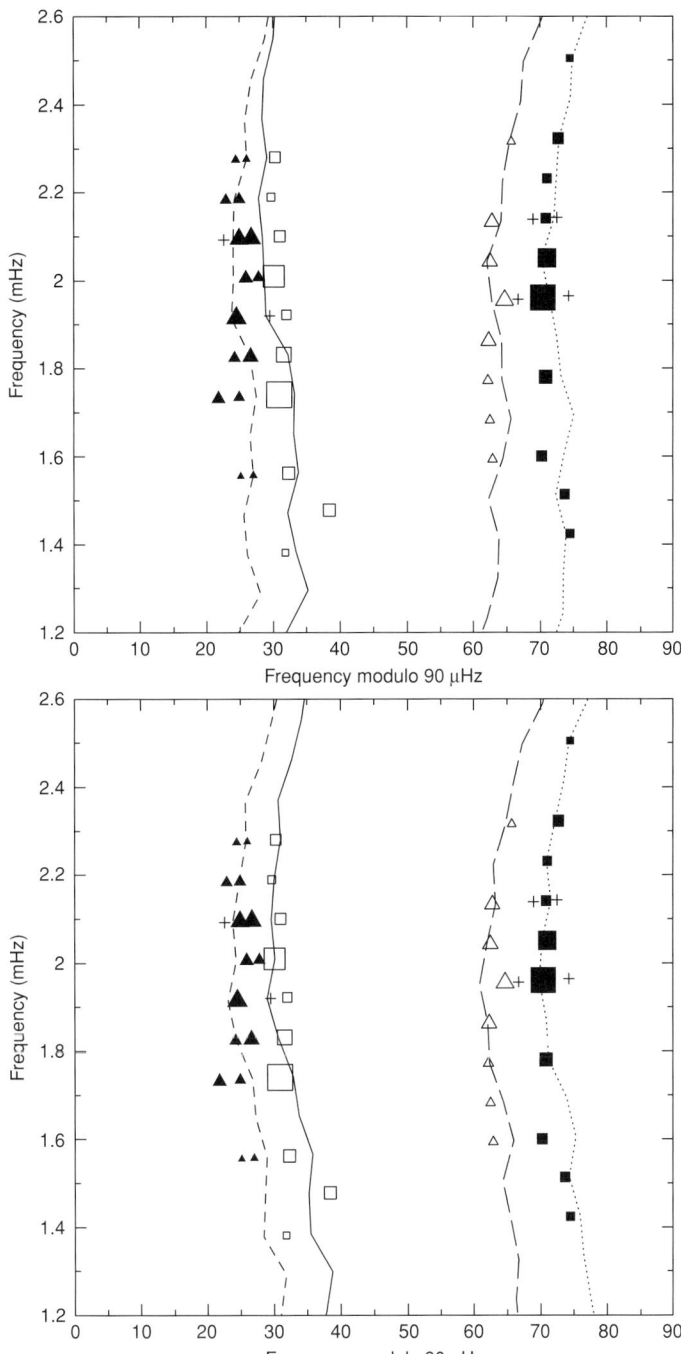

Figure 40.1. Echelle diagrams for two models of the star μ-Arae: one with primordial overmetallicity (top) and one with accretion (bottom). The ordinate represents the frequencies of the modes and the abseissa the same frequencies modulo the large separation, here 90 μHz. The lines represent the computations, from left to right $l = 2, 0, 3$, and 1, respectively, and the symbols represent the observations. After Bazot *et al.* (2005).

Table 40.1. *Observed parameters for ι-Hor, after Laymand & Vauclair (2006)*

T_{eff} (K)	log g	[Fe/H]	Reference
6,136 ± 34	4.47 ± 0.05	0.19 ± 0.03	González *et al.* (2001)
6,252 ± 53	4.61 ± 0.16	0.26 ± 0.06	Santos *et al.* (2004)
6,097 ± 44	4.34 ± 0.06	0.11 ± 0.03	Fischer & Valenti (2005)

targets of the COROT mission, an exoplanet-host star that will be observed for five consecutive months.

2 Modeling

In all the computations and comparisons presented below, the models were computed using the Toulouse–Geneva stellar-evolution code (TGEC), with the OPAL equation of state and opacities (Rogers & Nayfonov 2002; Iglesias & Rogers 1996) and the NACRE nuclear reaction rates (Angulo *et al.* 1999). In all models microscopic diffusion was included using the Paquette prescription (Paquette *et al.* 1996, Richard *et al.* 2004). The treatment of convection was done in the framework of the mixing-length theory and the mixing-length parameter was adjusted as in the Solar models ($\alpha = 1.8$), except in some specific cases that are specially discussed. Various kinds of models were computed, according to the initial assumptions: overmetallic models with two different initial helium values, and accretion models.

For the computations of overmetallic models, the helium value is crucial because differences in helium may lead to completely different evolutionary tracks. For the computations of the accretion models, how accretion occurred and the composition of the accreted matter are also subjects of debate. In the models, the simplest assumption was used: instantaneous accretion of matter with Solar composition for metals and no light elements, at the beginning of the main sequence. Neither extra mixing nor overshoot was included.

Adiabatic oscillation frequencies were computed using the PULSE code adapted from Brassard *et al.* (1992). For each evolutionary track, many models were computed inside the observed spectroscopic boxes (color–magnitude or log g–log T_{eff} diagrams, as described below). The most useful combinations of the oscillation frequencies (large and small separations, echelle diagrams) were computed for comparisons with observational data already available or to be obtained in future.

3 The example of μ-Arae

The exoplanet-host star μ-Arae is a G5V star with a visual magnitude of $V = 5.1$, a Hipparcos parallax $\pi = 65.5 \pm 0.8$ marcsec, which gives a distance to the Sun

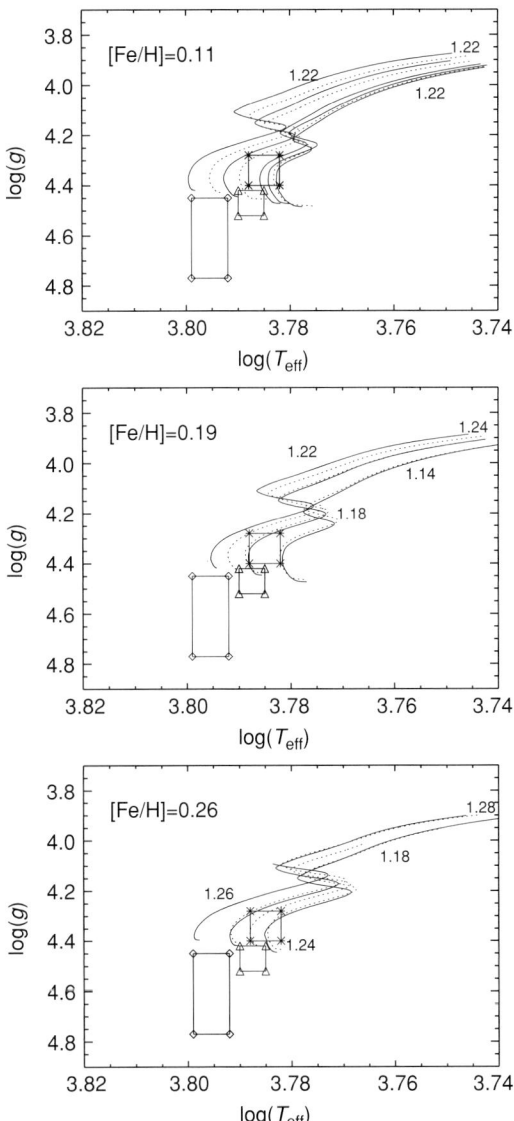

Figure 40.2. These graphs display the error boxes for the position of the star ι-Hor in the log g–log T_{eff} diagram, as given by the three groups that observed this star (Table 40.1). The author's boxes which correspond to the chosen metallicity, in each graph, are emphasized with boldface lines. After Laymand and Vauclair (2006).

398   *Metallicity of Solar-type main-sequence stars*

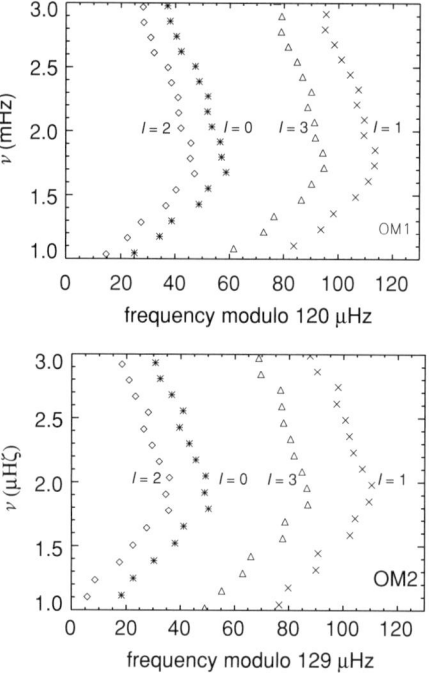

Figure 40.3. Echelle diagrams for two possible models (OM1 and OM2) of the exoplanet-host star ι-Hor, after Laymand and Vauclair (2006). Observations of this star were to be carried out with the HARPS spectrometer in November 2006.

of 15.3 pc, and a luminosity of $\log(L/L_\odot) = 0.28 \pm 0.012$. Spectroscopic observations by various authors gave five different effective temperatures and metallicities; see references in Bazot *et al.* (2005). The HARPS observations allowed the identification of 43 oscillation modes of degrees $l = 0$ to $l = 3$ (Bouchy *et al.* 2005). From the analysis of the frequencies and comparison with models, the values $T_{\rm eff} = 5813 \pm 40$ K and [Fe/H] $= 0.32 \pm 0.05$ dex were derived: these values, which lie inside the spectroscopic boxes, were obtained with a much better precision than is available from spectroscopy.

For each evolutionary track, many models were computed inside the observed spectroscopic boxes in the HR diagram, but only those which could reproduce the observed echelle diagram were kept for subsequent tests (Figure 40.1). For these models the large separation $\Delta\nu_l = \nu_{n+1,l} - \nu_{n,l}$ is exactly 90 µHz: a difference of only 0.5 µHz completely destroys the fit with the observations.

We can see a clear difference between the overmetallic and the accretion cases for the lines $l = 0$ and $l = 2$: in the overmetallic case they come closer at large frequencies and even cross around $\nu = 2.7$ mHz, which does not happen in the accretion case. This behavior clearly appears in the representations of the small

Table 40.2. *Observed parameters for HD 52265; see Soriano* et al. *(2006).*

T_{eff} (K)	log g	[Fe/H]	authors
6,103 ± 52	4.28 ± 0.12	0.23 ± 0.05	Santos *et al.* (2004)
6,162 ± 22	4.29 ± 0.04	0.27 ± 0.02	González *et al.* (2001)
6,076 ± 44	4.26 ± 0.06	0.19 ± 0.03	Fischer & Valentini (2005)
6,069 ± 15	4.12 ± 0.03	0.19 ± 0.03	Takeda *et al.* (2005)
6,179 ± 18	4.36 ± 0.03	0.24 ± 0.02	Gillon & Magain (2006)

separations; see specific figures in Bazot *et al.* (2005). I will come back to this important effect below, for another star. For μ-Arae it was not yet possible to decide which scenario was the best one, in spite of the very good data obtained, but we are still working on it.

4 Modeling of the particular star ι-Horologii

Among exoplanet-host stars, ι-Hor is a special case for several reasons. Three groups have published different stellar parameters for this star (Table 40.1). Meanwhile, Santos *et al.* (2004) suggested a mass of $1.32 M_\odot$, whereas Fischer and Valenti (2005) gave it a mass of $1.17 M_\odot$.

Laymand & Vauclair (2006) computed evolutionary tracks and models lying inside the error boxes given by the observers (Figure 40.2). The oscillation frequencies for several characteristic models, which can account for the spectroscopic observations, have been computed. Two of them are presented as echelle diagrams in Figure 40.3. From this preliminary study, some important conclusions have already been drawn: a metallicity larger than [Fe/H] = 0.20 dex is quite unprobable and the stellar mass cannot exceed $1.22 M_\odot$.

Some authors (Grenon 2000; Chereul and Grenon 2000; Kalas and Delorn 2006) pointed out that this star has the same kinematical characteristics as the Hyades: as with many other stars observed in the whole sky, it belongs to the "Hyades stream." This may be due to dynamical effects in the Galaxy, but it is also possible that the star formed together with the cluster and moved out due to evaporation. Models with the metallicity and the age of the Hyades are indeed quite realistic. Comparison of asteroseismic observations of this star, which will be done in November 2006 with the HARPS spectrometer in La Silla, with model predictions should allow us to evaluate whether this star formed together with the Hyades. While a negative answer would be interesting in itself, a positive answer would be taken as proof that the overmetallicity has a primordial origin.

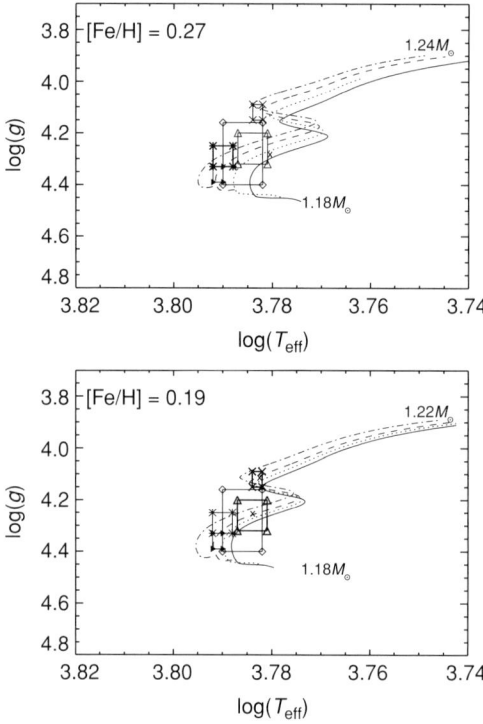

Figure 40.4. These graphs display the error boxes for the position of the star HD 52265 in the log g–log T_{eff} diagram, as given by the five groups that observed this star (Table 40.2). The author's boxes which correspond to the chosen metallicity, in each graph, are emphasized with boldface lines. After Soriano *et al.* (2006)

5 The "COROT star" HD 52265

For the "COROT star" HD 52265, which will be observed for five consecutive months, we expect very precise data, which it is hoped will lead to very good frequency determinations and mode identifications. We will use the same kind of tests as before, and we expect to be able to draw precise conclusions on the internal structure and past history of this exoplanet-host star.

Preliminary computations and modeling of HD52265 have been done using the same techniques as for μ-Arae, as a preparation for the future observations with COROT. Five groups of observers have given external parameters for HD 52265 (Table 40.2). Its luminosity is $L/L_\odot = 1.94 \pm 0.16$ (Soriano *et al.* 2006).

Evolutionary tracks have been computed for three different metallicities, as given from spectroscopic observations (Figure 40.4). Oscillation frequencies and seismic tests have been studied for specific models. Here I show only two extreme and interesting examples, for metallicities [Fe/H] = 0.19 and 0.27.

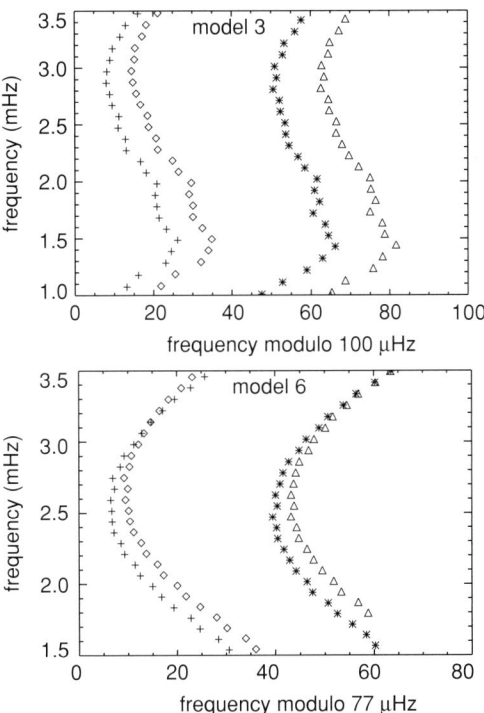

Figure 40.5. Echelle diagrams computed for two extreme models of HD 52265, the first one with metallicity 0.27 inside the González et al. (2001) error box (model 3) and the second with metallicity 0.19 inside the Takeda et al. (2005) error box (model 6).

The echelle diagrams for these two models are given in Figure 40.5. They present important differences. The most interesting feature is found in the Takeda et al. (2005) case. The $l = 0$ and $l = 2$ lines cross for a frequency around 3 mHz, and the $l = 1$ and $l = 3$ lines also come very close. This is related to the presence of a convective, helium-rich core, as discussed in Soriano et al. (2006).

6 Conclusion

Overmetallicity is a frequent phenomenon in stars of our Galaxy. The case of exoplanet-host stars is especially interesting in that respect, because most of these stars are observed to be overmetallic relative to the Sun, with an average value of 0.2 dex. This observed feature is certainly related not only to the planet formation but also to the reason why giant planets migrate towards the central star in some cases. The stars for which no planets have been detected may very well host giant planets that have not migrated, as in the Solar System. The connection between overmetallicity and planet migration may be related to the timescale of planet formation, decreasing with increasing metallicity: we may expect that planets that are formed rapidly can have time to migrate, due to friction with the remaining disc. In conclusion, asteroseismology of metal-rich stars, particularly exoplanet-host stars, is an important new subject, from which we expect many developments in the near future.

References

Angulo C., Arnould M., & Rayet M. (NACRE collaboration) (1999), *Nucl. Phys. A* **656**, 1 (http://pntpm.ulb.ac.be/Nacre/nacre.htm)
Bazot, M., & Vauclair, S. (2004), *A&A* **427**, 965
Bazot, M., Vauclair, S., Bouchy, F., & Santos, N. (2005), *A&A* **440**, 615
Bouchy, F., Bazot, M., Santos, N., Vauclair, S., & Sosnowska, D. (2005), *A&A* **440**, 609
Brassard, P. *et al.* (1992), *ApJS* **81**, 747
Chereul & Grenon, M. (2000), in *Dynamics of Star Clusters and the Milky Way*, eds Deiters *et al.* (San Francisco, CA, Astronomical Society of the Pacific)
Fischer & Valenti (2005), *ApJ*, **622**, 1102
Gillon & Magain (2006), *A&A* **448**, 341
González, G. (2003), *Rev. Mod. Phys.* **75**, 101
González, G., *et al.* (2001), *AJ* **121**, 432
Grenon, M. (2000), in *The Evolution of the Milky Way*, eds Matteucci & Giovanelli ed., p. 47
Iglesias, C. A., & Rogers, F. J. (1996), *ApJ* **464**, 943
Kalas & Deltorn (2006), submitted to *ApJ*
Laymand, M., & Vauclair, S. (2006), *A&A* (in press), astro-ph/0610501
Mayor, M., & Queloz, D. (1995), *Nature* **378**, 355
Paquette, C., Pelletier, C., Fontaine, G., & Michaud, G. (1986), *ApJS* **61**, 177
Richard, O., Théado, S., & Vauclair, S. (2004), *So Ph* **220**, 243
Rogers, F. J., & Nayfonov, A. (2002), *ApJ* **576**, 1064
Santos, N. C., Israelian, G., & Mayor, M. (2004), *A&A* **415**, 1153
Santos, N. C., Israelian, G., Mayor, M. *et al.* (2005), *A&A* **437**, 1127
Santos, N. C., Israelian, G., Mayor, M., Rebolo, R., & Udry, S. (2003), *A&A* **398**, 363
Soriano, M., Laymand, M., Vauclair, S., & Vauclair, G. (2006), in preparation
Takeda, Y., Ohkubo, M., Sato, B., Kambe, E., & Sadakane, K. (2005), *PASJ* **57**, 27
Vauclair, S. (2004), *ApJ* **605**, 874

41

Chemical-abundance gradients in early-type galaxies

Patricia Sánchez-Blázquez

University of Central Lancashire, Preston PR1 2HE, UK

We present long-slit spectra for 11 early-type galaxies observed with the Keck telescope. We measure rotation-velocity and velocity-dispersion profiles together with 20 Lick line-strength gradients. Gradients of indices are transformed into ages, metallicities and [α/Fe] using stellar-population models that take into account variations in chemical-abundance ratios. We find that the line-strength gradients are mainly due to radial variations of metallicity, although small gradients of [α/Fe] and age are also present. Contrary to what is expected in simple collapse models, galaxies in our sample have both positive and negative [α/Fe] profiles. This rules out a solely inside-out or outside-in formation mechanism for all early-type galaxies. Metallicity gradients correlate with the shape of the isophotes and the rotational velocity but do not correlate with the mass of the galaxies. Galaxies with younger populations in their centres have steeper metallicity gradients. Our results suggest a scenario whereby galaxies form through the merger of smaller structures and the degree of dissipation during those mergers increases when the masses of the progenitor galaxies decrease.

1 Introduction

The formation and evolution of massive, early-type galaxies constitutes a long-standing and crucial problem in cosmology. Studies of the stellar population in the centres of the galaxies have revealed that more-massive galaxies are older than less-massive ones and that they formed their stars on longer timescales. Modern versions of monolithic collapse and hierarchical models of galaxy formation can reproduce these star-formation histories by tuning their feedback prescriptions. However, any successful model of galaxy formation should predict not only the

The Metal-rich Universe, eds. G. Israelian and G. Meynet. Published by Cambridge University Press.
© Cambridge University Press 2008.

stellar population in the centre but also the radial variations. Studies of stellar population gradients have been sparse up to now due to the high signal-to-noise ratio necessary to reach the external parts of the galaxies. Here we present very high-quality line-strength gradients for a sample of 11 early-type galaxies covering a wide range of luminosities.

2 Data and analysis

The observations were made using LRIS (Oke *et al.* 1995) in long-slit mode using the Keck II telescope. Standard data-reduction procedures were performed. The error frames were treated in parallel with the science images and, therefore, an associated error was generated for each individual data spectrum. From each full galaxy frame, a final frame was created by extracting spectra along the slit, binning in the spatial direction to guarantee a minimum signal-to-noise ratio per Å of 50 in the spectral region of the Hβ index.

We derived the age, [Fe/H] and α-element abundance ratios for all the galaxies of the sample using the χ^2-minimisation technique detailed in Proctor & Sansom (2002) with the models published by Thomas *et al.* (2003, TMB03 hereafter) and 19 Lick indices. The α-enhancement is parametrised in TMB03 models by [E/Fe]: the abundance ratio of the α-elements O, Ne, Mg, Si, S, Ar, Ca and Ti plus the elements N and Na. (see the original references for details). To quantify the profiles we performed a linear fit weighting with the errors in the y-direction and excluding the central points in order to avoid seeing effects. We fitted only those regions of the profiles which had a linear variation.

3 Results and discussion

One way to determine the evolutionary paths of early-type galaxies is to study the relation between the metallicity gradients and other global properties of these systems, since different physical processes are expected to lead to different correlations. For example, dissipational processes are believed to create steeper gradients in more-massive galaxies (Larson 1974; Bekki & Shioya 1999), although this is sensitive to the feedback prescription adopted in the simulations (e.g. Bekki & Shioya 1999). Dissipationless mergers of galaxies, in contrarst, are expected to produce some dilution of the gradients in galaxies (White 1980), deleting or producing an inverse correlation among stellar population gradients and mass. Figure 41.1 shows the correlation of the metallicity with the central velocity dispersion (which is a proxy for virial mass) for our sample of galaxies. Although the sample is not very large, we confirm the lack of correlation previously noted by other authors using line-strength indices (e.g. Gorgas *et al.* 1990; Mehlert *et al.* 2003). It is not galaxies

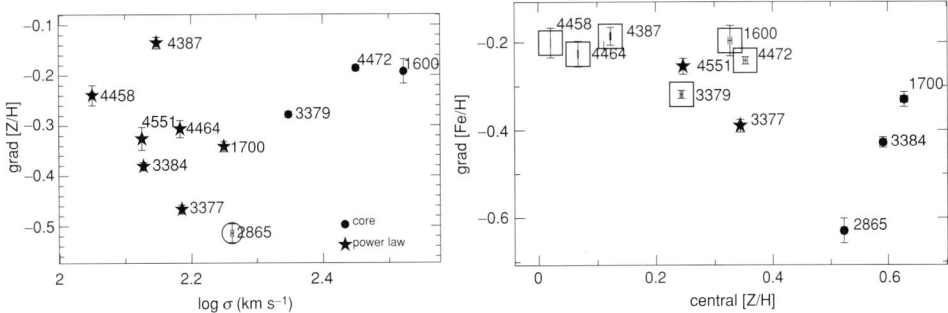

Figure 41.1. The relations between the metallicity gradients and the central velocity dispersion (left panel) and between the metallicity gradient and the metallicity (right panel).

with steeper gradients that are the most massive, but rather the ones with $\sigma \sim 120$ km s^{-1} (corresponding roughly to galaxies with $M_B \sim -20.5$, $M \sim 10^{11} M_\odot$). In the figure we also distinguish between galaxies with different shapes of the inner profile. It can be seen there that *power-law* and *core* galaxies follow different trends in this plane. The trends could be explained by assuming that more-massive galaxies formed via mergers without dissipation (and that more-massive galaxies have suffered more mergers) whereas less-massive ones have grown by gas-rich mergers.

The right panel in Figure 41.1 shows the relation between the central metallicity and the metallicity gradient. We have also separated the galaxies as a function of their central ages. There exists a trend for galaxies with higher central metallicities also to have steeper metallicity gradients, although there is considerable scatter in the relation. We also find a relation between the metallicity gradient and the central age, whereby galaxies with a younger age in their centres also have steeper metallicity gradients.

Finally, we study the relations between the metallicity gradient and some structural parameters of the galaxies. Figure 41.2 shows the relations between the metallicity gradient and the rotational velocity, the anisotropy parameter ($\log(v/\sigma)^*$), as defined by Bender (1990), and the $(a_4/a) \times 100$ parameter, which measures the deviation of the shape of the isophotes from a perfect ellipse. We could not find reference values of a_4 for three of our galaxies (NGC 3384, NGC 4458 and NGC 4464) and, therefore, they are not included in Figure 41.2. A non-parametric Spearman rank-order test gives a probability lower than 0.5% that the correlation observed in the figures could have been produced by chance for the three relations, although, in the case of the relation with the anisotropy parameter, the correlation is mainly driven by two galaxies (NGC 1600 and NGC 2865). The correlation between grad[Z/H] and $(a_4/a) \times 100$ (disky galaxies have stronger gradients and

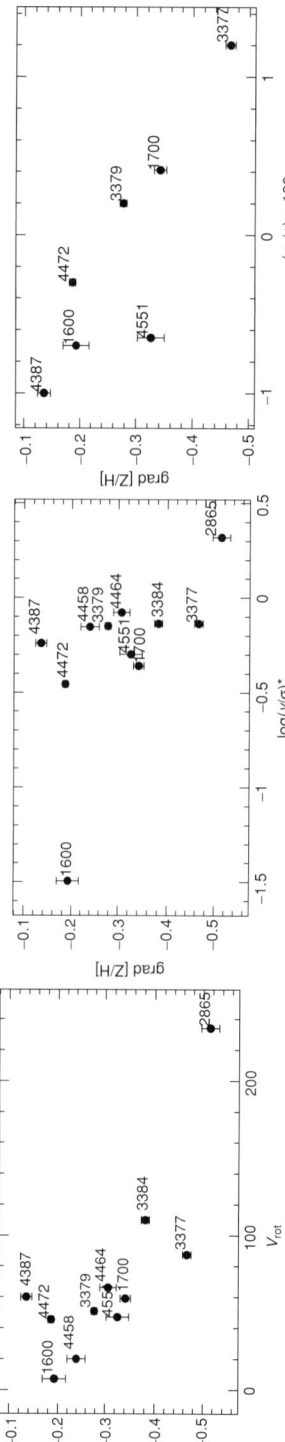

Figure 41.2. Relations between the metallicity gradient and other structural parameters. Left panel: the relation between the metallicity gradient and the rotational velocity. Central panel: the relation between the metallicity gradient and the anisotropy parameter normalised with respect to an isotropic rotator. Right panel: the relation between the metallicity gradient and the (a_4/a) ×100 parameter, extracted from Bender et al. (1989).

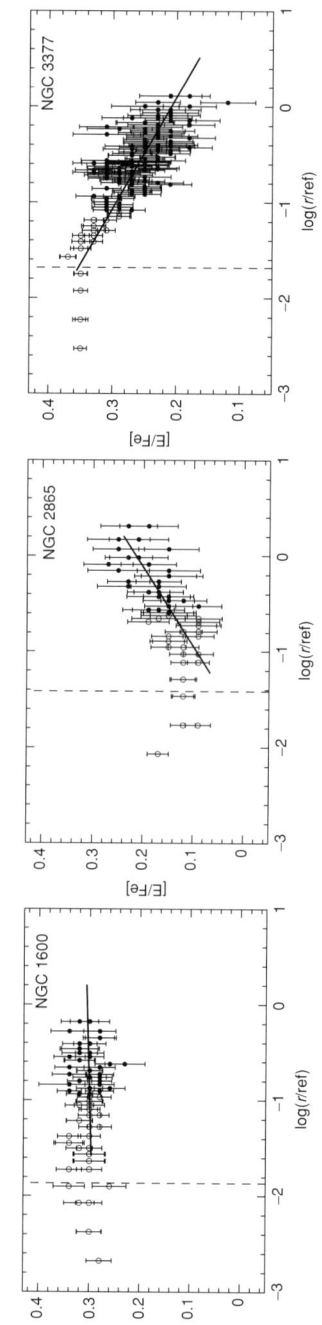

Figure 41.3. Examples of the [E/Fe] profile from the sample of galaxies presented here. The lines represent a linear fit to the data. Open symbols represent the points excluded from the fit.

the gradient decreases with the boxiness of the isophotes) is somehow surprising, because the a_4 parameter measured for simulated galaxies depends on the projection effects and, therefore, the same galaxy can have boxy and disky isophotes depending on the viewing angle (e.g. Stiavelli *et al.* 1991; Governato *et al.* 1993).

It has been suggested in the literature that the age spread in the stellar population of nearby galaxies found in several studies (e.g. González 1993; Trager *et al.* 2000; Sánchez-Blázquez *et al.* 2006) could be a consequence of a small frosting of young stars in the centre of the galaxies, but not connected with the main mechanism of galaxy formation. However, the results presented here show that the central populations are related to the metallicity gradients, which are related to the structural parameters and, therefore, probably connected with the formation mechanism of the galaxy.

Bekki & Shioya (1998) performed numerical simulations of mergers between gas-rich galaxies, studying the effect of star formation on the structural parameters of the remnant. They found that the rapidity of gas consumption by star formation greatly affects the isophotal shape of the merger remnant. Mergers with gradual star formation are more likely to form elliptical galaxies with disky isophotes, whereas those for which the star formation is more rapid are more likely to form boxy ellipticals (although this depends on the viewing angle and, therefore, they can be seen as disky too). This scenario could explain the relation between the metallicity gradient and the shape of the isophotes found here.

The variations in chemical-abundance ratio with radius give information about the timescales of star formation within the galaxies.[1] In the classical models of monolithic collapse stars form in essentially all regions during the collapse and remain in their orbits with little inward migration, whereas the gas, being continuously enriched by the evolving stars, dissipates inwards. In these scenarios the star formation is halted by the galactic winds, which initiated beforehand in the external parts of the galaxies due to their shallower potential well. This mechanism creates positive [α/Fe] gradients. The resultant [α/Fe] gradient of a merger remnant is more difficult to predict, since it depends on the initial gradient and gas fraction of the interacting galaxies and also on whether star formation occurs.

Figure 41.3 shows some examples of [E/Fe] for our sample of galaxies. As can be seen, there are both positive and negative cases. This rules out a simple inside-out or outside-in mechanism for all the galaxies. Furthermore, they are considerably shallower than the relationship predicted by simple models of gaseous collapse (e.g. Pipino *et al.* 2006).

---

[1] Assuming that there are no radial variations in initial mass function or selective mass losses.

4 Concluding remarks

In general, the gradients presented here and their correlations with other parameters are compatible with a scenario in which elliptical galaxies formed through the merger of smaller structures and the relative amount of dissipation experienced by the baryonic mass component along during stellar mass assembly of ellipticals decreases with increasing galaxy mass. Although the quality of the gradients analysed here is very high, the sample is not very large. Therefore, the conclusions presented here will need to be confirmed with larger samples.

References

Bekki, K., & Shioya, Y. (1998). *ApJ* **497**, 108
  (1999) *ApJ* **513**, 108
Bender, R. (1990), *A&A* **229**, 441
Bender, R. *et al.* (1989), *A&A* **217**, 35
González J. J. (1993), Ph.D. Thesis, University of Santa Cruz, California
Gorgas, J., Efstathiou, G., & Aragón-Salamanca, A. (1990), *MNRAS* **245**, 217
Governato, F., Reduzzi, L., & Rampazzo, R. (1993), *MNRAS* **261**, 379
Larson, R. B. (1974), *MNRAS* **166**, 585
Mehlert D., Thomas, D., Saglia, R. P., Bender R., & Wegner G. (2003), *A&A* **407**, 423
Oke J. B., Cohen, J. G, Carr, M. *et al.* (1995), *PASP* **107**, 375
Proctor, R. N., & Sansom, A. E. (2002), *MNRAS* **333**, 517
Sánchez-Blázquez, P., Gorgas, J., Cardiel, N., & González, J. J. (2006), *A&A* **457**, 809
Stiavelli, M., Londrillo, P., & Messina, A. (1991), *MNRAS* **251**, 57
Thomas, D., Maraston, C., & Bender, R. (2003), *MNRAS* **339**, 897
Trager, S. C., Faber, S. M., Worthey, G., & González, J. J. (2000), *ApJ* **120**, 165
White, S. D. M. (1980), *MNRAS* **191**, 1

42

Oxygen-rich droplets and the enrichment of the interstellar medium

Grażyna Stasińska,[1] Guillermo Tenorio-Tagle,[2] Mónica Rodríguez[2] & William J. Henney[3]

[1]*LUTH, Observatoire de Paris-Meudon, 5 Place Jules Jansen, 92195 Meudon, France*
[2]*Instituto Nacional de Astrofísica, Óptica y Electrónica, Apartado Postal 51, 72000 Puebla, Mexico*
[3]*Centro de Radioastronomía y Astrofísica, Universidad Nacional Autónoma de México, Campus Morelia, Apartado Postal 3–72, 58090 Morelia, Mexico*

> We argue that the discrepancies observed in H II regions between abundances derived from optical recombination lines (ORLs) and collisionally excited lines (CELs) might well be the signature of a scenario of the enrichment of the interstellar medium (ISM) proposed by Tenorio-Tagle (1996). In this scenario, the fresh oxygen released during massive supernova explosions is confined within the hot superbubbles as long as supernovae continue to explode. Only after the last massive supernova explosion does the metal-rich gas start to cool down and fall onto the galaxy in the form of metal-rich droplets. Full mixing of these metal-rich droplets and the ISM occurs during photoionization by the next generations of massive stars. During this process, the metal-rich droplets give rise to strong recombination lines of the metals, leading to the observed ORL–CEL discrepancy.

1 Introduction

There is no doubt that galaxies suffer chemical enrichment during their lives; see e.g. Cid Fernandes *et al.* (2007) for a recent systematic approach using a large database of galaxies from the Sloan Digital Survey Data Release 5 (Adelman-MacCarthy *et al.* 2007). The main source of oxygen production was since long ago identified as due to supernovae from massive stars (Type-II supernovae). Yet, the exact process by which chemical enrichment proceeds is poorly known (Scalo & Elmegreen 2004).

Ten years ago, Tenorio-Tagle (1996, hereafter T-T96) proposed a scenario in which the metal-enhanced ejecta from supernovae follow a long excursion in galactic halos before falling down onto the galaxies in the form of oxygen-rich droplets.

The Metal-rich Universe, eds. G. Israelian and G. Meynet. Published by Cambridge University Press.
© Cambridge University Press 2008.

In the present work, a fuller version of which has been published in *Astronomy & Astrophysics* (Stasińska *et al.* 2007), we suggest that the discrepancy between the oxygen abundances derived from optical recombination lines (ORLs) and from collisionally excited lines (CELs) in H II regions (e.g. García-Rojas *et al.* 2006 and references therein) might well be the signature of those oxygen-rich droplets. In fact, Tsamis *et al.* (2003) and Péquignot & Tsamis (2005) had already suggested that the ORL–CEL discrepancy in H II regions is the result of inhomogeneities in chemical composition of these objects. Our aim is to make explicit the link between the ORL–CEL discrepancy and the T-T96 scenario, and to check whether what is known of the oxygen yields allows one to explain the ORL–CEL discrepancy in a quantitative way.

2 The Tenorio-Tagle (1996) scenario

Figures 42.1–42.5 present the T-T96 scenario in cartoon format.

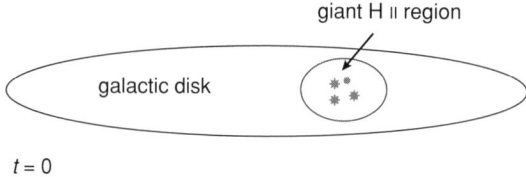

Figure 42.1. At time $t = 0$, a burst of star formation occurs and a giant H II region forms.

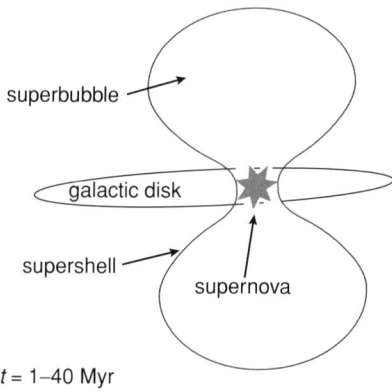

Figure 42.2. During the next $\sim$40 Myr, supernovae explode, creating a hot superbubble confined within a large expanding supershell that bursts into the galactic halo. The superbubble contains the matter from the oxygen-rich supernova ejecta mixed with the matter from the stellar winds and with the matter thermally evaporated from the surrounding supershell.

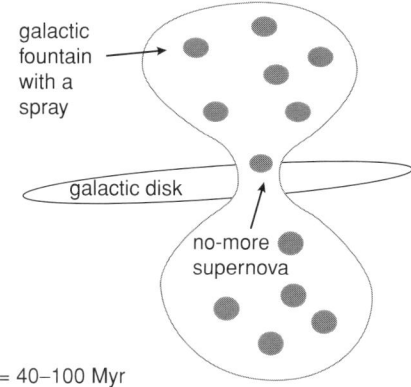

Figure 42.3. After the last supernova has exploded, the gas in the superbubble begins to cool down. Loci of higher densities cool down quicker. Owing to a sequence of fast repressurizing shocks, this leads to the formation of metal-rich cloudlets. The cooling timescale is of the order of 100 Myr.

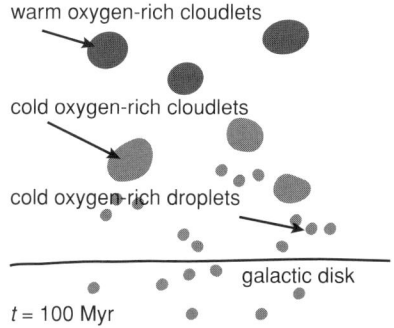

Figure 42.4. The now-cold metal-rich cloudlets fall onto the galactic disk. They are further fragmented into metal-rich droplets by Rayleigh–Taylor instabilities. This metal-rich rain affects a region whose extent is of the order of kiloparsecs, i.e. much larger than the size of the initial H II region.

3 The ORL–CEL discrepancy in the context of the T-T96 scenario

The details of the physical arguments concerning the amount of oxygen available in the droplets, the mixing processes and the simulation of the ORL–CEL discrepancy with a multi-zone photoionization model are described in Stasińska et al. (2007). Here, we simply give the most important conclusions.

Photoionization of the oxygen-rich droplets predicted by the T-T96 scenario can reproduce the observed abundance discrepancy factors (ADFs, i.e. the ratios of abundances obtained from ORLs and from CELs) derived for Galactic and

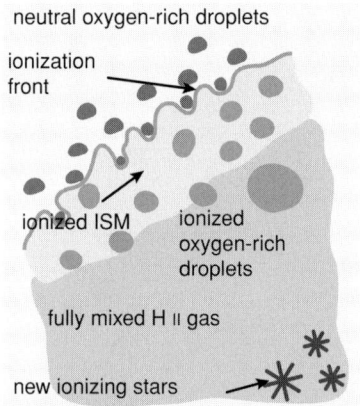

Figure 42.5. When a next generation of massive stars forms, they photoionize the surrounding interstellar medium (ISM), including the metal-rich droplets. It is only after the droplets have been photoionized that their matter is intimately mixed with the matter from the ISM and that proper chemical enrichment has occurred. The whole process since the explosion of the supernovae that provided fresh oxygen has taken at least 100 Myr.

extragalactic H II regions. The recombination lines arising from the highly metallic droplets thus show mixing at work.

We find that, if our scenario holds, the recombination lines strongly overestimate the metallicities of the fully mixed H II regions. The collisionally excited lines may also overestimate them, although to a much smaller extent. In the absence of any recipe to correct for these biases, we recommend that one should discard objects with large ADFs when one is attempting to probe the chemical evolution of galaxies.

To proceed further with this question of inhomogeneities, one needs as many observational constraints as possible. On the theoretical side, one needs robuster estimates of the integrated stellar yields as well as better knowledge of the impact of massive stars on the ISM and of the role of turbulence. All these issues are relevant to our understanding of the metal enrichment of the Universe.

References

Adelman-MacCarthy J. K. *et al.* (2007)
Cid Fernandes, R., Vala Asari, N., Sodré L. Jr *et al.* (2007), *MNRAS* **375**, L16
García-Rojas, J., Esteban, C., Peimbert, M. *et al.* (2006), *MNRAS* **368**, 253
Scalo, J., & Elmegreen, B. G. (2004), *ARA&A* **42**, 275
Stasińska, G., Tenorio-Tagle, G., Rodríguez, M., & Henney, W. J. (2007), *A&A* **471**, 193
Tenorio-Tagle, G. (1996), *AJ* **111**, 1641 (T-T96)
Tsamis, Y. G., Barlow, M. J., Liu, X.-W., Danziger, I. J., & Storey, P. J. (2003), *MNRAS* **338**, 687
Tsamis, Y. G., & Péquignot, D. (2005), *MNRAS*, **364**, 687

Part VII

Chemical and photometric evolution beyond
Solar metallicity

43

Models of the Solar vicinity: the metal-rich stage

Leticia Carigi

Instituto de Astronomía, Universidad Nacional Autónoma de México, Apartado Postal 70–264, México 04510, D.F., Mexico

I present a review of chemical-evolution models of the Solar neighborhood. I pay special attention to the ingredients necessary to reproduce the observed [Xi/Fe] ratios in nearby metal- and super-metal-rich stars, and to the chemical properties of the Solar vicinity, focusing on [Fe/H] ≥ -0.1. I suggest that the observed abundance trends are due to material synthesized and ejected by intermediate-mass stars with Solar metallicity in the AGB stage, and also by massive stars with (super)solar metallicity in the stellar wind and supernovae stages. The required tool to build chemical-evolution models that reach supersolar metallicities is the computation of stellar yields for stellar metallicities higher than the initial Solar value. With these models it might be possible to estimate the importance of merger events in the recent history of the Galactic disk as well as the relevance of radial stellar migration from the inner to the outer regions of the Galaxy. I also present a short review of the photospheric Solar abundances and their relation to the initial Solar abundances.

1 Introduction

The Solar neighborhood is an invaluable laboratory for the chemical-evolution models because the number of free parameters is similar to the number of observational constraints.

Certain assumptions are typically adopted in chemical-evolution models of a galactic zone: (i) the formation mechanism and formation time of the galactic zone; (ii) when, how many, and what types of stars are formed; (iii) when those stars die; and (iv) the chemical abundances of the material ejected during the life and death of the stars.

The Metal-rich Universe, eds. G. Israelian and G. Meynet. Published by Cambridge University Press.
© Cambridge University Press 2008.

Once a chemical-evolution model for the Solar vicinity satisfies the observational constraints, it is also possible to test its congruence with both the Galaxy-formation process and the properties of the underlying stellar populations. Therefore, the accuracies of the estimations of stellar and H II-region abundances define the strength of our tests.

For that reason, this review is to a large extent based on new data for the Solar neighborhood: (i) for H II regions, obtained by Esteban, Peimbert and collaborators (see Chapter 17 in this volume); and (ii) for F and G dwarf stars, obtained by Bensby and Feltzing (see Chapters 5 and 7 in this volume). Also I will compare these data with the abundances of other Galactic components and other galaxies, in particular bulge stars and extragalactic H II regions, in order to analyze the origin of the abundance trends at high metallicity.

2 Observational constraints

The definition of "Solar vicinity" has several interpretations, ranging from a zone including all low-redshift galaxies to a region as small as the one including the stars within 1 pc of the Sun. In the chemical-evolution context the Solar vicinity corresponds to a cylinder centered around the Sun, 8 kpc from the Galactic Center, that includes objects belonging to the Galactic halo and disk. The dimensions of this cylinder depend on the locations of the objects used as observational constraints. Typical adopted dimensions are 1 kpc radius and $\sim$3 kpc height, for a cylinder oriented orthogonally to the Galactic plane.

Since I am interested in (super)-metal-rich objects, I will focus on objects with metallicity near or higher than Solar.

2.1 Solar abundances

The photospheric Solar abundances provide the reference pattern for general abundance determinations in the Universe (for stars, ionized nebulae, galaxies) and the inferred initial Solar abundances correspond to the interstellar medium (ISM) of the Solar neighborhood 4.5 Gyr ago. During the last $\sim$20 years the observational estimations of photospheric Solar abundances regarding the most abundant heavy elements, such as C, N, O, and Ne, have decreased.

In Table 43.1 I show the chemical-abundance determinations for some common elements in the Solar photosphere computed by Anders & Grevesse (1989, AG89), Grevesse & Sauval (1998, GS98), and Asplund et al. (2005, AGS05). These abundances are expressed as $12 + \log(X/H)$ by number; I have added the values of the mass fractions of He and metals, Y and Z, respectively. I also present the decreasing factors in the abundance determinations between the Anders & Grevesse work and the Asplund et al. data and between the Grevesse & Sauval work and the Asplund et al. data.

Table 43.1. *Chemical composition in the Solar photosphere*

Element	AG89	GS98	AGS05	$\frac{(X_i/H)_{AGS05}}{(X_i/H)_{AG89}}$	$\frac{(X_i/H)_{AGS05}}{(X_i/H)_{GS98}}$
C	8.56 ± 0.04	8.52 ± 0.06	8.39 ± 0.05	0.68	0.74
N	8.05 ± 0.04	7.92 ± 0.06	7.78 ± 0.06	0.54	0.72
O	8.93 ± 0.04	8.83 ± 0.06	8.66 ± 0.05	0.54	0.68
Ne	8.09 ± 0.10	8.08 ± 0.06	7.84 ± 0.06	0.56	0.58
Fe[a]	7.51 ± 0.03	7.50 ± 0.01	7.45 ± 0.05	0.87	0.89
Y	0.2743	0.2480	0.2486	0.91	1.00
Z	0.0189	0.0170	0.0122	0.65	0.72

[a]Fe abundance in meteorites only for AG89.

Since Fe is one of the most common elements and the value of its Solar photospheric determination has remained almost constant during the last few years, I will compare the available stellar data on the basis of [Fe/H]. For H II regions, I will consider the O/H value determined for those nebulae as the reference ratio, since Fe is strongly depleted by accretion onto dust in ionized nebulae.

In order to reproduce the helioseismology observations, Sun models require more metals than the Solar Z value obtained by Asplund *et al.* (2005). Bahcall *et al.* (2006), by considering the photospheric metallicity of the Sun determined by Asplund *et al.* (2005), found that the initial Solar metallicity, Z_{in}, is 0.01405, 15% more metals than observed in the Solar photosphere. This difference is due to diffusive settling of the elements in the photosphere during the last 4.5 Gyr, the age of the Sun (Carigi & Peimbert 2008). This fact has implications for the chemical-evolution models, because the Solar abundances in the photosphere have been taken as representative of the abundances of the ISM when the Sun was born, but those photospheric Solar abundances should be corrected for Solar diffusion. Moreover, the diffusive-settling effect should be considered in the abundance determinations of other stars, taking into account that the amount of material settled depends on the stellar age.

During this review I assumed $Z_\odot = 0.012$, $Z_{in} = 0.014$, and $Z_{can} = 0.020$ as the photospheric, initial, and canonical Solar metallicities, respectively.

2.2 Abundances from H II regions

Abundance estimations in H II regions give us the present-day abundances, so they are very important for chemical-evolution models.

Esteban *et al.* (2005) derived the C/H and O/H values of eight Galactic H II regions from C and O recombination lines. They found higher C/H and O/H values than the photospheric Solar ones for the Orion nebula and five other Galactic H II regions closer to the Galactic Center (see Figure 43.2). These values are in agreement with the C/H estimations derived by Slavin & Frish (2006 and references therein), who find C/H = 8.78 ± 0.20, a value higher than Solar (photospheric and internal), along one line of sight of the Local Interstellar Cloud. Knowledge of H II-region gradients, particularly in the inner Galactic disk, might be useful for analyzing the chemical enrichment at high metallicities.

The abundances of Galactic and extragalactic metal-rich H II regions, and also the various methods of abundance determination, have been discussed by Bresolin (2007) and in Chapter 17 of this volume). Since the various methods used to determine abundances provide different chemical abundances, a consensus on the abundance determinations for H II regions is required.

2.3 The age–[Fe/H] relation

One of the most fundamental observational results in cosmochemistry is the age–[Fe/H] relation because it links the ages of dwarf stars to their chemical properties. This relation presents a large scatter throughout the metallicity range and metal-rich stars are no exception. According to Bensby *et al.* (2005) thin-disk stars with [Fe/H] > +0.2 present ages between 3 and 9 Gyr, but stars with 0 < [Fe/H] < +0.2 have ages between 0 and 6 Gyr. According to Soubiran & Girard (2006) the mean age of the thin-disk stars of [Fe/H] > +0.15 is 5 Gyr, with a dispersion of 3.4 Gyr, while the mean age of stars of 0 < [Fe/H] < +0.2 is 3.8 Gyr with a dispersion of 2.1 Gyr.

The dispersion is partly caused by stars born at other Galactocentric radii with different star-formation histories (SFHs) that migrated to the Solar vicinity. Since this stellar migration requires time, Rocha-Pinto *et al.* (2006) showed that the Solar neighborhood has been polluted by old and metal-poor stars from inner and outer radii (between 6 and 9.5 kpc). The age dispersion of the metal-rich stars might be explained by a superposition of young stars that were born in the Solar vicinity and old stars that were born at inner radii with an early and efficient star-formation rate (SFR) as chemical-evolution models of the Galactic disk predict (e.g. Carigi *et al.* 2005).

An alternative explanation for the age dispersion presented in the age–[Fe/H] relation is the possibility of mergers of one or several satellite galaxies with different SFHs.

2.4 The [X/Fe] versus [Fe/H] relations

Other important observational constraints are provided by the [X/Fe] versus [Fe/H] relations derived for dwarf stars in the Solar neighborhood. These relations give information from which to infer the past of the Solar vicinity and the properties of its stellar populations.

2.4.1 Alpha enhancement

Authors of some studies have found α enhancement in thick-disk stars relative to the thin-disk stars in the $-0.7 < $ [Fe/H] < -0.1 range (Feltzing in Chapter 5 of this volume and references therein). No chemical-evolution model that assumes a simple formation mechanism for the disk can reproduce that behavior, but in the literature there are some models with complex disk-formation histories that can explain the α enhancement.

(i) Nykytyuk & Mishenina (2006) suggest a two-zone model with different gas infalls and SFHs for the thick and thin disks. Their model can reproduce the α enhancement in the thick-disk stars relative to the thin-disk stars, but not the dispersion in the age–[Fe/H] relation. Chiappini (2001) and during this meeting (Chapter 47) has suggested a similar model assuming a double infall for the Galactic disk and that the thin disk formed with material from the thick disk and from the intergalactic medium.

(ii) Brook *et al.* (2005) suggest hierarchical mergers and fragmentation models. Specifically, the thick disk formed by multiple gas-rich mergers at early times (7.7 Gyr ago) at redshifts higher than ∼1. Part of the gas of the thick disk was left over by shock heating, then the thin disk formed from primordial infalling gas and the gas pre-enriched by the thick-disk stars that falls later onto the thin disk. Their results are in good agreement with α enhancement but in only partial agreement with the dispersion in the age–[Fe/H] relation.

2.4.2 Metal-rich disk stars

Bensby *et al.* (2005) determined chemical abundances in F and G dwarfs and found strong abundance trends, which are shown in Figure 43.1. As can be noted from this figure, the slope of [X/Fe] versus [Fe/H] relations for [Fe/H] > -0.1 (i) differ significantly for Na, Ni, and Zn, with Δ[X/H]/Δ[Fe/H] > 0, and for Ba, with Δ[X/H]/Δ[Fe/H] < 0; (ii) vary moderately for Mg, Al, Si, Ca, and Ti, being Δ[X/H]/Δ[Fe/H] ~ 0; and (iii) do not vary for O, Cr, and Eu.

Similar trends have been observed for the red giants of the Galactic bulge (Cunha & Smith 2006; Lecureur *et al.* 2007): (i) [Na/Fe] increases with [Fe/H]; (ii) [Na/Mg] increases with [Fe/H] > 0, but not as much as in the thin disk; and (iii) [O/Mg] decreases with [Fe/H] > 0 like in the disk stars.

Johnson *et al.* (2007) recently determined abundances for a super-metal-rich G dwarf of the Galactic bulge and found that the [α/Fe] ratios are subsolar, whereas ratios for the odd-Z elements are slightly supersolar. These values are in agreement with the trends seen in the more-metal-rich stars of the Galactic disk (see Figure 43.1).

3 Chemical-evolution models of the Solar vicinity

The goal of any chemical-evolution model is to explain the observed chemical properties. In the literature there are successful models that match the abundance trends for [Fe/H] ≤ 0; for example, see the excellent reviews by Gibson *et al.* (2003) and by Matteucci (2004). Those models are computed with different codes and assumptions. In Figure 43.1 I present the [Xi/Fe] versus [Fe/H] evolution predicted by some well-known models. If those theoretical trends are extrapolated to [Fe/H] ~ +0.5 no chemical-evolution model is able to reproduce the change in the [X/Fe] versus [Fe/H] relation for [Fe/H] > −0.1.

Therefore, I will study the dependences of [X/Fe] on the various ingredients of a chemical-evolution model, in order to find an explanation for the [X/Fe] trends presented by metal-rich stars of the Galactic thin disk.

It is well known that the [X/Fe] ratios depend on gas flows, SFRs, the initial mass function, and stellar yields, the last two being the most important factors.

3.1 Gas and star flows

Infalls and outflows change the [X/Fe] ratios depending on the abundances and the amount of gas of the flows. Rich outflows with SN II material reduce [X/Fe] as opposed to the increment required by most of the observed trends. Rich outflows with SN Ia material increase [X/Fe] but they also decrease [Fe/H], preventing the formation of metal-rich stars. Infall of a metal-rich gas overabundant in elements present in metal-rich stars, such as Na, can reproduce the rise in [X/Fe], but how does the infalling gas acquire those [X/Fe] values?

Brook *et al.* (2005) explain the chemical properties of thick- and thin-disk stars with models that assume mergers and infalls mainly at redshifts lower than 1, but their results do not predict the abundance trends for [Fe/H] > −0.1.

Reddy (in this meeting; see Chapter 8) showed a secondary peak in that [Fe/H] distribution for [Fe/H] > 0. Inclusion of a significant amount of stars (or gas that triggered the star formation) from a merger event could explain the secondary peak. If the thin-disk metal-rich stars formed in one or several Galactic satellites that settled in the Galactic disk, how did the stars of those satellites reach supersolar [Fe/H] with (super)solar [X/Fe]?

According to the merger scenario, the bulge was also formed by satellites that fell early on into the Milky Way, therefore the old and metal-rich stars of the bulge and

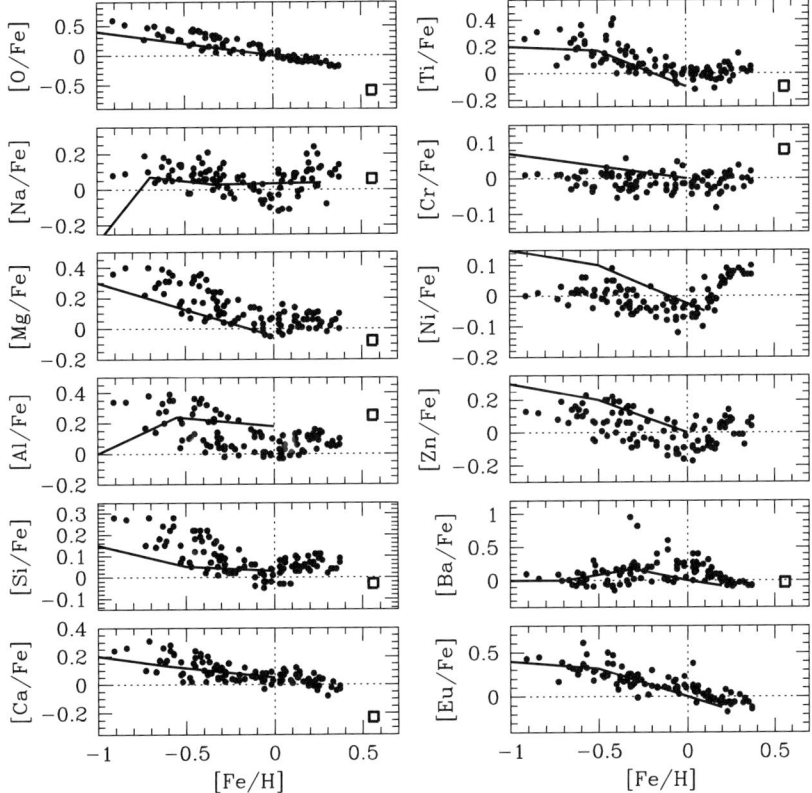

Figure 43.1. Evolution of [X/Fe] versus [Fe/H] predicted by various models: for Al by Timmes et al. (1995), for Ca by Portinari et al. (1998), for Ti, Ni, and Zn by François et al. (2004), for O by Gavilán et al. (2005), for Mg, Si, and Cr by Prantzos (2005), for Ba and Eu by Cescutti et al. (2007), and for Na by Izzard et al. (2007). Filled circles: F and G dwarf disk stars (Bensby et al. 2005). Open squares: the most-metal-rich G dwarf bulge star (Johnson et al. 2007). Corrections to [X/H] to take account of the different photometric Solar values assumed by the authors have not been made.

Galactic disk originated from small galaxies, or metal-rich stars form of material from small structures, but again, how did those structures reach [X/Fe] > 0?

According to another scenario the bulge could have been formed by the stars that were born in the inner Galactic disk and were dynamically heated by the bar (Colín et al. 2006). Moreover, the same bar could have been able to produce radial flows of stars from the inner to the outer part of the Galactic disk. Therefore the metal-rich stars of the bulge and the Solar neighborhood formed in the inner disk, but how did the inner disk reach [X/Fe] > 0?

Radial gradients can be powerful tools allowing one to decide whether metal-rich stars observed in the Solar vicinity and the bulge formed *in situ* or alternatively were formed with in the inner Galactocentric radius or originated from merged satellites. In the most complicated case (or the most realistic), a combination of these three possibilities should be the answer.

Therefore, stellar or gaseous infalls might explain the abundance trends observed for [Fe/H] > −0.1, but they pose the question of how these infalls could acquire those (super)solar [X/Fe] values.

3.2 The star-formation rate

Important changes in the SFR affect the [X/Fe] ratios mainly after a significant star-formation burst (e.g. Carigi *et al.* 1999, 2002; Chiappini 2001).

The spiral wave is the most important inner mechanism that triggers star formation and that could recently have formed stars from a metal-rich gas. Rocha-Pinto *et al.* (2000) and Hernández *et al.* (2000) inferred the SFH from the color-magnitude diagram for the Solar vicinity. They found (i) a decreasing exponential general behavior of the SFH during the last ∼10 Gyr, and (ii) variations from the general behavior related directly to the spiral-wave passages. These results indicate that there were no significant bursts of star formation within the last ∼6 Gyr, therefore it is unlikely that a burst could have modified the [X/Fe] slopes.

An important fact is that metal-rich stars with similar [X/Fe] values have been observed in the Solar neighborhood and in the bulge, galactic components with different SFHs. The Solar vicinity had a moderate SFR for 12 Gyr (e.g. Carigi *et al.* 2005) whereas the bulge formed very quickly, in less than 0.5 Gyr, with a high SFR (Ballero *et al.*, Chapter 48, and Matteucci, Chapter 44, in this volume).

Therefore, I discard changes in the SFR as the explanation of the change in the [X/Fe] slopes for [Fe/H] > −0.1.

3.3 The initial mass function

The initial mass function (IMF) gives the mass distribution of the stars formed in a star-formation burst. This function is parametrized by the slope for various mass ranges and by the lower and upper mass limits of the stars formed. Since [X/Fe] depends strongly on the IMF, a dependence of the IMF on metallicity, density, or gas mass could explain the change of the [X/Fe] slope.

According to Kroupa (Chapter 24 in this volume) there is no evidence that the IMF changes with Z. Nevertheless, Bonnell suggested (in this meeting; see Chapter 39) that the IMF changes with Z: for supersolar metallicities the slope for massive stars could be steeper and the upper limit could be lower than for subsolar metallicities, producing subsolar [X/Fe] values, in contradiction with the observed values.

In a metal-rich gas it is more difficult to create massive stars due to the dependence of the Jeans mass on $Z^{-2/3}$ and metal-rich stars truncate the star-formation process by the action of their stellar winds. This suggestion could explain the low ionization of the metal-rich H II regions compared with that of H II regions with subsolar metallicity: in a metal-rich gas the number of massive stars may be lower, leading to a smaller number of ionizing photons; on the other hand, the lower stellar temperatures of the metal-rich stars help to reduce the number of ionizing photons. Further work on this suggestion needs to be done.

The dependence of the IMF on the gas density should be less important, because the same abundance trends have been observed in Galactic components with different densities (bulge, open clusters, isolated dwarfs), but with a common property: supersolar metallicity.

According to Weidner & Kroupa (2005) the IMF depends on the gas mass. They found that the slope and the upper mass limit change with the gas mass available to form stars. In dwarf galaxies the upper mass limit is lower and the slope in the massive-star range is steeper, producing lower [X/Fe] values than those of normal galaxies for elements synthesized only by massive stars.

Moreover, Carigi & Hernández (2008) found important effects on the abundance ratios when the IMF is stochastically populated. The [O/Fe] values varied within three orders of magnitude for a stellar population of 500 $M_\odot$ enriching a gas mass of $10^4 M_\odot$. This effect could explain the dispersion observed in the abundance ratios, but not the abundance trends.

Therefore, possible modifications to the IMF cannot explain the abundance trends observed for [Fe/H] > −0.1.

3.4 Stellar yields

Since supersolar [X/Fe] values seem to be a common property of stars with [Fe/H] > 0 in Galactic components (thin disk, bulge, open clusters) with different formation histories, the abundance trends can be explained as being due to the stellar yields of (super) solar-metallicity stars. The observed abundances will provide strong constraints on the physical processes taking place in the stellar cores.

The models shown in Figure 43.1 consider different Z-dependent yields for massive stars, low- and intermediate-mass stars (LIMSs), and SN Ia. These stellar yields were computed for $Z \leq Z_{\text{can}}$ with the exception of the Portinari *et al.* (1998) yields, but those authors never used their yields for $Z = 0.05$ because they stopped their computations at [Fe/H] = 0. Cescutti *et al.* (2006) and François *et al.* (2004) modified the stellar yields obtained from stellar-evolution models in order to reproduce the observed trends for [Fe/H] < +0.1.

Edmunds (during this meeting; see Chapter 46) suggested that stellar yields increasing with Z raise the [X/Fe] values for [Fe/H] > −0.1. The [O, Mg/Fe] values for bulge and thin-disk stars indicate that there is a Z dependence in the ratio of the O to Mg yield.

Meynet *et al.* (Chapter 36 in this volume) show that non-rotating massive-stars with $Z = Z_{can}$ and a high mass-loss rate eject more C than O, and that these stars are an important source of He, C, Ne, and Al, but not so much of O. These facts could explain the decrease in [O/Fe] with increasing [Fe/H] while the [Al/Fe] values remain almost constant for [Fe/H] ≥ 0.

The significant change in the [Na/Fe] slope suggests that there is an extra source of Na production. Assuming that SN II and AGB stars produce Na, Izzard *et al.* (2007) reproduce the [Na/Fe] values for [Fe/H] < −0.2, but fail to reproduce the increase in [Na/Fe] for [Fe/H] > 0. They suggest that the change in the [Na/Fe] slope may be explained by invoking secondary Na produced by SN II. Nevertheless, according to Fröhlich (see Chapter 37) the effect of core-collapse supernovae cannot explain the increase in [Na/Fe] observed for [Fe/H] ≥ 0.

Another channel that contributes to the enrichment of a metal-rich gas is provided by SN Ia. According to Yoon (Chapter 38 in this volume) there are various scenarios for SN Ia with different time delays, but the amount of heavy elements ejected is similar for all of the scenarios. The role of rotation might be important in the production of chemical elements, but this effect has not been studied yet. Therefore, new stellar yields for massive stars and intermediate-mass stars of Solar and supersolar metallicity are required.

3.4.1 The importance of stellar winds in metal-rich stars

One of the most important problems in the chemical evolution of galaxies is that of C production. Carigi *et al.* (2005, 2006) have studied the contributions to C enrichment by massive stars and LIMSs in various types of galaxies. We have found that the massive stars have contributed 48% and 36% of the total C produced in the Solar neighborhood and in the dIrr galaxy NGC 6822, respectively. The difference is due to the effect of Z on the stellar winds of massive stars. Massive stars of Solar Z eject more C than do those of subsolar Z through stellar winds (see Meynet *et al.*, Chapter 36, and Crowther, Chapter 29, in this volume).

Carigi *et al.* (2005) made a chemical-evolution model of the Galaxy in which they assumed that the metal-rich stars behave like stars of Z_{can}. They are able to reproduce the C/O and O/H values of the Solar vicinity as well as the O/H and C/O gradients observed by Esteban *et al.* (2005) but cannot reproduce the decrease in [C/Fe] for [Fe/H] > −0.1 shown by Bensby & Feltzing (2006) and Allende-Prieto in Chapter 3 (see Figure 43.2).

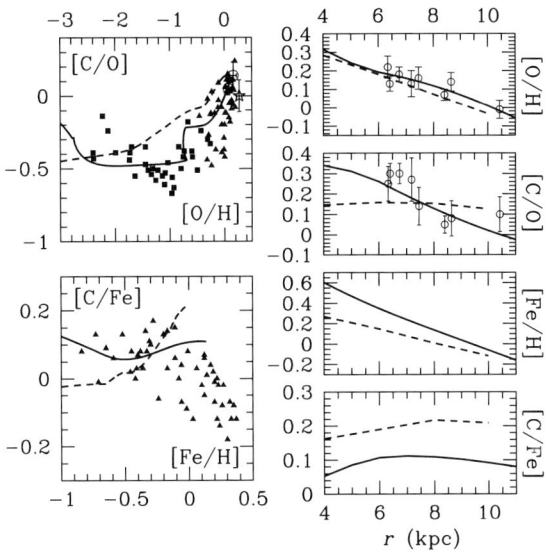

Figure 43.2. Model predictions by Carigi (2000) (dashed lines) and Carigi *et al.* (2005) (continuous lines) considering yields of metal-rich massive stars published by Portinari *et al.* (1998) and Maeder (1992), respectively. Left panels: [C/O, Fe] evolution with [O, Fe/H] in the Solar vicinity. Right panels: present-day ISM abundance ratios as a function of Galactocentric distance. Open circles: galactic H II regions, gas plus dust, by Esteban *et al.* (2005) and Carigi *et al.* (2005). Star: extragalactic H II region (H1013) in M101 (Bresolin 2007). Filled triangles: F and G dwarf disk stars (Bensby & Feltzing 2006). Filled squares: dwarf stars (Akerman *et al.* 2004). Photometric Solar values from AGS05 are considered, except for the data published by Bensby & Feltzing (2006) because they assumed their own Solar abundances.

Carigi (2000) made a model considering yields of massive stars published by Portinari *et al.* (1998) for $Z = 2.5 Z_{can}$ and predicted a C/O gradient flatter than the observed one. The flattening of the gradient is due to the increase in the mass-loss rate with Z ($\propto Z^{0.5}$) assumed by Portinari *et al.*; consequently metal-rich massive stars are stripped before C is synthesized and their C yields are lower than those of stars with $Z = Z_{can}$.

According to Meynet *et al.* (Chapter 36 in this volume) rotating stars with $Z = Z_{can}$ are more efficient at ejecting C and O than are rotating stars with $Z = 2Z_{can}$. This could explain the decrease in [C/Fe] exhibited by metal-rich stars of the thin disk, but the C contribution due to LIMSs must be included also in order to have a complete picture of the evolution at high Z.

In order to reproduce the high C/O values for inner Galactocentric radii the mass-loss rate for metal-rich stars has to be lower than that assumed by Portinari *et al.* (1998). On the other hand, to reproduce the low C/Fe values for [Fe/H] > 0 in the Solar neighborhood the mass-loss rate has to be higher than that assumed by Portinari *et al.* Puls (during this meeting; see Chapter 31) gave limits for the mass-loss rate, which varies as $Z^{0.62\pm0.15}$.

Consequently, there is an inconsistency between theory and observations regarding the behavior of C/O and C/Fe for high metallicities and a more complex explanation is needed.

4 Conclusions

Chemical-evolution models of the Solar vicinity for metal-rich stars are in the early stages of development. Nevertheless, I present the following conclusions.

- Models that assume hierarchical mergers and fragmentation explain most of the chemical and kinematic properties of thick and thin disks for [Fe/H] ≤ 0.
- Models that assume different SFHs and infalls for the thin and thick disks explain their chemical properties only for [Fe/H] ≤ 0.
- There is no published chemical-evolution model of the Solar vicinity that can reach the maximum [Fe/H] value observed in the thin disk, that is [Fe/H] $\sim +0.4$.
- Simple extrapolations of chemical-evolution models that assume [Fe/H] ≤ 0 to the [Fe/H] $\sim +0.4$ regime result in predictions that fail to match the chemical abundances found in (super)-metal-rich stars.
- The similar abundance-ratio trends observed for stars in the Solar neighborhood and the high-metallicity bulge stars suggest that both were created by the pollution of supersolar massive stars and Solar intermediate-mass stars. The stellar yields of these stars are required in order to study the metal enrichment of the ISM in the Solar vicinity and the bulge.
- No current chemical-evolution model includes adequately the contribution of metal-rich massive stars, because their yields have not been computed completely yet.
- The intermediate-mass stars in the AGB may produce an important amount of the chemical elements heavier than oxygen, but in the case of Na their contribution is not enough to explain the rise in [Na/Fe] for [Fe/H] > -0.1.
- The large dispersion in age of metal-rich stars in the Solar vicinity could be indicating an external origin such as merger events or stellar migrations from the inner disk.

In my opinion, we are entering a new phase of astronomy: from the past to the future of the Universe (before this meeting the emphasis was from the present to the past), the metal-rich past and present give us hints to the future.

Acknowledgment

I thank Garik Israelian and the Organizing Committee for kindly inviting me to give this review. I am grateful to Manuel Peimbert for several fruitful discussions and helpful suggestions, to Brad Gibson for sending me the paper by Izzard *et al.* in advance of publication, and to Octavio Valenzuela for a careful reading of the manuscript. I received partial support from the Spanish MCyT under project AYA2004–07466 to attend this meeting.

References

Akerman, C. J., Carigi, L., Nissen, P. E., Pettini, M., & Asplund, M. (2004), *A&A* **414**, 931–942
Anders, E., & Grevesse, N. (1989), *Geochimica et Cosmochimica Acta* **53**, 197–214
Asplund, M., Grevesse, N., & Sauval, A. J. (2005), *ASPC* **336**, 25–37
Bahcall, J. N., Serenelli, A. M., & Basu, S. (2006), *ApJS* **165**, 400–431
Bensby, T., & Feltzing, S. (2006), *MNRAS* **367**, 1181–1193
Bensby, T., Feltzing, S., Lundström, I., & Ilyin, I. (2005), *A&A* **433**, 185–203
Bresolin, F. (2007), *ApJ* **656**, 186–197
Brook, C. B., Gibson, B. K., Martel, H., & Kawata, D. (2005), *ApJ* **630**, 298–308
Carigi, L. (2000), *RMxAA* **36**, 171–184
Carigi, L., & Hernandez, X. (2008), *MNRAS* Submitted
Carigi, L., & Peimbert, M. (2008), *ApJ* Submitted (arxiv: 0801.2867)
Carigi, L., Colín, P., & Peimbert, M. (1999), *ApJ* **514**, 787–797
Carigi, L., Hernandez, X., & Gilmore, G. (2002), *MNRAS* **334**, 117–128
Carigi, L., Peimbert, M., Esteban, C., & García-Rojas, J. (2005), *ApJ* **623**, 213–224
Carigi, L., Colín, P., & Peimbert, M. (2006), *ApJ* **644**, 924–939
Cescutti, G., François, P., Matteucci, F., Cayrel, R., & Spite, M. (2006), *A&A* **448**, 557–569
Chiappini, C. (2001), *RMxAAC* **11**, 171–176
Colín, P., Valenzuela, O., & Klypin, A. (2006), *ApJ* **644**, 687–700
Cunha, K., & Smith, V. V. (2006), *ApJ* **651**, 491–501
Esteban, C., García-Rojas, J., Peimbert, M. *et al.* (2005), *ApJ* **618**, L95–L98
François, P., Matteucci, F., Cayrel, R. *et al.* (2004), *A&A* **421**, 613–621
Gavilán, M., Buell, J. F., & Mollá, M. (2005), *A&A* **432**, 861–877
Gibson, B. K., Yeshe, F., Agostino, R., Daisuk, K., & Hyun-chul, L. (2003), *PASA* **20**, 401–415
Grevesse, N., & Sauval, A. J. (1998), *Space Science Reviews* **85**, 161–174
Hernández, X., Valls-Gabaud, D., & Gilmore, G. (2000), *MNRAS* **316**, 605–612
Izzard, R. G., Gibson, B. K., & Stancliffe, J. (2007), *IAU Symp.* **238**, 121
Johnson, J. A., Gal-Yam, A., Leonard, D. C. *et al.* (2007), *ApJ* **655**, L33–L36
Lecureur, A., Hill, V., Zoccali, M. *et al.* (2007), *A&A* **465**, 799–814
Maeder, A. (1992), *A&A* **264**, 105–120
Matteucci, F. (2004), *ASPC* **317**, 337
Nykytyuk, T. V., & Mishenina, T. V. (2006), *A&A* **456**, 969–976
Portinari, L., Chiosi, C., & Bressan, A. (1998), *A&A* **334**, 505–539
Prantzos, N. (2005), *Nuclear Physics A* **758**, 249–258
Rocha-Pinto, H. J., Scalo, J., Maciel, W. J., & Flynn, C. (2000), *A&A* **358**, 869–885
Rocha-Pinto, H. J., Rangel, R. H. O., Porto de Mello, G. F. *et al.* (2006), *A&A* **453**, L9–L12
Slavin, J. D., & Frisch, O. C. (2006), *ApJ* **651**, L37–L40
Soubiran, L., & Girard, P. (2005), *A&A* **438**, 139–151
Timmes, F. X., Woosley, S. E., & Weaver, T. A. (1995), *ApJS* **98**, 617–658
Weidner, C., & Kroupa, P. (2005), *ApJ* **625**, 754–762

44

Chemical-evolution models of ellipticals and bulges

Francesca Matteucci

Departimento di Astronomia, Università di Trieste, Via G. B. Tiepolo 11, 34100 Trieste, Italy

We review some of the models of chemical evolution of ellipticals and bulges of spirals. In particular, we focus on the star-formation histories of ellipticals and their influence on chemical properties such as [α/Fe] versus [Fe/H], galactic mass and visual magnitudes. By comparing models with observational properties, we can constrain the timescales for the formation of these galaxies. The observational properties of stellar populations suggest that the more-massive ellipticals formed on a shorter timescale than less-massive ones, in the sense that both the star-formation rate and the mass-assembly rate, strictly linked properties, were greater for the most-massive objects. Observational properties of true bulges seem to suggest that they are very similar to ellipticals and that they formed on a very short timescale: for the bulge of the Milky Way we suggest a timescale of 0.1 Gyr. This leads us to conclude that the bulge evolved in a quite independent way from the Galactic disk.

1 Introduction

Galaxy formation is still an open subject and various scenarios for the formation of elliptical galaxies have been proposed. Originally, Toomre & Toomre (1972) suggested that ellipticals can form from major mergers of massive disk galaxies. Later on, Larson (1974), followed by many others (Arimoto & Yoshii 1987; Matteucci & Tornambè 1987; Bressan *et al.* 1994; Gibson 1995; Pipino & Matteucci 2004; Merlin & Chiosi 2006), suggested an early monolithic collapse of a gas cloud or early merging of lumps of gas, in which dissipation plays a fundamental role and star formation stops after the occurrence of a galactic wind, as the main mechanism for the formation of ellipticals. In the following years Bender *et al.* (1993) suggested

The Metal-rich Universe, eds. G. Israelian and G. Meynet. Published by Cambridge University Press.
© Cambridge University Press 2008.

early merging of lumps containing gas and stars in which some energy dissipation is present. In more recent times a great deal of interest was devoted to the bottom-up scenario for galaxy formation expected from the cold-dark-matter (CDM) model, in which ellipticals should form by merging of early-formed stellar systems in a wide redshift range and preferentially at late epochs with also some star formation associated with the merger events (e.g. Kauffmann *et al.* 1993). More recently, Bell *et al.* (2004) invoked dry merging at low redshift for the formation of ellipticals, namely merging of quiescent objects with no associated star formation. Bulges of spirals in most cases seem to have the same properties as ellipticals (see Chapter 27). In the following we will try to see whether any of these proposed scenarios can account for the majority of the observational constraints of ellipticals and bulges.

2 Observational properties of elliptical galaxies

From the observational point of view, elliptical galaxies are characterized by spanning a large range in luminosities and masses, and by containing mainly red giant stars and no gas. For a recent exhaustive review on elliptical galaxies we address the reader to Renzini (2006).

In addition, the following must be taken into account.

- Ellipticals are metal-rich galaxies with mean stellar metallicity in the range $\langle[Fe/H]\rangle_*$ $= -0.8$ to $+0.3$ dex (Kobayashi & Arimoto 1999) and they are characterized by having large $[\alpha/Fe]$ ratios ($\langle[Mg/Fe]\rangle_* > 0$, from 0.05 to 0.3 dex) in nuclei of giant ellipticals (Peletier 1989; Worthey *et al.* 1992; Weiss *et al.* 1995; Greggio 1997; Kuntschner *et al.* 2001; Chapter 16). This fact indicates that these galaxies and especially the most massive ones had a short duration of formation (~ 0.3–0.5 Gyr) (Matteucci 1994; Weiss *et al.* 1995): in fact, in order to have high $\langle[Mg/Fe]\rangle_*$ ratios in their dominant stellar population, the Type-Ia supernovae (SNe), which occur on a large interval of timescales, should not have had time to pollute the interstellar medium (ISM) significantly before the end of the star formation.
- Abundance gradients exist in ellipticals, with typical metallicity gradients of $\Delta[Fe/H]/\Delta \log r = -0.3$; see Kobayashi & Arimoto (1999), and references therein for a compilation of gradients. These abundance gradients in the stellar populations are well reproduced by "outside-in" models for the formation of ellipticals, as suggested by Martinelli *et al.* (1998), Pipino & Matteucci (2004, hereafter PM04), and Pipino *et al.*(2006). It is not clear whether there is a correlation between abundance gradients and galactic mass, see Ogando *et al.* (2005), for a recent paper, as required by the classic monolithic model of Larson.
- The tightness of the color–central-velocity-dispersion relation found for Virgo and Coma ellipticals (Bower *et al.* 1992) also indicated a short process of galaxy formation (~ 1–2 Gyr). Bernardi *et al.* (1998) extended this conclusion also to field ellipticals, but derived a slightly longer timescale of galaxy formation (~ 2–3 Gyr).

- The thinness of the fundamental plane seen edge-on (M/L versus M) for ellipticals in the same two clusters (Renzini & Ciotti 1993) indicates again a short process for the formation of stars in these galaxies.
- Another very interesting feature of ellipticals is the increase of the central $\langle[\mathrm{Mg/Fe}]\rangle_*$ ratio with velocity dispersion (galactic mass, luminosity), i.e. [Mg/Fe] versus σ_0 (Trager et al. 1993; Worthey et al. 1992; Matteucci 1994; Jørgensen 1999; Kuntschner et al. 2001), which indicates that more-massive objects evolve faster than less-massive ones.
- Lyman-break and SCUBA galaxies at $z \geq 3$, for which the star-formation rate is as high as $\sim(40-1000) M_\odot \mathrm{yr}^{-1}$, could be young ellipticals (Dickinson 1998; Pettini et al. 2002; De Mello et al. 2002; Matteucci & Pipino 2002).
- The existence of old fully assembled massive spheroidals already at $1.6 \leq z \leq 1.9$ (Cimatti et al. 2004) also indicates an early formation of ellipticals, at least at $z > 2$.
- Very recently the Hubble Space Telescope has provided evidence for the existence of old massive spheroids at very high redshift. In particular, Mobasher et al. (2005) reported evidence for a massive ($M = 6 \times 10^{11} M_\odot$) post-starburst galaxy at $z = 6.5$.

These observational facts, in particular the high-z old and massive early-type galaxies are challenging most N-body and semi-analytical simulations published so far, in which these galaxies are very rare objects. Differences between predictions and observations are as high as a factor of ten and increase with z (Sommerville et al. 2004). In addition, evidence for mass downsizing and "top-down" assembly of ellipticals (Cimatti et al. 2006; Renzini 2006) arises from a new analysis of the rest-frame B-band COMBO-17 and DEEP2 luminosity functions and from a photometric analysis of galaxies at $z = 1$ (Kodama et al. 2004). Therefore, all of these findings indicate the formation of ellipticals at very high redshift.

On the other hand, the arguments favoring the formation of ellipticals at low redshift can be summarized as follows.

- The relatively large values of the Hβ index measured for a sample of nearby ellipticals, which could indicate prolonged star-formation activity up to 2 Gyr ago (González 1993; Trager et al. 1998).
- The blue cores found in some ellipticals in the Hubble Deep Field (Menanteau et al. 2001), which indicate continuous star formation.
- The tight relations in the fundamental plane at low and higher redshift can be interpreted as due to a conspiracy of age and metallicity, namely to an age–metallicity anti-correlation: more-metal-rich galaxies are younger than less-metal-rich ones (e.g. Worthey et al. 1995; Trager et al. 1998, 2000; Ferreras et al. 1999).
- The main argument in favor of formation at low redshift was for years the apparent paucity of high-luminosity ellipticals at $z \sim 1$ relative to now (e.g. Kauffmann et al. 1996; Zepf 1997; Menanteau et al. 1999). However, more recently Yamada et al. (2005) found that 60%–85% of the local early-type galaxies were already in place at $z = 1$.

In the following we will show the predictions of various models for the formation and evolution of ellipticals and compare these predictions with the observations described above.

3 Models based on galactic winds

Monolithic models assume that ellipticals suffer a strong star formation and quickly produce galactic winds when the energy from SNe injected into the ISM equates to the potential energy of the gas. Star formation is assumed to halt after the development of the galactic wind and the galaxy is assumed to evolve passively afterwards. The original model of Larson (1974) suggested that galactic winds should occur later in more-massive objects due to the assumption of a constant efficiency of star formation in ellipticals of different masses and to the increasing depth of the potential well in more-massive ellipticals. Unfortunately, this prediction is at variance with the observation that the $\langle[Mg/Fe]\rangle_*$ ratio increases with galactic mass, which instead suggests a shorter period of star formation for larger galaxies. This was first suggested by Trager *et al.* (1993), and Matteucci (1994, hereafter M94), who also computed models for ellipticals by assuming a shorter period of star formation in big ellipticals. In order to obtain that, an increasing efficiency of star formation with galactic mass was assumed, with the consequence of obtaining a galactic wind occuring earlier in the massive than in the small galaxies. She called this process "inverse wind" and showed that such a model was able to reproduce the increase of $\langle[Mg/Fe]\rangle_*$ with galactic mass.

More recently, PM04 presented a revised monolithic model that allows for the formation of ellipticals by a fast merger of gas lumps at high redshift. The model is multi-zone and predicts that each elliptical forms "outside-in" (star formation stops in the outer before the inner regions owing to a galactic wind). In other words, the galactic wind develops outside-in. Following the original suggestion by M94, they assumed an increasing efficiency of star formation with the galactic mass. They also suggested a shorter timescale τ for the gas assembly with increasing galactic mass. In Figure 44.1 we show the predicted histories of star formation in the "inverse-wind scenario" of M94 and PM04. As one can see, the most-massive ellipticals undergo a shorter and more intense episode of star formation than the less-massive ones.

In Figure 44.2 we show the predictions of Matteucci (1994) concerning $\langle[Mg/Fe]\rangle_*$ versus the galactic mass (stellar) in the inverse-wind scenario, in Larson's classical scenario, and in the case of a variable initial mass function (IMF). The last case assumes that more-massive ellipticals should have a flatter IMF. However, this particular scenario requires too flat an IMF for massive ellipticals,

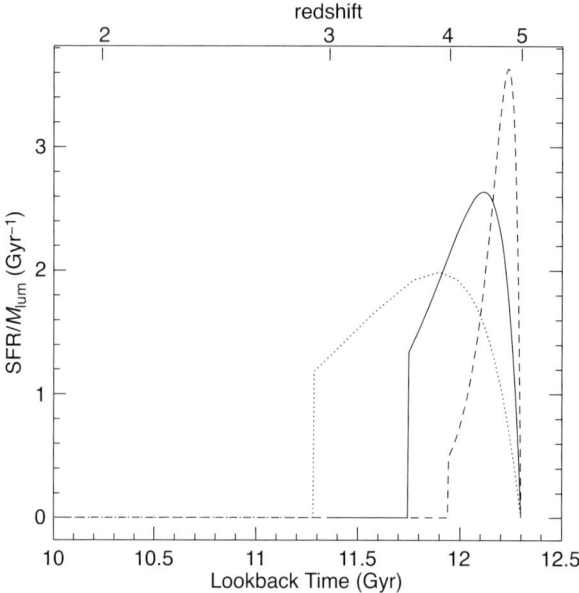

Figure 44.1. The predicted star-formation histories (star-formation rate per unit stellar mass) for galaxies of masses 10^{12} (dashed line), 10^{11} (continuous line), and $10^{10} M_\odot$ (dotted line). Such a behavior is obtained by assuming that the efficiency of star formation is increasing with galactic mass whereas the timescale for the assembly of the gaseous lumps giving rise to the galaxies is a decreasing function of mass (downsizing both in star formation and in mass assembly). In these models (PM04) the galactic wind occurs earlier in the more-massive galaxies than it does in less-massive ones.

which is at variance with observational properties (e.g. M/L ratio, colour–magnitude diagram).

PM04 recomputed the relation $\langle[\mathrm{Mg/Fe}]\rangle_*$ versus mass (velocity dispersion) and compared it with the data published by Thomas *et al.* (2002), who showed how hierarchical semi-analytical models cannot reproduce the observed $\langle[\mathrm{Mg/Fe}]\rangle_*$ versus velocity-dispersion trend, since in this scenario massive ellipticals have longer periods of star formation than smaller ones. In Figure 44.3, we have plotted the predictions of PM04 (continuous line) compared with data and hierarchical-clustering predictions.

More recently, Thomas *et al.* (2005) presented a suggestion about the star-formation histories in ellipticals in the cases of high- and low-density environment (clusters and field) and suggested that the formation of ellipticals in the field might have started 2 Gyr after that of ellipticals in clusters. This suggestion is based on recent data relating to [α/Fe] and ages in ellipticals and is shown in Figure 44.4. As one can see, their suggestion for the star formation in ellipticals in clusters is similar to that of PM04.

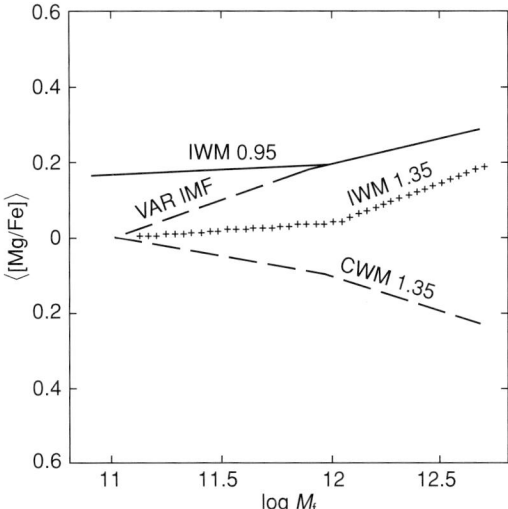

Figure 44.2. The predicted $\langle[Mg/Fe]\rangle_*$ versus $\log M_f$ (final mass) for ellipticals, under several of the assumptions considered by M94. The curves labeled IWM0.95 and IWM1.35 correspond to models with star-formation histories similar to those shown in Figure 44.1, the only difference being that the ellipticals are considered as closed-box systems until the occurrence of a galactic wind. The case IWM0.95 assumes for all galaxies an IMF with $x = 0.95$, whereas the case IWM1.35 assumes a Salpeter (1955) IMF. The curve labeled CWM indicates classic wind models in which the galactic wind occurs earlier in less-massive than it does in more-massive galaxies. Finally, the curve labeled VARIMF assumes that the IMF varies as a function of the galactic mass (see the text).

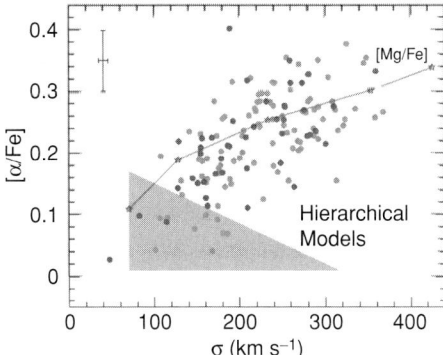

Figure 44.3. The [Mg/Fe] versus σ relationship for ellipticals (continuous line) as predicted by PM04 compared with observations (dots) and with the predictions of hierarchical semi-analytical models (shaded area). Figure adapted from Thomas et al. (2002).

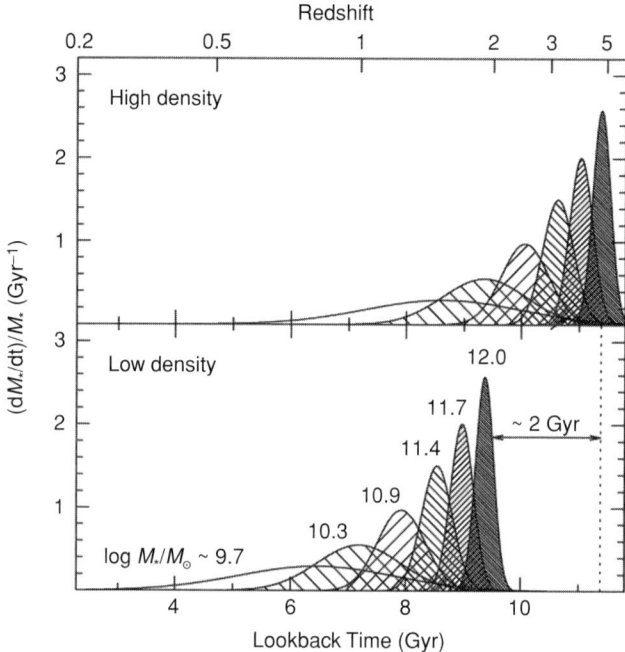

Figure 44.4. Thomas *et al.*'s view of the star-formation history (star-formation rate per unit mass of gas) in ellipticals of different masses and in different environments. Figure from Thomas *et al.* (2005).

4 Models with mergers

In order to check the effect of possible gaseous mergers triggering star formation during the lifetime of elliptical galaxies on their chemical and photometric properties, Pipino & Matteucci (2006) computed some cases with various merging epochs and amounts of accreted matter. In Figure 44.5 we show the predicted and observed $\langle[\text{Mg/Fe}]\rangle_V$ versus M_V (visual magnitude) relation (the equivalent of the $\langle[\text{Mg/Fe}]\rangle_*$ versus mass relation) for ellipticals. The quantity $\langle[\text{Mg/Fe}]\rangle_V$ represents the stellar [Mg/Fe] ratio averaged over M_V instead of over the mass.

Figure 44.5 contains the predictions of the best model of PM04 for galaxies of various masses, which lie well inside the area of existence of the observations, whereas the predictions of models with mergers tend to fall well outside the observed region unless the merger is unimportant. In particular, the agreement with observations worsens with increasing merged mass and consequent star formation.

Thomas *et al.* (1999) studied a scenario in which the formation of ellipticals occurs by merging of two spirals like the Milky Way. They concluded that this scenario fails to reproduce the α-element-enhanced abundance ratios in the metal-rich stars of ellipticals, unless the IMF is flattened during the burst ignited by the merger.

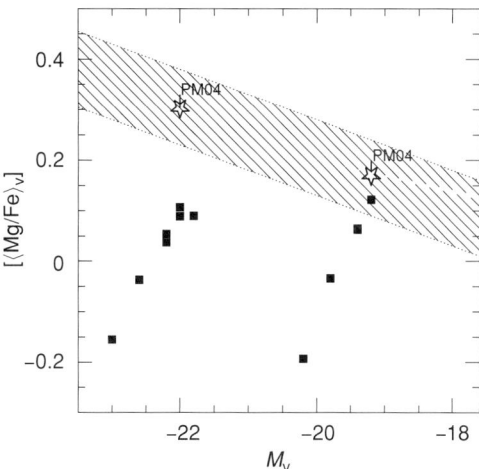

Figure 44.5. The observed $\langle[\mathrm{Mg/Fe}]\rangle_V$ versus M_V for ellipticals (shaded area) compared with models without and with mergers. The big empty stars represent the best model of PM04 whereas the filled squares represent models with mergers. The squares lying further from the shaded area represent models in which mergers with large amounts of gas and consequent star formation are allowed.

It is worth mentioning that recently De Lucia *et al.* (2006) studied the star-formation histories, ages, and metallicities of ellipticals by means of the Millennium Simulation of the concordance ΛCDM cosmology. They also suggested that more-massive ellipticals should have shorter star-formation timescales, but lower assembly (by dry mergers) redshift than less-luminous systems. This is the first hierarchical-paper admitting "downsizing" in the star-formation process in ellipticals. However, the lower assembly redshift for the most-massive system is still in contrast to what is concluded by Cimatti *et al.* (2006), who show that the downsizing trend should be extended also to the mass assembly, in the sense that the most-massive ellipticals should have assembled before the less-massive ones. This is in agreement with the model of PM04, which assumes an increasing timescale for the assembly of less-massive ellipticals.

5 The evolution of bulges

Regarding the bulges of spiral galaxies, one generally distinguishes between true bulges, hosted by S0–Sb galaxies, and the "pseudobulges" hosted in later-type galaxies (Renzini 2006). Generally, the properties (luminosity, colors, line strengths) of true bulges are very similar to those of elliptical galaxies. In the following, we will refer only to true bulges and in particular to the bulge of the Milky

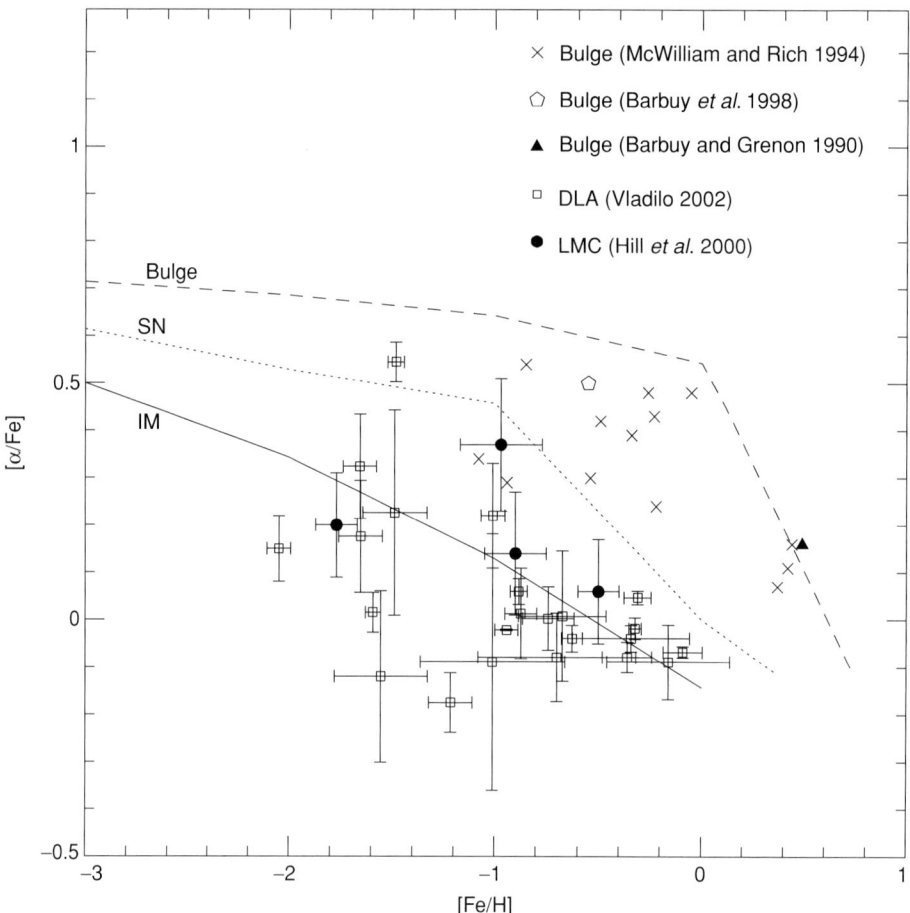

Figure 44.6. The predicted [α/Fe] versus [Fe/H] relations for the bulge (upper curve), the Solar vicinity (middle curve) and irregular galaxies (lower curve). Data for the bulge are shown for comparison. Data for the LMC and damped Lyman-α (DLA) systems are also shown for comparison, indicating that DLA systems are probably irregular galaxies. Figure from Matteucci (2001).

Way. The bulge of the Milky Way is, in fact, the best-studied bulge and several scenarios for its formation have been put forward in past years. As summarized by Wyse & Gilmore (1992), the proposed scenarios are (1) the bulge formed by accretion of extant stellar systems, which eventually settled in the center of the Galaxy; (2) The bulge was formed by accumulation of gas at the center of the Galaxy and subsequent evolution with either fast or slow star formation; and (3) The bulge was formed by accumulation of metal-enriched gas from the halo, thick disk, or thin disk in the center of the Galaxy.

The metallicity distribution of stars in the bulge and the [α/Fe] ratios greatly help in selecting the most probable scenario for formation of the bulge. In Figure 44.6

we present the predictions by Matteucci (2001) of the [α/Fe] ratios as functions of [Fe/H] for galaxies of various morphological types. In particular, for the bulge or an elliptical galaxy of the same mass, for the region of the Solar vicinity, and for an irregular Magellanic galaxy (the LMC and SMC). The underlying assumption is that different objects undergo different histories of star formation, this being very fast in the spheroids (bulges and ellipticals), moderate in spiral disks, and slow and perhaps gasping in irregular gas-rich galaxies. The effect of different star-formation histories is evident in Figure 44.6, where the predicted [α/Fe] ratios in the bulge and ellipticals stay high and almost constant over a large interval of [Fe/H]. This is due to the fact that, since star formation is very intense, the bulge very soon reaches Solar metallicity thanks only to the SNe II; then, when SNe Ia start exploding and restituting Fe into the ISM, the change in the slope occurs at larger [Fe/H] than that in the Solar vicinity. In the extreme case of irregular galaxies the situation is the opposite: here the star formation is slow and when the SNe Ia start exploding the gas is still very metal-poor. This scheme is quite useful since it can be used to identify galaxies merely by looking at their abundance ratios. A model for the bulge behaving as shown in Figure 44.6 is able to reproduce also the observed metallicity distribution of bulge stars (see Matteucci & Brocato 1990; Matteucci *et al.* 1999). The scenario suggested in these papers favors the formation of the bulge by means of a short and strong starburst, in agreement with Elmegreen (1999) and Ferreras *et al.* (2003). A similar model, although updated with the inclusion of the development of a galactic wind and more recent stellar yields, is presented in Chapter 48 by Ballero *et al.*, who show how a bulge model with intense star-formation (star-formation efficiency $\sim 20\,\mathrm{Gyr}^{-1}$) and rapid assembly of gas (within 0.1 Gyr) can best reproduce the most recent accurate data on abundance ratios and the metallicity distribution.

Previous attempts to model the galactic bulge were presented by Mollá *et al.* (2000) and by Samland *et al.* (1997). Both these models, although they differ in other respects, assumed a more prolonged period of star formation than in the models discussed above, which produces [α/Fe] ratios behaving more akin to those in the Solar vicinity. Very recently, Zoccali *et al.* (2006) derived oxygen and iron abundances for 50 K giants in the Galactic bulge. The spectra were taken with the UVES at the VLT and have quite a high resolution ($R = 45,000$). These data show a longer plateau for [O/Fe] than in the Solar vicinity, with a change in slope in the [O/Fe] versus [Fe/H] relation occurring at [Fe/H] ~ -0.2 dex, in very good agreement with the predictions of Ballero *et al.* in Chapter 48. Also dynamical models by Immeli *et al.* (2004), who simulated the formation of galaxies from clouds with various dissipation efficiencies, can explain the bulge abundance ratios by means of a short starburst occurring when the dissipation efficiency was quite high.

6 Conclusions

We have compared the observational properties of ellipticals and bulges with the model predictions and reached the following conclusions.

- The existence of old and massive galaxies at high redshift ($z > 3$) argues in favor of the scenario whereby ellipticals form very rapidly and at high redshift.
- The increasing $\langle[Mg/Fe]\rangle_*$ with galactic mass as well as most of the properties of ellipticals (mass–metallicity relation, color–magnitude diagram) can be well explained by a "quasi-monolithic" model whereby star formation stops first in more-massive ellipticals because of galactic winds induced by supernovae. This implies a downsizing process both in the star formation and in the mass assembly of these galaxies. Models of co-evolution QSO ellipticals also suggest a shorter period of star formation in the most-massive objects (Granato *et al.* 2001).
- Classic hierarchical models for galaxy formation cannot reproduce the trend of [Mg/Fe] since they allow star formation to continue until recently in ellipticals. Only the recent model of De Lucia *et al.* (2006) combines a hierarchical scenario with downsizing of star formation, but it does not allow for a mass downsizing, as suggested by the recent work of Cimatti *et al.* (2006).
- Major mergers occurring for elliptical galaxies at any redshift or late mergers with a variety of accreted masses, coupled with star formation, should be ruled out on the basis of the results of chemical and spectrophotometric models of ellipticals (Pipino & Matteucci 2006). Also mergers of spirals can be ruled out on the basis of chemical arguments (Thomas *et al.* 1999).
- True bulges seem to be very old systems with properties similar to those of elliptical galaxies. The most-recent Galactic models and accurate abundance data suggest that the bulge of the Milky Way must have formed at very early times on a very short timescale (<0.5 Gyr) and by means of a very intense burst of star formation with a high efficiency of star formation.

References

Arimoto, N., & Yoshii, Y. (1987), *A&A* **173**, 23
Barbuy, B., & Grenon, M. (1990), in *Bulges of Galaxies*, eds. B. J. Jarvis & D. M. Terndrup (Garching bei München, European Southern Observatory), p. 83
Barbuy, B., Ortolani, S., & Bica, E. (1998), *A&AS* **132**, 333
Bell, E. F., Wolf, C., Meisenheimer, K. *et al.* (2004), *ApJ* **608**, 752
Bender, R., Burstein, D., & Faber, S. M. (1993), *ApJ* **411**, 153
Bernardi, M., Renzini, A., da Costa, L. N. *et al.* (1998), *ApJ* **508**, L143
Bressan, A., Chiosi, C., & Fagotto, F. (1994), *ApJS* **94**, 63
Cimatti, A., Daddi, E., Renzini, A. *et al.* (2004), *Nature* **430**, 184
Cimatti, A., Daddi, E., & Renzini, A. (2006), *A&A* **453**, L29
Bower, R. G., Lucey, J. R., & Ellis, R. S. (1992), *MNRAS* **254**, 613
De Lucia, G., Springel, V., White, S. D. M., Croton, D., & Kauffmann, G. (2006), *MNRAS* **366**, 499
de Mello, D., Wiklind, T., Leitherer, C., & Pontoppidan, K. (2002), *Ap&SS* **281**, 549

Dickinson, M. (1998), in *The Hubble Deep Field*, eds. M. Livio, M. Fall, & P. Madau (Cambridge, Cambridge University Press), p. 219
Elmegreen, B. G. (1999), *ApJ* **517**, 103
Ferreras, I., Charlot, S., & Silk, J. (1999), *ApJ* **521**, 81
Ferreras, I., Wyse, F. G., & Silk, J. (2003), *MNRAS* **345**, 1391
Gibson, B. K. (1995), Ph.D. Thesis, University of British Columbia, Canada
González, J. J. (1993), Ph.D. Thesis. University of California, Santa Cruz
Granato, G. L., Silva, L., Monaco, P. et al. (2001), *MNRAS* **324**, 757
Greggio, L. (1997), *MNRAS* **285**, 151
Jørgensen, I. (1999), *MNRAS* **306**, 607
Kauffmann, G., Charlot, S., & White, S. D. M. (1993), *MNRAS* **283**, L117
    (1996), *MNRAS* **283**, 117
Kobayashi, C., & Arimoto, N. (1999), *ApJ* **527**, 573
Kodama, T., Yamada, T., Akiyama, M. et al. (2004), *ApJ* **492**, 461
Kuntschner, H., Lucey, J. R., Smith, R. J., Hudson, M. J., & Davies, R. L. (2001), *MNRAS* **323**, 625
Hill, V., François, P., Spite, M., Primas, F., & Spite, F. (2000), *A&A* **364**, L19
Immeli, A., Samland, M., Gerhard, O., & Westera, P. (2004), *A&A* **413**, 547
Larson, R. B. (1974), *MNRAS* **169**, 229
Martinelli, A., Matteucci, F., & Colafrancesco, S. (1998), *MNRAS* **298**, 42
Matteucci, F. (1994), *A&A* **288**, 57
    (2001), *The Chemical Evolution of the Galaxy*, (Dordrecht, Kluwer Academic Publisher)
Matteucci, F., & Brocato, E. (1990), *ApJ* **365**, 539
Matteucci, F., & Pipino, A. (2002), *ApJ* **569**, L69
Matteucci, F., Romano, D., & Molaro, P. (1999), *A&A* **341**, 458
Matteucci, F., & Tornambé, A. (1987), *A&A* **185**, 51
McWilliam, A., & Rich, R. M. (1994), *ApJS* **91**, 749
Menanteau, F., Ellis, R. S., Abraham, R. G., Barger, A. J., & Cowie, L. L. (1999), *MNRAS* **309**, 208
Menanteau, F., Jimenez, R., & Matteucci, F. (2001), *ApJ* **562**, L23
Merlin, E., & Chiosi, C. (2006), *A&A* **457**, 437
Mobasher, B., Dickinson, M., Ferguson, H. C. et al. (2005), *ApJ* **635**, 832
Mollá, M., Ferrini, F., & Gozzi, G. (2000), *MNRAS* **316**, 345
Ogando, R. L. C., Maia, M. A. G., Chiappini, C., Pellegrini, P. S., Schiavon, R. P., & da Costa, L. N. (2005), *ApJ* **632**, L61
Peletier, R. (1989), Ph.D. Thesis, University of Groningen, The Netherlands
Pettini, M., Rix, S. A., Steidel, C. C., Adelberger, K. L., Hunt, M. P., & Shapley, A. E. (2002), *ApJ* **569**, 742
Pipino, A., & Matteucci, F. (2004), *MNRAS* **347**, 968 (PM04)
    (2006), *MNRAS* **365**, 1114
Renzini, A. (2006), *ARAA* **44**, 141
Renzini, A., & Ciotti, L. (1993), *ApJ* **416**, L49
Salpeter, E. E. (1955), *ApJ* **121**, 161
Samland, M., Hensler, G., & Theis, C. (1997), *ApJ* **476**, 544
Sommerville, R. S., Moustakas, L. A., Mobasher, B. et al. (2004), *ApJ* **600**, L135
Thomas, D., Greggio, L., & Bender, R. (1999), *MNRAS* **302**, 537
Thomas, D., Maraston, C., Bender, R., & Mensez de Oliveira, C. (2005), *ApJ* **621**, 673
Thomas, D., Maraston, C., & Bender, R. (2002), *Reviews in Modern Astronomy* **15**, 219
Toomre, A., & Toomre, J. (1972), *ApJ* **178**, 623

Trager, S. C., Faber, S. M., González, J. J., & Worthey, G. (1993), *AAS* **183**, 4205
Trager, S. C., Faber, S. M., Worthey, G., & González, J. J. (2000), *AJ* **120**, 165
Trager, S. C., Worthey, G., Faber, S. M., Burstein, D., & González, J. J. (1998), *ApJS* **116**, 1
Vladilo, G. (2002), *A&A* **391**, 407
Weiss, A. Peletier, R. F., & Matteucci, F. (1995), *A&A* **296**, 73
Wyse, R. F. G., & Gilmore, G. (1992), *AJ* **104**, 144
Worthey, G., Faber, S. M., & González, J. J. (1992), *ApJ* **398**, 69
Worthey, G, Trager, S. C., & Faber, S. M. (1995), *Fresh Views of Elliptical Galaxies*, eds. A. Buzzoni, A. Renzini & A. Serrano (San Francisco, CA, Astronomical Society of the Pacific), p. 203
Yamada, T., Kodama, T., Akiyama, M. *et al.* (2005), *ApJ* **634**, 861
Zepf, S. E. (1997), *Nature* **390**, 377
Zoccali, M., Lecureur, A., Barbuy, B. *et al.* (2006), *A&A* **457**, L1

45

Chemical evolution of the Galactic bulge

Brad K. Gibson,[1] Angela J. MacDonald,[1] Patricia Sánchez-Blázquez,[1] & Leticia Carigi[2]

[1] University of Central Lancashire, Centre for Astrophysics, Preston PR1 2HE, UK
[2] Instituto de Astronomía, UNAM, México, D.F., Mexico

The chemical evolution of the Galactic bulge is calculated by adopting a single-zone framework, with accretion of primordial gas on a free-fall timescale, assuming (i) a correspondingly rapid timescale for star formation and (ii) an initial mass function biased towards massive stars. We emphasise here the uncertainties associated with the underlying physics (specifically, stellar nucleosynthesis) and how those uncertainties are manifested in the predicted abundance-ratio patterns in the resulting present-day Galactic-bulge stellar populations.

1 Background

In many respects, bulges of spiral galaxies are very similar to elliptical galaxies. Both adhere to many common scaling relations, including the fundamental plane, possess high stellar densities, have little in the way of gas and dust, and appear essentially old, with enhanced abundance ratios of α-elements with respect to iron. In our own Milky Way, the bulge accounts for $\sim$20% of the Galaxy's baryons – a factor of ten greater than the stellar halo. Despite this significance, relatively few detailed chemical-evolution models of the bulge exist,[1] due in part to a dearth of high-resolution spectroscopic studies of its individual stars. Having said that, in lieu of such data, notable exceptions have appeared in the literature, drawing upon extant metallicity-distribution functions (MDFs) and inferred abundance ratios derived from lower-resolution data.

---

[1] In contrast to the case of the halo, which, despite its trace baryonic contribution to the Milky Way, has had at least ten times the number of models published to explain its origin.

The Metal-rich Universe, eds. G. Israelian and G. Meynet. Published by Cambridge University Press.
© Cambridge University Press 2008.

Köppen & Arimoto (1989, 1990) assumed infall of primordial gas on a free-fall timescale (0.1 Gyr) and power-law initial mass functions (IMFs) of slope $x = 1.05$ (Köppen & Arimoto 1989) and $x = 1.30$ (Köppen & Arimoto 1990), over the mass range $0.05 < m/M_\odot < 60$. Rapid and efficient star formation ($10\,\text{Gyr}^{-1}$) was halted by a supernova-driven wind after 1 Gyr, the metal-enriched outflowing gas providing fuel to the Galactic disc for future star formation. The Köppen & Arimoto models successfully recovered the bulge's MDF, present-day gas mass fraction and enhanced [α/Fe], despite the (i) neglect of Type-Ia supernovae and (ii) use of the instantaneous-recycling approximation.

Matteucci & Brocato (1990) and, later, Matteucci *et al.* (1999) relaxed these two limitations of Köppen & Arimoto, also concluding that flatter-than-Salpeter IMFs ($1.1 \lesssim x \lesssim 1.3$, over the mass range $0.1 < m/M_\odot < 100$) in conjunction with (i) a Schmidt-like star-formation law, (ii) rapid infall of primordial gas on timescales of 0.01 Gyr (Matteucci & Brocato 1990) and 0.1 Gyr (Matteucci *et al.* 1999) and (iii) efficient star formation ($20\,\text{Gyr}^{-1}$) applied. Both models were successful in terms of recovering the bulge's MDF and enhanced [α/Fe].

Samland *et al.* (1997) suggested that the bulge's MDF is consistent with the use of a more traditional Salpeter IMF ($x = 1.35$, over the mass range $0.1 < m/M_\odot < 100$) and a more prolonged star-formation phase (with the bulge being 3–5 Gyr younger than the halo, a conclusion that is perhaps less secure), with 'breathing' phases of infall and outflow throughout the bulge's history. Mollá *et al.* (2000) also adopted the Salpeter IMF and assumed two infall phases ('bulge' and 'core', with a longer infall timescale for the dominant 'bulge' phase of 0.7 Gyr). As with Samland *et al.* (1997), infall and outflow lead to exchange of matter among halo, bulge and core, and ultimately to a predicted bulge MDF matching that observed.

All of the above models have their merits and detriments, but space precludes a detailed intercomparison. The prediction of α-enhanced abundance patterns *across the full range of bulge metallicities* ($-1 \lesssim$ [Fe/H] $\lesssim +0.5$) is somewhat unique to the 'Matteucci' models, for obvious reasons (IMF + star-formation efficiencies + timescales).

The recent appearance of spectacular high-resolution spectroscopic data for the bulge (e.g. Lecureur et al. 2007 and references therein), makes it timely not only to revisit the traditional [α/Fe] patterns predicted by chemical-evolution models, but also to begin to inspect individual α-to-α-element predictions, to seek further insights into bulge formation (and, as we will suggest, stellar evolution). Such a preliminary analysis was undertaken by Gibson (1995), but the data quality at the time made the conclusions speculative, at best. In this short contribution, we revisit the issue of bulge abundance patterns, concentrating instead on a previously (somewhat) ignored component of the models – specifically, the sensitivity to the compilation of the yield from Type-II supernovae adopted.

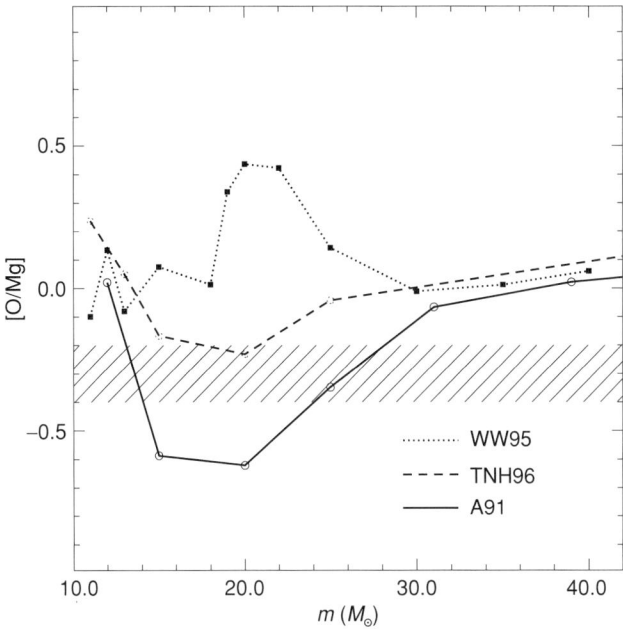

Figure 45.1. Nucleosynthetic abundance-ratio (oxygen-to-magnesium: [O/Mg]) patterns at $z = 0.02$ predicted by the Solar-metallicity Type-II-supernova models of Woosley & Weaver (1995, WW95), Thielemann *et al.* (1996, TNH96) and Arnett (1991, A91). The shaded region is representative of the subsolar [O/Mg] range encountered in the Galactic bulge: $-0.4 \lesssim$ [O/Mg] $\lesssim -0.2$ (McWilliam & Rich 1994; Lecureur *et al.* 2007).

2 Results and discussion

Many of the models described in Section 1 adopt nucleosynthetic yields from Woosley & Weaver (1995) or one of its predecessors. However, as discussed by Gibson *et al.* (1997) in a different context (clusters of galaxies), while stellar nucleosynthesis models may have identical global metal (i.e. Z) yields, the relative distribution of elements and isotopes therein may be quite different (driven by differences in the treatment of reaction rates, mass loss, convection, etc.). Figure 45.1, adapted from Gibson (1995), provides a graphic demonstration for the important α-element pair of oxygen and magnesium – a factor of 5–10 difference in O/Mg exists, for example, between Woosley & Weaver (1995) and Arnett (1991), at Solar metallicity, in the mass range $15 \lesssim m/M_\odot \lesssim 25$.

The shaded region highlights an interesting puzzle – as hinted at already in the seminal work of McWilliam & Rich (1994) and confirmed recently by Lecureur *et al.* (2007), O/Mg in the bulge appears to be a factor of two lower than that of the Sun, over the metallicity range $-0.5 \lesssim$ [Fe/H] $\lesssim +0.5$ (i.e. a range spanning the bulk of the stars in the bulge). As discussed by Gibson (1995), such subsolar [O/Mg] values are essentially impossible to recover with *any* chemical-evolution model employing the Woosley & Weaver (1995) yields, because not a single model in the grid lies within the shaded region (thus, no IMF that would lead to a model matching the data could be constructed *a posteriori*); the situation does not appear particularly tenable with the Thielemann *et al.* (1996) models either. What *is* interesting from Figure 45.1, though, is the location of the little-used Arnett (1991) yields in this particular plane – specifically, a natural byproduct of the models is the increased production of magnesium in the mass range $15 \lesssim m/M_\odot \lesssim 25$, shifting those models into the abundance-pattern regime populated by the stars of the Galactic bulge; (this is a not entirely surprising consequence, in the light of Gibson (1997, Figure 3).

Using the Arnett (1991; henceforth A91), Woosley & Weaver (1995, henceforth WW95) and Thielemann *et al.* (1996; henceforth TNH96) Type-II supernova yields, we have constructed representative models of the Galactic bulge using the one-zone infall precursor analogue (Gibson & Matteucci 1997) to GEtool (Fenner & Gibson 2003). Our fiducial bulge model is patterned after the $(x = 0.95, k = 20 \text{ Gyr}^{-1}, \pi = 0.1 \text{ Gyr})$ model of Matteucci *et al.* (1999), the primary differences being (i) the reduction of the upper mass limit of the IMF from $100 M_\odot$ to $35 M_\odot$ and (ii) the inclusion of three different Type-II supernova yield options. Note, in keeping with our philosophy, we do *not* perform any *a posteriori* normalisation of our models. As in Köppen & Arimoto (1989, 1990), we have stopped the simulations at 1 Gyr, although this does not alter the general thrust of our conclusions.

The predicted chemical evolution of this fiducial model, using the three different yield sources, is shown in Figure 45.2. It should come as little surprise (in the light of the discussion surrounding Figure 45.1) that only the model incorporating Arnett's (1991) yields successfully predicts the bulk of the bulge stars to have $-0.4 \lesssim [\text{O/Mg}] \lesssim -0.2$. Conversely, Matteucci *et al.* (1999, Figure 3) and Mollá *et al.* (2000, Figure 3) predicted $[\text{O/Mg}] \approx +0.05 \pm 0.05$ for the stars of the Galactic bulge. It should be stressed that this was a natural prediction of Arnett's yields, and required no *a posteriori* rescaling of the magnesium abundances, as is normally done when employing the method of Woosley & Weaver (1995).

Having said all this, it would be foolish to suggest that this is definitive proof in favour of the Arnett (1991) compilation; all of the caveats noted in Gibson *et al.* (1997) regarding their input physics remain valid today. More importantly, that which might ameliorate one important abundance-ratio problem in the bulge may also lead to irreparable consequences for other patterns, or which is more likely, problems for the Solar neighbourhood (although a cynic could turn the problem around and say that having to resort to a different IMF for the bulge, as opposed to the Solar neighbourhood, is not necessarily 'better'). Indeed, we suspect that a detailed accounting of *all* relevant observables will suggest that the A91 yields are

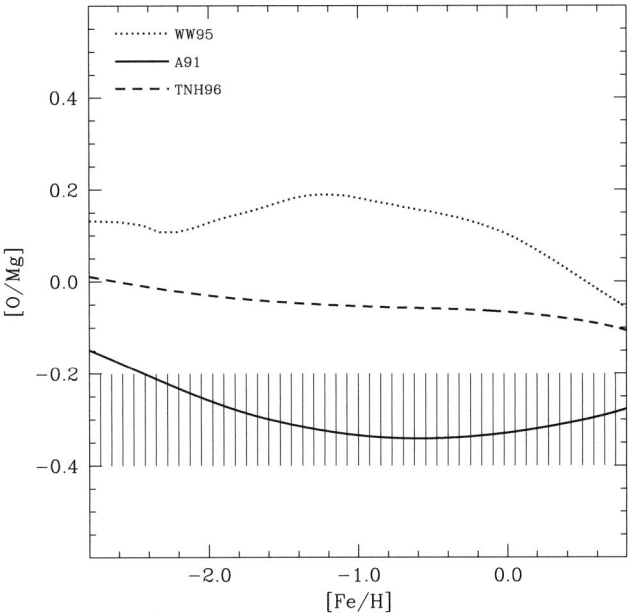

Figure 45.2. Predicted chemical-evolution trajectories in the [O/Mg]–[Fe/H] plane, for Type-Ia supernovae, under the assumption of a massive-star ($m_u = 35 M_\odot$)-biased initial mass function, $x = 0.95$, rapid infall (on a free-fall timescale) of primordial gas, and the three Type-II supernova yield compilations shown in Figure 45.1: WW95, Woosley & Weaver (1995); TNH96, Thielmann et al. (1996); and A91, Arnett (1991). As in Figure 45.1, the shaded region corresponds to the subsolar [O/Mg] range encountered in the Galactic bulge.

not a panacea for the chemical evolution of the bulge, but our goal here was not to prove (or disprove) that statement, but simply to remind the end-user of such yield tables that, while WW95 is an extraordinarily beautiful suite of models, one should be cautious in assuming that their use constitutes the elimination of nucleosynthesis as a significant systemtic uncertainty in models of galactic chemical evolution!

Acknowledgment

Simulations were performed at the University of Central Lancashire High Performance Computing Facility.

References

Arnett, D. (1991), in *Frontiers of Stellar Evolution*, ed. D. L. Lambert (San Francisco, CA, Astronomical Society of the Pacific), pp. 389–401 (A91)
Fenner, Y., & Gibson, B. K. (2003), *PASA* **20**, 189–195
Gibson, B. K. (1995), Ph.D. Dissertation, University of British Columbia, Canada
 (1997), *MNRAS* **290**, 471–489
Gibson, B. K., & Matteucci, F. (1997), *MNRAS* **291**, L8–L12
Gibson, B. K., Loewenstein, M., & Mushotzky, R. F. (1997), *MNRAS* **290**, 623–628

Köppen, J., & Arimoto, N. (1989), *ApSS* **156**, 47–50
 (1990), *A&A* **240**, 22–35
Lecureur, A., Hill, V., Zoccali, M., *et al.* (2007), *A&A* **465**, 799
Matteucci, F., & Brocato, E. (1990), *ApJ* **365**, 539–543
Matteucci, F., Romano, D., & Molaro, P. (1999), *A&A* **341**, 458–468
McWilliam, A., & Rich, R. M. (1994), *ApJS* **91**, 749–791
Mollá, M., Ferrini, F., & Gozzi, G. (2000), *MNRAS* **316**, 345–356
Samland, M., Hensler, G., & Theis, Ch. (1997), *ApJ* **476**, 544–559
Thielemann, F.-K., Nomoto, K., & Hashimoto, M.-A. (1996), *ApJ* **460**, 408–436 (TNH96)
Woosley, S. E., & Weaver, T. A. (1995), *ApJS* **101**, 181–235 (WW95)

46

How do galaxies become metal-rich? An examination of the yield problem

M. G. Edmunds

School of Physics and Astronomy, Cardiff University, Queens Buildings, 5, The Parade, Cardiff CF24 3AA, UK

Reports of high metallicities in galactic systems have always been controversial. I disuss whether observational claims both for nebulae and for stars are well-founded, and try to form a rational view of just how metal-rich some regions of galaxies do become. Metallicity is linked to the evolution of star formation in a galaxy through the yield, the mass of metals produced each time star formation locks up unit mass of interstellar material. The mechanisms by which real or apparent high yields might be achieved are examined – global and local gas flows, poor mixing, star formation and metallicity effects in stellar evolution. As perhaps expected, it turns out to be not so easy to 'get rich', quickly or otherwise – suggesting that sorting out the lingering uncertainties in the abundance analysis of H II regions and stars remains a priority.

1 Introduction

What is rich? I suspect that rather like our incomes, where we tend to define a rich person as anyone who earns more than twice as much as we do, 'metal-rich' is something with twice the metallicity of the objects you are used to working with! Richness tends to be relative. But what can we agree on? The following statement is probably not too contoversial: (1) in the central regions of our Galaxy there are stars with twice the Solar metal abundance and there is gas with twice the Solar abundance. The next statement might raise a few objections: (2) there are open clusters, and stars locally, which have at least three times Solar abundance (e.g. Twarog *et al.* 2003). A third statement probably cannot be confirmed or denied at present: (3) there are central regions of galaxies, known as quasars, where gas abundances reach at least five times Solar. There are two immediate problems. The

The Metal-rich Universe, eds. G. Israelian and G. Meynet. Published by Cambridge University Press.
© Cambridge University Press 2008.

first is that we do not know the *absolute* yield and hence whether twice Solar represents reaching the yield or having to reach double the yield. The second problem is asking whether (1) and (2) could in any event actually be consistent. Before tackling the subtleties of chemical evolution we ought to look at the accuracy of the abundance measurements.

I do not want to review in great detail the problem of measuring abundances from H II regions, particularly extragalactic ones, since this has already been done very well at this conference by F. Bresolin (see Chapter 17). I will note that calibrations of the high-end metallicity from strong-line methods of analysis have reduced very significantly (e.g. Bresolin *et al.* 2004; Pilyugin 2001; Pilyugin *et al.* 2004) from early calibrations like Edmunds and Pagel (1984). I do wonder whether use of recombination lines and the sort of effects outlined by Stasińska (2005) will result in the calibration ending up somewhere between the earlier and later values. But one thing really puzzles me – the old calibrations were based on computed H II-region models – and if the calibration *is* so wrong, then what is wrong with those models? I do not think I will be happy until consistency with theoretical H II-region models is attained. Whatever the outcome of calibration, it still seems as if galaxies do display abundance gradients – and the smaller magnitude of these gradients implied by recent calibrations may alleviate some of the problems (e.g. Edmunds and Greenhow 1995) of generating the largest gradients implied by the old calibrations.

Can we trust stellar abundances? The determination of element ratios appears to be very good nowadays, but lingering uncertainties persist in *absolute* abundances, e.g. Ayres *et al.*'s (2006) discussion of the innovative work of M. Asplund, and an open mind is still needed.

2 Models and yields

The formal definition of the 'yield', often denoted p, is the mass of metals produced per mass of material locked up in long-lived stars or remnants (e.g. Pagel 1997; Edmunds 1990). In a closed-box (or 'simple') model this gives the elementary relation that the gas abundance $Z = p \ln(1/f)$, where f is the 'gas fraction' – the mass of gas left in the system relative to the mass of stars plus gas. The mean metallicity in the stellar population, often denoted $\langle Z \rangle$, tends to the value p as the gas is exhausted. What we are really asking (and this is the yield problem(s)) is (1) whether the true yield p is equal to the Solar abundance, or some (small) multiple of it; (2) how far other processes can produce an 'effective' yield (e.g. one definition of which is the mean metallicity $\langle Z \rangle$ of a stellar population when gas is exhausted) that is 2, 3, 5 ... times p; (3) whether p might vary with the metallicity of the gas from which the stars formed or other conditions; and (4) whether there are regions that are too metal-rich to be explained by 'normal' galactic evolution. In the simple model, the gas metal abundance can become high (much greater than p) at small gas fractions, but one should be aware that at small gas fractions the release of low-metallicity gas from long-lived stars (which is not allowed for in the simple model) might in fact dilute the gas somewhat (Talbot & Arnett 1971). The mean metallicity $\langle Z \rangle$ of the stellar population will, however, tend stubbornly to the yield p. So how can the mean metallicity of a stellar population be varied? An obvious way is if evolution does not proceed very far and then all the remaining gas is lost. Another way is just to have steady mass loss at some constant times the star-formation rate. In this case the whole metallicity structure of the stellar population – i.e. the relative numbers of stars of various metallicities – is simply displaced to lower abundance (i.e. a reduced mean but with the same spread in logarithmic abundance). Arbitrary inflow of unenriched gas has a rather more subtle effect – the mean abundance (at gas exhaustion by star formation) would still tend to p, but the effect of inflow is to *narrow* the metallicity stucture. On defining an 'effective' yield for the gas by $p_{\text{eff}} = \langle Z \rangle / \ln(1/f)$, where f is the gas fraction, and that for the stars by $p_{\text{eff}} = \langle Z \rangle / p$, we could summarise by saying that outflow reduces both the gas and the stellar effective yield, while inflow of unenriched gas reduces the gas effective yield, but does not affect the stellar effective yield. So far we do not seem to be making much progress towards producing *increased* yields!

3 The initial mass function for star formation and yield variation

An obvious way to increase the yield is to favour the formation of massive, metal-synthesising stars; but I have always felt that resorting to altering the inital mass function (IMF) is a sign of moral weakness. At this conference P. Kroupa has acted like a saint, while I. Bonnell has been sorely tempted. I would not deny that at zero metallicity in the early Universe the IMF might have been very different, but once even a small metallicity has built up there seems reasonable observational evidence that the IMF is fairly constant on a large scale. But suppose that one persists in Sin and asks how IMF varition might affect yields. An obvious possibility is that increasing metallicity implies more dust and hence more opacity. This could affect the upper mass limit by radiation-pressure effects (but perhaps in the wrong way, if an increase in yield is wanted?), or perhaps the whole IMF by opacity effects on fragmentation. I am unaware of any reliable systematic theoretical study of the effect of IMF variations on yields – which probably requires better knowledge of mass cuts in supernovae anyway – but a very simple estimate with an IMF of the form $n(m) dm = k m^{-\alpha} dm$ at the high end and with metal production proportional to progenitor mass implies the possibility of a three-fold change of yield with variation of α between, say, 2.1 and 2.5. Statistical effects forced by the finite size of star clusters can also have some influence on the overall yield (Goodwin and Pagel 2005). Now, although I have a moral objection to IMF variation, this does not mean that the yield has to be a metal-independent constant. We know that stellar evolution and mass loss are sensitive to metallicity, so it is not unreasonable to consider yield variations, perhaps of linear, $p = p_0 + aZ$, or quadratic, $p = p_0 + bZ^2$, form. The metallicity effect might not show itself at low metal abundances, but as things become metal-richer 'to those that hath shall more be given', provided that a or b is positive. The effect of the linear variation on the gas metal abundance in a simple closed model is unsurprising. Perhaps more interesting is the effect on the metallicity structure of the population. This now spreads out and increases the relative numbers of stars at higher metallicities. If we also allow some inflow of unenriched gas, it appears, for certain combinations of parameters, Edmunds (2007), to magnify the effect on the yield, so this mechanism might well be a promising way of attaining apparent yields that are several times the yield p implied from lower-metallicity systems. This is pure speculation, so before taking it too seriously its relevance (or lack of relevance) for the local stellar metallicity distribution should be checked. A metallicity-dependent yield could also help in building abundance gradients. With a *constant* yield one can show (Edmunds and Greenhow 1995) that the mean stellar metal abundance with radial gas flow could reach $2p$ (i.e. twice that of the simple model) in a disc configuration and up to $3p$ in spherical/spheroidal geometry. Presumably a metal-dependent yield could increase this significantly, assuming that one means by this an increase over the p_0 yield measured at low metallicity. An astrophysical justification for a metallicity-dependent yield would be needed too – indeed, if one argues that increased metallicity leads to increased mass loss, a look at the oxygen yields of Maeder (1992) would imply a reduction in yield with increased mass loss – just the opposite of what we want here!

4 Metallicity-dependent yields and [Mg/Fe]

As discussed extensively at this conference, one of the most noticeable abundance-ratio behaviours is the high value of [Mg/Fe] in the stellar population of galactic bulges. It may be telling us something fundamental about galaxy formation. Is it just the SN II/SN Ia timescale that is involved, with a yield of the form $p_{\text{Fe}} = p_{\text{SN II}} + p_{\text{SN Ia}}$ (delayed), showing that the bulges were formed relatively fast? I would just point out that, if the magnesium yield had a suitable metallicity-dependent component $p_{\text{Mg}} = p_0 + aZ$, where Z is the overall metallicity (mainly oxygen; stellar-evolution specialists could no doubt invent a suitable mechanism!), then one could have an [Mg/Fe] that actually increased at high metallicity in a fast-forming system where star formation halts before the release of most of the SN Ia iron, while the [Mg/Fe] would be dragged down by the emerging SN Ia iron in a slowly-forming system. A more careful analysis of what happens to the mean metallicities in the stellar populations is needed here – particularly if inflowing gas moving into a central forming bulge might enhance any metallicity-dependent effects, followed by outflow to shed SN Ia low-[Mg/Fe] gas.

5 Metallicity–luminosity relations

I still do not understand metallicity–luminosity relations for galaxies, although I will give a conventional explanation (e.g. Garnett 2002; Pilyugin *et al.* 2004). Tremonti *et al.* (2004) give a fine plot of abundances versus mass for some 53 000 galaxies in the Sloan digital survey that is based on strong-line H II-region abundance indicators. It is not difficult to make an elementary model that fits the trend of metallicity with mass that is based on gas outflow whose extent is determined by the mass of the galaxy (Tremonti *et al.* 2004; Edmunds 2005). It would be interesting to see whether one could fit the mass–metallicity relation at a redshift of 2 shown by M. Pettini at this conference (Figure 20.5, taken from Erb *et al.* (2006)) with the same model, but simply at an earlier time (i.e. with star formation not having proceeded so far). I thought initially that this could be done, but I note that there are problems in relating the nearby and distant metallicities, depending on what calibration and strong-line indicator is used. Erb *et al.* suggest that the mass loss may have a different form early on, independently of galaxy mass and with quite a high true yield of four times Solar. Modelling of this is worth persuing, with special effort to make the abundance analysis right for both nearby and distant systems. If it can be shown that a single mass-loss mechanism is at work and that the redshift evolution is simply a matter of time and star-formation rate, then it perhaps reinforces my misgivings (or interest) – because just *how* do galaxies know at what rate to form stars? The metallicity–luminosity plot for a given epoch can be spread out all over the place if the star-formation rates in individual galaxies of a given mass vary by a factor of a few. So either this is the wrong explanation for the metallicity–mass relation, or somehow the local conditions that determine the star formation in a galaxy ARE tightly controlled by its mass.

To return briefly to the high gas abundances in quasar gas. The observational estimates may be wrong – analysis is a tricky business, given the problems we seem to have with fairly 'normal' metal-rich H II regions, let alone where there may be very strong and hard radiation fields from an active quasar nucleus. Perhaps there are almost 'pathological' models for disc evolution with multiple generations of star formation fed by an inflow of enriching gas that manages to build up to five or more times Solar abundance in the gas – this need not be impossible if the total mass of really enriched gas is relatively small; and it might be worth investigating just how contrived such models would have to be. An additional enhancement might occur if you could break up some of the long-lived stars and re-use them as gas in star formation. I suppose that the effect would be rather like having a lower 'lock-up' fraction α, so that the formal yield $p = p'/\alpha$ would increase. Reducing α by a factor of two would double the yield, but an environment that destroys half the stars does not sound like a particularly suitable place for the necessary subsequent fresh star formation and resulting nucleosynthesis.

6 Concluding remarks

I will not delve deeply into the question of whether local chemical inhomogeneity of the interstellar medium could lead to particularly metal-rich regions. I still see no reason to abandon the idea that the interstellar medium is locally very well mixed (e.g. Scalo and Elmegreen 2004; Karlsson and Gustaffson 2001, 2005). Indeed, there are indications from the constancy of element ratios that the interstellar medium must have been very well mixed even during quite early Galactic epochs (e.g. Arnone *et al.* 2005). Variation in abundances between stars and between star clusters must therefore indicate difference in epoch of formation and/or that they formed at well-separated radial distances in the Galaxy.

So we have not really made much progress in trying to produce convincing high yields. Much of the need for high yields might well go away if some re-calibrations of H II regions are right. Perhaps we are simply left with the obvious need to develop confidence in both H II-region and cluster-star analysis before invoking special mechanisms (or spurious IMF variations) to enhance the true yield. We are still not really sure whether we have a (low) true yield that needs to be increased in some situations or a (high) true yield that needs to be decreased in most other situations! Outflow (and, under certain conditions, inflow) can certainly reduce yields. Increases by factors of two to three can be accomodated by the effect of large-scale gas flows in a galaxy during its star-forming activity, but more enhancement than this is indeed problematic. Galactic nuclei may remain a special case.

References

Arnone, E. *et al.* (2005), *A&A* **430**, 507
Ayres, T. R. *et al.* (2006), *ApJS* **165**, 618
Bresolin, F., Garnett, D. R., & Kennicutt, R. C. (2004), *ApJ* **615**, 228
Edmunds, M. G. (1990), *A&A* **384**, 879–883
   (2005), *A&G* **4.2**.
   (2007), *MNRAS in preparation*.
Edmunds, M. G., & Greenhow, R. M. (1995), *MNRAS* **246**, 678
Edmunds, M. G., & Pagel, B. E. J. (1984), *MNRAS* **211**, 507
Erb, D. K. *et al.* (2006), *ApJ* **644**, 813
Garnett, D. R. (2002), *ApJ* **581**, 1019
Goodwin, S. P., & Pagel, B. E. J. (2005), *MNRAS* **359**, 707

Karlsson, T., & Gustafsson B. (2001), *A&A* **379**, 461
   (2005), *A&A* **436**, 879
Maeder, A. (1992), *A&A* **264**, 105
Pagel, B. E. J. (1997), *Nucleosynthesis and the Chemical Evolution of Galaxies* (Cambridge, Cambridge University, Press).
Pilyugin, L. S. (2001), *A&A* **369**, 594
Pilyugin, L. S., Vílchez, J. M., & Contini, T. (2004), *MNRAS* **425**, 849
Scalo, J., & Elmegreen, B. G. (2004), *ARAA* **42**, 275
Stasińska, G. (2005), *A&A* **434**, 507
Talbot, R. J. Jr, & Arnett, W. D. (1971), *ApJ* **170**, 409
Tremonti, C. A. *et al.* (2004), *ApJ* **613**, 898
Twarog, B. A., Anthony-Twarog, B. J., & de Lee, N (2003), *AJ* **125**, 1383

47

Abundance patterns: thick and thin disks

Cristina Chiappini[1,2]

[1]*Osservatorio di Trieste, INAF, Via Tiepolo 11, 34131 Trieste, Italy*
[2]*Geneva Observatory, 51 Chemin des Maillettes, CH-1290 Sauverny, Switzerland*

We study the effect of assuming different formation timescales for the thick and thin disks on the variation of the abundance ratios of several elements with metallicity. We show that, if the thin disk was formed on a longer timescale ($\simeq 7$ Gyr) than the thick disk ($\simeq 0.8$ Gyr), the abundance ratio shifts between the thick and thin disk, as a function of the metallicity, can be well explained. Moreover, these observations offer a powerful constraint on stellar yields in general (massive stars, low- and intermediate-mass stars, and SN Ia) and their dependence on metallicity.

1 Introduction

High-quality abundances for thick- and thin-disk stars are now available together with kinematic information, allowing the abundance patterns in the thick and thin disks to be studied separately (see Chapter 5). These abundance patterns reflect the different star-formation histories of these two components. Moreover, whereas for metallicities [Fe/H] < -2.5 the interstellar medium was enriched essentially by massive stars, for [Fe/H] > -1.5 up to $\simeq 0.5$ (the metallicity range of the thick/thin disks) the chemical enrichment is complex, involving not only the contribution of stars of all masses and of Type-Ia supernovae, but also metallicity effects that can modify the stellar yields. Clearly, the study of this problem requires the use of detailed chemical-evolution models in which the approximation of instantaneous recycling is not assumed.

In this work we assume that the two components form by gas accretion on different timescales. For the thin disk we adopt a model similar to that of Chiappini *et al.* (1997), but in this case we model the thin disk as an independent quantity,

The Metal-rich Universe, eds. G. Israelian and G. Meynet. Published by Cambridge University Press.
© Cambridge University Press 2008.

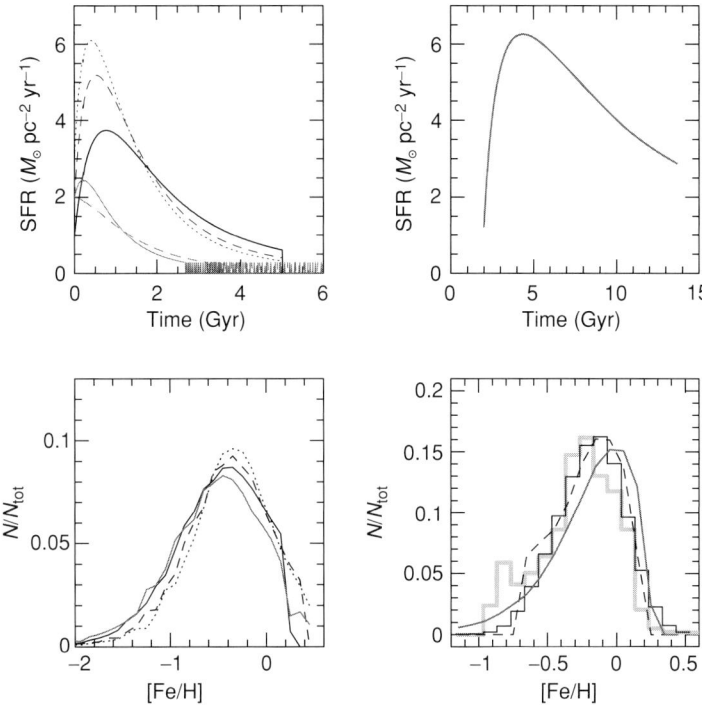

Figure 47.1. Variation with time of the star-formation rate (SFR) (top panels) and the predicted metallicity distribution (bottom panels) in the thick disk (left panels) and thin disk (right panels), from several models. The dashed line in the bottom-right panel shows, for comparison, the best model of Chiappini *et al.* (1997).

as did e.g. Pagel & Tautvaišienė (1995). In our models the thin disk forms by exponential infall of primordial gas with an e-folding time of ~ 7 Gyr in the Solar vicinity. For the thick disk we have fewer constraints, namely (a) its stars are older than ~ 10 Gyr, (b) the metallicity distribution has a peak around [Fe/H] = -0.5 and extends from about -1.5 to Solar or above, and (c) the thin disk is 4%–15% of the mass of the thin disk (Jurić *et al.* 2006). The above constraints can be satisfied if we assume a shorter timescale of gas accretion for the thick disk.

2 Results

Figure 47.1 shows the resulting star formation and metallicity distribution for the thick (left panel) and thin (right panel) disks. In the case of the thick disk we show several possible models computed with different assumptions regarding (a) its mass with respect to that of the thin disk, (b) the late evolution of the star formation (truncated, threshold, continuous), and (c) the infall timescale (0.4 or 0.8 Gyr). Given the uncertainty in all these quantities, we check the range of model parameters that could still explain the observed thick-disk metallicity distribution. In the next figures we will show the predictions corresponding to the model shown by a thick solid line in Figure 47.1 (which has a timescale of 0.8 Gyr and a normalization with respect to the thin disk of $\sim$10% – see details in Chiappini *et al.* (2008, in preparation). In what follows we will see that such a model naturally accounts for the observed abundance patterns in the thick disk. The chemical-evolution models for the thick and thin disks were computed assuming the same initial mass function and stellar yields for both components. For the stellar yields we adopted the Woosley & Weaver (1995) prescriptions for SN II, model W7 from Nomoto *et al.* (1997) for SN Ia, and the yields of van den Hoek & Groenewegen (1997) for low- and intermediate-mass stars. The yields for massive stars and SN Ia were modified according to the prescriptions given in François *et al.* (2004) (model A). Moreover, we also show the results obtained from models computed with the same prescriptions as detailed above, except for the massive stars for which we adopted the new stellar yields of Nomoto *et al.* (2006) (model B).

2.1 The α-elements

Figure 47.2 shows the predicted patterns for the α-elements for the thick (right panel) and thin disks (left panel) for models A (solid line) and B (dotted line) described above. Figure 47.3 (left panel) shows data for the thick- and thin-disk stars plotted together in an [X/O] versus [O/H] diagram, on which the differences, if they are indeed caused by different formation timescales, should be clearer. The solid curves correspond to the thick-disk predictions; the dashed ones to the predictions for the thin disk (for model A). It is clearly seen that, despite some absolute shifts (which will depend on the stellar yields adopted), the relative behavior of the two curves is in very good agreement with the data, namely (a) Mg/O relationships in the thick and thin disks trace each other; (b) larger differences are seen in the Fe/O ratio, as expected; (c) the α-elements do not have the same overabundances, because of the different amounts coming from massive stars but also due to the contribution of SN Ia for S, Ca, and Si; and (d) the variations of the shifts between the thick- and thin-disk patterns with metallicity are in good agreement with those predicted by the models. A detailed discussion of these results and their implications for the stellar yields of particular elements can be found in Chiappini *et al.* (2008, in preparation).

2.2 The iron-peak elements

Figure 47.3 (right panel) shows a comparison between our predictions (model A) for the variation of [X/Fe] versus metallicity, for X = Sc, Co, V, Ni, Cr, Cu, Mn,

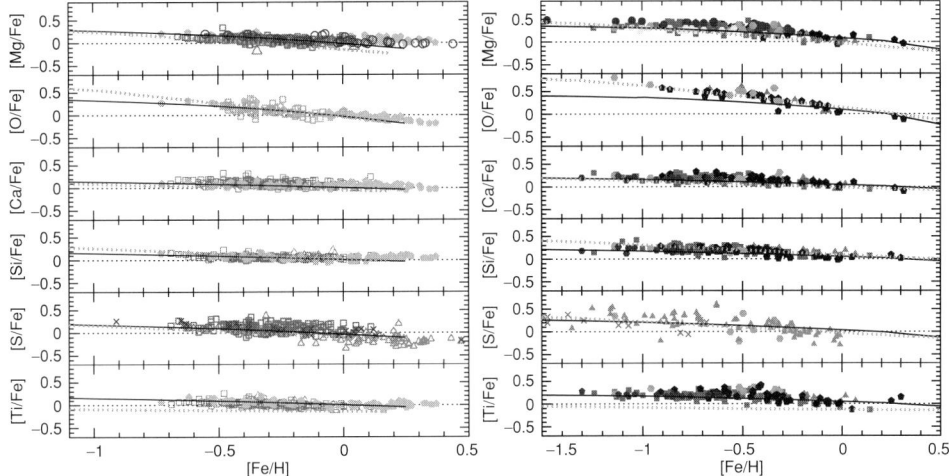

Figure 47.2. Plots of [α/Fe] versus [Fe/H] for a selected dataset for which it is possible to assign the stars to either the thick or the thin disk with more than 70% probability. See Chiappini *et al.* (2008 in preparation for the choice of the datasets included in this figure. The data are from Brewer & Carney (2006), Bensby *et al.* (2003, 2004, 2005), Caffau *et al.* (2006), Prochaska *et al.* (2000), Reddy *et al.* (2003), Fuhrman (2004), Meléndez *et al.* (2002), Nissen *et al.* (2004), and Reddy *et al.* (2006.) Left panel: thin disk. Right panel: thick disk. For explanation of models A (solid line) and B (large-dotted line) see the text. For color figures, see online version.

and Zn, for the thick (solid line) and thin (dashed line) disks. It can be seen that the two curves overlap for Co, V, Ni, Cr, and Zn. A small shift between the curves is obtained for Mn, Cu, and Sc. The data seem to behave in the same way, there being only little shifts for Sc, Mn, and, perhaps, Cu. The elements K, Sc, and V are odd-Z elements produced mainly by oxygen-burning (K in hydrostatic burning; Sc and V in explosive burning). The nucleosynthesis prescriptions for the last two elements are thus more uncertain. Here we adopted the prescriptions given in François *et al.* (2004), which assume that the stellar yields of the above elements are not metallicity-dependent. Although this seems to be the case for Cr and Ni, a certain dependence on metallicity is expected for Mn, Co, Cu, and Zn. For models including the metallicity dependencies of the latter elements, see Chiappini *et al.* (2008, in preparation).

3 Concluding remarks and open questions

- A model in which the thick and thin disks form on different timescales (around 0.8 and 7 Gyr, respectively) can well explain the observed differences between the abundance ratios of thick- and thin-disk stars, for several elements, as a function of metallicity.

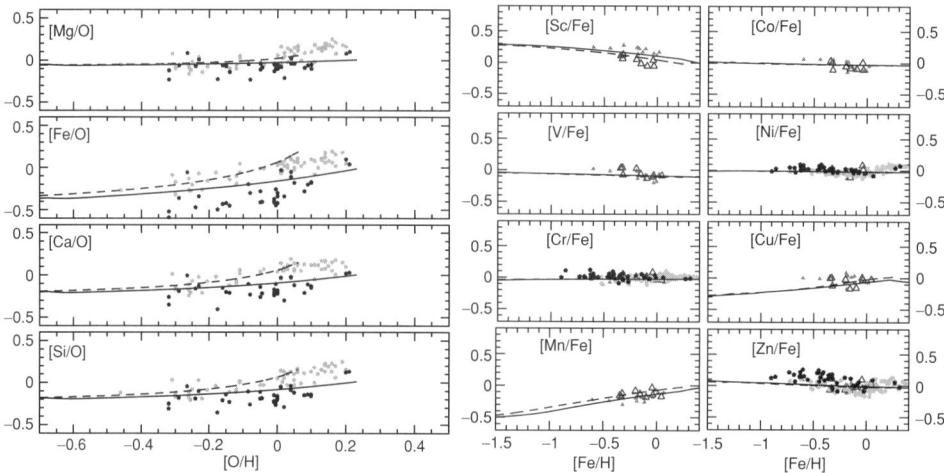

Figure 47.3. Comparing thick- and thin-disk abundance patterns predicted by model A. Left panel: [α/O] versus [O/H] diagrams showing data from Bensby *et al.* (2003, 2004, 2005) for thin (light pentagons) and thick (black pentagons) disks. Right panel: [iron-peak/Fe] versus [Fe/H] diagrams. The data are again from Bensby *et al.* (pentagons) and from Brewer & Carney (2006) (triangles – open for thin-disk and closed for thick-disk stars).

- There is still no agreement on how metal-rich the thick disk is (does it extend to metallicities above Solar?). The answer to this question is important in order to constrain the late phases of the star-formation history of the thick disk.
- Are the over-Solar-metallicity stars in the Bensby *et al.* sample thin-disk stars from the Solar vicinity? According to our models, with current stellar yields, and given the other constraints, the thin disk stars with [Fe/H] > 0.2 should have come from an inner region of the disk. A curve for 4 kpc would fit these data well – see Selwood & Binney (2002) and Grenon (1999) for dynamical effects that might cause radial migration of stars.
- Taking together the high-quality data now available for the thin- and thick-disk and halo stars, we have the opportunity to better constrain the stellar yields and their dependencies on metallicity.

Acknowledgment

This research was partially supported by the INAF PRIN grant CRA 1.06.08.02.

References

Bensby, T., Feltzing, S., & Lundström, I. (2003), *A&A* **410**, 527–551
 (2004), *A&A* **415**, 155–170
Bensby, T., Feltzing, S., Lundström, I., & Ilyin, I. (2005), *A&A* **433**, 185–203
Brewer, M., & Carney, B. W. (2006), *AJ* **131**, 431–454
Caffau, E., Bonifacio, P., Faraggiana, R. *et al.* (2006), *A&A* **441**, 533–548
Chiappini, C., Matteucci, F., & Gratton, R. (1997), *ApJ* **477**, 765–780
François, P., Matteucci, F., Cayrel, R. *et al.* (2004), *A&A* **421**, 613–621

Fuhrmann, K. (2004), *Astron. Nachr.* **325**, 3–80
Grenon, M. (1999), *Astrophys Space Sci.* **265**, 331–336
Jurić, M., Ivezić, Z., Brooks, A. *et al.* (2006), astro-ph/0510520
Meléndez, J., & Barbuy, B. (2002), *ApJ* **575**, 474–483
Nissen, P. E., Chen, Y. Q., Asplund, M., Pettini, M. (2004) *A & A* **415**, 993–1007
Nomoto, K., Tominaga, N., Umeda, H. *et al.* (2006), *Nucl. Phys. A* **777**, 424–458.
Nomoto, K., Hashimoto, M., Tsujimoto, T. *et al.* (1997), *Nucl. Phys. A* **616**, 79c
Pagel, B. E. J., & Tautvaišienė, G. (1995), *MNRAS* **276**, 505–514
Prochaska, J. X., Naumov, S. O., Carney, B. *et al.* (2000), *ApJ* **120**, 2513–2549
Reddy, E. B., Lambert, D. L., & Allende Prieto, C. (2006), *MNRAS* **367**, 1329–1366
Reddy, E. B., Tomkin, J., Lambert, D. L., & Allende Prieto, C. (2003), *MNRAS* **340**, 304–340
Sellwood, J. A., & Binney, J. J. (2002), *MNRAS* **336**, 785–796
van den Hoek, L. B., & Groenewegen, M. A. T. (1997), *A&AS* **123**, 305–328
Woosley, S. E., & Weaver, T. A. (1995), *ApJS* **101**, 181–235

48

Formation and evolution of the Galactic bulge: constraints from stellar abundances

Silvia K. Ballero,[1,2] Francesca Matteucci[1,2] & Livia Origlia[3]

[1] *Dipartimento di Astronomia, Università di Trieste, Via G. B. Tiepolo 11, I-34124 Trieste, Italy*
[2] *INAF – Osservatorio Astronomico di Trieste, Via G. B. Tiepolo 11, I-34121 Trieste, Italy*
[3] *INAF – Osservatorio Astronomico di Bologna, Via G. Ranzani 1, I-40127 Bologna, Italy*

We present results for the chemical evolution of the Galactic bulge in the context of an inside-out formation model of the Galaxy. A supernova-driven wind was also included in analogy with elliptical galaxies. New observations of chemical-abundance ratios and the metallicity distribution have been employed in order to check the model results. We confirm previous findings that the bulge formed on a very short timescale with quite a high star-formation efficiency and an initial mass function more skewed towards high masses than the one suitable for the Solar neighbourhood. A certain amount of primary nitrogen from massive stars might be required in order to reproduce the nitrogen data at low and intermediate metallicities.

1 Introduction

The issue of bulge formation and evolution has lately received renewed attention due to results from recent studies, which form the basis for a growing consensus that the bulge is old and that its formation timescale was relatively short (≤ 1 Gyr). We want to test the hypothesis of a quick dissipational collapse via the study of the evolution of the abundance ratios coupled with considerations on the metallicity distribution and to show that abundance ratios can provide an independent constraint for the bulge-formation scenario since they differ depending on the star-formation history (Matteucci, 2000). The α-elements in particular are of paramount importance in probing the star-formation timescale, since the signature of a very short burst of star formation must result in their enhancement with respect to iron.

The Metal-rich Universe, eds. G. Israelian and G. Meynet. Published by Cambridge University Press.
© Cambridge University Press 2008.

2 The chemical-evolution model and the data

The model we are adopting is described by Ballero *et al.* (2007) and belongs to the category of fast dissipational collapse prior to the settling of the disc. It follows the trend of previous models by Matteucci & Brocato (1990) and Matteucci *et al.* (1999). The star-formation rate is proportional to the gas surface mass density of the disc:

$$\psi(r, t) = \nu G(r, t), \tag{48.1}$$

where ν is the star-formation efficiency, i.e. the inverse of the star-formation timescale. The stellar initial mass function (IMF) is parametrised as a power law of logarithmic index x:

$$\phi(m) \propto m^{-(1+x)}, \tag{48.2}$$

which may differ among the various mass ranges. Finally, the bulge forms via infall of gas shed from the Galactic halo on a timescale τ:

$$\dot{G}_{\text{inf}}(t) \propto e^{-t/\tau}. \tag{48.3}$$

We updated the stellar lifetimes by adopting those of Kodama (1997) and we introduced the stellar yields published by François *et al.* (2004), which were calibrated in order to fit the chemical properties of the Solar neighbourhood. Following the photometric measurements of Zoccali *et al.* (2000) the IMF index for below $M_\odot$ was set to $x = 0.33$. The Type-Ia supernova rate was computed following the single-degenerate scenario, according to Matteucci & Recchi (2001). Finally, we considered primary production of nitrogen from massive stars, as computed by Matteucci (1986), as opposed to purely secondary production, which is usually adopted in chemical-evolution models. We also introduced a supernova-driven wind in analogy with elliptical galaxies. To develop it, we supposed that the bulge lies at the bottom of the potential well of the Galactic halo, whose mass is $M = 10^{12} M_\odot$ and whose effective radius is ~ 100 times the effective radius of the bulge's mass distribution. The binding energy of the bulge is then calculated following Bertin *et al.* (1992), who split the potential energy into two parts, one due to the interaction between the bulge gas and the bulge potential and the other to the interaction of the bulge gas with the potential well of the dark-matter halo of the Milky Way, which depends on the relative mass distribution. The thermal energy of the interstellar medium due to supernova explosions is given by

$$E_{\text{th,SN I}} + E_{\text{th,SN II}} = \int_0^t \epsilon(t - t') R_{\text{SN I/II}}(t') dt', \tag{48.4}$$

where R is the rate of Type-Ia and Type-II supernovae, t' is the explosion time and ϵ is the energy content of a supernova remnant, whose evolution is calculated following Cox (1972) and Cioffi & Shull (1998). When $E_{\text{th,SN I/II}} = E_{\text{b,gas}}$ a

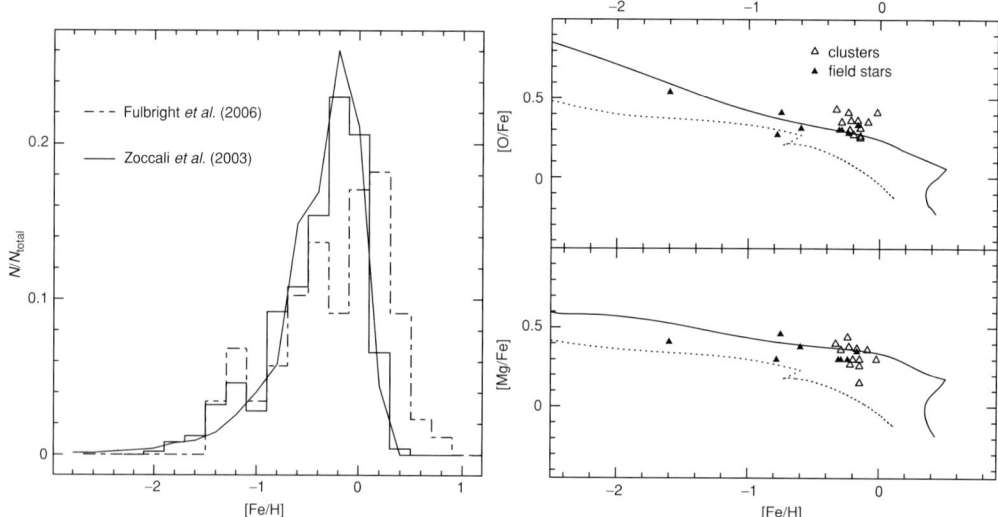

Figure 48.1. Left panel: the predicted metallicity distribution in our bulge reference model ($x = 0.95$ for $M > M_\odot$, $\nu = 20\,\text{Gyr}^{-1}$, $\tau = 0.1\,\text{Gyr}$) compared with the observed distributions (Zoccali et al. 2003; Fulbright et al. 2006). Right panel: the evolution of [O/Fe] and [Mg/Fe] versus [Fe/H] in the bulge for the reference model (solid line). A Solar-neighbourhood fiducial line (dotted line) is plotted for comparison. Data are taken from the infrared-spectroscopic database (Origlia et al., 2002; Origlia & Rich 2004; Origlia et al. 2005; Rich & Origlia 2005).

wind develops and deprives the bulge of all its gas, and the evolution is passive thereafter.

3 Discussion

We investigated variations of the parameters which influence the chemical evolution, namely of the IMF index, from 0.33 to the Scalo (1998) value; of the star-formation efficiency ν (from 2 to $200\,\text{Gyr}^{-1}$); and of the infall timescale τ (from 0.01 to 0.7 Gyr) and matched the results with the most recent data for the metallicity distribution and the evolution of [α/Fe] abundance ratios with metallicity (see Figure 48.1). We found out that in order to reproduce all the constraints it is necessary to adopt a short formation timescale (0.01–0.1 Gyr) and an intense efficiency of star formation (10–20 Gyr^{-1}), combined with an IMF flatter ($x = 0.95$ for $M > M_\odot$) than that suitable for the Solar neighbourhood. The reference model we plotted is also able to explain the different trends of [O/Fe] and [Mg/Fe] seen in the observations. Moreover, if we compare (Figure 48.2) the [N/O] versus [O/H] plot with the data from planetary nebulae of Górny et al. (2004), we see that in order to reproduce the average trend of observations it is necessary to assume a

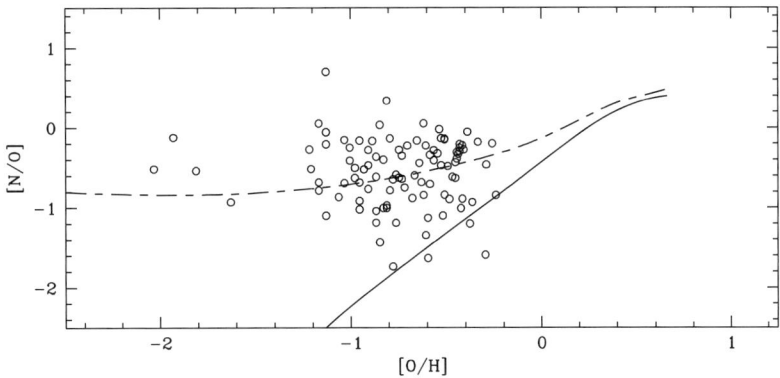

Figure 48.2. The evolution of [N/O] versus [O/H] in the Galactic bulge in our reference model (solid line) compared with a model (Matteucci 1986) in which primary production of N from massive stars is assumed (dashed line). If the average trend of these observations is representative of the pristine [N/O] values, the fit is achieved with the latter model. Data for N and O in bulge PNe are from Górny *et al.* (2004).

primary production of nitrogen from massive stars of all masses at every metallicity, in analogy with the Solar neighbourhood (Ballero *et al.* 2006). The most promising way of doing this seems to be constituted by use of the rotational yields of Meynet *et al.* (2006), as was shown in fact by Chiappini *et al.* (2006). In any case, no firm conclusions regarding nitrogen can be drawn due to the fact that nitrogen is self-enriched to some extent in planetary nebulae.

References

Ballero, S. K., Matteucci, F., & Chiappini, C. (2006), *NewA* **11**, 306–324
Ballero, S. K., Matteucci, F., Origlia, L., & Rich, R. M. (2007), *A&A* **467** 123
Bertin, G., Saglia, R. P., & Stiavelli, M. (1992), *ApJ* **384**, 423–432
Chiappini, C., Hirschi, R., Meynet, G., Ekström, S., Maeder, A., & Matteucci, F. (2006), *A&A* **449**, L27–L30
Cioffi, D. F., McKee, C. F., & Bertschinger, E. (1998), *ApJ* **334**, 252–265
Cox, D. P. (1972), *ApJ* **178**, 143–158
François, P., Matteucci, F., Cayrel, R., Spite, M., Spite, F., & Chiappini, C. (2004), *A&A* **421**, 613–621
Fulbright, J. P., McWilliam, A., & Rich, R. M. (2006), *ApJ* **636**, 821–841
Górny, S. K., Stasińska, G., Escudero, A. V., & Costa, R. D. D. (2004), *A&A* **427**, 231–244
Kodama, T. (1997), Ph.D. Thesis, University of Tokyo
Meynet, G., Ekström, S., & Maeder, A. (2006), *A&A* **447**, 623–639
Matteucci, F. (1986), *MNRAS* **221**, 911–921
Matteucci, F. (2000), in *The Evolution of the Milky Way: Stars versus Clusters*, eds F. Matteucci & F. Giovannelli (Dordrecht, Kluwer), pp. 3–12
Matteucci, F., & Brocato, E. (1990), *ApJ* **365**, 539–543
Matteucci, F., & Recchi, S. (2001), *ApJ* **558**, 351–358
Matteucci, F., Romano, D., & Molaro, P. (1999), *A&A* **341**, 458–468
Origlia, L., & Rich, R. M. (2004), *AJ* **127**, 3422–3430

Origlia, L., Rich, R. M., & Castro, S. (2002), *AJ* **123**, 1559–1569
Origlia, L., Valenti, E., & Rich, R. M. (2005), *MNRAS* **256**, 1276–1282
Rich, R. M., & Origlia, L. (2005), *ApJ* **634**, 1293–1299
Scalo, J. M. (1998), in *The Stellar Initial Mass Function*, eds. G. Gilmore & D. Howell (San Francisco, CA, Astronomical Society of the Pacific), pp. 201–236
Zoccali, M., Cassisi, S., Frogel, J. A. *et al.* (2000), *ApJ* **530**, 418–428
Zoccali, M., Renzini, A., Ortolani, S. *et al.* (2003), *A&A* **399**, 931–956

Summary

B. E. J. Pagel

Astronomy Centre, University of Sussex, Brighton, UK

Topics figuring in this conference include limits to high metallicity, metallicity characteristics of stellar populations, [M/Fe] in bulges and discs, effects of metallicity on star formation and the initial mass function, its relation to planet formation, effects of high metallicity on stellar evolution, yields and galactic chemical evolution, metal-rich H II regions and metallicities at high redshift.

1 Introduction: how metal-rich can you get?

This question (which Daniel Kunth and Mike Edmunds raised) seems like a good starting point for a meeting on the metal-rich Universe. Given a sufficiently strong initial burst, one can reach a metallicity in the gas phase of several times the yield for a short period, before it is diluted by mass loss from the first generation of stars in an isolated system (Talbot & Arnett 1971). This situation may be relevant to the broad-line regions of quasars, where metallicities (basically in terms of CNO) of several times Solar have been estimated (Hamann *et al.* 2004; Nagao *et al.* 2006). These estimates are somewhat model-dependent, but they need to be taken seriously.

High stellar abundances (based on SSP population-synthesis models), at least in terms of magnesium, are found in the inner regions of giant elliptical galaxies (e.g. Trager 2004), with direct implications for the yield(s). Francesca Matteucci gave a robust defence of monolithic models for galaxy formation, such as the ones presented here by Antonio Pipino and Silvia Ballero, while Patricia Sánchez-Blásquez and Brad Gibson considered merger models (either dissipative or 'dry')

The Metal-rich Universe, eds. G. Israelian and G. Meynet. Published by Cambridge University Press.
© Cambridge University Press 2008.

to be compatible with high α/Fe ratios and the relation with velocity dispersion under certain conditions. Leticia Carigi found that merger–fragmentation models work well for subsolar, but not for supersolar, metallicities.

2 Metallicities of stellar populations

Spinrad & Taylor (1969) famously (and at first controversially) drew attention to 'super-metal-rich' stars in the Solar neighbourhood (such as μ-Leo, since convincingly confirmed), which are now receiving much attention due to the correlation with planets, and in galactic bulges, as Mike Rich reminded us. In our own bulge, the (iron) metallicity distribution function (MDF) peaks at Solar (making the Mg MDF somewhat higher, as Jon Fulbright mentioned) and it is not clear that there is a 'G-dwarf problem'; however, the metal-rich and metal-poor components differ kinematically, with the former exhibiting a vertex deviation. In M31, by contrast, Pascal Jablonka sees a definite G-dwarf problem (as well as in the Galactic bulge itself), which is also found generally in 'hot' stellar systems (Bressan *et al.* 1994; Worthey 1994) and was strikingly illustrated at this meeting by Scott Trager's deep HR diagrams from the HST for M32 and NGC 5128.

How many metal-rich stars ([Fe/H] > 0) are there in the Solar neighbourhood, and what are their characteristics in age and kinematics? The local MDF is somewhat controversial, depending on adopted calibrations of the Strömgren colour indices and the weighting given to the thick disc, with some estimating the peak to be at Solar metallicity (Haywood 2002; Reid 2002), which would make nearly half the stars metal-rich, whereas most others, including Carlos Allende-Prieto and his Texas colleagues, place the peak 0.1–0.2 dex lower. Metal-rich stars are typically old, like those in NGC 6791 and other open clusters discussed by Angela Bragaglia and Sofia Randich, and members of the thin disc (Grenon 1999), but there is some overlap. Thomas Bensby and Alexandra Ecuvillon have studied the Hercules stream and other moving groups, which may combine high metallicity with thick-disc kinematics, maybe resulting from an origin in the inner Galaxy. However, we heard from Aurélie Lecureur that the α/Fe ratios in Galactic-bulge giants behave differently from those of the thick disc.

Nuno Santos and Doug Lin discussed the remarkable correlation between metallicity and the presence of known extrasolar planets, i.e. gas giants with close orbits. (The effect is absent among giants, according to Luca Pasquini.) Surface pollution now seems to be ruled out (Sylvie Vauclair described possible acid tests from stellar seismology), and it appears that intrinsically high metallicity somehow favours the formation of large planets by core accretion. Doug described numerical simulations in which dust in the proto-stellar disc is driven inwards by orbital decay to the 'snow line', where it can dry to form chondrules that stick together to form planetesimals. Hypothetical Earth-like planets would not be affected by this process.

3 M/Fe ratios

Alpha-element/Fe ratios are thought to indicate the relative roles of core-collapse and Type-Ia supernovae, with implications for yields and for the timescale of star formation. Sofia Feltzing compared relevant data for the thin and thick disks, finding an extended period of star formation and some metal-rich stars in the latter. Carla Fröhlich and Song-Chul Yoon gave theoretical accounts of the yields from the two types of supernovae, of which some details could be gleaned from the surface composition of stellar companions to compact X-ray sources as described by Rafael Rebolo. An intriguing feature of the M/Fe–Fe/H relationships, which is not readily accounted for by the models, is that O/Fe continues on down above Solar Fe/H, whereas Mg/Fe flattens, as was first shown by Lambert (1989). The behaviour of oxygen is understood in numerical models of Galactic chemical evolution like those of Matteucci and her colleagues (e.g. Chiappini *et al.* 1997) as a consequence of SN Ia forming continuously with the progress of time, increasing their relative contribution as star formation slows down or is interrupted, but in these models magnesium should behave in the same way, which it doesn't. The behaviour of magnesium is more readily understood in terms of our 'Mickey Mouse'-type analytical models in which SN Ia are assumed to appear at a fixed interval after star formation (Pagel & Tautvaišienė 1995).

Another contentious issue is whether there is a smooth transition from thick to thin disc or a discontinuity, revealed in the [α/Fe]–[Fe/H] diagrams. As was mentioned by Cristina Chiappini, Gratton *et al.* (1997) and Fuhrmann (1998) found a distinct gap in O/Fe and Mg/Fe, respectively, which receives new support from the Texas survey reported by Carlos at this meeting. Thick and thin discs overlap in Fe/H, but their [α/Fe] trends are different, having a constant value around 0.3 in the thick disc and a downward trend in the thin one. The implication is that the thick disc 'got rich quick' and the metallicity was actually lower when the thin disc came into being. Carlos flashed up on the screen (and immediately took down) a schematic model not unlike one that I put forward at a meeting some years ago (Pagel 2001). We agreed in private conversation that what you have to do (in Carlos's words) is 'to add water'.

4 Star formation and evolution at high metallicities

Metallicity strongly affects mass-loss rates from massive stars, making the properties of Wolf–Rayet (WR) stars (notably the incidence of WNL stars and the

WC/WN ratio) a powerful metallicity indicator, meaning iron metallicity according to Götz Gräfener. Paul Crowther described the effects observed in the inner Galaxy, M31 and M83; late WC stars, for instance, are seen only in environments with supersolar metallicity, with a combination of high mass loss and resulting relatively soft ionizing radiation. Francisco Najarro derived a Solar-like abundance in the Arches cluster from the incidence of WNL stars. In low-metallicity environments the binary channel is the predominant source of WR stars. However, many details are unclear, and Joachim Puls in a lively presentation drew attention to serious reservations about the actual mass-loss rates. Georges Meynet described the complications that could be introduced by rotation and magnetic fields.

Does metallicity affect the initial mass function (IMF)? Pavel Kroupa and Ian Bonnell gave theoretical reasons why it might, but nothing definite seems to be known. Peter Westera and João Souza Leão indicated that the question cannot be easily investigated observationally either.

5 H II regions

Stellar and gas-phase abundance estimates are beginning to converge, apart from those for elements depleted from the gas phase, as shown, for example, by the study of the Orion-nebula cluster reported by Sergio Simón-Díaz and the data for hot stars in general reported by Danny Lennon. Fabio Bresolin described the problems in abundance analysis for more-distant and fainter nebulae, for which temperature-sensitive lines are often unobservable, leading to the need to use 'empirical' methods based on strong lines, with their associated calibration problems, and there are still some niggling discrepancies even when they are detected and measured. Strong-line ('strong-arm') methods have often led to overestimates of abundance in the past, because not enough cooling processes were included in photoionisation models. Thus, from the work of Pilyugin and others, extrapolated central abundances in spiral galaxies are about Solar and no more. On the other hand, internal temperature gradients and fluctuations can, if neglected, lead to underestimates, which can now be checked using recombination lines of carbon and oxygen in regions of high surface brightness. Peimbert & Peimbert (2006) have found a typical difference of about 0.2 dex, making the largest oxygen abundance found in H II regions around 8.75, compared with the recent estimate of 8.67 for the Sun. Pepe Vílchez applied a similar correction for large spirals like M51, leading to 8.87 in the gas phase and 8.95 when corrected for depletion onto dust. However, Grażyna Stasińska pointed out that metal-rich droplets from a supernova superbubble can cool and dominate the recombination-line emission, in which case the upward correction to abundances deduced from collisionally excited lines would be overestimated. So the upper limit to H II-region abundances is still somewhat uncertain. Angeles Díaz drew attention

Table 1. *Baryon and metal budgets at $z = 0$, after Finuguenov et al. (2003)*

Component	Z (10^{-2})	Ω_Z (10^{-5})	Ω_b (10^{-2})
Stars	1.2	2.3–4.6 ⎱ Most	0.2–0.4
O VI absorbers	0.26	1.8–5.2 ⎰ metals	0.7–2.0 ⎱ Most
Lyman-α forest	0.01	0.1	1.2 ⎰ baryons
X-ray gas, clusters	0.7	1.4	0.2
Total	0.28	5.6–11.3‡	2.3 – 3.8
Predicted			3.9

to circum-nuclear H II regions or 'hot spots' where these highest abundances are most likely to be found.

Gerhard Hensler considered the possibility of local pollution from stellar winds and concluded that, with his choice of parameters, effects on N and O are negligible, although there would be a 20% enhancement of carbon. Observationally, however, there are well-authenticated effects, at least for nitrogen, in localised regions of NGC 5253 and a few other cases (Kobulnicky & Skillman 1996; Pustilnik et al. 2004).

6 How metal-rich was the Universe at $z \simeq 2$?

Max Pettini described how, apart from quasars, quite substantial abundances are found in Lyman-break galaxies (LBGs) at redshifts of 2–4, with a mass–metallicity (i.e. oxygen) relation falling short of the present-day one by 0.2–0.3 dex. Gas fractions are quite large, which is consistent with a supersolar yield plus outflow. Are these the progenitors of present-day elliptical galaxies? Damped Lyman-α absorption-line systems (DLAs) tend to be metal-poor, but Bryan Penprase described a study of DLAs in front of γ-ray-burst afterglows, which occasionally come close to Solar and can be quite dusty; Sara Ellison described one such system that has diffuse interstellar bands. The question of a selection bias in DLA surveys caused by dust is still controversial; no such bias has been revealed by the CORALS survey of radio-selected quasars.

7 Conclusion: how metal-rich is the Universe now?

A final question that seems appropriate to this meeting concerns the abundance and distribution of baryons and metals in the Universe as a whole, to which some tentative answers have been given by Persic & Salucci (1992), Fukugita *et al.* (1998) and Fukugita & Peebles (2004). Table 1 is largely based on their work and especially on that of Finuguenov *et al.* (2003), but quite similar numbers have also been given by Dunne *et al.* (2003). The table has some striking features: 'metals' are more or less equally divided between stars and tenuous intergalactic gas now beginning to be detected in O VI, still with substantial uncertainty in the quantitative aspect, which also shares the baryons predominantly with cold intergalactic gas revealed by the Lyman-α forest at low redshift. With the assumed numbers, the yield for the whole Universe is about 0.03, about twice Solar, but consistent with a Salpeter IMF between $0.1 M_\odot$ and around $100 M_\odot$ (Madau *et al.* 1996).

Acknowledgments

It remains for me to thank Garek Israelian and the Organising Committee for the privilege of attending this fascinating conference and the LOC for the friendly and efficient organisation in these splendid surroundings.

References

Bressan, A., Chiosi, C., & Fagotto, F. (1994), *ApJS* **94**, 63
Chiappini, C., Matteucci, F., & Gratton, R. (1997), *ApJ* **477**, 765
Dunne, L., Eales, S. A., & Edmunds, M. G. (2003), *MNRAS* **341**, 589
Finuguenov, A., Burkert, A., & Böhringer, H. (2003), *ApJ* **594**, 136
Fuhrmann, K. (1998), *A&A* **338**, 161
Fukugita, M., Hogan, C. J., & Peebles, P. J. E. (1998), *ApJ* **503**, 518
Fukugita, M., & Peebles, P. J. E. (2004), *ApJ* **616**, 643
Gratton, R., Carretta, E., Matteucci, F., & Sneden, C. (1997), unpublished preprint.
Grenon, M. (1999), *Astrophys. Space Sci.* **265**, 331
Hamann, F., Dietrich, M., Sabra, B. M., & Warner, C. (2004), in *Origin and Evolution of the Elements*, eds. A. McWilliam & M. Rauch (Cambridge, Cambridge University Press), p. 440
Haywood, M. (2002), *MNRAS* **337**, 151
Kobulnicky, H. A., & Skillman, E. D. (1996), *ApJ* **471**, 211
Lambert, D. L. (1989), in *Cosmic Abundances of Matter*, ed. C. J. Waddington (New York, American Institute of Physics), p. 168
Madau, P., Ferguson, H. C., Dickinson, M. E. *et al.* (1996), *MNRAS* **283**, 1388
Nagao, T., Marconi, A., & Maiolino, R. (2006), *A&A* **447**, 157
Pagel, B. E. J. (2001), in *Cosmic Evolution*, eds E. Vangioni-Flam, R. Ferlet & M. Lemoine (Singapore, World Scientific), p. 223
Pagel, B. E. J., & Tautvaišienė, G. (1995), *MNRAS* **276**, 505
Persic, M., & Salucci, P. (1992), *MNRAS* **258**, 4P
Pustilnik, S., Kniazev, A., Pramskij, A. *et al.* (2004), *A&A* **419**, 469
Peimbert, M., & Peimbert, A. (2006), *Revista Méxicana Astronomía y Astrofísica*, **26**, 163
Reid, I. N. (2002), *PASP* **114**, 306
Spinrad, H., & Taylor, D. J. (1969), *ApJS* **22**, 445
Talbot, R. J., & Arnett, W. D. (1971), *ApJ* **170**, 409
Trager, S. C. (2004), in *Origin and Evolution of the Elements*, eds. A. McWilliam & M. Rauch (Cambridge, Cambridge University Press), p. 388
Worthey, G. (1994), *ApJS* **95**, 107